Manual Therapy:

IMPROVE MUSCLE AND JOINT FUNCTIONING

Ad Warmerdam

Manual Therapy:

Improve Muscle and Joint Functioning

Ad Warmerdam
D.P.T., O.C.S., M.T.C.

Manual Therapy:
Improve Muscle and Joint Functioning

Ad Warmerdam
D.P.T., O.C.S., M.T.C.

Pine Publications
2947 Jerusalem Avenue
Wantagh, New York 11793-2020

Copyright 1999, Ad Warmerdam. All rights reserved.
Library of Congress Catalog Card No. 97-91614
ISBN No. 0-9657910-0-9

Editing, Design, and Type Composition by Barbara Field

Pine Publications
2947 Jerusalem Avenue
Wantagh, New York 11793-2020

Library of Congress Cataloging in Publication Data

    Warmerdam, Ad, 1958–
        Manual therapy: Improve muscle and joint functioning
        References, Index

No part of this book may be reproduced or transmitted in any form or by any means, electronic or mechanical, including photocopying, recording, or by any information storage and retrieval system, without written permission from the author.

This book has been designed to provide information in regard to the subject matter covered. The purpose of this book is to educate. Although every effort has been made to make this book as complete and as accurate as possible, the author and publisher assume no responsibility for errors, omissions, or misapplication or misuse of the information provided. The information given in this book is appropriate in most cases. However, everyone is different. The author and publisher disclaim all responsibility for any liability, damage, injury, loss, or risk, personal or otherwise, incurred as a consequence, directly or indirectly, of the use and application of any of the contents of this book.

Distributor, North America	Distributor, All Other Countries
OPTP	Almere Motion
P.O. Box 47009	Kweekgrasstraat 60
Minneapolis, MN 55447	1313 BT Almere
U.S.A.	Netherlands
Fax: 612-553-9355	Fax: 036-5304009
Tel: 612-553-0452	Tel: 036-5304010
Tel: 800-367-7393	

*For Caro, Anica,
and Marc*

Contents

Foreword xiii

Acknowledgments xv

Introduction 1

1. Muscle Weakness: An Overview 3
   - Inactivity 4
   - Immobilization 5
   - Decreased Circulation in the Muscle ... 6
   - Aging 7
   - Myopathy 7
   - Nociception 10
   - Disorders of Nerve Conduction 11
   - Antagonistic Tissue with Reduced Extensibility 13
   - Summary and Conclusions 18
   - References 19

2. Articular Receptors 29
   - Articular Nerves 29
   - Innervation of the Joint 30
   - Articular Receptors 31
   - Summary 38
   - References 39

3. Joint Angle and Ligamento-Capsular Tension 43
   - Summary 45
   - References 45

4. Position and Movement Sense 47
   - Articular Receptors 47
   - Spindle Cell 47
   - Muscle or Joint? 48
   - Gamma Motoneuron 48
   - Active Positioning 49
   - Type I and Type II Articular Receptors ... 49
   - Summary 50
   - References 50

5. Joint Pain 53
   - References 54

6. Articular Reflexes 57
   - Articular Nociceptive Reflexes 58
   - Articular Mechanoreceptor Reflexes .. 58
   - Final Common Input 59
   - Spindle Cell 60
   - Static and Dynamic Gamma Motoneurons 62
   - Static and Dynamic Spindle Responses ... 63
   - Tonic and Phasic Alpha Motoneurons and Muscle Fibers 63
   - Static and Dynamic Articular Reflexes ... 64
   - Receptor Hysteresis and Residual Muscle Tension 65
   - Summary 65
   - References 66

7. Articulo-Muscular Protective Reflex 71
   - Ankle Mechanoreceptor Reflexes 71
   - Reciprocal Coordination of Articular Reflexes 73
   - Joint Stability 74
   - Articulo-Muscular Protective Reflex .. 74
   - Gamma Motoneurons 75
   - Anterior Cruciate Ligament 76
   - Ankle Sprains 77
   - Elbow 77
   - Summary 78
   - References 78

8. Withdrawal Reflex 81
   Interneurons 82
   Interneurons in the Withdrawal Reflex ...82
   Reflex Reversal 83
   Prolonged Facilitation of the
   Withdrawal Reflex 84
   Summary and Conclusions 84
   References 84

9. Dysfunctions and Pathology 89
   Altered Articular Afference 89
   Aging 89
   Trauma 89
   Surgery 90
   Summary 90
   References 90

10. Arthritis 93
    Terminology 93
    Effusion 93
    Muscle Weakness 96
    Substance P 98
    The Nervous System and Arthritis 98
    Type IV Nociceptors 99
    Capsular Pattern 100
    Acute Arthritis 103
    Chronic Arthritis 103
    Summary 104
    References 104

11. Neurogenic Arthropathy 109
    References 110

12. Radiculopathy 111
    Summary 112
    References 113

13. Pseudoradiculopathy 115
    Referred Pain 115
    Facilitated Segment 116
    Pseudoradiculopathy 117
    Radiculopathy or Pseudoradiculopathy? ..118
    Summary 120
    References 120

14. Arthrogenic Muscle Dysfunction 123
    Summary 125
    References 125

15. Manual Muscle Testing 127
    Testing Movements 127
    Testing Individual Muscles 128
    Multijoint Muscles 128
    Isometric Versus Concentric Manual
    Muscle Testing 129
    Grading Muscle Strength 129
    Qualitative Muscle Testing Technique ...130
    Assessment of Muscle Strength 131
    Evaluation of Muscle Strength 132
    Loose-Packed Position 132
    Testing in Multiple Positions 134
    Barrier to Movement 134
    Joints Without Active Range 137
    Generalized Muscle Weakness 139
    Active Exercises 139
    Reliability and Validity 140
    Summary 142
    References 143

16. Evaluation of Spinal and Multiple
    Dysfunctions 145
    Evaluation 145
    Spinal Origin of Peripheral
    Dysfunctions 147
    Radiculopathy 150
    Pseudoradiculopathy 150
    Innervation of Extremity Muscles 152
    Evaluation of Weakness from Multiple
    Sources 154
    Summary 155
    References 155

17. Therapy 157
    Traditional Strengthening Exercises ...157
    Mobilization 158
    Immediate and Aftereffects of a
    Mobilization 159
    Sustained Versus Oscillatory
    Mobilizations 160
    Active Mobilizations 160
    Postmobilization Weakness 164
    High-Velocity Thrust Manipulation 165
    Pain Reduction 167
    Radiculopathy 167
    Hypertonic Muscles 168

Sequencing the Treatment in
Orthopedic Conditions 169
Home Exercises 169
Indications and Contraindications 170
Summary 171
References 172

18. Notation 175
Example I 175
Example II 176
Example III 176

19. Technique Procedures 177
Joints 177
Cervical Spine 180
Thoracic Spine 191
Rib Cage 198
Lumbar Spine 210
Sacroiliac Joints .218
Symphysis Pubis .237
Shoulder Girdle .239

Glenohumeral Joint 250
Elbow and Forearm 266
Wrist 273
Fingers 280
Thumb 286
Hip 293
Knee 303
Ankle and Tarsal Joints 316
Toes 329
Radiculopathy and
Pseudoradiculopathy 332
References 335

20. Case Studies 337
Case 1, Median Nerve Paresis 337
Case 2, Median Nerve Laceration 339
Case 3, Lyme Disease 340
Case 4, Drop Foot 343
Case 5, Low Back Pain 345

Index 351

Foreword

New approaches to patient treatment often arise from clinical experiences that occur on a daily basis. By questioning our current knowledge base and being critical of our own successes and failures with patient treatment, we allow ourselves to develop new ideas and concepts in patient care. At a time when many types of medical treatment are being challenged and are subject to scientific inquiry in research base study, our patients serve as single-subject case studies.

This book challenges currently held beliefs regarding muscle function, and particularly muscle strength. It asks the question why a muscle tests weak and, more important, whether or not this is true motor weakness or rather muscular inhibition secondary to joint dysfunction. As a manual therapist and critical colleague of Mr Warmerdam, I have had the opportunity to successfully apply his approach in my clinical practice, expanding my thinking regarding the treatment of muscle imbalances.

Mr. Warmerdam has attempted wherever possible to support his work with scientific evidence that is available to date. His concepts will most certainly stimulate much thought and discussion, and will need to be rigorously scientifically tested. He is to be congratulated for his perseverance, dedication, and enthusiasm, with which this book abounds.

MARK R. BOOKHOUT, M.S., P.T.

Acknowledgments

Writing a book is a lengthy process. The concepts in this book were developed over more than 17 years, during which many people have contributed to the theory and clinical applications it presents and many helped give physical shape to those ideas and applications. The process began with two people introducing me to a gentle way of manual muscle testing. Diane Propster and Sandy not only made me look at muscle testing in a different way, they also changed my belief that it required weeks to strengthen muscles. I am grateful for the seed they planted in me.

Many clinical applications in this book are influenced by the teachings of other clinicians. I am indebted to all of them, especially: Mark R. Bookhout, David S. Butler, Robert L. Elvey, Olaf Evjent, Freddy M. Kaltenborn, David Lamb, Geoff D. Maitland, Robin A. McKenzie, Fred L. Mitchell, Jr., and Stanley V. Paris. Most learning, however, was under the direct guidance of my patients. I thank them for teaching me the intimate interactions between muscles and joints. I have the utmost appreciation for the knowledge they displayed and for their patience in educating me.

I am also grateful to Barry Wyke, whose publications first suggested to me that there might be a neurophysiological explanation for the clinical phenomena my patients taught me. I also gratefully acknowledge Rolf Hoogland, who provided information about pseudoradiculopathy, and James Cyriax, in whose excellent publications I found support for my clinical observations and their neurophysiological basis.

Many thanks to all my students and colleagues, especially Anthony Davis and Jamie Rockwin, for their valuable suggestions toward literature and more effective evaluation and treatment techniques. I also thank those who understood the value of my work and stimulated me in developing it into continuing education courses: Bruce Franke, Barbara Goldfarb, Howard Makofsky, Atti Noordhof, Ron Peyton, Ann Pitmann, and Linda Tremain.

I thank Catherine Lang for retrieving the many publications that support the concepts in this book, and Henk Brandsma, whose enthusiastic teachings made writing the book easier. Thanks also to Karen Baxter and Chris Contino for their secretarial assistance, to Robert Benno, Larry Goldfarb, Linda Lotito-Finneran, and Howard W. Makofsky for their critical review of the first chapter and their suggestions for improvement, and to Marcel Ates and Thom Warmerdam II for reviewing the parts on physics. Any errors are mine. A grateful acknowledgement to Marloes van Rees for the photography, to Roel Klap for managing the light during the photo sessions, and to Caro van Iersel for modeling.

Special thanks to my colleagues and friends, Frans Abbink, Agatha de Jongh-Abbink, Roel Klap, Marloes van Rees, and Claire Marston, with whom I shared part of my American dream, and to Janet Anderson and Diane Propster, who made me want to have an American dream. Thanks to Greet and Ben van Iersel for always being available to help. Many thanks to Marian and Thom Warmerdam, Sr., for showing me that reaping a harvest comes only after sowing and working the land.

This book was written in partial fulfillment of the requirements for obtaining my doctorate in physical therapy at the University of St. Augustine for Health Sciences, Institute of Physical Therapy. I extend my greatest appreciation to three wonderful people, the members of my doctoral committee, who helped me get the best out of me and helped shape the contents of this book: Barbara H. Connolly, Mark R. Bookhout, and Stanley V. Paris. This work would not have come to fruition without your guidance.

Finally, my utmost appreciation to Caro, off whom I could bounce ideas, and who walked with me as those ideas slowly took shape in print.

Introduction

This book may shift the paradigm that muscle strengthening takes a long time. How long does muscle strengthening take? Weeks? Months? This book will propose that it is often possible to "strengthen" muscles in minutes. Muscle weakness is a symptom, one that can only be effectively treated by addressing the cause of the weakness. By finding the cause of the weakness and correcting it, muscles may be strengthened in a fraction of the time it took using traditional strengthening exercises.

A muscle's strength depends on three factors: use, the nervous system, and the joint it crosses.[1] When a muscle is paralyzed, every clinician is aware of the possibility of the nervous system causing the weakness, and that the nervous system may be addressed through neuro-developmental treatment, spinal traction for nerve root impingement, and so on. Why then is it that in the past, when a muscle was found weak, the cause was so frequently thought to reside in the muscle itself, without attending to the nervous system or the joint it crosses?

In 1981, I graduated as a physical therapist in the Netherlands. I consider it coincidence when early in my clinical practice I learned that mobilization of the sacroiliac joint could influence the strength of weak hip flexors. However, since then I have observed patients systematically to determine the laws governing the changes in muscle function following joint mobilization. I learned that the influence of joints on muscle tone applies not only to the sacroiliac joint but to all synovial joints in the body. In addition, the influence of joints on the muscles is not random; specific principles apply to the articular control of muscles. The nervous system mediates these muscle tone changes. Using neurophysiological terminology, the mechanisms that explain the altered muscle functioning in response to various joint positions and joint mobilization include: altered articular reflex activity, modulation of the withdrawal reflex, a changed articulo-muscular protective reflex, and inhibition of a facilitated segment. All these phenomena explain the influence of joints on muscle tone and functioning and will be presented in more detail in the following chapters.

This book is the result of 15 years of clinical research and literature studies. Many clinicians have communicated findings similar to mine, and where appropriate, I have included these communications. I believe this book contains what many manual therapists and physicians have experienced in the clinic but, until now, has never been systematically organized.

The muscle tests in this book are not intended to evaluate the maximum force a muscle can produce. This is merely a number and has little meaning. For instance, a child can produce less muscle force than an adult, but that does not mean the child's muscles are weak. Even in sex-, age-, and lean-body-mass-matched individuals, the maximum force a muscle can produce has little meaning. The muscle tests in this book evaluate the quality of a contraction. A muscle can have a good quality of contraction or a poor quality of contraction. A muscle with poor contraction quality needs to be further evaluated to find the cause of its poor performance. If the cause of a muscle weakness is found and corrected, muscle function should improve immediately in most cases, not some five weeks later.

Correction of muscle weakness by mobilization of a joint cannot replace other therapies. It is but another piece of the puzzle among other therapies. However, it is a valuable piece, for it may provide a link between other pieces of the puzzle. For instance, a lumbar evaluation and treatment might be essential in correcting a lower extremity dysfunction or a shoulder condition. The missing link between a correctly functioning shoulder and the lumbar spine might be a properly working latissimus dorsi.

The first chapter of this book provides an overview of the causes of muscle weakness with, at times, detailed reporting of research. The ensuing chapters explore the capacity of joints to cause muscle weakness and the role of the manual therapist in correcting joint dysfunction with the corresponding muscle weakness. Whereas the first half of the book presents mainly research on joint-muscle interactions, the second half provides the practical applications of these interactions. The practical applications are most clearly demonstrated in chapter 19, which presents techniques for evaluating and treating patients. In this chapter, the patient is of the female gender and the therapist is of the male gender, to make the description of the techniques easier to understand; the choice of gender is arbitrary. The last chapter, chapter 20, contains five case studies, integrating the information from the preceding chapters and giving an overview of the applications in a clinical setting.

By presenting both the research and the practical applications of the joint-muscle interactions, I intended to provide the reader with valuable insights into the function of the musculoskeletal system and, on a practical level, provide clinicians a better understanding of their patients' symptomatology. I trust that the information in this book gives clinicians a greater scale of evaluation and treatment options that will benefit their patients.

Reference

1 Cyriax J: Textbook of Orthopaedic Medicine, ed 8. London, Baillière Tindall, 1982, vol 1

1. Muscle Weakness: An Overview

When working with patients, therapists often encounter muscle weakness. This weakness may be the primary complaint of the patient or may be secondary to other conditions. Muscle weakness is typically treated using resistive (strengthening) exercises. The basis for this approach is that weakness is caused by atrophy of the muscle resulting from diminished use; however, a muscle may be weak for several reasons, such as circulatory disorders, aging, myopathy, nociception, disorders of nerve conduction, and inhibition by antagonistic tissue with reduced extensibility (tightness) (see Table 1.1). Each of these potential causes of muscle weakness has unique characteristics. By prescribing strengthening exercises without evaluating the cause of the muscle weakness, one may not be treating the condition appropriately.

Janda wrote in 1986 that muscle weakness has usually been considered a result of decreased activity.[109] Observing that minimal attention had been paid to the idea that decreased muscle activity and the resulting muscle weakness may be due to inhibition, he went on to state that inhibition may be an integral part of many, if not all, forms of weakness. Hurley shared Janda's viewpoint and wrote that muscle weakness may result from two factors: first, a decreased number or size of extrafusal muscle fibers, and second, a failure to activate all muscle fibers.[106] A decreased number or size of muscle fibers may be termed *atrophy*, whereas failure to activate all muscle fibers may be termed *inhibition*.[106,109] If a muscle is inhibited for a prolonged period, muscle fibers may atrophy.[106]

A key question is whether using strengthening exercises to correct muscle weakness addresses both atrophy and inhibition. Various clinical investigators have

Table 1.1 The potential causes of muscle weakness.

- Inactivity
- Immobilization
- Circulatory disorders
- Aging
- Myopathy
  - Muscular dystrophy
  - Inflammatory disorders
  - Endocrine myopathy
  - Metabolic myopathy
  - Toxin- and drug-induced myopathy
  - Infectious myopathy
- Nociception
- Disorders of nerve conduction
  - Upper motoneuron
  - Lower motoneuron
    -- Anterior horn
    -- Peripheral nerve
    -- Neuromuscular junction
- Tight antagonistic structures
  - Skin
  - Muscle
  - Neural tissue
  - Articular ligaments and capsule

reported that strengthening exercises have resulted in changes in the nervous system, such as learning the correct sequence of muscle contractions and/or activating motor units with a higher threshold, that improve the trainee's ability to perform the exercise.[123,125,149,150,157,183] These initial changes in response to strength training cause muscles to be stronger without a matching increase in muscle fiber size.[66,123,125,149,150,157,183] Only at later stages of training will muscle fibers gradually increase in size. These later stages of training may begin after about 4 to 12 weeks.[123,150,157] In addition to changes in the nervous system and in muscle fiber size, a third factor may be responsible for increased strength following a training program. Muscle-specific factors such as increased glycolytic and mitochondrial enzyme activity also may play a role in improved strength by improving the efficiency of force generation.[66,123]

Thus, strengthening exercises may address muscle atrophy, inhibition of the muscle, and muscle-specific factors. Does this make strengthening exercises the treatment of choice for all forms of muscle weakness? This chapter will present the various causes of muscle weakness and suggest treatment strategies for the manual therapist to address each of these causes.

Inactivity

Inactivity is defined here as long-term reduced activity of the muscle while range of motion is maintained. Inactivity weakness may occur during extended bed rest and extended stays in weightlessness (e.g., space flight).[19,21,54,83] Lower limb unloading, such as may occur during non-weight-bearing activities, also may cause inactivity weakness.[19,22,54] Berg et al. and Dudley et al. studied the effects of lower limb unloading on skeletal muscle mass and function and compared them with the effects of bed rest as reported in other studies.[19,54] They reported that lower limb unloading could simulate the effects of bed rest; the reduction in muscle mass and function during unloading was of similar magnitude to that produced by bed rest.[19,54] In the Berg et al. study, six healthy men ambulated for four weeks on crutches with one leg suspended in a harness that prevented weight bearing.[19] While the subjects were sitting or in bed, the harness was not worn and there was no restriction in ankle, knee, or hip mobility. Following the four weeks of unloading, the subjects were asked to resume normal physical activities and to perform 2 to 2.5 hours of resistive-type or aerobic exercises each week. Concentric and eccentric peak torque and angle-specific torque of the quadriceps were obtained with a dynamometer. These were measured before, during, and at various times after unloading. Cross-sectional area (CSA) and radiological density of the quadriceps were measured with CT scans. The study demonstrated marked decreases in muscle strength and CSA in response to unloading. Concentric and eccentric peak torque and angle-specific torque across varying speeds decreased ($p < .05$) by 22 percent and 16 percent, respectively. Radiological density and CSA decreased ($p < .05$) by 6 percent and 7 percent, respectively. Four days of resumed weight bearing partially restored muscle function. Seven weeks after normal activities were resumed, muscle strength and CSA measurements had returned to normal.

In the above study, in response to unloading, torque decreased more than radiological density and CSA. In another study, in which nine volunteers underwent five weeks of bed rest, LeBlanc and colleagues also reported a disproportionate loss in strength compared with the loss in CSA.[136] They found a 26 percent reduction ($p < .05$) in gastrocnemius and soleus strength as measured with a dynamometer and a 12 percent reduction ($p < .05$) in CSA as measured with the help of magnetic resonance imaging (MRI). The twofold reduction in strength as compared with CSA is a common finding in many studies on the effects of inactivity on muscles.[19,21,136]

Berg and Tesch observed 10 healthy men who ambulated with one leg non-weight bearing for 10 days.[22] Five of the men used a harness to unload the leg; the other five walked with the leg hanging freely while weight bearing was prevented by a 10-centimeter sole under the weight-bearing leg. The authors measured isometric quadriceps torque with a dynamometer and the concurrent quadriceps EMG activity. These measurements were taken during maximal voluntary contraction and during contraction at a constant submaximal load. Measurements were taken before unloading, after 10 days of unloading, and twice after the subjects resumed weight bearing. In response to unloading, maximal torque decreased ($p < .05$) by 13 percent, whereas EMG activity did not change. This suggested that inhibition was not responsible for the decreased maximal torque. At submaximal torques, EMG activity increased ($p < .05$) by 25 percent. This increase may have been required to maintain the specific submaximal force level.[21,22] It is postulated that the increased EMG activity compensated for loss of muscle mass, enabling the atrophied quadriceps to produce the same submaximal torque.[21,22] Torque and EMG activity recovered ($p < .05$) after four days of resumed weight bearing.[22] Without obtaining quadriceps cross-section measurements, but based on previous studies on unloading, Berg and Tesch concluded that the pronounced decrease and rapid recovery in maximal torque did not appear to result solely from changes in muscle mass.[22] Neither did inhibition seem responsible for the decreased maximal quadriceps torque after short-term unloading; EMG activity during maximal voluntary contraction had remained unchanged, and the EMG activity during submaximal loads was increased. The authors proposed that some unspecified muscle-specific factors were partly responsible for the reduced maximal torque.[22] They reported that recovery of muscle function after short-term unloading seems to be completed in a shorter time span than the preceding period of inactivity.[22]

In another study, Berg et al. took measurements of the quadriceps in seven healthy men before and after six weeks of complete bed rest.[21] Maximal voluntary isometric and concentric quadriceps torque decreased ($p < .05$) by 25 percent and 29 percent, respectively, as a result of the bed rest. Quadriceps CSA, measured by MRI, decreased ($p < .05$) by 14 percent. The greater loss in strength compared with the loss in quadriceps CSA suggested that inhibition and/or reduced efficiency in force generation contributed to the large decrease in voluntary strength.[21] EMG studies during maximal quadriceps activity before and after the bed rest showed inhibition to be present in addition to the muscle atrophy. Maximal quadriceps EMG activity decreased ($p < .05$) by 19 percent.

At fixed submaximal loads, a certain increase in EMG activity seems logical to compensate for the loss in quadriceps CSA.[21] In the Berg et al. study, EMG activity at submaximal loads (100 N-m) increased ($p < .05$) by 44 percent. The increase in EMG activity at this submaximal load far exceeded what would be expected to compensate for the decrease in quadriceps CSA.[21] This suggested that less efficient force generation was present in addition to atrophy and inhibition.[21] The authors reported that a decreased number of active cross bridges per volume of muscle may explain the less efficient force generation; however, impairment of other muscle-specific factors could not be ruled out.[21]

Muscle strength recuperates faster after shorter rest periods such as 10 days, whereas muscles take longer to regain strength after longer rest periods such as four weeks.[19,20,22] Studies on quadriceps weakness after inactivity suggest that resuming normal weight bearing alone, or resuming normal physical activities plus performing 2 to 2.5 hours of resistive-type or aerobic exercises weekly, may normalize quadriceps strength in four days to seven weeks.[20,21,22,54]

Intense muscle training may prevent the development of weakness in patients on bed rest.[9,10,83] Germain et al. placed 12 healthy male subjects in two groups. The six subjects in the first group performed maximal-intensity isometric and isokinetic leg exercises in the supine position while on 28 days of bed rest. The exercises were performed daily for 30 to 45 minutes during the last 20 days of the bed rest period. The second group was on bed rest as well but performed no exercises. Measurements on both groups were obtained with a dynamometer three days before and after the bed rest. Germain et al. observed a 10 percent reduction ($p \leq .0001$) in quadriceps strength in the control group, whereas no significant strength changes (+4 percent) were observed in the group that exercised.[83] Thus, daily maximal-intensity isometric and isokinetic exercises may prevent muscle weakness in the lower extremities of patients on bed rest.

A MEDLINE search provided no literature on the effects of inactivity on muscles other than those in the lower extremity. Research is needed to determine whether, and to what extent, the results of studies on the lower extremity may be generalized to other skeletal muscles in the body.

Immobilization

Immobilization, such as during casting or bracing, invariably causes muscle weakness. This immobilization-induced weakness may begin after about 1.5 to 5 weeks.[6,40,43,52,53,80,81,144,145,154] The extent to which inactivity during the immobilization is a contributing factor to muscle weakness is unclear; nor do we know the extent to which immobilization-induced stiffness of periarticular tissues contributes to the weakness.

MacDougall et al. placed seven healthy male subjects in two groups. Three subjects trained their triceps for five to six months and then had their upper extremity immobilized in an elbow cast for five to six weeks.[144] Four subjects did the reverse, starting with immobilization followed by resistance training. MacDougall et al. measured the effect of the immobilization and training on the triceps using strength and CSA measurements. Strength measurements were taken with a Cybex dynamometer, and CSA was calculated from cryostat sections of needle biopsies. When the results from all seven subjects were combined, the heavy resistance training resulted in a 98 percent increase ($p < .01$) in the strength of the triceps. The CSA of the muscle fibers increased as well, but to a lesser extent. Fast-twitch fiber CSA increased ($p < .05$) by 39 percent and slow-twitch fiber CSA increased ($p < .05$) by 31 percent. MacDougall et al. found no relationship between the amount of increased strength and the magnitude of the hypertrophy as exemplified by one subject who showed no increase in fast- or slow-twitch fiber size as a result of the training but who increased strength by 143 percent. Following immobilization of the seven subjects' elbows, strength decreased ($p < .05$) by 41 percent. Fast-twitch fiber CSA decreased ($p < .05$) by 30 percent and slow-twitch fiber CSA decreased ($p < .05$) by 25 percent. Again, MacDougall et al. found no relationship between the magnitude of the muscle fiber atrophy and the decreased strength. They stated that maximal elbow extension strength may be affected more by changes within the central nervous system (CNS) than by muscle fiber size changes. The authors did not state what the changes in the CNS entailed,[144] but referred to a study by Milner-Brown et al.[155]

In three studies performed by Milner-Brown et al., EMG activity of the first dorsal interosseus muscle in the hand was recorded.[155] The tendency of the motor units' activity to be grouped over a period of several milliseconds was calculated and termed *synchronization ratio*. In the first study, Milner-Brown et al. recorded data from seven weight lifters and seven control subjects.

More than 80 percent of the motor units in weight lifters had synchronization ratios greater ($p \leq .001$) than .2, whereas less than 20 percent of the motor units recorded from the control subjects had synchronization ratios greater ($p \leq .001$) than .2.

In the second study, four subjects performed six weeks of strengthening exercises on one hand. The exercises consisted of six or more maximal voluntary contractions per day against a spring held between the thumb and index finger. As a result of the exercises, the synchronization ratio increased significantly ($p \leq .01$) above .2, whereas no change was observed in the unexercised hand. In a subsequent six-week period without exercises, synchronization of motor unit activity decreased to slightly above control level.

In the third study, the reflex responses to median nerve stimulation of four weight lifters and five control subjects were studied. EMG was recorded at the thenar muscles. The recorded late reflex responses in the weight lifters suggested changes in supraspinal pathways. The authors postulated that as a result of exercising, direct supraspinal connections from the motor cortex to the spinal motoneurons may be affected in a manner that could increase synchronization of motor units.[155] In response to a lack of exercise, the reverse may have been responsible for the decreased synchronization.[155]

Davies et al. and Duchateau et al. found weakness after immobilization to be caused by muscle atrophy and unspecified changes within the CNS.[43,53] Other studies suggested that these changes within the CNS consisted of a reduced number of motor units being active and a decreased motor unit firing rate.[52,81] Duchateau et al. suggested that muscular intracellular factors such as decreased energy storage also contributed to weakness after immobilization.[53] Thus, intracellular muscle changes and changes in motor unit function may be responsible for the decreased strength after immobilization that muscle atrophy alone cannot account for.[52,53,81,154]

Heavy resistance training such as weight lifting and training on an isokinetic station may undo the effects of immobilization by increasing muscle fiber size and creating changes within the CNS leading to increased strength.[81,144] The changes within the CNS leading to increased strength are increased motor unit activation and possibly increased synchronization of motor unit activity. Exercise may also undo the effects of immobilization by increasing the reduced intracellular energy storage, but few data are available.[53]

Decreased Circulation in the Muscle

Insufficient blood circulation in a muscle may cause weakness and pain.[38,82,97,103,176,216,217] When circulation is insufficient to meet the muscle's demands during repeated contractions, intermittent claudication in the leg or arm may result.[38,65,97,102,103,176] Intermittent claudication is caused by occlusion of the major arteries or, less frequently, occlusion of the major veins supplying an extremity.[38,65,101–103,176,216,217] Causes of reduced circulation are thoracic outlet compression syndrome in the upper extremity,[216] atherosclerosis in the lower extremity,[38,65,97,101–103,130] and progressive arterial vascular diseases such as thromboangiitis obliterans (Buerger's disease) and Raynaud's disease, which may affect both the upper and lower extremities.[49,89,130]

A strong and painless contraction becoming painful and/or weak on repeated contractions or persistent exertion suggests insufficient circulation to the contracting muscle.[38,216] In the lower extremity, pain is usually felt in the calf, but can also occur in the thigh or buttock.[38,65,103,130] If arterial occlusion is the cause of pain, the site of occlusion is proximal to the area of pain.[97,130] Arterial pulses may be absent distally in the extremity.[38,130] Other symptoms of reduced blood circulation may be decreased skin temperature, numbness, and paresthesia.[97,130]

Treatment for muscle weakness secondary to decreased circulation may consist of reduction of compression, as in the manual therapy or surgical treatment of thoracic outlet compression syndrome, or endurance-type exercises to improve the muscle's metabolism.[65,82,102,130,216,217] The optimum endurance-type exercise program should be supervised to maximize the patient's compliance.[65,102,103,175,176] The exercises should be done at a high intensity (i.e., fast walking to moderate or maximal claudication pain) and regularly for at least two months.[38,65,82,102] Walking exercises may improve blood flow in the lower extremities, enhance metabolic activity in the muscle, and improve gait.[65,102,130,176] Exercise therapy is contraindicated when the patient has ischemic ulcers; blood flow in the skin may be reduced during exercises as resistance to blood flow in the exercising muscle decreases.[97,130] Other than exercises, cessation of smoking may help, if applicable, since smoking has a vasoconstrictive effect.[38,65,97,130]

Aging

Advancing age may be accompanied by muscle atrophy and weakness.[25,30,56,67,68,74,76,100,148,169,170,178,202,205,228,229] The increasing muscle atrophy associated with advancing age is termed *sarcopenia*.[67] In men, the age-related decline in muscle strength begins at age 60; in women, it begins with the onset of menopause, unless hormone-replacement therapy is used.[169] Besides muscle atrophy, other causes that may be responsible for declining muscle strength include malnutrition, immobility, reduction in the muscle's metabolic capacity, reduced motor unit discharge rate, and a decrease in the number of motor units.[25,30,68,74,76,138,202,205,228]

Treatment of muscle weakness in the elderly may depend on the cause of weakness. For example, malnutrition may require a different intervention than inactivity.[74] Fiatarone et al. randomly placed 100 frail nursing home residents (mean age of 87 years) into four groups.[76] The different groups received 10 weeks of exercise training, 10 weeks of multinutrient supplementation, both interventions, or neither. The exercise training consisted of hip and knee extensor isokinetic training at 80 percent of the one-repetition maximum. The training was done on alternate days, and each individual session lasted 45 minutes. The residents who underwent exercise training increased muscle strength by 113 percent as compared with 3 percent in the nonexercising residents ($p < .001$). Gait velocity, measured over a six-meter course, increased by 12 percent in the exercising residents and was reduced by 1 percent in the nonexercising residents ($p = .02$). Stair-climbing power, measured on a four-riser staircase with banisters, improved by 28 percent in the exercising residents as compared with 4 percent in the nonexercising residents ($p = .01$). The nutritional supplements showed no effect on any primary outcome measure. The population may have had no nutritional deficits, or the nutritional supplements may not have been sufficient to augment muscle function.[76] Fiatarone et al. concluded that high-intensity resistance exercise training is an effective method of counteracting muscle weakness in very elderly people.[76] Other authors confirmed that high-intensity resistance exercise training leads to improved muscle strength and may compensate for the aging-associated loss of skeletal muscle mass.[74,75,100,148,202,205]

Myopathy

Myopathy is defined as any disease or abnormal condition of striated muscle.[201] Myopathies tend to create symmetrical weakness affecting the proximal muscles to a greater extent than the distal muscles.[41,64,199] In the upper extremities, the muscles inserting on the scapula may be weak while the hand muscles remain unaffected.[64] In the lower extremities, weakness of the hip flexors and extensors is often conspicuous.[64] Myopathy may be associated with normal sensation, and pain is not a primary feature, although it may be secondary.[199]

When a therapist suspects a myopathy is causing a patient's muscle weakness, determining if the patient has had an appropriate medical workup is important. Physical therapy may not be able to address the primary cause of myopathy; however, it may be invaluable in improving the patient's functional status through conditioning exercises, or in addressing symptoms associated with myopathy, such as joint stiffness (reduced extensibility of tissue).

The following myopathies are accompanied by muscle weakness: muscular dystrophy, inflammatory disorders, endocrine myopathy, metabolic myopathy, toxin- and drug-induced myopathy, and infectious myopathy.[55,64,89,195,199]

MUSCULAR DYSTROPHY

Muscular dystrophies are genetically determined disorders that often start in the first two decades of life.[64,163,182] In children, weakness may be suspected due to altered movement patterns and loss of developmental milestones. Three examples of muscular dystrophies are Duchenne's, fascioscapulohumeral, and myotonic muscular dystrophy.

Patients with muscular dystrophy may benefit from physical therapy to maintain range of motion (e.g., stretching fibrotic muscles in patients with Duchenne's muscular dystrophy) and to provide strengthening exercises.[195] Strengthening exercises may have to be done with the assistance of the therapist if the patient is too weak to perform them in a full arc of motion independently. The goals of physical therapy are to maintain or improve the range of motion, strength, and functional skills of the patient.

Duchenne's Muscular Dystrophy—Duchenne's muscular dystrophy affects predominantly males since it is transmitted as an X-linked recessive trait. The weakness in Duchenne's muscular dystrophy starts in the legs (pelvic muscles) at an early age and progresses relatively rapidly to include most muscles.[55,182,195] Pseudohypertrophy of the muscles due to infiltration of fat and connective tissue is often present.[64,182] The connective tissue infiltration decreases muscle extensibility, often creating tightness of the triceps surae and the tensor fascia lata.[195] Weakness of muscles antagonistic to the tight muscles further reduces range of motion. The patient may lose the ability to walk at the age of 9 or 10.[195] The life span of patients with Duchenne's muscular dystrophy varies from the late teens to the early thirties secondary to respiratory or cardiac failure.[195]

Fascioscapulohumeral Muscular Dystrophy—Fascioscapulohumeral muscular dystrophy is inherited as an autosomal dominant or recessive trait and affects males and females equally.[182,195] The disorder is slowly progressive and typically affects the facial and shoulder girdle muscles first,[55,182] hence the name "fascioscapulohumeral muscular dystrophy." At early stages, the patient may be unable to purse the lips or elevate the arms.[195] If onset of the disorder is in early childhood, the patient may lose the ability to walk independently by the age of nine).[55,195] Onset at a later age (15 to 35) provides a much slower progression of the disorder.[55] In both early and late onset, the patient may need a wheelchair for locomotion as the disorder progresses and extremity muscles become weak.[195] Contractures are rare.[195] No specific prognostic information on the longevity of patients with childhood onset of the disorder is available.[195]

Myotonic Muscular Dystrophy—Myotonic muscular dystrophy is inherited as an autosomal dominant trait and affects males and females equally.[182,195] The weakness in myotonic muscular dystrophy is insidious and slowly progressive, affecting predominantly cranial muscles and muscles distally in the extremities (hands and feet).[182,195] In addition to weakness, the muscles display delayed relaxation after a vigorous voluntary contraction, which is termed *myotonia*.[55,182] Clinically, this impedes the patient's ability to cease contracting the muscle and gain full range of motion in the opposite direction. Contractures may develop in patients with this disorder.[195]

INFLAMMATORY DISORDERS

Myopathy caused by inflammatory disorders include polymyositis, dermatomyositis, and systemic sclerosis (scleroderma). The muscle weakness in these disorders follows the typical pattern for myopathy: symmetrical weakness affecting the proximal muscles to a greater extent than the distal muscles.

Polymyositis and Dermatomyositis—As the name "polymyositis" implies, it is a disease causing simultaneous inflammation of multiple muscles.[49,89] The disease results in damage to connective tissue of muscles as well as connective tissue of skin.[89] Patients with polymyositis may have increased muscle enzymes in the serum.[27,47,89] When the disease is accompanied by a red skin rash, it is known as "dermatomyositis."[27,49,89] The skin rash typically occurs on the upper eyelids, often with edema, but red, scaly thickening of the skin also may occur over bony prominences such as elbows, knees, and knuckles.[89]

Proximal symmetrical muscle weakness is a clinical hallmark of polymyositis and dermatomyositis.[27,89] Involvement of facial and distal muscles is rare.[199] Other clinical signs and symptoms are respiratory muscle weakness, dysphagia (secondary to muscle weakness), symmetrical muscle pain or tenderness, arthritis, and weight loss.[89]

Systemic Sclerosis (Scleroderma)—Hardening of connective tissue in patients with scleroderma may cause tightness of the skin, but also may cause tightness of connective tissue in other parts of the body, such as the fascia around muscles.[89,129] Atrophy and weakness of proximal muscles is a common finding in patients with scleroderma.[39,132] In addition to skin tightness and muscle weakness, clinical signs and symptoms include joint and muscle pain, arthritis, Raynaud's phenomenon (intermittent pallor of the digits brought on by cold or stress), and GI tract involvement (e.g., bloating, intermittent diarrhea, and weight loss).[89,132] Flexion contractures due to arthritis or skin tightness are common.[89,132]

ENDOCRINE MYOPATHY

Proximal muscle weakness may be an early manifestation of endocrine myopathy such as hyperthyroidism, hypothyroidism, hyperparathyroidism, acromegaly (hypersecretion of growth hormone), Cushing's syndrome (hyperfunction of the adrenal cortex), and diabetes mellitus (insufficient or defective insulin production by the pancreas).[89,104,199] An imbalance of hormones produced by certain endocrine organs may interfere with the normal growth and development of connective tissue structures in the musculoskeletal system and thus lead to muscle weakness.[89] In addition to proximal muscle weakness, clinical signs and symptoms include pain in muscles, carpal tunnel syndrome, spondyloarthropathy, periarthritis, and fatigue.[89,104,199]

METABOLIC MYOPATHY

A variety of metabolic disorders may cause muscle weakness, including renal tubular acidosis, metabolic alkalosis, hemochromatosis (increased iron storage in the body), and magnesium, phosphate, potassium, and calcium electrolyte imbalance.[89,199] Two examples, hypokalemia and osteomalacia, will be discussed here. Osteomalacia is the result of hypocalcemia.

Hypokalemia—An abnormal low plasma level of potassium (hypokalemia) could manifest itself as neuromuscular disorders ranging from weakness to paralysis.[31–33,49,50,230] After rapid loss of plasma potassium, muscle weakness may develop within 24 hours.[230] Hypokalemia may result from vomiting, diarrhea, kidney disorders, medication, or malnutrition.[31–33,41,50,230]

Human cells have different ion concentrations inside their cell membrane compared with outside, creating the membrane potential. The membrane potential of a nerve cell at rest depends primarily on a high concentration of potassium inside the cell membrane compared with a low potassium concentration outside.[41] When reduced amounts of potassium are present extracellularly (hypokalemia), the concentration difference of potassium inside versus outside the cell increases.[41] As a result, the nerve cell's membrane potential increases and it hyperpolarizes.[41] A hyperpolarized nerve is less excitable because the membrane potential is further removed from its threshold, making it harder to elicit action potentials.[31,41] Thus, it will be harder for the nerve to conduct impulses, and muscle weakness may result.

Osteomalacia—One of the characteristics of osteomalacia or "rickets" is muscle weakness (predominantly proximal and occasionally selective or asymmetrical).[89,185] Although weakness of neck muscles may be present, facial muscles are spared.[185] Other characteristics may be slight muscle atrophy, bone pain or tenderness, fractures, and skeletal deformities.[89,185] Osteomalacia may be caused by a vitamin D deficiency.[185] Vitamin D allows calcium absorption from the gastrointestinal tract.[199] Renal failure, leading to a reduced plasma level of calcium, may also result in osteomalacia.[185]

Calcium within the muscle cell is a necessary element in a series of events that lead to contraction of the cell. Intracellular calcium that attaches to the surface of actin filaments allows the actin and myosin filaments to interact and the sarcomeres to shorten.[41,84,148] More precisely, calcium binds to troponin on the actin filament, which then produces exposure of a receptor site for the myosin head.[84] The myosin head attaches to and actively draws the actin filament so that the muscle fiber may shorten.[84] Thus, calcium is an essential component of the process that allows a muscle to contract. Insufficient amounts of intracellular calcium reduce the actin/myosin interaction and cause varying degrees of muscle weakness.[185]

TOXIN- AND DRUG-INDUCED MYOPATHY

A variety of toxins and drugs may cause myopathy, possibly by harming specific or all body tissues.[27,199] These toxins and drugs include alcohol, cimetidine, clofibrate, colchine, and D-penicillamine. Corticosteroids are other examples of drugs that may cause muscle weakness.[45,89,135,171,177,199,203,207] Corticosteroids are used in the treatment of neuromuscular diseases such as multiple sclerosis, rheumatic and collagen disorders such as rheumatoid arthritis,[23,199] and chronic obstructive pulmonary disease.[45] Although corticosteroids are used to help the patient, the medication may contribute to the weakness accompanying these diseases.[45,89] It is not known at what dosage corticosteroids contribute to muscle weakness.[45,89,135,171,177,199,203,207]

INFECTIOUS MYOPATHY

This category includes certain viral, bacterial, parasitic, and fungal infections that cause myopathy.[199] Examples of viral causes of myopathy are human immunodeficiency virus (HIV), influenza, rabies, and rubella. One example of a bacterial cause of myopathy is salmonella. A parasitic cause of myopathy is toxoplasmosis, and candida tropicalis may be a fungal cause of myopathy. These viral, bacterial, parasitic, and fungal infections may cause muscle weakness by creating inflammation of muscles.[199] Although therapists may not generally treat patients with overt expressions of generalized infections such as fever, treatment of patients with chronic infections such as

HIV may be commonplace. Patients with HIV complain of muscle weakness[135] and may benefit from physical therapy interventions.

Nociception

The clinically familiar changes in posture and movement that accompany pain could be the result of nociceptive reflexes.[225–227] Examples of nociceptive reflexes may be an antalgic gait and involuntary guarding to protect a painful body part. Involuntary guarding, often called "muscle spasm," could be an expression of nociceptive reflexes facilitating muscle tone.[8,166] Besides increasing muscle tone,[8,41] nociception may inhibit muscle tone and cause weakness.[7,26,41,109,297]

Arvidsson et al. examined the effect of nociception on quadriceps strength in 10 subjects, 17 to 38 years of age, who had undergone reconstructive surgery for chronic anterior cruciate ligament (ACL) insufficiency.[7] The researchers took integrated EMG recordings of the quadriceps during maximal voluntary contractions the day after the surgery. The EMG recordings were taken before and at specified times after an epidural injection with an anesthetic. As pain gradually disappeared in response to the epidural injection, EMG recordings of the quadriceps contractions showed increased activity. At the final recording, 20 to 25 minutes after injection, the average integrated EMG was 2,728 percent of the preinjection level. In contrast to these 10 subjects, two healthy volunteers demonstrated a largely unchanged quadriceps function after an epidural injection, as did three other healthy volunteers after anesthetic injection into the distal quadriceps. The authors suggested that pain relief plays a large role in the ability to activate the quadriceps normally after reconstructive surgery for chronic ACL insufficiency.[7]

Janda stated that nociception may inhibit muscles by altering the timing and degree of muscle activation.[109] He described these phenomena as twofold:

1. The start of muscle activation is delayed. This results in a changed activation order of the individual muscles, with activation of the noninhibited muscles occurring earlier.
2. The inhibited muscle displays decreased activity in general. In extreme cases, the muscle may remain nearly silent on EMG.

The effect of nociception on muscle function is thought to occur through polysynaptic projections of nociceptive afferents on alpha motoneurons.[79,224,227] Through these polysynaptic projections, nociception may inhibit or facilitate muscle tone. Theoretically, the nociceptive reflexes result in inhibition of muscles whose contraction increases nociception and in simultaneous reciprocal facilitation of their antagonists (the muscles whose contraction decreases nociception).[7,26,87,109]

If nociception is responsible for muscle weakness, treatment of the weak muscles should address this, for example, by providing pain relief. Manual therapy may provide pain relief.[137,227] This phenomenon is explained by the "gate-control theory" of pain,[137,227] according to which nociception is conveyed by small-diameter myelinated and unmyelinated afferent fibers to transmitting neurons (t cells) within the spinal cord.[174] The activity of these t cells is also under the influence of large-diameter myelinated fibers that carry information from, among others, low-threshold mechanoreceptors.[112,137,152,158,174,211,212] Both the large-diameter afferent fibers and the small-diameter nociceptive fibers activate the t cells. In addition to activating t cells, the large-diameter myelinated fibers can also activate inhibitory interneurons.[112,174] These inhibitory interneurons inhibit the t cells and, in this way, inhibit transmission of nociception. In other words, large-diameter myelinated afferents from low-threshold mechanoreceptors can inhibit transmission of nociception.

The gate-control theory provides a basic explanation for pain reduction in response to mechanoreceptor stimulation[112,211]; however, pain modulation is much more complex than the original gate-control theory proposed.[174] For example, information from some mechanoreceptors is transmitted by large-diameter as well as small-diameter afferents.[174] In addition, the gate-control theory does not explain how local mechanical stimuli that block the pain experience have long-lasting effects beyond the time of mechanoreceptor stimulation.[174] Another criticism of the gate-control theory is that "gating" of nociception occurs not only at the segmental level of afferent input, but also at other, more cranial, levels.[137,174] Thus, although the gate-control theory is correct in its premise that transmission of nociception is under the control of low-threshold afferents, it has its limitations.[112,174,211,212,215] "Sensory gating" may be a better description of the process by which perception of one sensory modality (e.g., nociception) may be reduced by a concomitant stimulation of another (e.g., proprioception).[137]

If we assume that a reduction of nociception, and consequently a reduction of nociceptive reflexes, leads to normalization of nociception-induced muscle tone changes, treatment of such muscle tone changes (including weakness) should focus on reducing noxious stimuli and/or gating of nociception. Gating of nociception is best achieved by stimulation of low-threshold mechanoreceptors located close to the area where nociception originates.[137,166,212] If nociception originates from a muscle, massage of the muscle, passive movements to rhythmically lengthen and shorten the muscle, and rhythmic voluntary contractions are treatment options.[137,227] When nociception originates from a joint, rhythmic, oscillatory joint movements can be used.[137,227] In general, dynamic stimulation may produce more sensory gating than static events, and active movements may produce more gating than passive movements.[137]

Essential to the successful gating of nociception is that the gating stimulus be pain free or that the gating movements take place within a pain-free range.[137]

Disorders of Nerve Conduction

Conduction disorders of nerve fibers may cause muscle weakness by altering the efferent output to muscles. The affected nerves responsible for the muscle weakness may be upper or lower motoneurons.[182] Upper motoneuron lesions are those of the motor cortex or the descending pathways above the cranial motor nuclei and anterior horn cells.[1,34,182] Lower motoneuron lesions are those of the motoneurons of the spinal cord and brain stem that directly innervate skeletal muscles[182] and include lesions of the anterior horn, the peripheral nerve, and the neuromuscular junction. Depending on the location of the lesion, a characteristic pattern of muscle weakness and other manifestations may be present.[127,182]

UPPER MOTONEURON

Upper motoneuron lesions are characterized by increased muscle tone (spasticity), hyperactive tendon reflexes, Babinski's sign, and muscle weakness.[1,34,35,41,64,85,91,139] These characteristics are due to a lack of modulation of lower motoneuron activity by higher centers in the CNS (i.e., the upper motoneurons). In other words, muscle weakness in upper motoneuron lesions may result from an inability to coordinate a movement rather than an inability to generate muscle strength.[24,36]

Increased muscle tone in patients with upper motoneuron lesions may be present in the antigravity muscles: the extensors in the lower extremity and the flexors in the upper extremity.[36,41,139] In the lower extremity, weakness may be more pronounced in the flexors than the extensors.[36,41,64] In the upper extremity, weakness is generally more pronounced in the extensors than in the flexors[36,41,64]; however, this was disputed by Colebatch and colleagues,[34,35] who studied the pattern of weakness in 16 subjects with unilateral arm paresis (n = 10) or paralysis (n = 6) resulting from an upper motoneuron (cerebral) lesion.[34] In the control group consisting of 14 healthy subjects, strength was measured bilaterally with a myometer. In the subjects with paralysis of the arm, strength was only measured on the unaffected side. Colebatch et al. found that the upper extremity extensors were not weaker than the flexors in subjects with hemiparesis; the flexors of the wrist and fingers were on average the most severely affected, whereas the elbow extensors and shoulder adductors and abductors were among the least affected muscles. These two rankings were highly correlated (r = .85, p < .01). This study also found extremity weakness in the contralateral (unaffected) side. Most muscles on the "unaffected" side of the subjects with hemiplegia were weaker than those of the control subjects (p < .01, mean reduction 12 percent).[34] The weakness on the unaffected side may be explained by diminished use or by a bilateral projection from the cortex to extremity muscles.[1,34,35]

In a second study, Colebatch et al. studied maximal voluntary strength (torque) of elbow flexors and extensors in 56 normal subjects and 18 subjects with hemiparesis.[35] The flexor muscles were relatively more affected (inhibited) (p < .01) than the extensors in nearly all subjects with hemiparesis. This study also demonstrated a weakness of muscles on the clinically unaffected side in the subjects with hemiparesis as a group. The weakness was mild, and only four subjects had values more than two standard deviations from the mean for either of the muscles.[35]

Muscle weakness in the clinically unaffected side also may be present in the lower extremity of patients with an upper motoneuron lesion.[1] In a study with 22 control subjects, 16 subjects with unilateral leg paresis, 4 subjects with severe unilateral paralysis, and 5 subjects with paraparesis, Adams et al. showed that strength of the muscles on the clinically unaffected side was reduced compared with the control subjects.[1] The mean strength of muscles on the unaffected side in the subjects with hemiparesis was 93 percent of the muscle strength in the control subjects (p < .01). In the subjects with hemiplegia, the mean strength on the unaffected side was 74 percent compared with the control subjects (p < .01).

In the extremities, the distal muscles may be more affected than the proximal muscles.[1,34,41] In patients with hemiparesis and hemiplegia, muscles of the upper face, the muscles of mastication, and the trunk and respiratory muscles usually are spared, possibly because these muscles are activated bilaterally.[34,214]

Testing of muscles in patients with upper motoneuron lesions is difficult due to the presence of hyperactive tendon reflexes, which increase resistance to passive movements.[41] The amount of resistance by spastic muscles varies with the speed of the movement.[41,139] Movement with higher speed causes spasticity to increase[139]; thus, spastic muscles resist movements that stretch them and limit excursion in this direction. This may interfere with testing of their antagonistic muscles. The therapist may have to adjust standard muscle testing techniques (concentric versus isometric in different parts of the range) to elicit the most accurate measure of strength. It is postulated that recovery of strength in patients with upper motoneuron lesions is influenced by the extent to which other parts of the CNS take over the function that was lost due to the lesion.[85]

ANTERIOR HORN

Anterior horn disorders tend to create weakness and atrophy of the muscles innervated by the involved segment.[41,182] Poliomyelitis is one example in which weakness is found in myotomal distribution.[109] Shingles, or herpes zoster, serves as a second example. Herpes zoster may be accompanied by asymmetrical weakness in the myotome muscles, thus resembling a radiculopathy[153]; however, the herpes zoster virus does not always affect the myotome. It may also create weakness in the distribution area of a peripheral nerve.[153]

PERIPHERAL NERVE

Depending on the site of the peripheral nerve lesion, a specific pattern of affected muscles will be present. All muscles innervated by the affected nerve may be weak, whereas other muscles innervated by more proximal parts of the nerve or by other nerves are strong.[127,216,217] Since most peripheral nerves contain both motor and sensory nerve fibers, affections of peripheral nerves often create motor as well as sensory disturbances.[41,163,216,217] Tendon reflexes are reduced, and Tinel's sign may be positive (i.e., lightly tapping the site of the nerve lesion creates unpleasant sensations in the distribution area of the nerve).[182]

Guillain-Barré syndrome is an example of a peripheral nerve disorder.[89] This syndrome, a demyelinative disease of peripheral nerves, is believed to be an autoimmune disease and is often triggered by surgery, immunization, or viral infection.[41,89,182] The accompanying bilateral muscle weakness usually advances from the legs to the arms, then to the neck and chest.[89] Other clinical signs include sensory disturbances (paresthesia) and diminished tendon reflexes.[89] The patient may complain of nausea and malaise.[89]

NEUROMUSCULAR JUNCTION

The neuromuscular junction is the site where the motor nerve terminal meets the muscle fiber.[41] Under physiological conditions, an action potential over the motor nerve causes the motor nerve terminal to release stored acetylcholine,[41] which binds to receptors at the membrane of the muscle fiber to generate a muscle fiber action potential. Disorders of the neuromuscular transmission will interfere with this process and include disorders of the production and release of acetylcholine, binding of acetylcholine with the receptor, and the receptor response.[41] In myasthenia gravis, the number of functional receptors is reduced.[41,181]

Myasthenia gravis means "severe muscle weakness."[181] In most cases, an autoimmune mechanism blocks or destroys the acetylcholine receptors in the membrane of the muscle fiber.[41,89] The muscles most affected are often the extraocular muscles, producing diplopia (double vision) and ptosis (drooping of the eyelid).[89] Myasthenia gravis affects proximal muscles more than distal muscles. Weakness varies over the course of a day, from day to day, or over longer periods.[181] Muscles fatigue easily, and the weakness is aggravated by exertion.[41,89] Medication that prolongs the effect of acetylcholine at the neuromuscular junction (e.g., pyridostigmine or neostigmine) improves muscle strength.[23,181,203] Corticosteroids may be prescribed to suppress the autoimmune response.[23,135,203]

COMPRESSION

Patients with compression of nerve fibers that impairs conduction of the nerve, such as occurs in radiculopathy, are frequently encountered by physical therapists. Depending on the location, the compression may impair conduction in upper and/or lower motoneurons.[160–162,216,217] The nerve fiber itself or the blood vessels supplying the nerve fiber may be compressed.[29,196] As blood vessels such as venules or capillaries have a lower resistance to compression than nerve fibers, compression is likely to reduce blood flow before it deforms the nerve fiber.[29,196] Thus, compression may cause ischemia of nerve fibers before it deforms the fibers themselves.[29,196] Examples of compression of blood vessels reducing circulation to nerve fibers are the vertebral artery insufficiency syndrome, the thoracic outlet compression syndrome, and other peripheral compression neuropathies.[160–162,216,217] Weakness resulting from compression of blood vessels may alter with positions that increase or decrease the compression. In addition, weakness resulting from compression of blood vessels is present throughout the full range of the muscle unless the magnitude of compression changes within this range. If compression has deformed the nerve fibers, they (and the muscles they innervate) will not respond readily to positions that decrease pressure.

Whereas some neurological disorders are impossible to treat causally with physical therapy, neurological disorders resulting from compression may respond well to physical therapy. Treatment options include reducing compression (e.g., by positional distraction) and mobilizing the nerve.[29,165] Van Meeteren demonstrated that rats with a sciatic nerve lesion caused by compression have a greater functional recovery when subjected to increased physical activity with positive reinforcement.[208] Increased physical activity may improve recovery of nerves by possibly increasing the endoneural blood flow.[208] Van Meeteren made a case for more research to determine if increased physical activity also leads to improved functional recovery in humans with a peripheral nerve lesion.

Antagonistic Tissue with Reduced Extensibility

Tissues with reduced extensibility may inhibit muscle contraction and thus produce muscle weakness. The tissues capable of inhibiting muscles are those that become stretched when the muscle contracts concentrically. These tissues antagonistic to the muscle include skin, muscle, neural tissue, and articular ligaments and capsule. The capability of antagonistic tissues to inhibit muscles may increase with depth; skin may cause little muscle tone change, whereas articular tissue may have more influence on muscle tone.[37,38,210,222]

STRETCH WEAKNESS

Kendall et al. reported the possibility of muscle weakness in response to long-duration lengthening of a muscle, such as may occur in people with chronic abnormal postures.[127] They called this weakness "stretch weakness." Kendall et al. described three patients with stretch weakness superimposed on different types of neurological disorders[127]: a patient with a radial nerve lesion (18 months post-onset), a patient with poliomyelitis and a foot drop (four years post-onset), and a 12-year-old patient with congenital hemiplegia with a wrist drop. These patients were treated by casting the weak muscles in a shortened position for a minimum of one month. Kendall reported that all three patients displayed improved muscle strength as measured by manual muscle tests; the patient with the radial nerve lesion improved strength from grade poor for the wrist and finger extensors to being able to play the piano and type again. The patient with the foot drop improved from trace strength of the anterior tibialis to grade good, and the patient with the wrist drop improved strength of the extensor carpi radialis (grade poor) and the extensor carpi ulnaris (grade fair) to grade good.[127] It is doubtful that casting the weak muscles in a shortened position improved nervous system function, but the effects of previous long-duration lengthening of the muscles by chronic abnormal postures (causing stretch weakness) may have been corrected.[127]

Loubert proposed that the muscle with stretch weakness would exhibit weakness only in a standard testing position.[141] The muscle would generate its greatest force and test strong at its resting length, the length used in the abnormal posture.[91,141,198,213] Loubert based his challenge on morphological and physiological changes during length adaptations of the soleus muscle in rats.[141] Literature suggests that human muscle responds similarly to muscle in experimental animals; in response to maintaining a lengthened position, muscles increase their resting length, producing their greatest force at this length.[91] The muscles would exhibit weakness in a shortened position only.[141]

Kendall et al. recommended treatment of the muscle with stretch weakness by maintaining it in a shortened position and stretching antagonistic tissues with reduced extensibility.[127] Maintaining the muscle in a shortened position causes it to adopt a new resting length and generate its peak tension at approximately the point in the range at which the muscle maintained its length.[91,213] In addition, stretching antagonistic tissues (skin, muscle, neural tissue, and/or joint capsule and ligaments) with reduced extensibility may remove their possible inhibitory effect on the muscle.

SKIN

Afferent input from cutaneous and subcutaneous receptors exerts polysynaptic reflex effects on the spinal gamma motoneuron.[13,93,99] By influencing the gamma motoneuron, afferent input may bias the muscle spindle, influence the alpha motoneuron, and finally, alter muscle tone. Researchers have demonstrated the effect of stimulation of the skin on muscle tone.[46,51,95,140,172,179] Although skin stimulation does not always cause observable responses in muscles, it usually has a subthreshold effect on motoneuron activity.[90]

Clinical examples of the effect of skin stimulation on motoneuron activity are the "local reflexes." These reflexes are called "local" because tactile stimulation of the skin causes local effects, usually a contraction of the muscles beneath the stimulated skin.[41,57,58,95,179] Examples of local reflexes are the abdominal reflex, in which stroking the abdominal skin causes a reflex contraction of the abdominal muscles, and the cremaster reflex, in which stroking the upper inner aspect of the thigh causes a reflex contraction of the cremasteric muscle.[41,90] A third example is Babinski's sign, in which noxious stimulation of the plantar surface of the foot causes flexion of the toes.[41] In addition to tactile stimulation of the skin, stretching the skin may cause contraction of the underlying muscle.[57,58,95]

Skin stimulation may have more widespread effects on muscles as well. Babinski's sign again serves as an example. In the presence of an upper motoneuron lesion, the same stimulus that causes a local reflex in healthy people causes extension of the big toe with a flexion reflex of the entire lower extremity.[41]

The effects of skin stimulation on motoneurons are reciprocally coordinated; specific motoneurons are facilitated while, simultaneously, motoneurons of antagonistic muscles are inhibited.[90,95] For example, the flexion reflex in response to noxious stimulation of the skin causes facilitation of flexor muscles in the limb with simultaneous inhibition of the extensor muscles.

Nicholas et al. demonstrated inhibition of the shoulder abductors after tactile (scratch) stimulation of the skin overlying the pectoralis major.[159] The abductors in both arms were tested using a handheld dynamometer and a Cybex dynamometer. Measurements of strength on the dominant side were obtained after tactile stimulation of the skin while measurements of strength on the nondominant side served as controls. The subjects were a random population of 23 males and females and a group of 17 athletic men. Measurements with the handheld dynamometer following the scratch demonstrated a 19 percent decrease ($p < .05$) in abductor strength in the random population and a 17 percent decrease ($p < .05$) in the athletic population. Measurements with the Cybex dynamometer demonstrated an 8 percent strength decrease ($p < .05$) in the random population and a 4 percent decrease ($p < .05$) in the athletic population.

The relationship between skin receptor activity and muscle tone changes has not yet been researched in depth. If stretching tight skin (e.g., scarred skin) causes increased activity of skin receptors, increased muscle tone changes may be expected. Improving the extensibility of the tight skin such that lengthening causes less skin receptor activity may normalize muscle tone (i.e., tone of the underlying muscle decreases and tone of its antagonist increases). Connective tissue massage or myofascial release applied to the tight skin may be the techniques of choice to increase extensibility.

MUSCLE

Tight antagonistic muscles have an inhibitory influence on their agonists.[109,110,126] Janda found that tight antagonistic muscles may result in delayed activation and weakness of agonists.[109] He called this type of weakness "pseudoparesis."[109,110] Pseudoparesis may be explained by Sherrington's law of reciprocal innervation, which explains how the nervous system inhibits a muscle (group) while its antagonist contracts.[49,109,110,126]

Specific muscles such as the short cervical flexors, the rhomboids, the lower trapezius, and the glutei are more likely to develop weakness or pseudoparesis.[109,110,126] While assessing certain functional movements, inhibition of these muscles may be observed as delayed and/or decreased activity of the muscle.[109] Function of the weak muscles may be replaced by noninhibited stronger muscles, which often are synergists of the inhibited muscle.[109]

Janda suggested treating the weak muscle by stretching the tight antagonists.[109] He proposed that achieving normal length of these tight antagonistic muscles removes the inhibition of the weak muscle and improves its activity. Stretching the tight antagonists restores strength of the inhibited, weakened muscle, often immediately, without any other treatment.[110]

Janda stated that it does not seem reasonable to start treatment of the weak muscle with strengthening exercises.[110] Attempting to strengthen an inhibited muscle with resistance training may have the opposite effect and increase the inhibition. Strengthening exercises intended for the inhibited, weak muscle may result in a contraction of the tight antagonist, since this muscle is easier to activate.[110] For example, Janda described a patient with chronic low back pain who had weak abdominal muscles and tight lumbar extensors.[110] When this patient was given strengthening exercises for the abdominals, her condition worsened due to the lumbar extensors contracting during the sit-up exercises. Thus, the extensors became tighter and the abdominals became weaker. Only after the lumbar extensors were stretched did the abdominals function better, thereby improving her condition.[110] For correction of muscle weakness due to tight antagonists, normalizing the movement pattern and ensuring that the inhibited muscle contracts normally may be essential.[109] After the muscle is "disinhibited," strengthening exercises could be initiated, if needed.

NEURAL TISSUE

Neural tissue is defined here as a single nerve fiber, or a bundle of nerve fibers, with its connective tissue and blood vessels. Limited extensibility of neural tissue itself, and limited mobility of neural tissue in relation to its surrounding structures, has been suggested as potentially reducing joint excursion.[12,28,29,59,62,63,77,96,188] Antagonistic neural tissue lacks adequate research as a potential cause of restricted joint excursion but is nonetheless gaining interest among clinicians.[12,28,29,59,62,63,77,96,131,188] Although neural tissue is capable of restricting joint excursion, whether the reduced joint excursion is caused by mechanical limitation of the neural tissue or whether tension on the neural tissue initiates a protective reflex is not known.[77] Muscles may change tone to protect neural tissue.[8,60,61,63,77,87,96,105] Neural tissue may be responsible for inhibition of muscles, regardless of the nerve fibers' capacity to conduct impulses.

Balster et al. separated 20 asymptomatic subjects into two groups of 10 based on results of the brachial plexus tension test; the subjects with more extensible neural tissue were assigned to one group and those with less extensible neural tissue to the other group.[8] EMG recordings were taken of the upper trapezius muscle during the brachial plexus tension test. Both groups demonstrated increased activity of the upper trapezius muscle as neural tension developed; however, the group with less extensible neural tissue showed increased EMG activity ($p < .001$) compared with the more extensible group. Increased trapezius activity provided increased resistance to lengthening of the brachial plexus. The trapezius activity could be the result of either a stretch reflex protecting the brachial plexus and/or a nociceptive-mediated flexor withdrawal reflex.[8]

Stretching the sciatic nerve, such as is done with straight leg raises or the slump test, may cause activity of the hamstrings.[77,86–88,96] The hamstrings' activity may be a protective response to prevent excessive tension on the sciatic nerve's intra- and/or extraneural connective tissue.[29,87] Hall et al. found hamstrings tone during straight leg raises to be greater ($t = 2.58$, $p = .01$) in subjects ($n = 20$) with an L5 or S1 radiculopathy than in an age- and sex-matched control group of healthy individuals.[96]

These studies suggested that increased neural tension causes increased activity of muscles antagonistic to the movement, possibly as a means of protecting neural tissue against stretching injury. Although not documented in the literature, this increased activity of antagonistic muscles may be accompanied by inhibition of agonistic muscles, which would also prevent increased tension on the neural tissue. Inhibition of agonistic muscles could occur for two reasons: first, the agonists could be inhibited as part of a nociceptive reflex preventing tension on the stretched neural tissue; second, the increased activity of the antagonistic muscles could be expected to inhibit the agonist through reciprocal inhibition.

ARTICULAR LIGAMENT AND CAPSULE

As a joint moves toward its end range, tension in its ligamento-capsular tissue increases, resulting in increased firing rates of embedded receptors and activation of silent receptors.[69,72,157,184] For example, dorsiflexion of the ankle stretches the posterior part of the ankle capsule and stimulates the posterior articular mechanoreceptors.[78] Sufficient activity of low-threshold articular mechanoreceptors may result in articular reflexes. Stimulation of the posterior articular receptors in the ankle joint may cause facilitation of the gastrocnemius with simultaneous reciprocal inhibition of the anterior tibialis.[78,223] The reverse could occur during plantar flexion: the gastrocnemius may become inhibited; muscle tone in the anterior tibialis may increase.

As end range approaches during movement, the articular reflexes prevent potentially excessive joint movement by increasing tone in the stretched muscle and decreasing tone in the shortened muscle.[78,98,146,147,180,200,206] Thus, in response to stimulation of articular mechanoreceptors, the changes in muscle tone limit the joint's movement toward end ranges and can prevent joint damage by limiting excessive excursions.[114,116,119]

These articular reflexes contribute to the functional stability of a joint.[2,78,114,122,167,168,173,191,192]

The reciprocally coordinated muscle tone changes in response to stimulation of articular receptors do not result from the articular receptors directly influencing alpha motoneurons. The articular receptors influence muscle tone via the gamma motoneurons (see Fig. 1.1).[13,70,93,113,115,117–120,122,187,189,190,197] Through them, articular afferent input influences the gamma bias of the spindle cells in the muscle and subsequently the alpha motoneurons.

Enhanced articular reflexes may result from joint effusion, causing increased inhibition of the muscle that moves the joint out of the loose-packed position.[44,69,107,108,111,128,137,193,197,219] For example, effusion of the knee joint may cause increased discharge rates of certain articular receptors, resulting in inhibition of the quadriceps.[71,73,107,128,193,194,219–221] The inhibition of the quadriceps may be present when the knee is extended but absent when the joint is in a loose-packed position.[107,128,137,194,219] The inability of the quadriceps to fully extend the knee, even when normal passive extension is available, has been termed *quadriceps lag*.[194] Stratford studied one group of subjects (n = 8) who had one knee joint effused and another group consisting of sex- and age-matched subjects (n = 8) with normal knees.[194] Bilateral quadriceps (vastus medialis, vastus lateralis, and rectus femoris) activity was recorded by EMG in both groups. EMG measurements during maximal voluntary contraction were obtained at 0 and at 30 degrees of knee flexion. EMG recordings of the quadriceps acting over the normal knees (of subjects in both groups) demonstrated normal activity at 0 and 30 degrees. However, in studying the quadriceps acting over the effused knee joints, the EMG demonstrated decreased ($p < .001$) activity in the vastus medialis, lateralis, and rectus femoris at 0 degrees compared with 30 degrees.

As part of the articular reflexes, receptors in ligaments may signal tension and evoke muscle tone changes, which may serve to reduce tension in the ligament.[186] Ligamentous injury may alter the reflex activity.[48] An example of this concept is ACL damage with its resulting muscle tone changes. In physiological situations, stretch of the intact ACL (by anterior translation of the tibia) causes increased activity of ACL-embedded receptors, which results in a reflex contraction of the hamstrings.[11,93,113,133,164,191,192,218] This reflex contraction reduces the load on the ligament. In patients with ACL-deficient knees, researchers often observed a delayed contraction of the hamstrings in response to anterior translation of the tibia.[191,218] The fact that the hamstrings still contracted suggested that receptors other than those in the ACL were activated during the anterior translation of the tibia.[191] Receptors in other ligaments of the body likely serve a purpose similar to those in the ACL in reducing tension in the respective ligament and protecting the related joint from excessive excursions.[122,186]

In a joint with stiffness in part of its ligamento-capsular tissue, end range will be reached earlier and tension will occur earlier in the range. The increased tension in the ligamento-capsular tissue at end-range positions of the joint may increase the activity of articular mechanoreceptors embedded in the ligamento-capsular tissue along with enhanced articular reflexes: increased facilitation of the muscle antagonistic to the

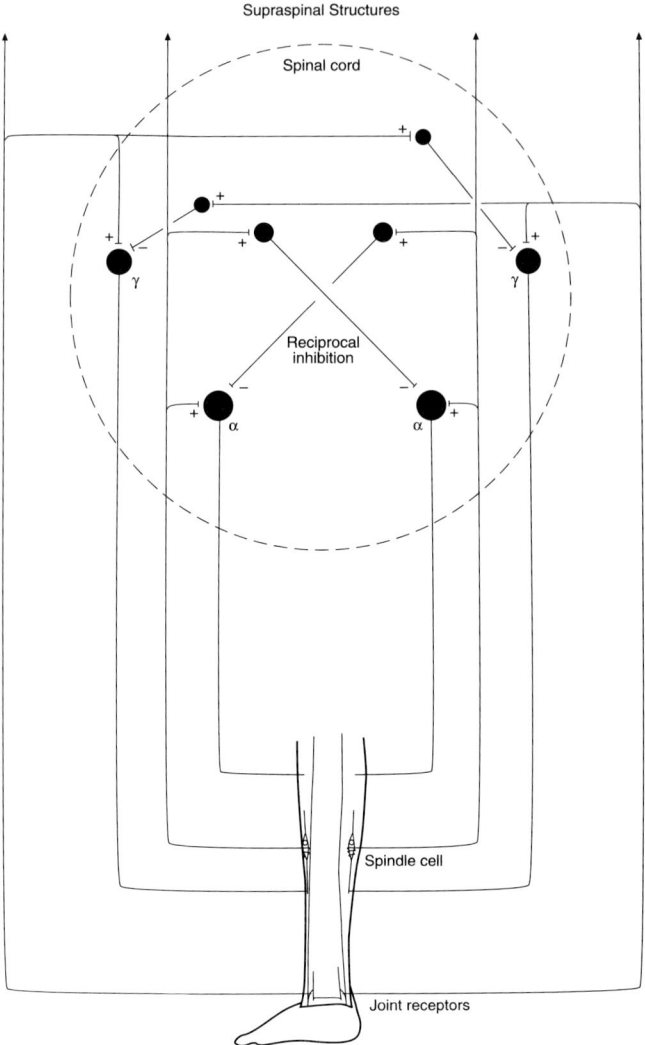

Fig. 1.1 Freeman and Wyke's proposed mechanism for the influence of the low-threshold articular afferents on the muscles.

movement with simultaneously increased inhibition of the agonistic muscle.[14,16–18,78,114,115,118,143,189,197]

When articular ligamento-capsular tissue with reduced extensibility is responsible for inhibition of the agonistic muscle, the muscle is expected to be inhibited only in those positions of the joint where the "inhibiting" tissue is placed under tension. In all other positions of the joint the muscle may have normal strength. A pattern of muscle strength that changes throughout the joint range indicates which tissue with limited extensibility is responsible for the inhibition.

One treatment for muscle weakness caused by inhibition from tight ligamento-capsular tissue is mobilization. The mobilization may range from a direct manipulative technique to an indirect technique such as a Functional Technique or Strain-Counterstrain. The direct manipulative technique increases tissue extensibility and therefore reduces articular receptor firing, whereas the indirect technique will silence the excessive articular afferent input by reducing the length of the ligamento-capsular tissue.[92,124,134]

INTERNEURONS

Interneurons in the spinal cord receive a wide and diverse convergence of afferents from various peripheral tissues.[42,204] For example, a single interneuron may receive afferents from the skin, from several synergistic muscles, and from joints (see Fig. 1.2).[143] In addition to peripheral afferent input, most interneurons receive input from supraspinal sources.[42] The extensive peripheral and descending input to an interneuron has important consequences. Depolarization of the interneuron does not depend on input from a single tissue (e.g., skin), but on the total afferent input converging on this interneuron. Thus, the sum of all inhibitory and facilitory influences on an interneuron determines whether it will depolarize. Input from a single tissue will depolarize the interneuron, provided it receives enough background facilitation from other peripheral and descending sources.

Through these interneurons, afferent input from the skin, muscles, and joints and input from supraspinal sources influence the gamma motoneurons, the muscle spindles, and the alpha motoneurons.[3–5,13,15,42,78,115,117,121,142,209] Thus, afferent input from skin, muscles, and joints has access to alpha motoneurons via interneurons.[15,17,42,142] For example, when a joint flexes while a stiffness is present on the extensor side of the joint, receptors in the skin, muscle, and ligamento-capsular tissue on that side become activated at increased rates.

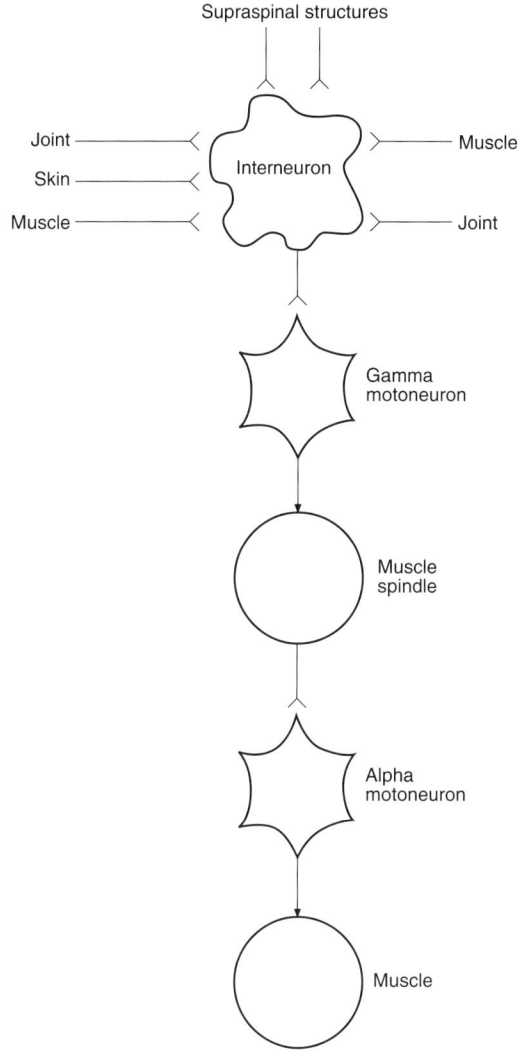

Fig. 1.2 Convergence of information from supraspinal structures and afferents from joints, skin, and muscles onto an interneuron in the spinal cord. The interneuron influences muscle tone via the gamma motoneuron, the muscle spindle, and the alpha motoneuron, respectively.

The sum of the increased afferent input from these tissues influences the interneurons and may influence the alpha motoneurons to inhibit the flexor muscle. Reducing stiffness by stretching any of the tissues with reduced extensibility will reduce the amount of afferent input to the interneurons and may increase strength of the inhibited muscle. Thus, muscle weakness secondary to inhibition by antagonistic tissue, whether it is primarily skin, muscle, or ligamento-capsular tissue, may be treated by addressing any of these tissues, provided their afferent input contributed to the weakness. The concept that multiple tissues may contribute to inhibition of the agonist may explain why biasing a treatment toward a single tissue may be effective for treatment of the whole complex. For example, a technique biased toward muscle, such as Muscle Energy, may be effective for stiffness that seems primarily located in other tissues (e.g., in the joint).[156] Any reduction in dysfunctional afferent input may normalize muscle tone and allow further movement into the range.

RELATIVE IMPORTANCE OF AFFERENT INPUT

Nociception from deep tissues produces more extensive sensory and motor disturbances than nociception from more superficial tissues.[37,38,210,222] Injury to a joint tends to produce more widespread and more prolonged sensory and motor disturbances than muscle lesions, and muscle lesions tend to produce more disturbances than skin lesions.[37,38,210,222] For example, a twisted ankle, which involves relatively minor destruction of joint tissue, produces more tenderness, pain, and changes in reflexes and gait than localized skin damage.[37,222] Reduction of nociception from deep tissues may have greater influence on muscle tone disturbances than reduction of nociception from more superficial tissues. In other words, when therapy has the potential to address a number of tissues (skin, muscle, joint) to normalize muscle disturbances, working at the level of the joint may be most effective.

Summary and Conclusion

Addressing muscle weakness in a patient can be challenging. This chapter categorized and discussed different causes of muscle weakness with the intent of assisting the therapist in the evaluation and treatment of patients with this condition.

Finding the cause, or causes, of decreased strength and providing appropriate therapeutic intervention requires detailed evaluation and frequent reevaluation of the patient. A detailed evaluation allows a treatment tailored to the specific type of weakness. Ideally, the result of a detailed evaluation is twofold. First, by categorizing the muscle weakness appropriately and treating it accordingly, improvement in muscle strength will be optimal. Second, patients with muscle weakness due to causes that are outside the scope of physical therapy will be recognized, thus avoiding unnecessary and ineffective treatment.

The potential causes of muscle weakness may include inactivity or immobilization of muscles, circulatory disorders, aging, myopathy, nociception, disorders of nerve conduction, and inhibition by antagonistic tissue with reduced extensibility (tightness). Along with the diversity of causes of muscle weakness, there is a diversity of treatment approaches. The most commonly used approach, strengthening exercises, addresses atrophy, inhibition, and/or muscle-specific factors that accompany muscle weakness and may be helpful when the weakness is caused by inactivity, immobilization, or aging (see Table 1.2).

Table 1.2 Causes of muscle weakness and potential interventions.

Cause	Intervention
Inactivity	• Strengthening exercises
Immobilization	• Strengthening exercises
Circulatory disorders	• Reducing compression • Endurance exercises • Cessation of smoking
Aging	• Strengthening exercises
Myopathy	• Medical workup • Treating associated symptoms such as poor general condition or stiffness
Nociception	• Reducing noxious stimuli • Gating of nociception
Disorders of nerve conduction	• Reducing compression, if possible • Mobilizing the nerve • Increasing physical activity?
Tight antagonistic structures	• Lengthening of tissues (e.g., joint mobilization)

Strengthening exercises are not appropriate and/or sufficient for other causes of weakness, however. Weakness caused by circulatory disorders may require reduction of the compression impeding the circulation, endurance-type exercises, and/or cessation of smoking. Patients suspected to have weakness caused by myopathy may need medical workup. Physical therapy may improve strength in these patients by addressing symptoms associated with the myopathy, such as poor general condition or joint stiffness. Weakness caused by nociception may be best treated by reducing noxious stimuli and/or gating the nociception. Decreased muscle strength caused by disorders of nerve conduction may, at times, be impossible to treat causally with physical therapy; however, if it results from compression of nerve fibers, the patient may respond well to physical therapy. Treatment options include reducing compression and mobilizing the nerve. Peripheral nerve lesions may have a greater functional recovery in response to increased physical activity of the patient. Muscle weakness caused by antagonistic tissues with reduced extensibility may require lengthening of these tissues. Treating the deeper structures (i.e., the joint) may have a more profound therapeutic effect.

References

1. Adams RW, Gandevia SC, Skuse NF: The distribution of muscle weakness in upper motor lesions affecting the lower limb. Brain 113 (Pt 5):1459-1476, 1990

2. Andersson S, Stener B: Experimental evaluation of the hypothesis of ligamento-muscular protective reflexes: II. A study in cat using the medial collateral ligament of the knee joint. Acta Physiol Scand 48 (suppl 166):27-48, 1959

3. Appelberg B, Hulliger M, Johansson H, Sojka P: Fusimotor reflexes in triceps surae elicited by natural stimulation of muscle afferents from the cat ipsilateral hind limb. J Physiol 329:211-229, 1982

4. Appelberg B, Hulliger M, Johansson H, Sojka P: Fusimotor reflexes in triceps surae muscle elicited by extension of the contralateral hind limb in the cat. J Physiol 355:99-117, 1984

5. Appelberg B, Johansson H, Sojka P: Fusimotor reflexes in triceps surae muscle elicited by stretch of muscles in the contralateral hind limb of the cat. J Physiol 373:419-441, 1986

6. Appell H-J: Muscular atrophy following immobilisation: A review. Sports Med 10(1):42-58, 1990

7. Arvidsson I, Eriksson E, Knutsson E, Arner S: Reduction of pain inhibition on voluntary muscle activation by epidural analgesia. Orthopedics 9(10):1415-1419, 1986

8. Balster SM, Jull GA: Upper trapezius muscle activity during the brachial plexus tension test in asymptomatic subjects. Manual Therapy 2(3):144-149, 1997

9. Bamman MM, Clarke MS, Feeback DL, Talmadge RJ, Stevens BR, Lieberman SA, et al: Impact of resistance exercise during bed rest on skeletal muscle sarcopenia and myosin isoform distribution. J Appl Physiol 84(1):157-163, 1998

10. Bamman MM, Hunter GR, Stevens BR, Guilliams ME, Greenisen MC: Resistance exercise prevents plantar flexor deconditioning during bed rest. Med Sci Sports Exerc 29(11):1462-1468, 1997

11. Baratta R, Solomonow M, Zhou BH, Letson D, Chuinard R, D'Ambrosia R: Muscular coactivation: The role of the antagonist musculature in maintaining knee stability. Am J Sports Med 16:113-122, 1988

12. Barton S, Brunsdon J, Bleicher K: An investigation of different methods of stretching the structures which limit straight leg raise. Proceedings of the International Federation of Orthopaedic Manipulative Therapists, Cambridge, Sept 4-9, 1988, pp 7-8

13. Baxendale RH, Davey NJ, Ellaway PH, Ferrell WR: The interaction between joint and cutaneous afferent input in the regulation of fusimotor neurone discharge. In Jami L, Pierrot-Deseilligny E, Zytnicki D (eds): Muscle Afferents and Spinal Control of Movement. IBRO Symposium at the College de France, Paris, France, September 16-19, 1991. New York, Pergamon Press, 1992, pp 95-104

14. Baxendale RH, Ferrell WR: Modulation of transmission in forelimb flexion reflex pathways by elbow joint afferent discharge in decerebrate cats. Brain Res 221:393-396, 1981

15. Baxendale RH, Ferrell WR: Ascending and descending effects of joint afferent discharge on forelimb and hindlimb flexion reflex excitability in decerebrate cats. Brain Res 332:394-396, 1985

16. Baxendale RH, Ferrell WR: Modulation of transmission in flexion reflex pathways by knee joint afferent discharge in the decerebrate cat. Brain Res 202:497-500, 1980

17. Baxendale RH, Ferrell WR: The effect of elbow afferent discharge on transmission in forelimb flexion reflex pathways to biceps and triceps brachii in decerebrate cats. Brain Res 247:57-63, 1982

18. Baxendale RH, Ferrell WR: The effect of knee joint afferent discharge on transmission in flexion reflex pathways in decerebrate cats. J Physiol 315:231-242, 1981

19. Berg HE, Dudley GA, Häggmark T, Ohlsén H, Tesch PA: Effects of lower limb unloading on skeletal muscle mass and function in humans. J Appl Physiol 70:1882-1885, 1991

20. Berg HE, Dudley GA, Hather B, Tesch PA: Work capacity and metabolic and morphologic characteristics of the human quadriceps muscle in response to unloading. Clin Physiol 13:337-347, 1993

21. Berg HE, Larsson L, Tesch PA: Lower limb skeletal muscle function after 6 weeks of bed rest. J Appl Physiol 82(1):182-188, 1997

22. Berg HE, Tesch PA: Changes in muscle function in response to 10 days of lower limb unloading in humans. Acta Physiol Scand 157:63-70, 1996

23. Berkow R (ed): The Merck Manual of Medical Information. West Point, PA, Merck Research Laboratories, 1997, pp 229-238, 318-322, 333

24. Bobath B: Adult hemiplegia: Evaluation and Treatment, 2nd ed. London, W. Heinemann Medical Books Ltd, 1985, pp 18-19

25. Brooks SV, Faulkner JA: Skeletal muscle weakness in old age: Underlying mechanisms. Med Sci Sports Exerc 26:432-439, 1994

26 Brügger A: Die Erkrankungen des Bewegungsapperates und seines Nervensystemes. New York, Gustav Fischer, 1977, pp 6-31, 712-795

27 Bunch TW: Polymyositis: A case history approach to the differential diagnosis and treatment. Mayo Clin Proc 65:1480-1497, 1990

28 Butler DS, Shacklock MO, Slater H: Treatment of altered nervous system mechanics. In Boyling JD, Palastanga N (eds): Grieve's Modern Manual Therapy, 2nd ed. New York, Churchill Livingstone, 1994, pp 693-703

29 Butler DS: Mobilisation of the Nervous System. New York, Churchill Livingstone, 1991, pp 57-62, 185-201, 247-258

30 Carmeli E, Reznik AZ: The physiology and biochemistry of skeletal muscle atrophy as a function of age. Proc Soc Exp Biol Med 206(2):103-113, 1994

31 Chhabra A, Patwari AK, Aneja S, Chandra J, Anand VK, Ahluwalia TP: Neuromuscular manifestations of diarrhea related hypokalemia. Indian J Pediatr 32:409-415, 1995

32 Chhabra A, Patwari AK, Aneja S, Cjandra J, Anand VK, Ahluwalia TP: Neuromuscular manifestations of diarrhea related hypokalemia. Indian J Pediatr 32:409-415, 1995

33 Chu CC, Huang CC, Chu NS: Recurrent hypokalemic muscle weakness as an initial manifestation of Wilson's disease. Nephron 73:477-479, 1996

34 Colebatch G, Gandevia SC: The distribution of muscular weakness in upper motor neuron lesions affecting the arm. Brain 112:749-763, 1989

35 Colebatch JG, Gandiva SC, Spira PJ: Voluntary muscle strength in hemiparesis: Distribution of weakness at the elbow. J Neurol Neurosurg Psychiatry 49:1019-1024, 1986

36 Cranenburgh B van: Inleiding in de toegepaste neurowetenschappen: Opvattingen over hersenlatsel en hemiplegie/functieherstel en revalidatie. Lochem, Netherlands, Tijdstroom, 1983, vol 2, pp 213-226

37 Cyriax J: Cyriax on Orthopaedic Medicine: Study Guide: The Diagnosis and Treatment of the Soft Tissue Lesions. London, OM Publications, 1981, pp 10-12

38 Cyriax J: Textbook of Orthopaedic Medicine, 8th ed. London, Baillière Tindall, 1982, vol 1, pp 34-35, 60, 419-421

39 D'Angelo WA, Fries JF, Masi AT, Shulman LE: Pathologic observations in systemic sclerosis (scleroderma): A study of fifty-eight autopsy cases and fifty-eight matched controls. Am J Med 46:428-440, 1969

40 Dastur DK, Gagrat BM, Manghani DK: Fine structure of muscle in human disuse atrophy: Significance of proximal muscle involvement in muscle disorders. Neuropathol Appl Neurobiol 5(2):85-101, 1997

41 Daube JR, Reagan TJ, Sandok BA, Westmoreland BF: Medical Neurosciences, 2nd ed. Boston, Little, Brown, 1986, pp 77-85, 177-180, 276-279, 282-294

42 Davidoff RA: Skeletal muscle tone and the misunderstood stretch reflex. Neurology 42:951-963, 1992

43 Davies CTM, Rutherford IC, Thomas DO: Electrically evoked contractions of the triceps surae during and following 21 days of voluntary leg immobilization. Eur J Appl Physiol 56:306-312, 1987

44 DeAndrade JR, Grant C, Dixon AStJ: Joint distension and reflex inhibition in the knee. J Bone Joint Surg Am 47:313-322, 1965

45 Decramer M, Lacquet LM, Fagard R, Rogiers P: Corticosteroids contribute to muscle weakness in chronic airflow obstruction. Am J Respir Crit Care Med 150:11-16, 1994

46 Delwaide PJ, Crenna P, Fleron MH: Cutaneous nerve stimulation and motoneuronal excitability: I. Soleus and tibialis anterior excitability after ipsilateral and contralateral sural nerve stimulation. J Neurol Neurosurg Psychiatry 44:699-707, 1981

47 DeVere R, Bradley WG: Polymyositis: Its presentation, morbidity and mortality. Brain 98:637-666, 1975

48 Di Fabio RP, Graf B, Badke MB, Breunig A, Jensen K: Effect of knee laxity on long-loop postural reflexes: Evidence for a human capsular-hamstring reflex. Exp Brain Res 90:189-200, 1992

49 Dorland's Iillustrated Medical Dictionary, 25th ed. London, Saunders, 1974

50 Dowd JE, Lipsky PE: Sjögren's syndrome presenting as hypokalemic periodic paralysis. Arthritis & Rheumatism 36(12):1735-1738, 1993

51 Drew T, Rossignol S: A kinematic and electromyographic study of cutaneous reflexes evoked from the forelimb of unrestrained walking cats. J Neurophysiol 57(4):1160-1184, 1987

52 Duchateau J, Hainaut K: Effects of immobilization on contractile properties, recruitment and firing rates of human motor units. J Physiol (Lond) 422:55-65, 1990

53. Duchateau J, Hainaut K: Electrical and mechanical changes in immobilized human muscle. J Appl Physiol 62(6):2168-2173, 1987

54. Dudley GA, Adams GR, Belew AH: Adaptations to unilateral lower limb suspension in humans. Aviation, Space, and Environmental Medicine 63:678-783, 1992

55. Dutkowsky JP: Neuromuscular disorders. In Canale ST, Beaty JH (eds): Operative Pediatric Orthopaedics, 2nd ed. New York, Mosby, 1995, pp 774-803

56. Dutta C, Hadley EC: The significance of sarcopenia in old age. J Gerontol 50A (special issue):1-4, 1995

57. Eldred E, Hagbarth KE: Facilitation and inhibition of gamma efferents by stimulation of certain skin areas. J Neurophysiol 17:59-65, 1954

58. Eldred E: Posture and locomotion. In Field J, Magoun HW, Hall VE (eds): Neurophysiology. Washington, American Physiological Society, 1960, sec 1, vol 1, pp 1067-1087

59. Elvey RL: Abnormal brachial plexus tension & shoulder joint limitation. In Gilraine F, Sweeting L (eds): Proceedings of the International Federation of Orthopaedic Manipulative Therapists, Vancouver, June 25-29, 1984, pp 132-139

60. Elvey RL: Clinical aspects of adverse neural tension. In Paris SV (ed): Proceedings of the International Federation of Orthopaedic Manipulative Therapists, Vail, CO, June 1-5, 1992, pp 45-46

61. Elvey RL: Nerve tension signs. In Paris SV (ed): Proceedings of the International Federation of Orthopaedic Manipulative Therapists, Vail, CO, June 1-5, 1992, pp 85-86

62. Elvey RL: The clinical relevance of signs of adverse brachial plexus tension. Proceedings of the International Federation of Orthopaedic Manipulative Therapists, Cambridge, Sept 4-9, 1988, pp 14-21

63. Elvey RL: The investigation of arm pain: Signs of adverse responses to the physical examination of the brachial plexus and related neural tissues. In Boyling JD, Palastanga N (eds): Grieve's Modern Manual Therapy, 2nd ed. New York, Churchill Livingstone, 1994, pp 577-585

64. Epstein O, Perkin GD, de Bono DP, Cookson J: Clinical Examination. St. Louis, Mosby, 1992, pp 1142-1145

65. Ernst E, Fialka V: A review of the clinical effectiveness of exercise therapy for intermittent claudication. Arch Intern Med 153(20):2357-2360, 1993

66. Esselman PC, de Lateur BJ, Alquist AD, Questad KA, Giaconi RM, Lehman JF: Torque development in isokinetic training. Arch Phys Med Rehabil 72:723-728, 1991

67. Evans WJ: Exercise, nutrition, and aging. Clin Geriatr Med 11:725-734, 1995

68. Faulkner JA, Brooks SV, Zerba E: Muscle atrophy and weakness with aging: Contraction-induced injury as an underlying mechanism. J Gerontol A Biol Sci Med Sci 50:124-129, 1995

69. Ferrell WR, Nade S, Newbold PJ: The interrelation of discharge, intra-articular pressure, and joint angle in the knee of the dog. J Physiol 373:353-365, 1986

70. Ferrell WR, Rosenberg JR, Baxendale RH, Halliday D, Wood L: Fourier analysis of the relation between the discharge of quadriceps motor units and periodic mechanical stimulation of cat knee receptors. Exp Physiol 75:739-750, 1990

71. Ferrell WR, Wood L: The effect of increased intra-articular volume on the discharge of stretch receptors in the cat knee joint. J Physiol (Lond) 329:59P-60P, 1982

72. Ferrell WR: The adequacy of stretch receptors in the cat knee joint for signaling joint angle throughout a full range. J Physiol 299:85-99, 1980

73. Ferrell WR. The effect of acute joint distension on mechanoreceptor discharge in the knee of the cat. Quarterly Journal of Experimental Physiology 72:493-499, 1987

74. Fiatarone MA, Evans WJ: The etiology and reversibility of muscle dysfunction in the aged. J Gerontol 48 (special issue):7-83, 1993

75. Fiatarone MA, Marks EC, Ryan ND, Meredith CN, Lipsitz LA, Evans WJ: High-intensity strength training in nonagenarians: Effects on skeletal muscle. JAMA 263(22):3029-3034, 1990

76. Fiatarone MA, O'Neill EF, Ryan ND, Clements KM, Solares GR, Nelson ME, et al: Exercise training and nutritional supplementation for physical frailty in very elderly people. N Eng J Med 330:1769-1775, 1994

77. Fidel, C, Martin E, Dankaerts W, Allison G, Hall T: Cervical spine sensitizing maneuvers during the slump test. J Manual Manipulative Ther 4(1):16-21, 1996

78. Freeman MAR, Wyke B: Articular reflexes at the ankle joint: An electromyographic study of normal and abnormal influences of ankle-joint mechanoreceptors upon reflex activity in the leg muscles. Br J Surg 54:990-1001, 1967

79 Freeman MAR, Wyke B: The innervation of the knee joint: An anatomical study in the cat. J Anat 101(3):505-532, 1967

80 Fuglevand AJ, Bilodeau M, Enoka RM: Short-term immobilization has a minimal effect on the strength and fatigability of a human hand muscle. J Appl Physiol 78:847-855, 1995

81 Fuglsang-Frederiksen A, Scheel U: Transient decrease in number of motor units after immobilisation in man. J Neurol Neurosurg Psychiatry 41:924-929, 1978

82 Gardner AW: Dissipation of claudication pain after walking: Implications for endurance training. Med Sci Sports Exerc 25(8):904-910, 1993

83 Germain P, Guell A, Marini JF: Muscle strength during bedrest with and without muscle exercise as a countermeasure. Eur J Appl Physiol 71:342-348, 1995

84 Ghez C: Muscles: Effectors of the motor system. In Kandell ER, Schwartz JH, Jessell TM (eds): Principles of Neural Science, 3rd ed. New York, Elsevier, 1991, pp 548-563

85 Ghez C: The control of movement. In Kandell ER, Schwartz JH, Jessell TM (eds): Principles of Neural Science, 3rd ed. New York, Elsevier, 1991, pp 533-547

86 Goeken LN, Hof AL: Instrumental straight-leg raising: A new approach to Lasegue test. Arch Phys Med Rehabil 72:959-966, 1991

87 Goeken LN, Hof AL: Instrumental straight-leg raising: Results in healthy subjects. Arch Phys Med Rehabil 74:194-203, 1993

88 Goeken LN, Hof AL: Instrumental straight-leg raising: Results in patients. Arch Phys Med Rehabil 75:470-477, 1994

89 Goodman CC, Snyder TEK: Differential Diagnosis in Physical Therapy, 2nd ed. London, W.B. Saunders, 1990, pp 60-61, 111-112, 325-326, 340-345, 373-374, 380, 473-477, 479-482, 488-493

90 Gordon J: Spinal Mechanisms of Motor Coordination. In Kandell ER, Schwartz JH, Jessell TM (eds): Principles of neural science, 3rd ed. New York, Elsevier, 1991, pp 581-595

91 Gossman MR, Sahrmann SA, Rose SJ: Review of length-associated changes in muscles. Phys Ther 62(12):1799-1808, 1982

92 Greenman PE: Principles of Manual Medicine, 2nd ed. London, Williams & Wilkins, 1996, pp 99-143

93 Grillner S, Hongo T, Lund S: Descending monosynaptic and reflex control of g-motoneurones. Acta Physiol Scand 75:592-613, 1969

94 Gruber J, Wolter D, Lierse W: Der vordere Kreuzbandreflex (LCA-Reflex). Unfallchirurgy 89:551-554, 1986

95 Hagbarth KE: Excitatory and inhibitory skin areas for flexor and extensor motoneurons. Acta Physiol Scand 26 (suppl 94):1-58, 1952

96 Hall T, Zusman M, Elvey R: Adverse mechanical tension in the nervous system? Analysis of straight leg raise. Manual Therapy 3(3): 140-146, 1998

97 Hammond MC, Merli GJ, Zierler RE: Rehabilitation of the patient with peripheral vascular disease of the lower extremity. In DeLisa JA (ed): Rehabilitation Medicine: Principles and Practice. Philadelphia, Lippincott Co, 1993, pp 1082-1086

98 He X, Proske U, Schaible HG, et al: Acute inflammation of the knee joint in the cat alters responses of flexors motoneurons to leg movements. J Neurophysiol 59:326-34,01988

99 Henneman E: Spinal reflexes and the control of movement. In Mountcastle VB (ed): Medical Physiology, 13th ed. St. Louis: C.V. Mosby Co, 1974, pp 651-667

100 Herbison GJ, Graziani V: Neuromuscular disease: Rehabilitation and electrodiagnosis: 1. Anatomy and physiology of nerve and muscle. Arch Phys Med Rehabil 76:S3-S9, 1995

101 Hiatt WR, Regenstein JG, Hargarten ME, Wolfel EE, Brass EP: Benefit of exercise conditioning for patients with peripheral arterial disease. Circulation 81(2):602-609, 1990

102 Hiatt WR, Regensteiner JG, Wolfel EE, Carry MR, Brass EP: Effect of exercise training on skeletal muscle histology and metabolism in peripheral arterial disease. J Appl Physiol 81(2):780-788, 1996

103 Hiatt WR, Wolfel EE, Meier RH, Regenstein JG: Superiority of treadmill walking exercises versus strength training for patients with peripheral arterial disease: Implications for the mechanism of the training program. Circulation 90(4):1866-1874, 1994

104 Houtman PM: Endocriene stoornissen en spier- en gewrichtsklachten. Nederlands Tijdschrift voor Manuele Therapie 14:70-76, 1995

105 Hu JW, Vernon H, Tatourian I: Changes in neck electromyography associated with meningeal noxious stimulation. J Manipulative Physiol Ther 18:577-581, 1995

106 Hurley MV: The effects of joint damage on muscle function, proprioception and rehabilitation. Manual Therapy 2(1):11-17, 1997

107 Iles JF, Stokes M, Young A: Reflex actions of knee joint afferents during contraction of the human quadriceps. Clin Physiol 10:489-500, 1990

108 Iles JF, Stokes M, Young A: Reflex actions of knee-joint receptors on quadriceps in man. J Physiol 360:48P, 1984

109 Janda V: Muscle weakness and inhibition (pseudoparesis) in back pain syndromes. In Grieve GP (ed): Modern Manual Therapy of the Vertebral Column. New York, Churchill Livingstone, 1986, pp 197-201

110 Janda V: Muscles, central nervous motor regulation and back problems. In Korr IM (ed): The Neurobiologic Mechanisms in Manipulative Therapy. New York, Plenum Press, 1978, pp 27-41

111 Jayson MIV, Dixon AStJ: Intra-articular pressure in rheumatoid arthritis of the knee: III. Pressure changes during joint use. Ann Rheum Dis 29:401-408, 1970

112 Jessell TM, Kelly DD: Pain and analgesia. In Kandell ER, Schwartz JH, Jessell TM (eds): Principles of Neural Science, 3rd ed. New York, Elsevier, 1991, pp 385-399

113 Johansson H, Lorentzon P, Sjolander P, Sojka P: The anterior cruciate ligament. Neuro-Orthopedics 9:1-23, 1990

114 Johansson H, Sjolander P, Sojka P: A sensory role for the cruciate ligaments. Clin Orthop 268:161-178, 1991

115 Johansson H, Sjolander P, Sojka P: Actions on g-motoneurones elicited by electrical stimulation of joint afferent fibers in the hind limb of the cat. J Physiol 375:137-152, 1986

116 Johansson H, Sjolander P, Sojka P: Activity in receptor afferents from the anterior cruciate ligament evokes reflex effects on fusimotor neurones. Neurosci Res 8:54-59, 1990

117 Johansson H, Sjolander P, Sojka P: Fusimotor reflex profiles of individual triceps surae primary muscle spindle afferents assessed with multi-afferent recording techniques. J Physiol Paris 85:6-19, 1991

118 Johansson H, Sjolander P, Sojka P: Fusimotor reflexes in triceps surae muscle elicited by natural and electrical stimulation of joint afferents. Neuro-Orthopedics 6:67-80, 1988

119 Johansson H, Sjolander P, Sojka P: Receptors in the knee joint ligaments and their role in the biomechanics of the joint. Crit Rev Biomed Eng 18:341-368, 1991

120 Johansson H, Sjolander P, Sojka P: Reflex actions on the g-muscle-spindle systems of muscles acting at the knee joint elicited by stretch of the posterior cruciate ligament. Neuro-Orthopedics 8:9-21, 1989

121 Johansson H: Reflex integration in the g-motor system. In Boyd IA, Gladden MH (eds): The Muscle Spindle. London, MacMillan Press Ltd, 1985, pp 297-301

122 Johansson H: Role of knee ligaments in proprioception and regulation of muscle stiffness. J of Electromyography and Kinesiology 1:158-179, 1991

123 Jones DA, Rutherford OM, Parker DF: Physiological changes in skeletal muscle as a result of strength training. Quarterly Journal of Experimental Physiology 74:233-256, 1989

124 Jones LH: Strain and counterstrain. Indianapolis, IN, American Academy of Osteopathy, 1981

125 Joynt RL, Findley TW, Boda W, Daum MC: Therapeutic exercise. In DeLisa JA (ed): Rehabilitation medicine: Principles and practice. Philadelphia: Lippincott Co, 1993, pp 526-554

126 Jull GA, Janda V: Muscles and motor control in low back pain: Assessment and management. In Twomey LT, Taylor JR (eds): Physical Therapy of the Low Back. New York, Churchill Livingstone, 1987, pp 253-278

127 Kendall FP, McCreary EK, Provance PG: Muscles: Testing and Function, 4th ed. Baltimore, Williams & Wilkins, 1993, pp 334-335, 394-403

128 Kennedy JC, Alexander IJ, Hayes KC: Nerve supply of the human knee and its functional importance. Am J Sports Med 10:329-335, 1982

129 King LE Jr, Olsen NJ, Puett D, Vital TL, Schulman M, Park JH: Quantitative evaluation of muscle weakness in scleroderma patients using magnetic resonance imaging and spectroscopy. Arch Dermatol 129:246-247, 1993

130 Kisner C, Colby LA: Therapeutic Exercise: Foundations and Techniques, 3rd ed. Philadelphia, F.A. Davis Co, 1996, pp 629-648

131 Kornberg C, Lew P: The effect of stretching neural structures on grade one hamstring injuries. J Orthop Sports Phys Ther 10:481-487, June 1989

132 Korst JK van de: Gewrichtsziekten. Utrecht, Netherlands, Bohn, Scheltema & Holkema, 1980, pp 203-207

133 Krauspe R, Schmidt M, Schaible HG: Sensory innervation of the anterior cruciate ligament. J Bone Joint Surg Am 74:390-397, 1992

134 Kusunose RS, Wendorff R: Strain and Counterstrain Syllabus. Encinitas, Jones Institute, 1990

135 Larson DE (ed): Mayo Clinic Family Health Book, 2nd ed. New York, William Morrow and Co, 1996, pp 919, 969-970, 1061

136 LeBlanc A, Gogia P, Schneider V, Krebs J, Schonfeld E, Evans H: Calf muscle area and strength changes after five weeks of horizontal bed rest. Am J Sports Med 16(6):624-629, 1988

137 Lederman E: Fundamentals of Manual Therapy: Physiology, Neurology and Psychology. New York, Churchill Livingstone, 1997, pp 139-143

138 Lillegard WA, Terrio JD: Appropriate strength training. Med Clin North Am 78(2):457-477, 1994

139 Little JW, Massagli TL: Spasticity and associated abnormalities of muscle tone. In DeLisa JA (ed): Rehabilitation Medicine: Principles and Practice. Philadelphia, Lippincott Co, 1993, pp 666-667

140 Loeb GE, Marks WB, Hoffer JA: Cat hindlimb notoneurons during locomotion: IV. Participation in cutaneous reflexes. J Neurophysiol 57(2):563-573, Feb 1987

141 Loubert PV: Morphological and Physiological Aspects of Length Adaptations of the Soleus Muscle in Spinalized Rats. Thesis. Ann Arbor, MI, University of Michigan, 1986

142 Lundberg A, Malmgren K, Schomburg ED: Reflex pathways from group II muscle afferents. Exp Brain Res 65:271-281, 1987

143 Lundberg A, Malmgren K, Schomburg ED: Role of joint afferents in motor control exemplified by effects on reflex pathways from Ib afferents. J Physiol 284:327-343, 1978

144 MacDougall JD, Elder GCB, Sale DG, Moroz JR, Sutton JR: Effects of strength training and immobilization on human muscle fibers. Eur J Appl Physiol 43:25-34, 1980

145 MacDougall JD, Ward GR, Sale DG, Sutton JR: Biochemical adaptation of human skeletal muscle to heavy resistance training and immobilization. J Appl Physiol 43(4):700-703, 1977

146 Magnus R: Zur regelung der Bewegungen durch das zentralnervensystem: II. Mitteilung. Pflügers Arch ges Physiol 134:253-269, 1909

147 Magnus R: Zur regelung der Bewegungen durch das zentralnervensystem: III. Mitteilung. Pflügers Arch ges Physiol 134:545-583, 1910

148 McArdle WD, Katch FI, Katch VL: Exercise Physiology, 2nd ed. Philadelphia: Lea & Febiger, 1986, pp 282-283, 289-304, 563-590

149 McComas AJ: Human neuromuscular adaptations that accompany changes in activity. Med Sci Sports Exerc 26(12):1498-1509, 1994

150 McDonagh MJN, Davies CTM: Adaptive response of mammalian skeletal muscle to exercise with high loads. Eur J Appl Physiol 52:139-155, 1984

151 McLain RF: Mechanoreceptor endings in the human cervical facet joints. Spine 19:495-501, 1994

152 Melzack R, Wall P: Pain mechanisms: A new theory. J Science 150:971-979, Nov 1965

153 Merchut MP, Gruener G: Segmental zoster paresis of limbs. Electromyogr Clin Neurophysiol 36:369-375, 1996

154 Miles MP, Clarkson PM, Bean M, Ambach K, Mulroy J, Vincent K: Muscle function at the wrist following 9 d of immobilization and suspension. Med Sci Sports Exerc 26:615-623, 1994

155 Milner-Brown HS, Stein RB, Lee RG: Synchronization of human motor units: Possible roles of exercise and supraspinal reflexes. Electroencephalogr Clin Neurophysiol 38:245-254, 1975

156 Mitchell FL Jr: Ostéopathie: Muscle Energy. Course Manual, College d'Etude d'Osteopathique, Montreal, Canada, Feb 1993, p 28

157 Moritani T, de Vries HA: Neural factors versus hypertrophy in the time course of muscle strength gain. Am J Phys Med 58(3):115-130, 1979

158 Nathan PW. The gate-control theory of pain: A critical review. Brain 99:123-158, 1976

159 Nicholas JA, Melvin M, Saraniti AJ: Neurophysiologic inhibition of strength following tactile stimulation of the skin. Am J Sports Med 8(3):181-186, 1980

160 Oostendorp RAB, Bernards ATM, Meldrum HA, Querido C, Hagenaars LHA: Neurologie en manuele therapie: De vertebrobasilaire insufficientie (deel III), Nederlands Tijdschrift voor Manuele Therapie 4:18-39, 1985

161 Oostendorp RAB, Bernards ATM, Querido C, Hagenaars LHA, Meldrum HA: Neurologie en manuele therapie: De vertebrobasilaire insufficientie (deel I, II), Nederlands Tijdschrift voor Manuele Therapie 3:48-73, 1984

162 Oostendorp RAB, Hagenaars LHA, Meldrum HA, Bernards ATM, Querido C: Neurologie en manuele therapie: De vertebrobasilaire insufficientie (deel IV), Nederlands Tijdschrift voor Manuele Therapie 5:16-45, 1986

163 Oosterhuis HJGH: Klinische neurologie: een beknopt leerboek, 5th ed. Utrecht, Netherlands, Bohn, Scheltema & Holkema, 1978, pp 213-218

164 Osternig LR, Caster BL, James CR: Contralateral hamstring (biceps femoris) coactivation patterns

and anterior cruciate ligament dysfunction. Med Sci Sports Exerc 27:805-808, 1995

165 Paris SV, Nyberg R, Irwin M: S2: Course notes. St. Augustine, FL, Institute of Physical Therapy, 1991, pp 163-165

166 Paris SV: Foundations of Clinical Orthopaedics: Course Notes. St. Augustine, FL, Institute Press, 1990, pp 255-257, 293

167 Partridge EJ: Joints: The limitation of their range of movement, and an explanation of certain surgical conditions. J Anat 58:346-354, 1924

168 Peterson I, Stener B: Experimental evaluation of the hypothesis of ligamento-muscular protective reflexes: III. A study in man using the medial collateral ligament of the knee joint. Acta Physiol Scand 48 (suppl 166):51-61, 1959

169 Phillips SK, Rook KM, Siddle NC, Bruce SA, Woledge RC: Muscle weakness in women occurs at an earlier age than in men, but strength is preserved by hormone replacement therapy. Clin Sci (Colch) 84:95-98, Jan 1993

170 Phillips SK, Wiseman RW, Woledge RC, Kushmerick MJ: Neither changes in phosphorus metabolite levels nor myosin isoforms can explain the weakness in aged mouse muscle. J Physiol (Lond) 463:157-167, 1993

171 Physicians' Desk Reference, 52nd ed. Montvale, NJ, Medical Economics Comp, 1998, p 1798

172 Pierrot-Deseilligny E, Bergego C, Katz R: Reversal in cutaneous control of Ib pathways during voluntary contraction. Brain Res 233:400-403, 1982

173 Pope MH, Johnson RJ, Brown DW, Tighe C: The role of the musculature in injuries to the medial collateral ligament. J Bone Joint Surg Am 61:398-402, 1979

174 Porth CM: Pathophysiology: Concepts of Altered Health States, 4th ed. Philadelphia: J.B. Lippincott Co, 1994, pp 983, 994-995

175 Regenstein JG, Steiner JF, Hiatt WR: Exercise training improves functional status in patients with peripheral arterial disease. J Vasc Surg 23:104-115, 1996

176 Regensteiner JG, Gardner A, Hiatt WR: Exercise testing and rehabilitation for patients with peripheral arterial disease: Status in 1997. Vasc Med 2:147-155, 1997

177 Reynolds JEF (ed): Martindale: The Extra Pharmacopoeia, 30th ed. London: Pharmaceutical Press, 1993, p 712

178 Roos MR, Rice CL, Vandervoort AA: Age-related changes in motor unit function. Muscle Nerve 20:679-690, 1997

179 Rossi A, Mazzocchio R: Cutaneous control of group I pathways from ankle flexors to extensors in man. Exp Brain Res 73:8-14, 1988

180 Rossignol S, Gauthier L: An analysis of mechanisms controlling the reversal of crossed spinal reflexes. Brain Res 182:31-45, 1980

181 Rowland LP: Diseases of chemical transmission at the nerve: Myasthenia gravis. In Kandell ER, Schwartz JH, Jessell TM (eds): Principles of Neural Science, 3rd ed. New York, Elsevier, 1991, pp 235-243

182 Rowland LP: Diseases of the motor unit. In Kandell ER, Schwartz JH, Jessell TM (eds): Principles of Neural Science, 3rd ed. New York, Elsevier, 1991, pp 244-257

183 Rutherford OM, Jones DA: The role of learning and coordination in strength training. Eur J Appl Physiol 55:100-105, 1986

184 Schaible HG, Schmidt RF: Responses of fine medial articular nerve afferents to passive movements of knee joint. J Neurophysiol 49:1118-1126, 1983

185 Schott GD, Wills MR: Muscle weakness in osteomalacia. Lancet 1:626-629, 1976

186 Schultz RA, Miller DC, Kerr CS, Micheli L: Mechanoreceptors in human cruciate ligaments. J Bone Joint Surg Am 66:1072-1076, 1984

187 Sjolander P, Johansson H, Sojka P, Rehnholm A: Sensory nerve endings in the cat cruciate ligaments: A morphological investigation. Neurosci Lett 102:33-38, 1989

188 Slater H, Butler DS, Shaklock MO: The dynamic central nervous system: Examination and assessment using tension tests. In Boyling JD, Palastanga N (eds): Grieve's Modern Manual Therapy, 2nd ed. New York, Churchill Livingstone, 1994, pp 587-606

189 Sojka P, Johansson H, Sjolander, Djupsjobacka M: Fusimotor neurones can be reflexly influenced by activity in receptor afferents from the posterior cruciate ligament. Brain Res 483:177-183, 1989

190 Sojka P, Sjolander P, Johansson H, Djupsjobacka M: Influence from stretch-sensitive receptors in the collateral ligaments of the knee joint on the g-muscle-spindle systems of flexor and extensor muscles. Neurosci Res 11:55-62, 1991

191 Solomonow M, Baratta R, Zhou BH, Shoji H, Bose W, Beck C, et al: The synergistic action of the anterior cruciate ligament and thigh muscles in main-

taining joint stability. Am J Sports Med 15:207-213, 1987

192 Solomonow M, D'Ambrosia R: Neural reflex arcs and muscle control of the knee stability and motion. In Scott WN (ed): Ligament and Extensor Mechanisms of the Knee. St. Louis, C.V. Mosby, 1991, pp 389-400

193 Spencer JD, Hayes KC, Alexander IJ: Knee joint effusion and quadriceps reflex inhibition in man. Arch Phys Med Rehabil 65:171-177, 1984

194 Stratford P: Electromyography of the quadriceps femoris muscles in subjects with normal and acutely effused knees. Phys Ther 62:279-283, 1981

195 Stuberg WA: Muscular dystrophy and spinal muscular atrophy. In Campbell SK (ed): Physical Therapy for Children. London, Saunders, 1995, pp 295-324

196 Sunderland S: Traumatized nerves, roots and ganglia: Musculoskeletal factors and neuropathological consequences. In Korr IM (ed): The Neurobiologic Mechanisms in Manipulative Therapy. New York, Plenum Press, 1977, pp 137-152

197 Swearingen RL, Dehne E: A study of pathological muscle function following injury to a joint. J Bone Joint Surg Am 46:1364, 1964

198 Tamai K, Kurokawa T, Matsubara I: In situ observation of adjustment of sarcomere length in skeletal muscle under sustained stretch. Nippon Seikeigeka Gakkai Zasshi 63(12):1558-1563, 1989

199 Tan JC: Practical Manual of Physical Medicine and Rehabilitation. New York, Mosby, 1998, pp 124-125, 345-346, 380-381

200 Tatton WG, Bawa P: Input-output properties of motor unit responses in muscles stretched by imposed displacements of the monkey wrist. Exp Brain Res 37:439-457, 1979

201 Thomas CL (ed): Taber's Cyclopedic Medical Dictionary, 17th ed. Philadelphia: F.A. Davis Co, 1993

202 Thompson LV: Effects of age and training on skeletal muscle physiology and performance. Phys Ther 74(1):71-81, 1994

203 Tierney LM Jr, McPhee SJ, Papadakis MA (eds): Current Medical Diagnosis & Treatment, 37th ed. Stamford, CT, Appleton & Lange, 1998, pp 969-970, 1094

204 Tracey DJ: Joint receptors and the control of movement. Trends Neurosci 3:253-255, 1980

205 Tseng BS, Marsh DR, Hamilton MT, Booth FW: Strength and aerobic training attenuate muscle wasting and improve resistance to the development of disability with aging. J Gerontol A Biol Sci Med Sci 50:113-119, 1995

206 Uexküll J v: Die ersten Ursachen des Rhythmus in der Tierreihe. Ergebn Physiol 3:1-11, 1904

207 United States Pharmacopeia: Complete Drug Reference. Yonkers, NY, Consumer Reports, 1997, pp 595-605

208 Van Meeteren N: Modulation of Peripheral Nerve Repair by Exercise Training and Chronic Stress in the Rat. Thesis. Utrecht, The Netherlands, University of Utrecht, 1994

209 Wadell I, Johansson H, Sjolander P, Sojka P, Djupsjobacka M, Niechaj A: Fusimotor reflexes influencing secondary muscle spindle afferents from flexor and extensor muscles in the hind limb of the cat. J Physiol Paris 85:223-234, 1991

210 Wall PD, Woolf CJ: Muscle but not cutaneous C-afferent input produces prolonged increases in the excitability of the flexion reflex in the rat. J Physiol 356:443-458, 1984

211 Wall PD: Introduction. In Wall PD, Melzack R (eds): Textbook of Pain, 2nd ed. New York, Churchill Livingstone, 1989, pp 11-12

212 Wall PD: The gate-control theory of pain mechanisms: A re-examination and re-statement. Brain 101:1-18, 1978

213 Williams PE, Goldspink G: Changes in sarcomere length and physiological properties in immobilized muscle. J Anat 127(3):459-468, 1978

214 Willoughby EW, Anderson NE: Lower cranial nerve function in unilateral vascular lesions of the cerebral hemisphere. Br Med J 289:791-794, 1984

215 Wilson PR, Lamer TJ: Pain mechanisms: Anatomy and physiology. In Raj PP: Practical Management of Pain, 2nd ed. London, Mosby-Yearbook, 1992, pp 65-78

216 Winkel D, Aufdemkampe G, Meijer OG: Orthopedische geneeskunde en manuele therapie: deel 2b. Diagnostiek extremiteiten. Houten, Netherlands, Bohn Stafleu Van Lochum, 1992, pp 9-30

217 Winkel D, Aufdemkampe G, Meijer OG: Orthopedische geneeskunde en manuele therapie: deel 2c. Diagnostiek extremiteiten. Houten, Netherlands, Bohn Stafleu Van Lochum, 1992, pp 228-229

218 Wojtys EM, Huston LJ: Neuromuscular performance in normal and anterior cruciate ligament deficient lower extremities. Am J Sports Med 22:45-50, 1994

219 Wood L, Ferrell WR, Baxendale RH: Pressures in normal and acutely distended human knee joints

and effects on quadriceps maximal voluntary contractions. Quarterly Journal of Experimental Physiology 73:305-314, 1988

220 Wood L, Ferrell WR: Fluid compartmentation and articular mechanoreceptor discharge in the cat knee joint. Quarterly Journal of Experimental Physiology 70:329-335, 1985

221 Wood L, Ferrell WR: Response of slowly adapting articular mechanoreceptors in the cat knee joint to alterations in intra-articular volume. Ann Rheum Dis 43:327-332, 1984

222 Woolf CJ, Wall PD: Relative effectiveness of C primary afferent fibers of different origins in evoking a prolonged facilitation of the flexor reflex in the rat. J Neurosci 6:1433-1442, 1986

223 Wyke B: Articular neurology: A review. Physiotherapy 58(3):94-99, 1972

224 Wyke B: Clinical significance of articular receptor systems. Ann R Coll Surg Engl 60:137, 1978

225 Wyke B: Neurology of the cervical spinal joints. Physiotherapy 65:73-76, 1979

226 Wyke B: The neurology of joints: A review of general principles. Clinics in Rheumatic Diseases 7:223-239, 1981

227 Wyke BD: Articular neurology and manipulative therapy. In Glasgow EF, Twomey LT, Scull ER, Kleynhans AM (eds): Aspects of Manipulative Therapy. Edinburgh, Scotland, Churchill Livingstone, 1985, pp 72-77

228 Young A, Stokes M, Crowe M: The relationship between quadriceps size and strength in elderly women. Clin Sci 63:35-36, 1982

229 Young A, Stokes M, Crowe M: The size and strength of the quadriceps muscles of old and young men. Clin Physiol 5:145-154, 1985

230 Zimhony O, Sthoeger Z, Ben David D, Bar Khayim Y, Geltner D: Sjögren's syndrome presenting as hypokalemic paralysis due to distal renal tubular acidosis. J Rheumatol 22:2366-2368, 1995

2. Articular Receptors

Joints influence muscle tone and therefore muscle function. The capability of a joint to alter muscle function is mediated by, among others, the articular receptors. The articular receptors can inhibit or facilitate muscle tone. Facilitation of a muscle might reveal itself as increased muscle tone, strength on manual muscle tests, and decreased extensibility (i.e., the muscle is shortened).[6] An example of facilitated muscle tone might be the muscle guarding associated with painful joint disorders. Inhibition of a muscle might reveal itself as decreased muscle tone on palpation, weakness, and increased muscle length.[6] An example of inhibited muscle tone might be the weakness and muscle atrophy associated with painful joint conditions.[10]

The inhibition and facilitation of muscle tone by articular receptors are joint-position dependent. For example, the quadriceps might have normal tone and normal strength on a manual muscle test when tested in 90 degrees of knee flexion; however, the muscle might become weak when the examiner tests the muscle in full knee extension. In this case, the articular receptors in the knee inhibited the strength of the quadriceps when tested in full knee extension. The alteration in contraction output often is not caused by the shortened position of the muscle. For instance, a biarticular muscle such as the hamstrings can be kept at the same length by changing the position of both the hip and knee joints. The hamstrings might have normal tone and test fully strong when tested with the hip and knee in 90 degrees of flexion; however, the muscle might become weak when the examiner tests the muscle in full knee flexion, regardless of whether the hip is flexed. In this case, the articular receptors in the knee cause an inhibition of the hamstrings' strength. The hamstrings do not lose their strength because they are tested in a shortened position; hip flexion can compensate for the length lost over the knee.

A second phenomenon will further demonstrate the influence of the joint on muscle strength. If a joint with restricted movement is mobilized such that normal joint movement is restored, a previously inhibited and weak muscle will test stronger immediately after the mobilization. Using the above example, a flexion mobilization of a stiff knee joint to allow it to bend further reduces the inhibition, and the hamstrings will test stronger immediately after the mobilization.[82] The flexion mobilization of the knee shortens the hamstrings muscle. Despite its reduced length, the muscle will test stronger. This phenomenon can easily be demonstrated throughout the body using various joints and the muscles that cross them.

To comprehend precisely the mechanism by which joints alter muscle tone, an understanding of articular receptor anatomy and physiology is necessary. It is at the level of the articular receptors that the nervous system starts its involvement in altering muscle activity in response to joint movement and position.

Articular Nerves

Joints are supplied by two different types of nerves: the primary and the accessory articular nerves. The primary articular nerves are the articular branches that originate directly from a main nerve trunk.[70,91] For example, the medial knee joint is partly innervated by the rami articularis, directly branching from the tibial nerve.[70]

The accessory articular nerves are branches of nerves that innervate nearby muscles, skin, or periost (see Fig. 2.1).[70,91] An example of an articular branch from a nerve innervating the skin is the ramus infrapatellaris.[70] The skin overlying the medial inferior knee joint is innervated by the ramus infrapatellaris from the saphenous nerve, which also innervates part of the knee joint. The nerves innervating periost also innervate the attachment of the joint capsule to the bone.[70] In addition, nerves innervating muscles give off small articular branches. For instance, the medial section of the knee joint is partly innervated by a nerve coming off the branch supplying the vastus medialis muscle.[70] In 1863, Hilton described joint innervation in what later became known as Hilton's law. He observed that a joint is innervated by the same nerve trunk as the one that innervates the overlying muscle.[40,17]

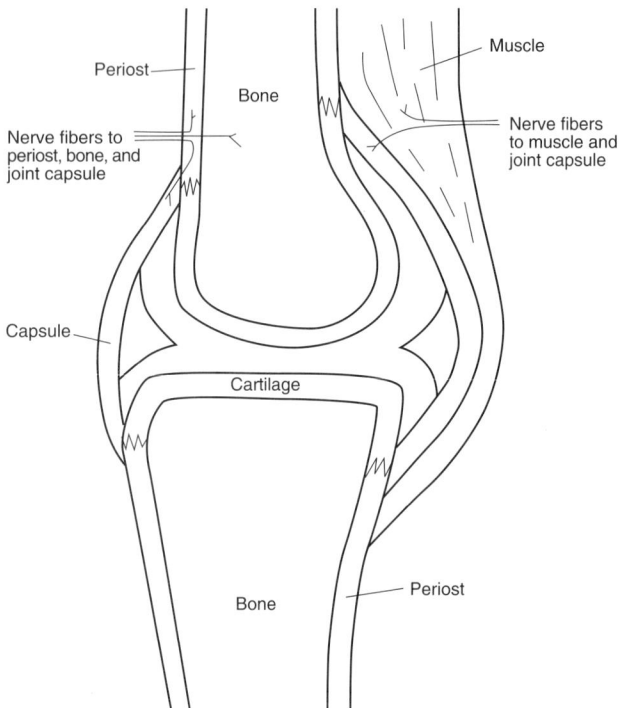

Fig. 2.1 Joint innervation.

The primary and accessory articular nerves often enter the periarticular tissue accompanied by blood vessels.[70,91] Branching of the articular nerve bundles follows that of the blood vessels.[31,36,70,81] Thus, the articular nerves and blood vessels innervate the joint in neurovascular bundles.[17,18,70,91] Once inside the joint capsule, the nerve bundles branch out to form plexuses.[70] Some nerve fibers, such as the unmyelinated efferents, innervate the smooth muscles of the blood vessels themselves, controlling their diameter, while others branch off into the surrounding tissues.[70]

The unmyelinated efferent fibers innervating the smooth muscles in the blood vessels are sympathetic fibers measuring less than 2 μm in diameter.[31,84,91] Thus, the sympathetic fibers control the diameter of articular blood vessels and therefore influence the amount of blood flow though articular tissues.[86,91]

Nerve endings are present in all synovial joints, including the facet, costotransverse, costovertebral, and temporomandibular joints.[1,22,23,45,48,87,95] All synovial joints are innervated multisegmentally. No synovial joint in the body is innervated by nerves that originate from only one spinal segment.[16,21,62,91] Therefore, a radiculopathy with complete obstruction to nerve conduction can never abolish all innervation to a joint. Even the spinal facet joints and the costotransverse and costovertebral joints are innervated by two or more spinal nerves.[29,68,87,90,91]

Innervation of the Joint

The joint capsule consists of two parts: the fibrous capsule and the synovial membrane (see Fig. 2.2).[39,60,83] The fibrous capsule is the superficial or outer layer and consists of parallel and interlacing bundles of connective tissue fibers interwoven with articular blood vessels and nerves. Local thickening of parallel connective tissue forms the capsular ligaments. Examples of capsular ligaments are the glenohumeral ligaments at the shoulder and the iliofemoral ligament at the hip.

In addition to capsular ligaments, accessory ligaments can be located inside or outside the fibrous capsule. Examples of intracapsular ligaments are the cruciate ligaments of the knee and the ligament of the head of the femur. Examples of extracapsular ligaments are the coracoclavicular ligament at the shoulder and the interspinous ligament between the spinous processes. The fibrous capsule and the ligaments assist in holding together the articular bones and in controlling joint motion.[39] Every ligament, for instance, becomes taut at the normal limit of some particular movement.[83]

The fibrous capsule is lined on its inside by the synovial membrane (see Fig. 2.2). The synovial membrane also lines all the intraarticular ligaments and tendons, such as the tendon of the long head of the biceps inside the shoulder joint.[39,83] The only parts inside a joint that are not covered by the synovial membrane are those involved in compression contact during activities of the joint, such as the articular cartilage or the menisci.

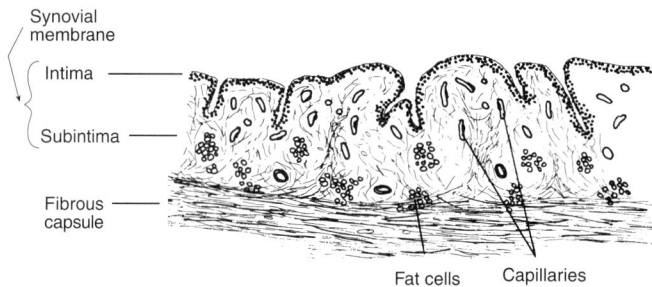

Fig. 2.2 Joint capsule.
Reprinted from Morree JJ de: Dynamiek van het Menselijke Bindweefsel, ed 2. Houten, Netherlands, Bohn Stafleu Van Lochem, 1993, with permission

The synovial membrane consists of two parts: the intima and the subintima. The subintima is located between the intima and the fibrous capsule. The subintima consists of loose connective tissue with a rich supply of capillaries and lymph vessels. Fat cells are part of the subintima but may also be located at the border between the subintima and the fibrous layer.[37] In contrast to the subintima, the intima is a thin layer consisting of only one to four synovial cells.

Nerve endings (sympathetic and afferent) are found in the fibrous capsule, including the capsular ligaments, in the intracapsular and extracapsular ligaments, and in the subintima, including the adipose (fat) tissue.[19,30,41,46,47,49,55,92] In the joint capsule, the greatest number of receptors is found in those parts of the joint most subject to variation in tension during movement.[22,70,86,91] For instance, in the knee and elbow, the greatest number of receptors is found anteriorly and posteriorly. The medial and lateral sides of these joints contain the fewest receptors. In the cruciate ligaments of the knee, receptors are found both in the ligaments themselves and in the synovial layer surrounding the cruciate ligaments.[46]

Until 1982, publications still reported the complete absence of any type of receptor in articular cartilage, synovial membrane, intraarticular menisci, and intervertebral discs in the joints of adults.[9,23,88,91,92,93] Recently, however, researchers have found receptors in the synovial membrane and the annulus fibrosis of the intervertebral discs.[4,5,20,31,46,47,55,58,72,81,93] The intima, which has some blood vessels running through it, was found to be innervated.[47,93] Small-diameter sensory nerves and possibly sympathetic nerves were found around the blood vessels extending through the intima, with some sensory fibers almost as superficial as the synovial surface.[47]

O'Connor and Zimny et al. reported an innervation of the knee menisci, finding the outer and middle third of knee menisci to contain receptors (see Fig. 2.3).[65,96] The anterior and posterior horns of menisci are innervated, possibly even more densely, and the more dense innervation of the anterior and posterior horns may be related to their tissue structure, which resembles ligaments.[65] Articular cartilage, the inner third of the menisci, and the nucleus pulposus of the intervertebral discs are still considered aneural, making it improbable for these tissues to be a direct source of pain.

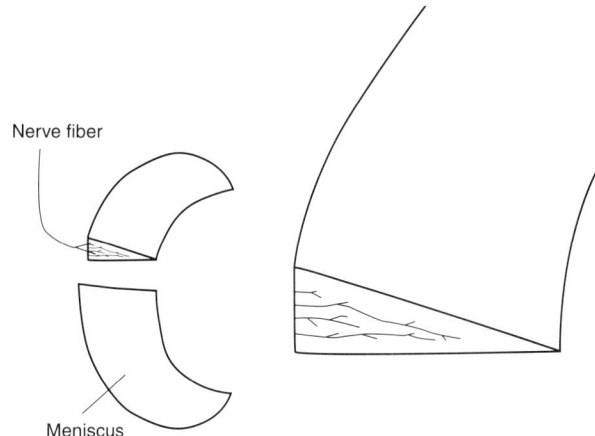

Fig. 2.3 Innervation of the knee meniscus. A portion of the meniscus (left) is enlarged (right). The nerve innervates two-thirds of the meniscus' width.

Articular Receptors

A single afferent fiber may supply one or more receptors. Depending on the number of receptors it supplies, a single afferent fiber may respond to stimulation of a larger or smaller tissue area. The area that, when stimulated, causes a response in the afferent fiber is termed a *receptive field*.[72] The receptive field of an afferent neuron with multiple receptors tends to be larger than that of an afferent neuron with a single receptor.[72] Single-receptor afferents, providing the central nervous system with information regarding a smaller receptive field, have the advantage of being spatially more specific to stimuli.[84] A large receptive field is less specific as to the location of a stimulus.

Stretching of periarticular tissue could cause a change in potential over the articular receptor's membrane. This change in potential is called the *receptor* or *generator* potential of the nerve ending. When the receptor potential reaches a certain value, the axonal portion of the receptor is stimulated, eliciting a barrage of impulses (action potentials) in the parent axon (the afferent fiber). The rate of impulse production in the axon is called the *discharge rate* or the *response frequency*.[72]

Different directions of joint movement stretch different parts of the joint's periarticular tissue and might depolarize articular receptors within the stretched tissue. Thus, receptors can be named by the movement that depolarizes them. For instance, articular receptors located in the front of the knee might depolarize during knee flexion, making these receptors *flexion receptors*.[72] Similarly, hip joint receptors activated by internal rotation of the femur might be termed *internal rotation receptors,* and receptors in the costovertebral joints activated by inspiration might be termed *inspiration receptors*.[22,72]

Freeman and Wyke categorized the articular receptors into four different types: Types I, II, III, and IV.[15,88] Their classification system proved to be useful and has been adapted by more recent researchers and authors.[4,11,12,32,41,57,83] In this book I also will use their classification system. The Type I, II, and III articular receptors are corpuscular mechanoreceptors and are therefore sometimes called the articular mechanoreceptors.[93] The Type IV variety is a nonencapsulated nociceptor. All four types of articular receptors are usually found near blood vessels, some on the vessels themselves.[15,34,36,37,69,79,91,92,93]

All articular receptors are anchored in their surrounding connective tissue.[37,79] Types I through III, the corpuscular receptors, are penetrated by collagen fibers.[35,37,79] Thus, collagen fiber movement can change the shape of the nerve ending.[35]

The four types of articular receptors have different functions. Each is stimulated in a unique way, and each receptor responds uniquely to stimulation. The Type I and II mechanoreceptors function as physiological receptors that are active during normal movement.[48] In contrast, the Type III and IV receptors normally are completely inactive and are only activated at the extremes of movement.[48] Type III and IV receptors might thus function under pathological conditions.[48]

TYPE I ARTICULAR RECEPTOR

Anatomy—The Type I articular receptor, like the Type III, is a Ruffini corpuscle (see Fig. 2.4).[37,86] The existence of such a corpuscle in the joint was first described by Angelo Ruffini, an Italian anatomist, in 1894.[12,17,73] In the periarticular tissues, the Ruffini corpuscles (the Type I and III receptors) are found in the

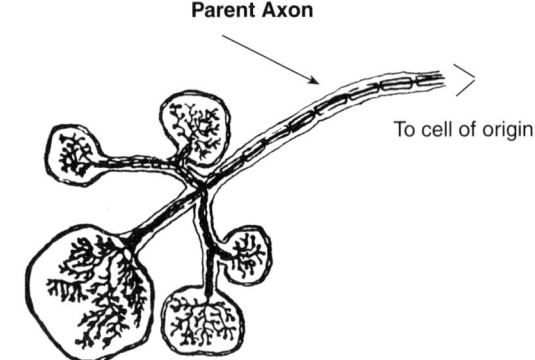

Fig. 2.4 Type I receptor.
Reprinted from Newton, R.A., Joint receptor contributions to reflexive and kinesthetic response. *Physical Therapy*, 62, p. 23, with permission of the APTA.

fibrous layers of the joint capsule and within the articular ligaments.[34,36,37] Halata and other researchers described various types of Ruffini corpuscles in the joint capsule.[37,75,77,97] One type had a well-developed perineural capsule and was located in connective tissue structures with a parallel orientation of the collagen fibers, such as ligaments. These large Ruffini corpuscles resembled Golgi tendon organs and were possibly identical to the Type III articular receptors described by Freeman and Wyke and other researchers.[15,37,88] Besides these larger Ruffini corpuscles, Halata also described a smaller type of Ruffini corpuscle without a perineural capsule. These smaller corpuscles were located in the connective tissue structures with no predominant orientation of the collagen fibers, like a capsule without ligamentous reinforcement. These Ruffini corpuscles with no capsule seem identical to the Type I articular receptors described by Wyke.[91,93]

Although the Ruffini-type corpuscles can be divided into a Type I and a Type III articular receptor, this division is arbitrary. A continuous spectrum of varieties of spray endings is present.[44] Thus, besides receptors with and without a perineural capsule, receptors also might be thinly encapsulated.[57]

In addition to being present in the fibrous capsule, authors have described the Type I receptors in the outer two-thirds of knee menisci, the anterior and posterior cruciate ligaments, and the capsular ligaments.[3,12,46,75,77,83,96,97] Grigg et al. and McLain described what are possibly Type I receptors in the synovial membrane close to the fibrous capsule.[28,57]

In the extremities, the proximal joints such as the shoulder and the hip contain more Type I receptors than the distal joints.[11,85,86,91] The number of Type I receptors decreases in a distal direction such that few are present in the interphalangeal joints of the hand and foot. In the facet joints of the spine, the population density is greatest in the cervical spine and decreases in a caudal direction.[85,86,91]

The Type I mechanoreceptor consists of a bundle of small, round corpuscles similar to bunches of grapes.[93] Up to eight corpuscles can be present within a cluster, however, McLain described them as single in the cervical spine.[12,57,91]

The afferent nerve of the Type I receptor is a medium-sized myelinated nerve fiber. The afferent nerve's diameter is described as 3 to 6 μm by Halata, as 5 to 8 μm by Clark and Wyke, and as 6 to 9 μm by Wyke.[9,34,37,85] Table 2.1 summarizes the anatomy of the articular receptors.

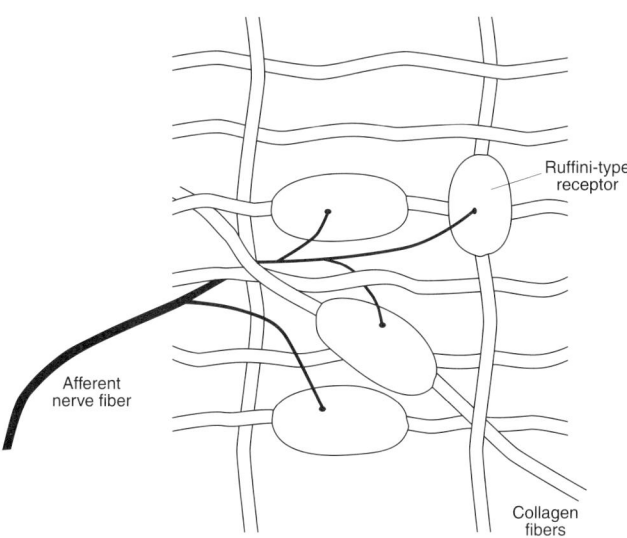

Fig. 2.5 Ruffini receptor endings penetrated by collagen fibers.[84]

Table 2.1 Articular receptor anatomy.

Type		Location	Afferent Fiber Size
I	Small Ruffini	Capsule (fibrous capsule and part of synovial membrane)	3–9 μm
		Capsular and cruciate ligaments	
		Knee menisci	
II	Pacini	Fibrous capsule	4–10 μm
		Adipose tissue	
		Ligaments	
		Knee menisci	
		Annulus fibrosus cervical discs	
III	Large Ruffini (Golgi tendon organ)	Ligaments (capsular, extracapsular, and intracapsular)	13–18 μm
		Capsule (between fibrous and synovial layers)	
		Menisci	
		Cervical discs	
IV	Free nerve endings	Capsule (fibrous, intima)	1–5 μm
		Fat pads	
		Ligaments	
		Menisci	

Physiology—The Type I receptor is a stretch receptor with a very low threshold, responding to lengthening of those collagen fibers that penetrate it.[23,25,26,34,35,37,48,83,84,91,93] The smallest change in position of the collagen fibers will cause a change in the receptor's shape.[35] Several receptors, all belonging to one afferent nerve, can depolarize under different stresses, depending on the directions of the collagen fibers that penetrate them (see Fig. 2.5).[25,26]

The discharge rate of the receptor is dependent on the amount of stretch of the surrounding connective tissue.[22,93] The maximum discharge rate always occurs at the extremes of movement and never at "loose-packed" positions.[22] Movement will cause increased tension of the capsule on one side of the joint and decreased tension on the opposite side (see Fig. 2.6). Similarly, Type I articular receptor activity increases on the side of the increased tension, and receptor activity will decrease in the opposite side of the capsule, where tension decreases.[22,91,93] Thus, a Type I flexion receptor, located on the extensor side of the joint capsule, is most active in a flexed position of the joint (i.e., when the Type I's surrounding connective tissue is stretched).[22,93] If a receptor is most active in flexion of a joint, it is least active in extension.[22] Similarly, an extension receptor is most active in extension and least

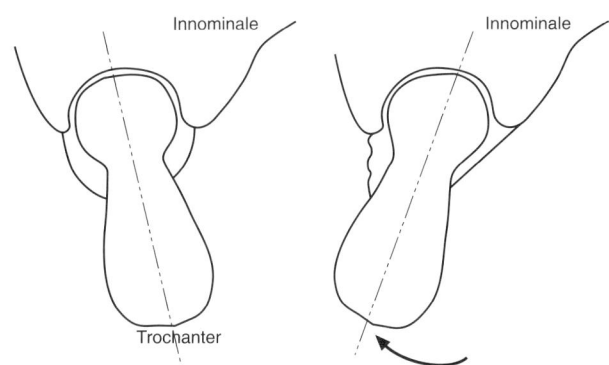

Fig. 2.6 Schematic representation of the hip joint as seen from above. As the femur rotates internally, the anterior capsule relaxes while the posterior capsule tenses up.

Adapted from Rossi A, Grigg P: Characteristics of the hip joint mechanoreceptors in the cat. J Neurophysiol 47: pp. 1029–1042, 1982, with permission from The American Physiological Society.

active in flexion.[22] The Type I receptors in other antagonistic parts of the joint respond in a similar manner (e.g., shoulder abductor and adductor receptors and internal and external rotation receptors).

Joint position is likely signaled based on tension in different regions of the soft tissues of a joint.[22] Type I receptors have a near-linear relationship between their discharge rate and joint position or stress on the surrounding connective tissue (see Fig. 2.7).[22,25,27] Godwin-Austen found that for every position of the joint there is a characteristic receptor discharge rate.[22] The near-linear relationship between Type I receptor activity and joint angle may be secondary to the fact that the Type I receptor adapts only slowly to any increased tension in its surrounding connective tissue.[93] Mountcastle went even further and called the Type I receptor an absolute detector of position.[61]

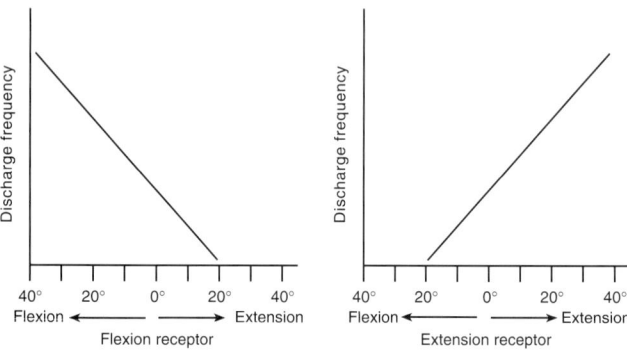

Fig. 2.7 The relationship between joint position and Type I discharge.

The Type I receptor's low threshold causes it to respond to very small changes of tension in its surrounding connective tissue.[91,93] In fact, some Type I receptors have such a low threshold that they signal joint position, even when the joint is immobile and in a loose-packed position.[86,91,93] These slowly adapting mechanoreceptors are active throughout the range of movement in response to the small static tension that prevails throughout the capsule.[11,22,42,91,93] In loose-packed positions, the passive tension in the capsule is caused by the relative vacuum within the joint attempting to "suck in" the capsule. In addition, the pull of muscles attaching to the capsule creates tension in the capsule.[91] An increase in joint capsule tension by active or passive movement, or by mobilization or manipulation, causes these receptors to discharge at a higher frequency and causes additional Type I receptors with a higher threshold to become active.[86,93]

During movement, the Type I receptor's discharge frequency changes. The speed of change of Type I receptor discharge indicates the velocity of joint movement.[22,63,93] As a result of the coordinated changes of the antagonistic receptors' discharge, the Type I receptor signals direction, velocity, and amplitude of a movement.[86,91,93] Again, Mountcastle termed the Type I receptor an absolute detector of direction, speed, and amplitude of movement.[61]

Distraction (an attempt to separate the two surfaces of a joint) stimulates Type I mechanoreceptors on all sides of the joint.[93] The discharge frequency of agonistic and antagonistic Type I receptors is proportional to the amount of traction force.[93] If the traction force is sustained, the frequency of the Type I receptor discharge remains proportional to the traction force.[93]

The Type I receptor influences the gamma motoneuron.[14,92] More specifically, evidence suggests that the Type I parent axon projects, via interneurons located in lamina IV of the spinal gray matter, to tonic gamma motoneurons.[91] The Type I articular receptor influences the tonic motor unit activity of muscles, probably via these tonic gamma motoneurons.[66,67,91] The following chapters will describe in detail the influence of articular receptors on muscle tone. The function of articular receptors is summarized in Table 2.2.

Table 2.2 Articular receptor function.

Type	Threshold	Adapting	Detector of
I	Low	Slowly	Joint position
			Joint movement
II	Low	Rapidly	Joint movement
III	High	Slowly	Large amounts of tension
			Position and movement, at end-range positions?
IV	High	Slowly	Noxious chemical and mechanical stimuli

TYPE II ARTICULAR RECEPTOR

Anatomy—The Type II articular receptor, a large laminated structure (see Fig. 2.8), is similar to the Pacinian receptor found elsewhere in the connective tissue of the body.[83] Its relatively large size might be the reason it was described earlier in joints than the Ruffini corpuscle.[12,18,50,57,64,71] Rauber was possibly the first to describe the articular Pacinian corpuscle, in

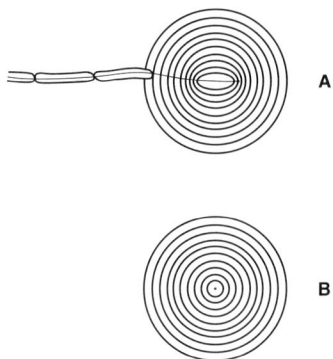

Fig. 2.8 Longitudinal section (A) and cross section (B) of the Type II receptor.

1867.[71] He described corpuscle nerve endings as varying in size from 0.5 to 4 mm in length, suggesting that the size of the Pacinian receptor varies from about the size of the dot of an *i* to a size slightly larger than a capital *O*.

The Type II receptor is located within the fibrous capsule, between the fibrous capsule and the synovial membrane, within the subintima and the ligaments, and inside adipose tissue.[3,12,15,33,35,37,38,41,51,83,91,93] It has been described extensively in the cruciate ligaments.[36,46,69,75,77,96,97] Zimny et al. described a Paciniform (Type II) receptor in the outer two-thirds of the knee menisci.[96] In the cervical spine, the receptor was found in the annulus fibrosis of the intervertebral discs.[58]

Within the joint capsule, the Type II receptor more densely populates those parts that undergo more variations in tension, similar to the Type I receptor.[91] However, the population density in the extremity joints is the reverse of that of the Type I receptor.[91] In the extremities, the population density of the Type II receptor increases in a distal direction.[11,85,86] The interphalangeal joints contain relatively more Type II receptors than proximal joints such as the hip or shoulder.[11] In the spinal facet joints, the population density is greatest cranially and decreases in a caudal direction.[91] Thus, Type II receptors are more numerous in the cervical spine compared with the lumbar spine.

The Type II receptor is individually (or in groups of up to four corpuscles) attached to its afferent nerve fiber.[12,86,91] Freeman and Wyke reported the afferent nerve fiber to be a relatively fast-conducting, medium-sized nerve of 9 to 12 μm.[15,90] However, more recent research found a smaller diameter of 4 to 10 μm for human Type II afferents.[36,37,41] The larger diameter found by Freeman and Wyke is possibly due to their research being done on cats.[37] Fiber diameter differences have been found between species.[70]

The Type II receptor itself consists of a terminal nerve ending surrounded by 1 to 30 concentric lamellae of perineural capsule.[37] Extracellular fluid separates the lamellae.[84] Similar to the Type I receptor, collagen fibers penetrate the receptor.[79]

Physiology—Like the Type I receptor, the Type II has a low threshold; however, unlike the Type I, it rapidly adapts to stimuli.[11,48,63,91] When the collagen fibers that penetrate the receptor shorten or lengthen, the Type II receptor (with its lamellae) is deformed. The lamellae are separated from each other by fluid, and slow deformation of the lamellae allows this fluid to move away from regions with high pressure.[8,52,53] This movement of the interlamellar fluid allows the outer lamellae to deform while the inner lamellae retain their original shape. Consequently, during slow deformation of the Type II receptor, this damping effect of the lamellae and the interlamellar fluid prevents the generation of a potential at the nerve ending. Only fast deformation of the Type II receptor produces a generator potential because it prevents the interlamellar fluid, which has a certain viscosity, from moving away in time. However, even when a generator potential is produced at the core of the receptor, the interlamellar fluid keeps moving until a new balance is created and the receptor becomes quiet again. Thus, although the receptor is deformed, and the tissue outside the receptor might be under tension or be deformed as well, the receptor has adapted and is "silent." Removal of tissue pressure or tension again causes a deformation of the receptor with movement between the lamellae and a possible depolarization.[53,56] Thus, the Type II receptor does not distinguish between "tension on" and "tension off." During movement, both agonistic and antagonistic Type II receptors will deform. The agonistic receptor might be deformed by increased tension in its surrounding connective tissue, whereas the antagonistic receptor will be deformed simultaneously by decreased tension in its surrounding tissue.

As a result of the Type II receptor's fast adaptation, it only responds to rapid changes in tension, adapting in less than half a second to any new tension in the capsule.[11,56,70,89,90,92,93] During active or passive movement of a joint, or during the application of traction, the Type II briefly discharges at the moment of sudden increased tension in its surrounding connective tissue.[91,93] The receptor becomes quiet during sustained positions or mobilizations and during immobilization.[91,93] On cessation of a mobilization or traction, the Type II receptor fires again briefly.[93] The receptor may generate a maximum number of impulses during rapid oscillations.[89,90,92]

Godwin-Austen researched the response properties of rapidly adapting articular receptors in the costovertebral joints. During normal breathing, no activity could be detected in these rapidly adapting receptors; only rapid movements of the ribs caused receptor activity, and discharges ceased altogether within 50 µs of movement termination.[22] These rapidly adapting receptors, regardless of their location in the costovertebral joint, discharged in response to either inhalation or exhalation.[22] Burgess and Clark reported similar transient responses to joint movements, regardless of whether this position was achieved through an agonistic or an antagonistic movement.[7]

In contrast to the Type I receptor that signals direction, velocity, and amplitude of movement, the Type II receptor merely signals that a movement is taking place.[22,93] Slow movements do not elicit Type II receptor activity. The speed of a movement determines the number of potentials generated. Therefore, the Type II receptor may be a rate-of-motion detector.[76]

Because of its low threshold, the Type II receptor can detect small movements, provided they are fast; however, despite its low threshold, it is completely useless for postural or kinesthetic sensation.[63] The receptor does not contribute to postural or kinesthetic sensation because it is inactive when the joint is immobile and because it has a short discharge duration during movement.[11,89]

The Type II receptor projects polysynaptically to the gamma motoneuron.[92] Evidence suggests that the Type II's parent axon projects via interneurons located in lamina IV to phasic gamma motoneurons, possibly influencing the phasic motor units of muscles.[66,67,91] Through its influence on gamma motoneurons, Type II receptor activity might be responsible for brief muscle activity at the onset of movement.[11] The Type II receptor may behave as a "booster" mechanoreceptor, briefly evoking supplementary motor unit activity in agonists at the initiation of a reflex, with simultaneous brief inhibition of the antagonist, to overcome the inertia of the part to be moved.[85] Once the reflex causes or stops a movement, the Type II activity ceases, leaving only Type I activity.

TYPE III ARTICULAR RECEPTOR

Anatomy —The Type III receptor is the largest of the articular corpuscles and is identical in structure and function to the Golgi tendon organ (see Fig. 2.9).[83,86,91] Both receptors may serve a protective function; on reaching the limit of joint range of motion, both tend to cause reflex inhibition of the agonist muscles.[59] The difference between the Golgi tendon organ and the Type III articular receptor is their location. The Golgi tendon organ is located in tendons, whereas the Type III receptor is located in the joint ligaments. Both receptors are situated in environments with a similar parallel arrangement of collagen fibers.[11,42,83]

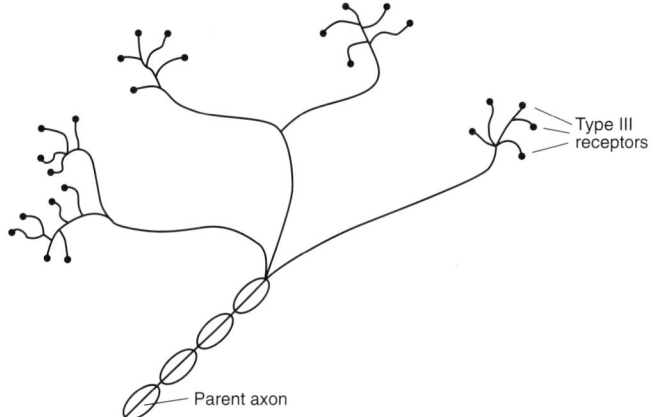

Fig. 2.9 Type III receptor.

The Type III receptor is a Ruffini corpuscle with a well-developed perineural capsule that is penetrated by collagen fibers.[37] Inside the Type III capsule, the nerve terminal is anchored in the connective tissue belonging to the ligament.[37]

In addition to the Type III receptor being the largest of the articular receptors, its afferent neuron is also the largest of the four receptor afferents. One to maximally three Type III corpuscles contact a single afferent nerve fiber. In cats, the Type III's afferent nerve fiber was described as being 13 to 17 µm in diameter.[12,91] Griffin and Harris described the afferent nerve fiber to the human temporomandibular joint as 15 µm in diameter.[24] Its large size makes this axon the fastest conducting articular afferent.

Capsular, intracapsular, and extracapsular ligaments contain Type III receptors.[12,34,46,74,75,77,91,97] Wyke described the Type III receptors as absent from the spine, although in an older publication he had described them in the facet joints.[86,91,93] McLain described Type III receptors in the cervical spine between the fibrous capsule and the subsynovial tissue, and Mendel et al. found receptors that resemble Type III's in the cervical discs.[57,58] Zimny described Golgi tendon organs (Type III receptors) in the outer two-thirds of menisci.[96]

Physiology—Because of its structural similarities to the Type I articular receptor, the Type III receptor shares a similar function. Both are slowly adapting mechanoreceptors, but unlike the Type I (and the Type II), the Type III receptor has a high threshold.[11,37,70,91,93] The Type III receptor responds only to high tensions generated in its surrounding connective tissue and is inactive during active and passive movements in loose-packed positions of a joint.[2,42,80,91,93] Consequently, the Type III articular receptor becomes active only at the extremes of joint movement, or when strong mobilization or traction forces are applied, such that high tensions are developed in the joint ligaments.[2,42,80,91,93] The Type III receptor is quiet during traction or mobilization of lesser intensity.[93]

If the extreme joint displacement or the considerable traction or mobilization force is maintained, the receptor adapts only very slowly.[86] Because of its slow adaptation, the Type III's discharge frequency is a continuous function of the magnitude of stress on the surrounding connective tissue.[93]

When the joint is at an end-range position, the Type III receptor could detect position and direction of movement.[63,78] Skoglund wrote that the receptor's low sensitivity makes it less sensitive to signal speed of movement.[78] The function of Type III receptors, when activated by extreme movement, is to inhibit the tone of the prime movers of the joint.[11] For example, stimulation of Type III flexion receptors by extreme flexion of a joint would cause inhibition of the flexors. Despite the Type III's seemingly simple effect on muscles (i.e., to inhibit them), its true function might be more complex. The inhibitory function could be compared with a car's brake pedal. The brake can be applied to slow the car (inhibit the muscle strength), or released to speed it up.[11,54,59] Thus, the result of Type III afferent discharge could be an inhibition or a facilitation of muscles. Which of these occurs depends on concurrent information arriving at interneurons from other Type III receptors, or from Golgi tendon organs, as well as information from low-threshold articular and skin receptors.

TYPE IV ARTICULAR RECEPTOR

Anatomy—The Type IV articular receptor is not a corpuscle but a free nerve ending without a perineural capsule (see Fig. 2.10). The terminal parts of the free nerve endings are anchored in the surrounding connective tissue.[37] The Type IV receptor is a nociceptor.[37,93] Its activity is often the reason patients consult us. The location of Type IV receptors is of particular interest to the medical profession because it answers the question, "What tissues can hurt?"

Wyke described the Type IV articular receptor as located in the entire thickness of the fibrous capsule, in the articular fat pads, and in the articular ligaments.[91,92,93] The presence of the nociceptor in the fibrous layer and within the ligaments has been confirmed by recent authors.[3,4,13,36,37,46,75,77,97]

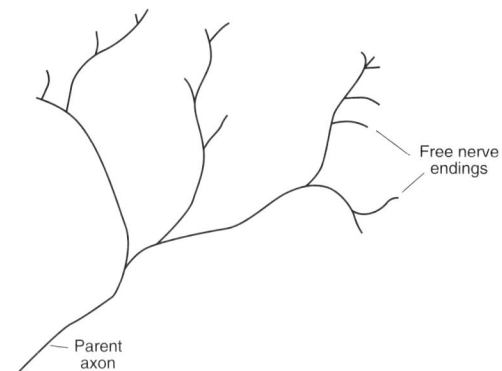

Fig. 2.10 Type IV receptor.

Wyke reported that the nociceptor is completely absent from the synovial layer, intraarticular menisci, and articular cartilage; however, more recently, researchers have located Type IV receptors in the outer two-thirds of knee menisci and at the border between the synovial and fibrous layers.[4,20,36,37,57,93,96] The Type IV receptor may even be located in the intima, almost as far as the synovial surface.[47]

In the joint capsule, the Type IV receptors form a feltlike network of unmyelinated nerve fibers.[84,92] This network is mainly located around the small articular blood and lymphatic vessels.[15,36,37,79,92] Type IV receptors attached to a single nerve fiber may terminate simultaneously at the blood vessel wall and in the surrounding tissue.[70] This means that the nerve itself branches to end on two different structures: on a blood vessel and on a structure near it. Such endings may be the morphological basis of the axon reflex (see chapter 10, "Substance P").[70]

The afferent nerve fibers of Type IV receptors are very small myelinated (2–5 μm) or unmyelinated (less than 2 μm) fibers, according to Wyke.[91] More recent publications have reported that the afferent nerves of the free nerve endings are unmyelinated or myelinated fibers measuring 1 to 2 μm in diameter.[36,37] These small afferents slowly conduct nociceptive impulses to the dorsal horn within the spinal cord. Gillette and colleagues found unilateral Type IV afferent fibers from a cat spinal facet joint to run to the bilateral dorsal horns.[21] Gillette also observed afferent fibers to run to more cranial and caudal segments. For instance, the afferent fibers innervating a unilateral L5 facet joint in the cat might run to the bilateral dorsal horns of levels L1 through S1. Even though the afferent fibers run to dorsal horns bilaterally, more fibers run to the ipsilateral dorsal horns. Similarly, most afferent fibers run to the segment the facet joint is part of, with fewer fibers running to the more cranial and caudal segments. Gillette and colleagues wrote that, as the innervation of the facet joints is multisegmental and bilateral, facet joint pain must be broadly distributed, poorly localized, and/or felt as radiating pain in extremities.[21]

Within the dorsal horn, the Type IV afferents terminate on interneurons. Some interneurons contact neurons that run to the brain, whereas others project polysynaptically through interneurons to motoneurons located in the anterior horn.[94] The neurons that run to the brain might produce pain, whereas the synapses on the anterior motor cells might produce changes in muscle tone.[94]

Physiology—Normally, the Type IV articular receptor is completely inactive.[92] It becomes active only when mechanical and/or chemical irritation stimulates the unmyelinated nerve endings.[91,92] Mechanical irritation is caused when abnormally high tensions develop in the articular tissue.[91] Examples of abnormal tension causing joint irritation are dislocation or subluxation of a joint, swelling (hemarthrosis or effusion), and prolonged distorted posture.[91]

The Type IV articular receptors are bathed in tissue fluid. Alteration of the chemical composition of this fluid produces chemical irritation of Type IV articular receptors.[91] High concentrations of irritant chemical substances such as lactic acid, K+ ions, prostaglandin E, and histamine may depolarize Type IV receptors.[91,93] These irritant chemicals are released by traumatized, ischemic, or inflamed tissues.[91]

Under normal circumstances, the chemical composition of the synovia is in equilibrium with that of the tissue fluid in the joint capsule.[91] When part of the joint capsule becomes inflamed, irritating chemicals in the synovial fluid increase, with a similar increase of these chemicals in the tissue fluid of the entire capsule.[91] Therefore, a local joint capsule irritation might be followed by an irritation of capsular Type IV receptors in the entire joint.[91]

Chemical irritants can sensitize the normally high-threshold Type IV nociceptors. Sensitization means that the Type IV's threshold is reduced. Whereas under normal circumstances strong mechanical stimuli would be required to excite Type IV receptors, sensitization can cause light stimuli to evoke action potentials. A patient with a chemical sensitization of a joint (for instance, due to an inflammation) might experience pain to light touch or with minimal movement.

Thus, abnormal chemical and/or mechanical tissue changes are responsible for depolarization of the Type IV receptor. This depolarization of the Type IV receptor and the conduction of the input over its afferent nerve is termed *nociception*. Nociception is not identical to pain. Only when nociception reaches the limbic regions of the cerebral cortex is it called pain.[90,93,94]

The intensity of a pain experience is not directly related to the intensity of the peripheral nociceptor irritation. The afferent nociceptive information en route to the cerebral cortex is modulated.[94] Modulation of nociception occurs at multiple levels due to inhibitory and facilitory influences.[94] One such inhibitory influence at the spinal level is from the peripheral mechanoreceptors.[93] In the spinal gray matter, peripheral mechanoreceptor activity modulates transmission of impulses from the Type IV afferents onto ascending neurons.[93] This modulation of nociception at the spinal level is termed *gating of nociception*.[93] The articular mechanoreceptors (i.e., the Type I, II, and III receptors and other nonarticular receptors) can gate nociception.[90] The inhibitory effect of articular mechanoreceptors on nociception will be discussed further under the topic of "gate control" (see chapter 5).

Summary

The Type I, II, and III articular receptors are corpuscular receptors that function as mechanoreceptors. The low-threshold Type I receptor is a detector of position and movement. The Type II receptor also has a low threshold and is active only briefly in response to movement, possibly functioning as a booster receptor to the Type I receptor. The Type III receptor is located predominantly in the articular ligaments and has a high threshold, which makes it respond only to large amounts of stress in its parent tissue. The Type IV receptor is a free nerve ending and functions as a nociceptor. All four types of articular receptors influence motoneuron function and therefore muscle tone. The Type I and II articular receptors influence muscle tone via the gamma motoneurons.

References

1. Abe K, Takata M, Kawamura Y: A study on inhibition of masseteric α-motor fibre discharges by mechanical stimulation of the temporomandibular joint in the cat. Arch Oral Biol 18:301-304, 1973

2. Aloisi AM, Carli G, Rossi A: Responses of hip joint afferent fibers to pressure and vibration in the cat. Neurosci Lett 90:130-134, 1988

3. Babu KS, Devanandan S: Sensory receptors situated in the interphalangeal joint capsule of the fingers of the Bonnet Monkey (Macaca radiata). Acta Anat 153:49-56, 1995

4. Biedert RM, Stauffer E, Friederich NF: Occurrence of free nerve endings in the soft tissue of the knee joint. Am J Sports Med 20:430-433, 1992

5. Bogduk N: The innervation of the intervertebral discs. In Boyling JD, Palastanga N: Grieve's Modern Manual Therapy, ed 2. Churchill Livingstone, New York, 1994, pp 149-161

6. Brügger A: Die Erkrankungen des Bewegungsapparates und seines Nervensystems. New York, Gustav Springer Verlag, 1980

7. Burgess PR, Clark FJ: Characteristics of knee joint receptors in the cat. J Physiol 203:317-335, 1969

8. Catton WT, Petoe N: A visco-elastic theory of mechanoreceptor adaptation. J Physiol 187:35-49, 1966

9. Clark RKF, Wyke BD: Contributions of temporomandibular articular mechanoreceptors to the control of mandibular posture: An experimental study. J Dent 2:121-129, 1974

10. deAndrade JR, Grant C, Dixon ASJ: Joint distention and reflex muscle inhibition in the knee. J Bone Joint Surg Am 47:313-322, 1965

11. Dee R: Mechanoreceptors in hip joint capsule and ligamentum capitis femoris and their reflex contribution to posture. In Symposium on Osteoarthritis. St. Louis, C.V. Mosby Co, 1976, pp 52-65

12. Dee R: The innervation of joints. In Sokoloff L (ed): The Joints and Synovial Fluid. London, Academic Press, 1978, pp 177-205

13. El-Bohy A, Cavanaugh JM, Getchell ML, et al: Localization of substance P and neurofilament immunoreactive fibers in the lumbar facet joint capsule and the supraspinous ligament of the rabbit. Brain Res 460:379-382, 1988

14. Freeman MAR, Wyke B: Articular reflexes at the ankle joint: An electromyographic study of normal and abnormal influences of ankle-joint mechanoreceptors upon reflex activity in the leg muscles. Br J Surg 54:990-1001, 1967

15. Freeman MAR, Wyke B: The innervation of the knee joint: An anatomical and histological study in the cat. J Anat 101:505-532, 1967

16. Gardner E: The innervation of the knee joint. Anat Rec 101:109-130, 1948

17. Gardner E: Physiology of movable joints. Physiol Rev 101:127-176, 1950

18. Gerlach F: Die Endkörperchen der sensiblen Nerven in der Gelenkkapsel. Jahresb u d Fort Anat u Physiol 18:381-382, 1889

19. Giles LGF, Harvey AR: Immunohistochemical demonstrations of nociceptors in the capsule and synovial folds of human zygapophyseal joints. Br J Rheumatol 26:362-364, 1987

20. Giles LGF, Taylor JR: Innervation of lumbar zygapophyseal joint synovial folds. Acta Orthop Scand 58:43-46, 1987

21. Gillette RG, Kramis C, Roberts WJ: Spinal projections of cat primary afferent fibers innervating lumbar facet joints and multifidus muscle. Neurosci Lett 157:67-71, 1993

22. Godwin-Austen RB: The mechanoreceptors of the costo-vertebral joints. J Physiol 202:737-753, 1969

23. Greenfield BE, Wyke B: Reflex innervation of the temporo-mandibular joint. Nature 211:940-941, 1966

24. Griffin CJ, Harris R: Innervation of the temporomandibular joint. Australian Dental Journal 20:78-85, 1975

25. Grigg P, Hoffman AH: Properties of Ruffini afferents revealed by stress analysis of isolated sections of cat knee capsule. J Neurophysiol 47:41-54, 1982

26. Grigg P, Hoffman AH: Ruffini mechanoreceptors in isolated joint capsule: Responses correlated with strain energy density. Somatosensory Research 2:149-162, 1984

27. Grigg P, Hoffman A: Calibrating joint capsule mechanoreceptors as in vivo soft tissue load cells. J Biomech 22:781-785, 1989

28 Grigg P, Hoffman AH, Fogarty KE: Properties of Golgi-Mazzoni afferents in cat knee joint capsule, as revealed by mechanical studies of isolated joint capsule. J Neurophysiol 47:31-40, 1982

29 Groen GJ: De innervatie van de wervelkolom bij de mens. Nederlands Tijdschrift voor Manuele Therapie 10:48-60, 1991

30 Grönblad M, Konttinen YT, Korkala O, et al: Neuropeptides in synovium of patients with rheumatoid arthritis and osteoarthritis. J Rheumatol 15:1807-1810, 1988

31 Grönblad M, Korkala O, Konntinen YT, et al: Silver impregnation and immunohistochemical study of nerves in lumbar facet joint plical tissue. Spine 16:34-38, 1991

32 Guido J, Voight ML, Blackburn TA, et al: The effects of chronic effusion on knee joint proprioception: A case study. J Orthop Sports Phys Ther 25:208-212, 1997

33 Halata Z: The ultrastructure of the sensory nerve endings in the articular capsule of the joint of the domestic cat (Ruffini corpuscles and Pacinian corpuscles). J Anat 124:717-729, 1977

34 Halata Z: Ruffini corpuscle - a stretch receptor in the connective tissue of the skin and locomotion apparatus. In Hamann W, Iggo A (eds): Progress in Brain Research 74:221-229, 1988

35 Halata Z, Badalamente MA, Dee R, et al: Ultrastructure of sensory nerve endings in monkey (Macaca fascicularis) knee joint capsule. J Orthop Res 2:169-176, 1984

36 Halata Z, Haus J: The ultrastructure of sensory nerve endings in human anterior cruciate ligament. Anat Embryol 179:415-421, 1989

37 Halata Z, Rettig T, Schulze W: The ultrastructure of sensory nerve endings in the human knee joint capsule. Anat Embryol 172:265-275, 1985

38 Hashimoto T, Hamada T, Sasaguri Y, et al: Immunohistochemical approach for the innervation of nerve distribution in the shoulder joint capsule. Clin Orthop 305:273-282, 1994

39 Hettinga DL: Inflammatory response of synovial joint structure. In Gould JA, Davies GJ (eds): Orthopaedics and Physical Therapy. Toronto, C.V. Mosby Co, 1985, vol 2

40 Hilton J: On the Influence of Mechanical and Physiological Rest in the Treatment of Accidents and Surgical Diseases, and the Diagnostic Value of Pain: A Course of Lectures, Delivered at the Royal College of Surgeons of England in the Years 1860, 1861, and 1862. London, Bell and Dalby, 1863, pp 166-167

41 Jerosch J, Steinbeck J, Clahsen H, et al: Function of the glenohumeral ligaments in active stabilisation of the shoulder joint. Knee Surg Sports Traumatol Arthrosc 1:152-158, 1993

42 Johansson H: Role of knee ligaments in proprioception and regulation of muscle stiffness. Journal of Electromyography and Kinesiology 1:158-179, 1991

43 Johansson H, Sjölander P, Sojka P: A sensory role for the cruciate ligaments. Clin Orthop 268:161-178, 1991

44 Johansson H, Sjölander P, Sojka P: Receptors in the knee joint ligaments and their role in the biomechanics of the joint. Crit Rev Biomed Eng 18:341-368, 1991

45 Kawamura Y, Abe K: Role of sensory information from temporomandibular joint. Bull Tokyo Med Dent Univ 21(suppl):78- 82, 1974

46 Kennedy JC, Alexander IJ, Hayes KC: Nerve supply of the human knee and its functional importance. Am J Sports Med 10:329-335, 1982

47 Kidd BL, Mapp PI, Blake DR, et al: Neurogenic influences in arthritis. Ann Rheum Dis 49:649-652, 1990

48 Klineberg IJ, Greenfield BE, Wyke BD: Contributions to the reflex control of mastication from mechanoreceptors in the temporomandibular joint capsule. Dental Practitioner 21:73-83, 1970

49 Konttinen YT, Grönblad M, Hukkanen M, et al: Pain fibers in osteoarthritis: A review. Semin Arthritis Rheum 18:35-40, 1989

50 Krause W: Nervenendigung in Gelenke. Zentralbl Med Wiss 12:211-212, 1874

51 Krenn V, Hofmann S, Engel A: First description of mechanoreceptors in the corpus adiposum infrapatellare of man. Acta Anat 137:187-188, 1990

52 Loewenstein WR, Mendelson M: Components of receptor adaptation in a Pacinian corpuscle. J Physiol 177:377-397, 1965

53 Loewenstein WR, Skalak R: Mechanical transmission in a Pacinian corpuscle: An analysis and a theory. J Physiol 182:346-378, 1966

54 Lundberg A, Malmgren K, Schomberg ED: Convergence from Ib, cutaneous and joint afferents in reflex pathways to motoneurones. Brain Res 87:81-84, 1975

55 Mapp P, Kidd B, Merry P, et al: Neuroanatomical features of the synovial membrane. Br J Rheumatol 27:92, 1988

56 Martin JH: Coding and processing of sensory information. In Kandel ER, Schwartz JH, Jessell TM: Principles of Neural Science, ed 3. New York, Elsevier, 1991, pp 337-338

57 McLain RF: Mechanoreceptor endings in human cervical facet joints. Spine 19:495-501, 1994

58 Mendel T, Wink CS, Zimny ML: Neural elements in human cervical intervertebral discs. Spine 17:132-135, 1992

59 Moore JC: The Golgi tendon organ: A review and update. Am J Occup Ther 38:227-236, 1984.

60 Morree JJ de: Dynamiek van het Menselijk Bindweefsel: Functie, Beschadiging en Herstel, ed 2. Houten, The Netherlands, Bohn Stafleu Van Loghum, 1993

61 Mountcastle VB: Sensory receptors and neural encoding: Introduction to sensory processes. In Mountcastle VB (ed): Medical Physiology. St. Louis, C.V. Mosby Co, 1974, vol 1, pp 285-306

62 Nade S, Bell E, Wyke B: Articular neurology of the feline lumbar spine. J Bone Joint Surg Br 60:292 (abstract), 1978

63 Newton RA: Joint receptors contributions to reflexive and kinesthetic responses. Phys Ther 62:22-29, 1982

64 Nicoladoni C: Untersuchungen über die Gelenknerven aus der Kniegelenkskapsel. Wien Med Wochenschr 23: 647, 1873

65 O'Connor BL: The histological structure of dog knee menisci with comments on its possible significance. Am J Anat 147:407-418, 1976

66 Oostendorp RAB: De musculaire dysbalans bij de patiënt met lage rugklachten. Nederlands Tijdschrift voor Fysiotherapie 90:82-90, 1980

67 Oostendorp RAB, Sande JAW van de: Arthrokinetische reacties en musculaire stabiliteit. Nederlands Tijdschrift voor Fysiotherapie 93:63-72, 1983

68 Paris SV: Anatomy as related to function and pain. Orthop Clin of North Am 14:475-489, 1983

69 Poláček P: Differences in the structure and variability of encapsulated nerve endings in the joints of some species of mammals. Acta Anat (47):112-124, 1961

70 Poláček P: Receptors of the joints: Their structure, variability, and classification. Acta Facultatis Medicae Universitatis Brunensis 23:1-107, 1966

71 Rauber A: Untersuchungen über das Vorkommen und die Bedeutung der Vater'schen Körperchen. Zentralbl Med Wiss 5:661-662, 1867

72 Rowinski MJ: Afferent neurobiology of the joint. In Gould JA, Davies GJ (eds): Orthopaedic and Sports Physical Therapy. St. Louis, C.V. Mosby Co, 1985, pp 50-64

73 Ruffini A: Sur un nouvel organe nerveux terminal et sur la présence des corpuscules Golgi-Mazzoni dans le conjunctif sous-cutané de la pulpe des doight de l'homme. Arch Ital Biol 21:249-265, 1894

74 Schultz RA, Miller DC, Kerr CS, et al: Mechanoreceptors in human cruciate ligaments: A histological study. J Bone Joint Surg Am 66:1072-1076, 1984

75 Schutte MJ, Dabezies EJ, Zimny ML: Neural anatomy of the human anterior cruciate ligament. J Bone Joint Surg Am 69:243-247, 1987

76 Schutte MJ, Happel LT: Joint innervation in joint injury. Clin Sports Med 9:511-517, 1990

77 Sjölander P, Johansson H, Sojka P: Sensory nerve endings in the cat cruciate ligaments: a morphological investigation. Neurosci Lett 102:33-38, 1989

78 Skoglund S: Joint receptors and kinaesthesia. In Iggo A (ed): Handbook of Sensory Physiology. Berlin, Springer, 1973, pp 111-136

79 Strasmann T, Wal J van der, Halata Z, et al: Functional topography and ultrastructure of periarticular mechanoreceptors in the lateral elbow region of the rat. Acta Anat 138:1-14, 1990

80 Tracey DJ: Joint receptors and the control of movement. Trends Neurosci 3:253-255, 1980

81 Vandenabeele F, Creemers J, Lambrichts I, et al: Fine structure of vesiculated nerve profiles in the human lumbar facet joint. J Anat 187:681-692, 1995

82 Warmerdam A: Arthrokinetic Therapy™: Improving muscle performance through joint manipulation. Proceedings of the 1992 Conference of the International Federation of Orthopaedic Manipulative Therapists, Vail, Colorado, pp 204-207

83 Williams PL, Warwick R, Dyson M, et al (eds): Gray's Anatomy, ed 36. New York, Churchill Livingstone, 1980

84 Willis WD jr, Grossman RG: Medical Neurobiology: Neuroanatomical and Neurophysiological Principles Basic to Clinical Neuroscience, ed 3. London, C.V. Mosby Co, 1981, pp 91-143

85 Wyke B: The neurology of joints. Ann R Coll Surg Engl 41:25-50, 1967

86 Wyke B: Articular Neurology - A review. Physiotherapy 58:94-99, 1972

87 Wyke B: Morphological and functional features of the innervation of the costovertebral joints. Folia Morphol (Warsz)(23):296-305, 1975

88 Wyke B: Clinical significance of articular receptor systems. Ann R Coll of Surg Engl 60:137, 1978

89 Wyke B: Conference on the ageing brain: Cervical articular contributions to posture and gait: Their relation to senile disequilibrium. Age Ageing (8): 251-258, 1979

90 Wyke B: Neurology of the cervical spinal joints. Physiotherapy 65:72-76, 1979

91 Wyke B: The neurology of joints: A review of general principles. Clinics in Rheumatic Diseases 7:223-239, 1981

92 Wyke B: Receptor systems in lumbosacral tissues in relation to the production of low back pain. In White AA, Gordon SL (eds): The American Academy of Orthopaedic Surgeons: Symposium on Idiopathic Low Back Pain. St. Louis, C.V. Mosby, 1982, pp 97-107

93 Wyke BD: Articular neurology and manipulative therapy. In Glasgow EF, Twomey LT, Scull ER, et al (eds): Aspects of Manipulative Therapy, ed 2. New York, Churchill Livingstone, 1985, pp 72-77

94 Wyke B: The neurology of low back pain. In Jayson MIV (ed): The Lumbar Spine and Back Pain, ed 3. New York, Churchill Livingstone, 1987, pp 56-99

95 Yamashita T, Cavanaugh JM, El-Bohy AA, et al: Mechanosensitive afferent units in the lumbar facet joint. J Bone Joint Surg Am 72:865-870, 1990

96 Zimny ML, Albright DJ, Dabezies E: Mechanoreceptors in the human medial meniscus. Acta Anat 133:35-40, 1988

97 Zimny ML, Schutte M, Dabezies E: Mechanoreceptors in the human anterior cruciate ligament. Anat Rec 214:204-209, 1986

3. Joint Angle and Ligamento-Capsular Tension

During the past century, researchers have attempted to answer the question: "Can articular mechanoreceptors signal joint angle?" A prerequisite for signaling joint angle is that every joint position has mechanoreceptor activity. Some researchers reported that most of the slowly adapting articular receptors (Types I and II) were active only near the end range of the joint.[5,6,14,28,29,30] Burgess, Grigg, and colleagues found only a few or no receptors signaling in midrange or "loose-packed positions."[3,12,13]

The discharge frequency of Type I receptors is linear with the amount of tension in the capsule.[10,14,15] Since Type I receptors detect capsular stress, they provide "a signal that indicates that the joint is at or near the limit of its range of movement . . . ."[14, p. 53] Rossi and colleagues wrote that the Type I articular receptors might therefore only function as limit detectors.[29,30]

In contrast to the previous researchers, who found articular receptors mostly active near the extremes of range, other researchers found these receptors (low-threshold, slowly adapting receptors: Type I) to be active in any position of the joint, including midranges.[1,4,7,10,16,18,19,25] Johansson and colleagues wrote that some articular receptors are always active in every position of a joint, since some part of the capsule and ligaments is stretched in every position of the joint.[17] Similarly, Fuss found that in the knee, different fibers of both cruciate ligaments are in constant tension throughout the range.[9] Considering that different fibers of the ligamento-capsular tissue are taut throughout the range, it seems likely that slowly adapting articular receptors are able to signal the joint angle adequately throughout the full range of movement.[9,16,20]

It may be that the two viewpoints, one being that articular receptors are mostly active at end-range positions and the other that articular receptors signal throughout the range, are both correct. The discharge pattern of articular receptors is probably that of some receptors firing at a low frequency at midrange positions but more receptors becoming active at the ends of the range.[7,8,21,31,32] These receptors fire with a higher frequency as the joint moves toward its end range.[7,8,31,32] Schaible and Schmidt suggested that the increased firing of receptors as the joint is about to leave its normal working range functions as a warning signal.[31] The warning signals increase in urgency as units with a high threshold to movement (Types III and IV) come into play. These signals probably induce motor reflexes that counteract the excess movement, thus preventing joint damage.[31]

The firing rate of the articular receptors resembles a U-shaped parabola (see Fig. 3.1).[8,26] The bottom of the U is the joint's "loose-packed position." The *maximally* loose-packed position combines the loose-packed positions of all the joint's degrees of freedom. For instance, the maximally loose-packed position of the shoulder joint is a combination of the loose-packed positions of flexion and extension, of external and internal rotation, and of abduction and adduction. A joint's maximally loose-packed position is located somewhere in its midrange and has the most ligamento-capsular relaxation (see Table 3.1).[22,33] Articular receptor activity is least in this position. Because of the ligamento-capsular tissue relaxation, the maximally loose-packed position is also the position where the largest amount of movement in any one plane is available. When a joint moves in a direction away from the maximally loose-packed position, the excursion in the other planes reduces.[2,11,23,24] For instance, in the shoulder's maximally loose-packed position, almost 160 degrees of internal-external rotation is available; however, after maximal flexion of the glenohumeral joint, the internal-external rotation range has reduced to approximately 30 degrees.

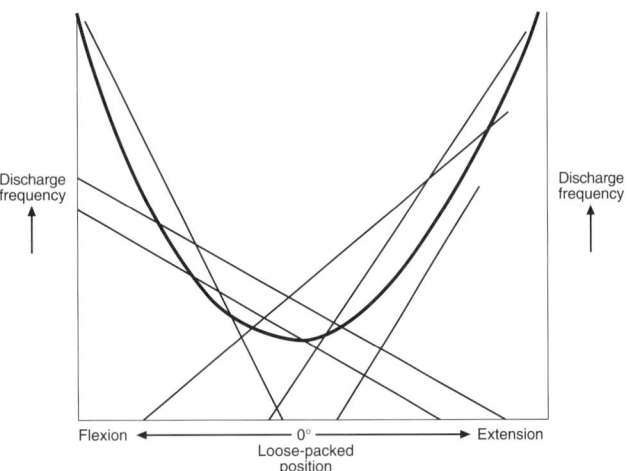

Fig. 3.1 Discharge rate of individual flexion and extension receptors as a function of joint angle. The *U*-shaped curve is the average discharge rate of the receptors.

The two legs of the *U* signify the greatest amount of firing due to ligamento-capsular tensioning. Although both legs represent ligamento-capsular tightening, one of the legs is, or is part of, the position wherein all the joint's capsular structures are maximally tightened. The maximally close-packed position of a joint is the position where maximal congruency of joint surfaces exists with maximal tightening of the ligamento-capsular tissues. Maximal tightening of the ligamento-capsular tissues is achieved by bringing the joint into a specific end-range position by combining movements in different planes (see Table 3.1). The maximally close-packed position for many distal extremity joints is achieved by movement in one plane only, since many distal extremity joints have only one degree of freedom (Fig. 3.2).[22,27,33] Proximal extremity joints have more degrees of freedom.[27] Ball-and-socket joints, such as the shoulder or hip joints, have three degrees of freedom. A ball-and-socket joint has two degrees of freedom for angular movements and one degree of freedom for spin movements. A spin movement is a rotation of one of the articular bones around an axis perpendicular to its joint surface (Fig. 3.3). For example, in the shoulder, the two degrees of freedom that allow angular movements permit abduction and adduction in the frontal plane and internal and external rotation in the transverse plane. The third degree of freedom comes from

Table 3.1 Maximally close-packed and maximally loose-packed positions.[22, 33, 34, 35]

Joint	Maximally Loose-Packed Position	Maximally Close-Packed Position
Spinal facet	Midposition	Maximal extension
Shoulder	Semi-abduction, -flexion, and -internal rotation	Maximal abduction with maximal external rotation and horizontal abduction
Ulnohumeral	Semi-flexion	Maximal extension
Radioulnar joints	Semi-flexion, -supination	—
Wrist	Semi-flexion, -ulnar deviation	Maximal dorsiflexion
Carpometacarpal I	Semi-flexion, -abduction	Maximal opposition
Metacarpophalangeal	Semi-flexion	Maximal flexion
Interphalangeal fingers	Semi-flexion	Maximal extension
Hip	Semi-flexion, -abduction, and -external rotation	Maximal extension with maximal internal rotation
Knee	Semi-flexion	Maximal extension (with coupled external rotation)
Ankle	Semi-plantar flexion	Maximal dorsiflexion
Subtalar joint	Midposition between inversion and eversion	Maximal inversion
Midtarsal joints	Semi-pronation	Maximal supination
Metatarsophalangeal	Semi-extension	Maximal extension
Interphalangeal toes	Semi-flexion	Maximal extension

the convex humeral head spinning on the glenoid. This degree of freedom permits flexion and extension of the shoulder joint.

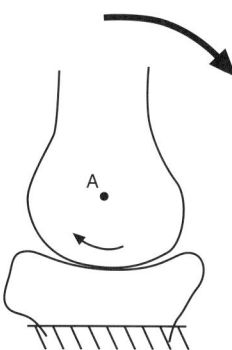

Fig. 3.2 Angular movement of the joint around one axis (A). The convex joint partner moves in one plane only.

Reprinted with permission from Bijl G van der: Het individuele functiemodel in de manuele therapie. Lochem, Netherlands, De Tijdstroom, 1986.

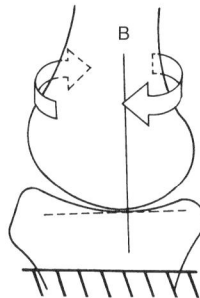

Fig. 3.3 The convex joint partner rotates around axis B, which is located perpendicular to the joint surface at the point of contact between the two bones.

Reprinted with permission from Bijl G van der: Het individuele functiemodel in de manuele therapie. Lochem, Netherlands, De Tijdstroom, 1986.

The maximally close-packed position of a joint is achieved by positioning the articulation at the end of its available angular range, combined with positioning it at the end of its "spin range," provided spin movements are available. If spin movements are available, a spin movement is always part of the movements to achieve a maximally close-packed position. The spin tightens ligamento-capsular tissue all around the joint. The close-packed position of a joint is the position of maximal afference (the greatest number of receptors firing at a maximum firing frequency).

Articular afferents are usually sensitive to displacements along more than one axis.[4] For instance, articular receptors located in the anterior capsule of the glenohumeral joint might be stretched by externally rotating the humerus, but possibly also by horizontally abducting the humerus or by a spin movement (flexion or extension of the humerus). Therefore, individual receptors are not reliable detectors of position, since the same discharge frequency may be reached in more than one position of the joint.[4] Therefore, position sense improves when the whole population, rather than individual receptors, is taken into account.[4] Glenohumeral external rotation stretches the anterior capsule but relaxes the posterior capsule. In contrast, horizontal abduction stretches the inferior capsule as well. Spin movements tend to stretch the whole capsule. The distinction between flexion and extension could be made by the receptors located in capsular diagonal fibers tightening with only one type of spin.

Effusion of a joint causes increased ligamento-capsular tension. An effusion changes the response of the articular receptors and might increase their discharge rate.[8] Even in the maximally loose-packed position an altered firing pattern of the receptors may take place.[8]

Summary

Articular receptors are capable of signaling joint position. In all positions of a joint, some Type I articular receptors are discharging. Afference from all Type I receptors in a joint is least in the maximally loose-packed position and greatest in the maximally close-packed position. During movement, antagonistic pairs of Type I receptors provide reciprocally coordinated information to the central nervous system, thereby providing the nervous system with accurate information about joint position.

References

1 Andrew BL, Dodt E: The deployment of sensory endings at the knee joint of the cat. Acta Physiol Scand 28:287-296, 1953

2 Bourdillon JF, Day EA, Bookhout MR: Spinal manipulation, ed 5. Boston, Butterworth-Heinemann, 1992

3 Burgess PR, Clark FJ: Characteristics of knee joint receptors in the cat. J Physiol 203:317-335, 1969

4 Carli G, Fontani G, Meucci M: Static characteristics of muscle afferents from gluteus medius muscle: Comparison with joint afferents of hip in cats. J Neurophysiol 45:1085-1095, 1981

5 Clark FJ: Information signaled by sensory fibers in medial articular nerve. J Neurophysiol 39:1462-1472, 1975

6. Clark FJ, Burgess PR: Slowly adapting receptors in cat knee joint: Can they signal joint angle? J Neurophysiol 38:1448-1463, 1975

7. Ferrell WR: The adequacy of stretch receptors in the cat knee joint for signalling joint angle throughout a full range of movement. J Physiol 299:85-99, 1980

8. Ferrell WR, Nade S, Newbold PJ: The interrelation of discharge, intra-articular pressure, and joint angle in the knee of the dog. J Physiol 373:353-365, 1986

9. Fuss FK: Anatomy of the cruciate ligaments and their function in extension and flexion of the human knee joint. Am J Anat 184:165-176, 1989

10. Godwin-Austen RB: The mechanoreceptors of the costo-vertebral joints. J Physiol 202:737-753, 1969

11. Greenman PE: Principles of Manual Medicine, ed 2. London, Williams & Wilkins, 1996

12. Grigg P: Mechanical factors influencing response of joint afferent neurons from cat knee. J Neurophysiol 38:1473-1484, 1975

13. Grigg P, Greenspan BJ: Response of primate joint afferent neurons to mechanical stimulation of knee joint. J Neurophysiol 40:1-8, 1977

14. Grigg P, Hoffman AH: Properties of Ruffini afferents revealed by stress analysis of isolated sections of cat knee capsule. J Neurophysiol 47:41-54, 1982

15. Grigg P, Hoffman AH: Calibrating joint capsule mechanoreceptors as in vivo soft tissue load cells. J Biomech 22:781-785, 1989

16. Johansson H: Role of knee ligaments in proprioception and regulation of muscle stiffness. Journal of Electromyography and Kinesiology 1:158-179, 1991

17. Johansson H, Sjölander P, Sojka P: Receptors in the knee joint ligaments and their role in the biomechanics of the joint. Crit Rev Biomed Eng 18:341-368, 1991

18. Klineberg I: Structure and function of temporomandibular joint innervation. Ann R Coll Surg Engl 49:268-288, 1971

19. Marshall KW, Tatton WG: Joint receptors modulate short and long latency muscle responses in the awake cat. Exp Brain Res 83:137-150, 1990

20. McCall WD, Farias MC, Williams WJ, et al: Static and dynamic responses of slowly adapting joint receptors. Brain Res 70:221-243, 1974

21. McLain RF: Mechanorecptor endings in human cervical facet joints. Spine 19:495-501, 1994

22. Mink AJF, Veer HJ ter, Vorselaars JACT: Extremiteiten: Functie-onderzoek en Manuele Therapie. Houten, The Netherlands, Bohn Stafleu Van Loghum, 1990

23. Mitchell FL jr: Elements of Muscle Energy technique. In Basmajian JV, Nyberg R (eds): Rational manual therapies. London, Williams & Wilkins, 1993

24. Mitchell FL jr: The Muscle Energy manual: Concepts & mechanisms, the musculoskeletal screen, cervical region evaluation and treatment. East Lansing, Michigan, Met Press, 1995, vol 1

25. Mountcastle VB: Sensory receptors and neural encoding: Introduction to sensory processes. In Mountcastle VB (ed): Medical Physiology. St. Louis, C.V. Mosby Co, 1974, vol 1, pp 285-306

26. Nade S, Newbold PJ, Straface SF: The effects of direction and acceleration of movement of the knee joint of the dog on medial articular nerve damage. J Physiol 388:505-519, 1987

27. Peck D, Brower TD: Algorithms for the segmental motor innervation of the extremities. Am Surg 53:270-273, 1987

28. Proske U, Schaible HG, Schmidt RF: Joint receptors and kinaesthesia. Exp Brain Res 72:219-224, 1988

29. Rossi A, Grigg P: Characteristics of hip joint mechanoreceptors in the cat. J Neurophysiol 47:1029-1042, 1982

30. Rossi A, Rossi B: Characteristics of the receptors in the isolated capsule of the hip in the cat. Int Orthop 9:123-127, 1985

31. Schaible HG, Schmidt RF: Responses of fine medial articular nerve afferents to passive movements of knee joint. J Neurophysiol 49:1118-1126, 1983

32. Skoglund S: Joint receptors and kinaesthesia. In Iggo A (ed): Handbook of Sensory Physiology. Berlin, Springer, 1973, pp 111-136

33. Williams PL, Warwick R (eds): Gray's Anatomy, ed 36. New York, Churchill Livingstone, 1980

34. Winkel D, Aufdemkampe G, Meijer OG, et al: Orthopedische Geneeskunde en Manuele Therapy: Diagnostiek Extremiteiten. Houten, The Netherlands, Bohn Stafleu Van Loghum bv, 1993, Vol 2b

35. Winkel D, Aufdemkampe G, Meijer OG, et al: Orthopedische Geneeskunde en Manuele Therapy: Diagnostiek Extremiteiten. Houten, The Netherlands, Bohn Stafleu Van Loghum bv, 1992, Vol 2c

4. Position and Movement Sense

Articular Receptors

What is the precise role of articular receptors in position and movement sense (kinesthesia)? Some researchers believe that position and movement sense derive exclusively from articular receptors, whereas others believe kinesthesia to derive from muscle receptors. Whether articular, muscle, or even skin receptors are most important for position and movement sense has been a topic of debate and research for many years.[1,11,40,45,50,61,62]

One group of authors stated that kinesthesia derives, in large part or exclusively, from articular receptors and not from muscles.[1,2,19,22,30,37,49,54] Henneman wrote in 1974, "It is now generally recognized that stretch receptors in muscle do not contribute to conscious perception; i.e., they do not furnish information regarding the position of a joint or a limb."[37, p. 634] "None of the information emerging from the spindle is perceived by the sensory cortex. . . ."[59, p. 390] Burke added to this the limited ability of spindle cells to signal joint position.[10]

Boyd and Berts measured the afferent discharges from the posterior articular nerve of the cat knee joint in response to movement and noticed that these afferent discharges were of two types.[9] The first type was a rapidly adapting response, and the second, a slowly adapting response. The rapidly adapting response to stretch of the posterior knee capsule was similar to the response from Pacinian corpuscles (Type II receptors). The slowly adapting response was similar to that from Ruffini endings (Type I receptors). They concluded that the slowly adapting receptors (Type I) can provide accurate information about the position of the joint.[9] Other researchers had confirmed that the Type I receptor contributes significantly to kinesthetic sensation.[15,17,65] From the fact that Type I receptors were found to signal midrange positions, some researchers concluded that these receptors are ideally suited to signal the joint angle and may play an important, if not exclusive, role in position sense.[2,19]

Spindle Cell

In 1977, Adams considered articular receptors to be the basis of position sense.[1] A year later, in response to Adams's article, Kelso wrote an article for the same journal that contradicted Adams's theory. Kelso wrote that muscle, and not articular afferents, provide a significant contribution to position and movement sense.[45] Besides Kelso, other authors also believed the muscle's spindle cell to be the basis for kinesthesia.[12,13,14,33,35,56] Even *Dorland's Medical Dictionary* chose sides and defined *kinesthetic* as "pertaining to kinesthesia or the *muscular* sense" (italics are mine).[18]

Cross and McCloskey and Grigg and colleagues found that good kinesthetic sensation remained without articular receptors.[16,34] In their studies, patients underwent a joint replacement with removal of structures such as the joint capsule. Kinesthetic sensation from the replaced joints remained intact.[16,34] Barrack and colleagues came to a similar conclusion; they found no difference in kinesthesia before and after an anesthetic injection into the knee joint.[5]

Muscle or Joint?

What forms the basis of position and movement sense, and is it possible to settle the question of the role of articular receptors in kinesthesia?[56] Afferent mechanoreceptor information from the joints seems to contribute to proprioception. The loss of articular receptor afference produces kinesthetic disturbances. For instance, joint conditions such as degenerative, inflammatory, and traumatic lesions, joint displacement, and arthroplasty can impair postural and movement sensation.[8,24,36,65,67,68] Barrack and colleagues found a marked decline in kinesthesia in patients with degenerative joint disease or complete tears of the anterior cruciate ligament.[6,7] Guido and colleagues found decreased position sense in a patient with a chronically effused knee joint.[36] Aspiration of the joint fluid improved position sense in passive positioning of the knee. Twenty-four hours after aspiration, the effusion had returned, and position sense had declined to pre-aspiration levels.

The loss of position and movement sense in patients with osteoarthritic joints may result from laxity of the capsule and ligaments caused by loss of cartilage and bone height. In addition, the degeneration may also damage the articular receptors directly. Barrett and colleagues provided some proof that the loss of joint space height causes impaired kinesthesia. In their study, patients who underwent replacement arthroplasty had slightly better proprioception than patients with osteoarthritis.[8] The arthroplasty restored alignment and "joint space height." However, Barrack and colleagues did not confirm Barrett's findings and found that arthroplasty on osteoarthritic knees did not alter kinesthesia.[7]

Both Type I and Type II receptors project polysynaptically to neurons in the parietal and paracentral regions of the cerebral cortex, where conscious perception takes place.[54,63,64,65,66,67] The afferent information travels cranially via dorsal and dorsolateral spinal columns.[66,67] However, the articular mechanoreceptors are not the only receptors that project to these cortical neurons; cutaneous mechanoreceptors and muscle spindle afferents also project to the same cortical sectors.[55,66] Kinesthetic sensation is therefore the cortical interpretation of information simultaneously received from articular, muscle, and cutaneous afferents.[21,23,26,27,31,47,52,53,57,60,66]

Most likely, no one group (articular, muscle, or skin afferents) can adequately signal the joint angle under all conditions.[21] Optimal kinesthesia requires analysis by the central nervous system of all available afferent sources.[21] During movement, mechanoreceptors in all three structures discharge, and all three contribute to kinesthetic sensation.[58,66] The central nervous system determines the position of a joint by comparing the information it receives from opposing parts of the joint. Flexion of the knee, for example, will relax the posterior capsule and skin and relax the hamstrings when it places tension on the anterior capsule, the skin, and the quadriceps.

Thus, articular afference aids in posture and movement sense but is not solely responsible. For instance, loss of articular mechanoreceptor function, through local anesthetic infiltration or surgical removal of the joint capsule, resulted in varying degrees of impairment but never abolished postural sensation.[34,46,66] Joint trauma also produced varying degrees of reduced postural sensation but never complete loss.[66]

Ferrell and colleagues found that patients with rheumatoid arthritis affecting the proximal interphalangeal joint have impaired position sense.[25] This loss of position sense was not due to neuropathy, a frequent complication of rheumatoid arthritis, especially if the patient used disease-modifying drugs, such as gold, for a long time. Loss of the articular mechanoreceptor afference was secondary to rheumatic destruction of the joint, which caused impaired (but not absent) position sense.[25]

The answer to the question of whether muscle or articular receptors are most important to kinesthesia was reported by Millar, Schutte, and Happel.[52,58] Joints and the muscles around them should be considered complementary components of a single afferent system, and they should be viewed as a unit instead of looking at the functions of the individual components.[52,58]

Gamma Motoneuron

Many researchers have suggested that articular afferents contribute to position and movement sense through the gamma loop system.[3,4,29,31,38,39,40,42,43,44,51] Description of the gamma loop begins with the gamma motoneuron, whose activity controls the length of the intrafusal fibers within the spindle cell. The spindle cell compares the length of the intrafusal fibers with that of the extrafusal muscle fibers (i.e., the muscle fibers proper). Information about these relative lengths is conveyed to the central nervous system for kinesthesia and influences the alpha motoneuron.

Afferent input, including that from joints, that could attribute to kinesthesia is not directly relayed to the cortex but goes through the gamma motoneuron. The reflex effects of the skin, articular, and muscle afferents on the gamma motoneurons are potent enough to modify the signaling of the muscle spindle's primary afferents.[3,4,39,40,41,42,43,51] This means the information conveyed by the muscle spindle's primary afferents is polymodal (i.e., the spindle cell integrated the muscle fibers' relative length with afferent information from the skin, joints, and muscles).[3,4,39,40,41,42,43,51] Of course, primary afferents from muscles do not project to gamma motoneurons; only afferents from Golgi tendon organs do so.[53] Thus, kinesthesia does not derive solely from the gamma loop itself, but is modified.[31]

Marsden and colleagues conducted a clinical study that supports the above theory.[48] One might simply think of the stretch reflex of a muscle as the spindle cell afferents signaling lengthening of the extrafusal muscle fibers, stimulating alpha motoneurons, and causing contraction of the extrafusal muscle fibers. However, the research by Marsden and colleagues showed that afference from the thumb skin and joints cooperate in the expression of the stretch reflex of the flexor pollicis longus. In their experiment, they blocked the afferents from the skin and the thumb joints using local anesthetic, resulting in the stretch reflex being almost completely abolished.[48] Similar results were obtained in studies of the ankle joint.[28,67]

Active Positioning

The position of a joint may be more accurately judged after active positioning, compared with passive positioning. Active positioning is the body's own muscles positioning its joint(s). Passive positioning is positioning by an outside force (e.g., an examiner positioning someone's relaxed limb). Grigg and colleagues proposed that the contraction of muscles by their attachments to the capsule would stimulate capsular mechanoreceptors that assist in kinesthesia.[32,33,35] This theory seemed to have merit, because muscular contractions powerfully activated articular neurons; however, it was rejected by the same authors, because the contraction needed to stimulate the articular receptors was rather high and could not explain why kinesthesia improved after active positioning with low force.[35]

Ferrell found that tetanic contraction of various muscles around the cat knee joint could alter the discharge of articular receptors, but only if the joint was near an end-range position.[22] The muscles used in the experiment were the quadriceps, the hamstrings, the medial and lateral gastrocnemius, and the popliteus. The altered discharge of the articular receptors in response to muscle contraction was not due to movement of the tibia or femur, as these were securely fixated. Possibly, the end-range position of the joint had the capsule prestretched, and the muscle contraction added or decreased tension in the capsule. Only if the joint capsule is already under pre-tension, as happens at the extremes of movement, could a muscular contraction increase the firing rate of the receptors. In loose-packed positions, contraction of muscles acting over a joint does not significantly increase tension in the joint capsule, and an altered discharge rate of the articular receptors was not observed in response to muscle contractions.[22]

A better explanation for why active positioning improves proprioception over passive positioning might be that the muscle receptors are allowed to play a larger role during active positioning.[46] For instance, passively approximating the origin and insertion of a relaxed muscle might not result in shortening of the intrafusal muscle fibers. However, if the intra- and extrafusal muscle fibers contract during muscle shortening, the altered muscle length can be perceived by the spindle cell.

Type I and Type II Articular Receptors

Of the four articular receptors, only the Type I articular receptor contributes to the conscious perception of posture.[66,67] Type I articular mechanoreceptors are the only mechanoreceptors active in healthy immobile joints. Type II receptors are not suitable for postural sensation because of their rapid adaptation to positions. Neither is the Type III suitable because of its high threshold and because it does not seem to project to the cortex.[66]

Besides position, the Type I receptor signals movement. During movement, the discharge of the Type I receptors in the stretched and destretched regions of the capsule alters to match the joint's changing positions.[66] Simultaneously, the brief discharges of the Type II receptors in the same regions of the capsule are superimposed on the Type I afference.[66]

The Type II receptor afference merely signals that movement is taking place. For example, active or passive knee flexion creates increased firing of anterior Type I and II receptors and decreased firing of posterior Type I receptors. The posterior Type II receptors will produce a brief burst of action potentials during the movement due to the change in tension in the posterior capsule. The afferent discharges from the Type I and II receptors on both sides of the joint can cause movement sensation.[66]

Summary

Articular receptors, together with skin and muscle receptors, contribute to position and movement sense via the gamma loop system. Information from these receptors modifies the spindle cell's primary afferents, which then provide information to the central nervous system about position and movement. The articular contribution to position sense derives from Type I articular receptors, whereas the articular contribution to movement sense derives from Type I and II receptor activity. The Type I receptor provides accurate information regarding position and movement. The Type II receptor can merely signal that a movement is taking place.

References

1. Adams JA: Feedback theory of how joint receptors regulate the timing and positioning of a limb. Psychol Rev 84:504-523, 1977
2. Andrew BL, Dodt E: The deployment of sensory endings at the knee joint of the cat. Acta Physiol Scand 28:287-296, 1953
3. Appelberg B, Hulliger M, Johansson H, et al: Reflex activation of dynamic fusimotor neurons by natural stimulation of muscle and joint receptor afferent units. In Taylor A, Prochazka A (eds): Muscle Receptors and Movement. New York, Oxford University Press, 1981, pp 149-161
4. Appelberg B, Johansson H, Sojka P: Fusimotor reflexes in triceps surae muscle elicited by stretch of muscles in the contralateral hind limb of the cat. J Physiol 373:419-441, 1986
5. Barrack RL, Skinner HB, Brunet ME, et al: Functional performance of the knee after intraarticular anesthesia. Am J Sports Med 11:258-261, 1983
6. Barrack RL, Skinner HB, Buckley SL: Proprioception in the anterior cruciate deficient knee. Am J Sports Med 17:1-6, 1989
7. Barrack RL, Skinner HR, Cook SD, et al: Effect of articular disease and total knee arthroplasty on knee joint-position sense. J Neurophysiol 50:684-687, 1983
8. Barrett DS, Cobb AG, Bentley G: Joint proprioception in normal, osteoarthritic and replaced knees. J Bone Joint Surg Br 73:53-56, 1991
9. Boyd IA, Berts TDMR: Proprioceptive discharges from stretch-receptors in the knee-joint of the cat. J Physiol 122:38-58, 1953
10. Burke D: Muscle spindle function during movement. Trends Neurosci 3:251-253, 1980
11. Carli G, Fontani G, Meucci M: Static characteristics of muscle afferents from gluteus medius muscle: Comparison with joint afferents of hip in cats. J Neurophysiol 45:1085-1095, 1981
12. Clark FJ: Information signaled by sensory fibers in medial articular nerve. J Neurophysiol 39:1462-1472, 1975
13. Clark FJ, Burgess PR: Slowly adapting receptors in cat knee joint: Can they signal joint angle? J Neurophysiol 38:1448-1463, 1975
14. Clark FJ, Horch KW, Bach SM, et al: Contributions of cutaneous and joint receptors to static knee-position sense in man. J Neurophysiol 42:877-888, 1979
15. Clark RKF, Wyke BD: Contributions of temporomandibular articular mechanoreceptors to the control of mandibular posture: An experimental study. J Dent 2:121-129, 1974
16. Cross MJ, McCloskey DI: Position sense following surgical removal of joints in man. Brain Res 55:443-445, 1973
17. Dee R: Mechanoreceptors in hip joint capsule and ligamentum capitis femoris and their reflex contribution to posture. In Symposium on Osteoarthritis. St. Louis, C.V. Mosby Co, 1976, pp 52-65
18. Dorland's Illustrated Medical Dictionary, ed 25. Philadelphia, W.B. Saunders Co, 1974
19. Ferrell WR: The adequacy of stretch receptors in the cat knee joint for signalling joint angle throughout a full range. J Physiol 299:85-99, 1980
20. Ferrell WR: The response of slowly adapting mechanoreceptors in the cat knee joint to tetanic contraction of hind limb muscles. Quarterly Journal of Experimental Physiology 70:337-345, 1985
21. Ferrell WR: Discharge characteristics of joint receptors in relation to their proprioceptive role. In Hnik P, Soukop T, Vejsada R, et al (eds): Mechanoreceptors, Development, Structure and Function. New York, Plenum Press, 1988, pp 383-388
22. Ferrell WR, Baxendale RH, Carnachan C, et al: The influence of joint afferent discharge on locomotion, proprioception and activity in conscious cats. Brain Res 347:41-48, 1985
23. Ferrell WR, Craske B: Contribution of joint and muscle afferents to position sense at the human

proximal interphalangeal joint. Exp Physiol 77:331-342, 1992

24 Ferrell WR, Crighton A, Sturrock RD: Age-dependent changes in position sense in human proximal interphalangeal joints. Neuroreport 3:259-261, 1992

25 Ferrell WR, Crighton A, Sturrock RD: Position sense at the proximal interphalangeal joint is distorted in patients with rheumatoid arthritis of finger joints. Exp Physiol 77:675-680, 1992

26 Ferrell WR, Gandevia SC, McCloskey DI: The role of joint receptors in human kinaesthesia when intramuscular receptors cannot contribute. J Physiol 386:63-71, 1987

27 Ferrell WR, Smith A: The effect of loading on position sense at the proximal interphalangeal joint of the human index finger. J Physiol 418:145-161, 1989

28 Freeman MAR, Wyke B: Articular reflexes at the ankle joint: An electromyographic study of normal and abnormal influences of ankle-joint mechanoreceptors upon reflex activity in the leg muscles. Br J Surg 54:990-1001, 1967

29 Gandevia SC, McCloskey DI: Joint sense, muscle sense, and their combination as position sense measured at the distal interphalangeal joint of the middle finger. J Physiol 260:387-406, 1976

30 Glencross D, Thornton E: Position sense following joint injury. Journal of Sports Medicine 21:23-27, 1981

31 Goodwin GM, McCloskey DI, Matthews PBC: The persistence of appreciable kinesthesia after paralysing joint afferents but preserving muscle afferents. Brain Res 37:326-329, 1972

32 Grigg P: Mechanical factors influencing response of joint afferent neurons from cat knee. J Neurophysiol 38:1473-1484, 1975

33 Grigg P: Response of joint afferent neurons in cat medial articular nerve to active and passive movements of the knee. Brain Res 118:482-485, 1976

34 Grigg P, Finerman GA, Riley LH: Joint-position sense after total hip replacement. J Bone Joint Surg Am 55:1016-1025, 1973

35 Grigg P, Greenspan BJ: Response of primate joint afferent neurons to mechanical stimulation of knee joint. J Neurophysiol 40:1-8, 1977

36 Guido J, Voight ML, Blackburn TA, et al: The effect of chronic effusion on knee joint proprioception: A case study. J Orthop Sports Phys Ther 25:208–212, 1997

37 Henneman E: Peripheral mechanisms involved in the control of muscle. In Mountcastle VB (ed): Medical Physiology, ed 13. St. Louis, C.V. Mosby Co, 1974, vol 1, pp 617-635

38 Jerosch J, Castro WHM, Halm H, et al: Does the glenohumeral joint capsule have proprioceptive capability? Knee Surg Sports Traumatol Arthrosc 1:80-84, 1993

39 Johansson H: Reflex integration in the γ-motor system. In Boyd IA, Gladden MH (eds): The Muscle Spindle. London, The MacMillan Press Ltd, 1985, pp 297-301

40 Johansson H: Role of knee ligaments in proprioception and regulation of muscle stiffness. Journal of Electromyography and Kinesiology 1:158-179, 1991

41 Johansson H, Sjölander P, Sojka P: Actions on γ-motoneurones elicited by electrical stimulation of joint afferent fibers in the hind limb of the cat. J Physiol 375:137-152, 1986

42 Johansson H, Sjölander P, Sojka P: A sensory role for the cruciate ligaments. Clin Orthop 268:161-178, 1991

43 Johansson H, Sjölander P, Sojka P: Receptors in the knee joint ligaments and their role in the biomechanics of the joint. Crit Rev Biomed Eng 18:341-368, 1991

44 Johansson H, Sojka P: Actions on γ-motoneurones elicited by electrical stimulation of cutaneous afferent fibers in the hind limb of the cat. J Physiol 366:343-363, 1985

45 Kelso JAS: Joint receptors do not provide a satisfactory basis for motor timing and positioning. Psychol Rev 85:474-481, 1978

46 Konradsen L, Ravn JB, Sorensen AI: Proprioception at the ankle: The effect of anaesthetic blockade of ligament receptors. J Bone Joint Surg Br 75:433-436, 1993

47 Kuno M, Muñoz-Martinez, Randic M: Sensory inputs to neurones in Clarke's column from muscle, cutaneous and joint receptors. J Physiol 228:327-342, 1973

48 Marsden CD, Merton PA, Morton HB: Servo action and stretch reflex in human muscle and its apparent dependence on peripheral sensation. J Physiol 216:21-23P, 1971

49 Matthews PBC: Mammalian Muscle Receptors and their Central Actions. London, Edward Arnold Ltd, 1972

50 Matthews PBC: Where does Sherrington's "muscular sense" originate? Muscles, joints, corollary discharges? Annu Rev Neurosci 5:189-218, 1982

51 McCloskey DI: Kinesthetic sensibility. Physiol Rev 58:763-820, 1978

52 Millar J: Joint afferent fibres responding to muscle stretch, vibration and contraction. Brain Res 63:38-383, 1973

53 Millar J: Convergence of joint, cutaneous and muscle afferents onto cuneate neurons in the cat. Brain Res 175:347-350, 1979

54 Mountcastle VB, Powell TPS: Central nervous mechanisms subserving position sense and kinesthesis. Johns Hopk Hosp Bull 105:173-200, 1959

55 Newton RA: Neural systems underlying motor control. In Montgomery PC, Connolly BH (eds): Motor Control and Physical Therapy: Theoretical Framework and Practical Applications, ed 1. Hixson, Tennessee, Chattanooga Group, 1991

56 Proske U, Schaible HG, Schmidt RF: Joint receptors and kinaesthesia. Exp Brain Res 72:219-224, 1988

57 Rossi A, Rossi B: Characteristics of the receptors in the isolated capsule of the hip in the cat. Int Orthop 9:123-127, 1985

58 Schutte MJ, Happel LT: Joint innervation in joint injury. Clin Sports Med 9:511-517, 1990

59 Solomonow M, D'Ambrosia RD: Neural reflex arcs and muscle control of knee stability and motion. In Scott WN (ed): Ligament and Extensor Mechanisms of the Knee. St. Louis, C.V. Mosby Co, 1991, pp 389-400

60 Swash M: J Neurol Neurosurg Psychiatry 49:100-106, 1986

61 Taylor JL, McCloskey DI: Ability to detect angular displacements of the fingers made at an imperceptibly slow speed. Brain 113:157-166, 1990

62 Williams WJ: A systems-oriented evaluation of the role of the joint receptors and other afferents in position and motion sense. Crit Rev Biomed Eng 7:23-77, 1981

63 Wyke B: Articular neurology - A review. Physiotherapy 58:94-99, 1972

64 Wyke B: Clinical significance of articular receptor systems. Ann R Coll Surg Engl 60:137, 1978

65 Wyke B: Conference on the Ageing Brain: Cervical articular contributions to posture and gait: Their relation to senile disequilibrium. Age Ageing 8:251-258, 1979

66 Wyke B: The neurology of joints: A review of general principles. Clinics in Rheumatic Diseases 7:223-239, 1981

67 Wyke BD: Articular neurology and manipulative therapy. In Glasgow EF, Twomey LT, Scull ER, et al (eds): Aspects of Manipulative Therapy, ed 2. New York, Churchill Livingstone, 1985, pp 72-77

68 Wyke B: The neurology of low back pain. In Jayson MIV (ed): The Lumbar Spine and Back Pain, ed 3. New York, Churchill Livingstone, 1987, pp 56-99

5. Joint Pain

Pain from a joint depends on Type IV afference. The afferent information from Type IV articular receptors and other nociceptors projects to the spinal cord. In the spinal cord, the Type IV afferents give off collateral branches that synapse with neurons located in the basal nucleus of the spinal gray matter.[13] Axons of these neurons ascend via the anterolateral spinal tract into the brain.[13] On reaching the cortex of the brain, the nociception produces the experience of pain.

Not all nociceptive information entering the spinal cord results in the experience of pain. In 1965, Melzack and Wall proposed the gate-control theory to explain the discrepancy between nociception and pain.[6] This theory explained why at times a minor trauma can cause much pain or a major trauma can cause only minor pain. These authors proposed that the transmission of nociception from the periphery onto cells within the central nervous system was under a "control." This "control," or "gate," for nociception was, among other factors, influenced by peripheral afferents.

The mechanism by which this control is achieved remains completely unknown. In their original paper, Melzack and Wall proposed detailed mechanisms of this gate; however, little has been proven to date, other than that the gate control mechanism exists.[8,10] Recent publications take a prudent view of the gating theory.[3] The articular mechanoreceptors (Type I, II, and III receptors) and other nonarticular receptors are capable of gating nociceptor afferent impulses.[13] Whether the large-diameter afferents from these receptors produce a presynaptic and/or postsynaptic inhibition of the nociceptive information is unknown.[3,8,10] Some publications, probably influenced by Melzack and Wall's original gate-control theory, still give detailed descriptions of the mechanism of the theory pertaining specifically to joints.[11,13,14] Lacking proof of such detailed mechanisms of the gate-control theory, it seems prudent to report only that the mechanism exists.

Essential to the experience of joint pain is Type IV receptor activity.[12] Concurrent activity in the mechanoreceptor afferents from the same or related tissues inhibits nociception.[11,14] The probability of Type IV nociceptive afference evoking an experience of pain is inversely related to the concurrent activity in these mechanoreceptor afferent nerves.[11,14] This means that the amount of joint pain experienced depends on the amount of chemical or mechanical irritation of a joint and the amount of afferent activity from articular and other related mechanoreceptors. Thus, a pain experience might result from nociception but might also result from insufficient gating of nociception. Decreased articular mechanoreceptor afferent input (e.g., guarding of a joint, or a diminished number of mechanoreceptors and/or mechanoreceptor afferent fibers) results in diminished gating of nociception. Degenerative, inflammatory, or traumatic affections of a joint result in a loss of the joint's mechanoreceptors or mechanoreceptor afferent nerves.[13] An example of a diminished number of mechanoreceptor afferents is postherpetic neuralgia.[14] The herpes virus selectively destroys the dorsal root ganglion of the large-diameter mechanoreceptor afferents in specific peripheral nerves, resulting in loss of gating of the nociceptor transmission. Another example would be a joint sprain.[5] The joint injury produces tissue damage and results in loss of mechanoreceptors.[11] Unfortunately, a progressive degenerative loss of mechanoreceptor afferents in all peripheral nerves (innervating both extremity and spinal joints) is an inevitable consequence of advancing age.[1,4,13,14] Interestingly enough, Type IV receptors are less affected by this degenerative process, or by inflammatory or traumatic processes.[13] The phylogenetic older age of the thin nociceptive afferents makes them less vulnerable and faster to recuperate than large-diameter mechanoreceptor afferents.[2]

Mechanoreceptor afferent fibers from a variety of tissues are capable of gating nociception.[9,10,13] Massage stimulates predominantly cutaneous receptors, whereas movements, active or passive, stimulate predominantly articular receptors. Active or passive movements to a joint are possibly more effective in relieving joint pain, because it derives from the same tissue as the nociception and enters the spine at the same cord level as nociception from that joint. This makes active or passive movement (including manual therapy) of the joint more specific in gating articular nociception than cutaneous therapeutic applications, such as massage, TENS, or heat.[9]

In the joint, the mechanoreceptors responsible for nociceptive gating are the Type I, II, and III receptors. The low threshold of the Type I and II articular receptors make them most suitable for this task. Their low threshold makes it very easy for a clinician to stimulate them and provide gating of nociception. Oscillations, gentle mobilizations, and active or passive movements stimulate the Type I and II receptors. Codman's exercise, whereby the arm is gently swung back and forth, is an example of an exercise that could inhibit shoulder nociception. Any technique that stimulates the slowly or rapidly adapting articular mechanoreceptors could block the transmission of nociception to the cortex, preventing the perception of pain.[14]

Oscillations of the joint seem most appropriate in relieving joint pain.[9,12] Oscillatory movements are more effective then static positions in closing the nociceptive gate. During oscillations, the otherwise relatively quiet Type II receptors constantly discharge. The rapidly adapting Type II receptors continuously signal altering tension in the joint tissue. The Type II activity added to the constantly present Type I discharges is more effective in closing the gate than Type I discharges alone.

The direction of the oscillation is clinically important, as it determines which joint structures are repeatedly loaded and unloaded. An oscillation of the shoulder joint in the direction of abduction/adduction will alter tension predominantly in the superior and inferior capsule, whereas an oscillation in the direction of internal/external rotation will alter tension predominantly in the anterior and posterior part of the joint. Similarly, the position of the joint will also determine the part of the capsule affected by the oscillations. For instance, an oscillatory movement of the lower leg with the knee in relative extension will affect predominantly the posterior knee capsule. Depending on what part of the capsule is affected by the oscillations, different spinal levels are influenced, and the effectiveness of the treatment is determined. Different parts of a joint might be innervated by different spinal segments.[7] For instance, the hip joint capsule is anteriorly and cranially innervated by L(1)2-4, whereas the posterior capsule is innervated by L(4)5-S3.[7] Thus, depending on the oscillations used, different parts of the capsule are stretched, different mechanoreceptors are stimulated, and different spinal segments are influenced.

In summary, pain results from nociception or its lack of gating. Mechanoreceptor afferent impulses are capable of gating nociceptive transmission in the spinal cord. Articular nociception might be most effectively gated by input from mechanoreceptors in the same joint or the same part of the joint as the origin of the nociception. Oscillatory movements of the joint, stimulating the low-threshold Type I and II articular receptors, seem the most effective method of blocking nociception.

References

1. Barrack RL, Skinner HB, Cook SD, et al: Effect of articular disease and total knee arthroplasty on knee joint-position sense. J Neurophysiol 50:684-687, 1983

2. Cranenburgh B van: Inleiding in de Toegepaste Neurowetenschappen: Opvattingen over zenuwstelsel en hersenen. Lochem, The Netherlands, De Tijdstroom, 1987, vol 1

3. Cranenburgh B van: Inleiding in de Toegepaste Neurowetenschappen: Pijn. Gent, Belgium, De Tijdstroom, 1987, vol 3

4. Ferrell WR, Crighton A, Sturrock RD: Age-dependent changes in position sense in human proximal interphalangeal joints. Neuroreport 3:259-261, 1992

5. Freeman MAR, Dean MRE, Hanham IWF: The etiology and prevention of functional instability of the foot. J Bone Joint Surg Br 47:678-685, 1965

6. Melzack R: The Puzzle of Pain. Harmondsworth, Middlesex, England, Penguin Books, 1973

7 Mink AJF, ter Veer HJ, Vorselaars JACT: Extremiteiten: Functie-onderzoek en Manuele Therapie, ed 6. Houten, The Netherlands, Bohn Stafleu Van Loghum, 1993

8 Nathan PW: The gate-control theory of pain: A critical review. Brain 99:123-158, 1976

9 Paris SV, Loubert PV: Foundation of Clinical Orthopaedics. St. Augustine, Florida, Institute Press, 1990

10 Wall PD: The gate control theory of pain mechanisms: A re-examination and re-statement. Brain 101:1-18, 1978

11 Wyke B: Neurology of the cervical spinal joints. Physiotherapy 65:73-76, 1979

12 Wyke B: The neurology of joints: A review of general principles. Clinics in Rheumatic Diseases 7:223-239, 1981

13 Wyke BD: Articular neurology and manipulative therapy. In Glasgow EF, Twomey LT, Scull ER, et al (eds): Aspects of Manipulative Therapy, ed 2. New York, Churchill Livingstone, 1985, pp 72-77

14 Wyke B: The neurology of low back pain. In Jayson MIV (ed): The Lumbar Spine and Back Pain, ed 3. New York, Churchill Livingstone, 1987, pp 56-99

6. Articular Reflexes

Joint afferents enter the spinal cord via the dorsal roots.[53,56] Within the gray matter of the spinal cord, the joint afferents project to the same cord segment and to more cranial and caudal segments. From these segments, the joint afferents, directly or through interneurons, reach neurons at all levels of the central nervous system.[67] For instance, the afferent information might reach the cortex, where it produces conscious perception of articular events.

The central effects of joint receptor activity can be divided into four categories:

- Perception of joint position and movement,
- Perception of joint pain,
- Articular nociceptive reflexes,
- Articular mechanoreceptor reflexes.

Perception of joint position, movement, and pain was discussed in the previous two chapters. The articular reflexes are the muscle tone changes resulting from Type IV, or articular mechanoreceptor, activity.

The articular receptor afferents terminate in laminas I, V, VI, and VII of the posterior horn (see Fig. 6.1). Within these laminas, the joint receptor afferents make extensive synaptic connections to interneurons. Some interneurons finally synapse in the anterior horn with the motoneurons associated with the musculature.[64] All articular reflexes are polysynaptic; that is, through interneurons.[19,75] This means that no joint afferent directly synapses on a motoneuron. Through their polysynaptic influence on motoneurons, the four types of joint receptors can produce reflex muscle tone changes.

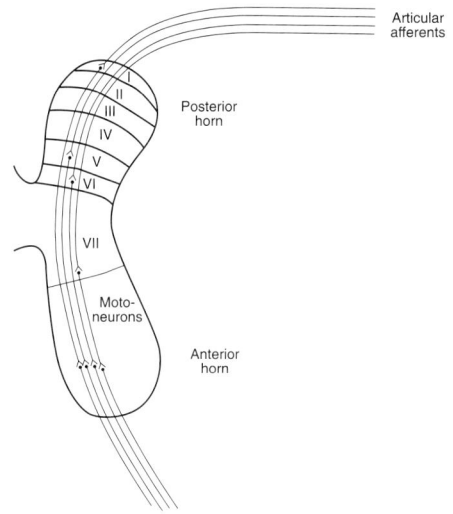

Fig. 6.1 Spinal cord lamina.

Researchers have described these articular reflex muscle tone changes in the extremity and spinal joints, the temporomandibular joint (TMJ), and the costovertebral joints.[12,43,52,61,69]

The joint afferents not only influence motoneurons of the same segment they entered, but motoneurons at different segments as well, because they ascend and descend to other cord segments and make extensive synaptic connections.[47,64,66,76] Thus, the four types of joint receptors, through their influence on the motoneurons of multiple segments, can influence tone of the muscles that act over the same or even remote joints.[78]

All four joint receptors produce articular reflexes through different pathways. Some joint receptors polysynaptically influence the alpha motoneuron, whereas other joint receptors polysynaptically influence the gamma motoneuron. If the gamma motoneuron is influenced, subsequently the alpha motoneuron is indirectly influenced.

Articular Nociceptive Reflexes

Articular nociceptive reflexes originate in Type IV receptor activity.[10,78] Polysynaptically, the Type IV receptors project to the alpha motoneurons of muscles related to the joint.[23,75,78,80] The afferents also project polysynaptically to alpha motoneurons of the contralateral extremity.[78,] Through its influence on the alpha motoneurons, Type IV activity may produce muscle tone changes.[77,80] The clinically familiar distortions of postures and movements associated with pain result from articular nociceptive reflexes.[77,78,80]

The articular mechanoreceptors (i.e., Type I, II, and III receptors) are responsible for the articular mechanoreceptor reflexes, which are active throughout our activities of daily living. However, the articular nociceptive reflexes from the Type IV receptors possibly override these mechanoreflexes, causing reflex disorders of posture and movement.[77,78,80] Clinically, the patient might walk with an antalgic gait instead of a normal gait, or reflexly guard the shoulder instead of freely swinging the arm.

Similar to the above mechanism of articular nociception interfering with normal articular reflexes, Type IV afference also impairs postural and movement sensation.[77,78,80] In other words, joint pain could prevent the patient from experiencing other sensations from the joint.

The reverse is true as well. Mechanoreceptor afference can inhibit Type IV nociceptive afference. On the perceptual level, mechanoreceptor afference can inhibit the perception of pain (gate-control theory). On the reflex level, mechanoreceptor afferents might also inhibit articular nociceptive reflexes. Willer found that large-diameter nerve afferents (for instance, from Type I, II, or III receptors) have an inhibitory effect on both pain and the withdrawal reflex.[72] The articular nociceptive reflex is likely the articular form of the withdrawal reflex. Willer's study suggested a relationship between pain perception and the withdrawal reflex, finding that initiation of the flexion withdrawal reflex in the biceps femoris coincided with the perception of pain in normal volunteers.[72] Both the pain and the withdrawal reflex could be inhibited by mechanoreceptor activity.

Articular Mechanoreceptor Reflexes

In contrast to the Type IV nociceptor, the Type I and Type II receptors operate through the gamma motoneuron system.[23,29,35,38,39,41,46,59,62,63,68,75,78,80] They influence the gamma motoneuron polysynaptically.[8,21,32,78,80] Through the gamma motoneuron, the Type I and Type II joint receptors produce static and dynamic articular reflexes. The dynamic articular reflexes are produced during joint movement, whereas the static articular reflexes are produced in the absence of movement.[78] The static articular reflexes are the product of the discharge of the Type I receptors.[78] The dynamic articular reflexes result from the combined effect of the altered Type I discharge and the brief burst of impulses from the Type II receptors.[78] Thus, whereas the Type I receptor is responsible for both the static and dynamic articular reflexes, the Type II is only responsible for the dynamic reflexes.[10,78]

Besides the Type I articular receptor, the Type III articular receptor is responsible for the static articular reflexes.[10] The Type I and Type III receptors are similar in that they slowly adapt. The Type III receptor has a higher threshold than the Type I receptor. If a joint displacement is of sufficient intensity to activate Type III receptors, the normal articular reflexes will be distorted by the superadded effect of the Type III receptor discharges.[78]

In contrast to the Type I projecting to the gamma motoneuron, Freeman and Wyke suggested that the Type III receptor exerts a polysynaptic reflexogenic influence on muscle tone via the alpha motoneuron.[23,75,78] However, their suggestion that the Type III receptor projects to the alpha motoneuron may not be correct.

Both the Type I and Type III receptors are slowly adapting Ruffini-type receptors.[31,57,62,83] The Type III receptor has a well-developed perineural capsule and is located in structures with a parallel orientation of the collagen fibers, such as ligaments.[31] The Type I lacks a perineural capsule and is located in connective tissue structures without a predominant orientation of the collagen fibers, such as a capsule without ligamentous reinforcement. This division is arbitrary, since many receptors exist with intermediate properties.[41,49] A continuous spectrum of varieties of Ruffini endings are present.[41,49]

Thus, it seems odd that one end of this continuous spectrum of receptors would project to the gamma motoneuron, whereas the other end would project to the alpha motoneuron. The influence of the Type I joint receptors on the gamma motoneuron is well established.[23,29,35,38,39,41,46,62,63,68,75,78,80] However, Freeman and Wyke's suggestion that the Type III receptor influences the alpha motoneuron seems unconfirmed by other research. The influence of the Type III receptor on alpha motoneurons may be indirect in that it influences the gamma motoneuron first. The direct polysynaptic influence of the Type III receptor on the motoneuron needs more research, since little is known about the reflex effects of this receptor.

Collateral branches from the articular mechanoreceptor afferents (Types I, II, and III) project to other segments and to the contralateral side.[1,80] The collateral projections explain the muscle tone changes, for instance, in the contralateral extremity in response to articular mechanoreceptor stimulation.[26,44,73] Besides projecting to the local spinal motoneurons, the articular mechanoreceptors project to the brain stem motoneurons and to neurons in various parts of the cerebellum, brain stem reticular formation, and thalamus.[78] The collateral branches of the articular mechanoreceptors projecting to the cerebellum ascend in the posterior spinal columns.[76] Articular mechanoreceptor information arriving in the cerebellum, as well as in all other nervous system structures, causes coordinated reflex effects on muscles that contribute to the control of posture and movement.[20,76,78]

Following are two examples of articular mechanoreceptor reflexes on muscles. The first example is a study of the temporomandibular joint, and the second example discusses the tonic neck reflex in the cat.

Kawamura and Abe measured the responses of slowly and rapidly adapting TMJ receptor afferents in response to movements of the condylar head. They concluded that "afferent discharges from the articular mechanoreceptors in the TMJ capsule make a significant contribution to the reflex coordination of the masticatory muscle activities during the jaw movement. Especially in the phase of jaw-opening, the mechanoreceptors in the TMJ modify the excitatory level of the masseter and digastric motoneurons and make jaw opening smooth *(p. 81).*"[42]

In the cervical spine, the articular receptors seem responsible for the asymmetrical tonic neck reflex.[66] These receptors exert powerful reciprocally coordinated facilitory and inhibitory reflex effects on motor unit activity of the upper and lower extremities.[52] According to Wyke, the low-threshold articular receptor (Type I and Type II) afferents in the cervical spine project polysynaptically to motoneurons of the entire neck and upper and lower extremity muscles.[76]

McCough and colleagues studied the response to rotation and tilting of the cervical spines of cats.[47] The results of their study suggested that the receptors responsible for the tonic neck reflex are articular receptors located in the joints of the upper cervical spine, especially the atlantoaxial and atlantooccipital joints.[47] They found that the tonic neck reflex in the cat remained unimpaired when deprived of information from muscle or skin receptors. However, with intact muscle and skin receptors, but lacking information from the upper cervical joints, the tonic neck reflex was abolished.

A study by Sheer may be related to the tonic neck reflex.[60] This study of 11 patients with recent onset of neck pain showed weight-bearing changes after a cervical mobilization or manipulation. Sheer measured the weight distribution on the legs on two scales before and after an atlantoaxial mobilization or manipulation. Of the 11 subjects, 9 were asymmetrical in weight bearing before intervention. In seven of these patients, Sheer found an immediate change toward weight-bearing equalization after an "appropriate manual therapy technique." The weight bearing in two patients did not change. Sheer concluded that the Type I and II receptors in the cervical joints and periarticular tissue appear to influence the locomotor system.[60]

Final Common Input

The low-threshold articular receptor afferents influence the gamma motoneurons frequently and powerfully.[23,29,35,37,39,41,62,63,68,75,78,80] Low mechanical stimulation intensities seem to have strong effects on the gamma motoneurons.[38,39,46,62,63] The gamma motoneurons are highly responsive to a variety of afferent articular information. For instance, the gamma motoneurons were found to respond strongly to electrical stimulation of ipsilateral knee joint nerves, as well as to passive movements of the contralateral joints, pressure on the knee joint capsule, and stretch of the cruciate and collateral ligaments.[2,36,37,39,41,63] Only one study was unable to obtain clear evidence that low-threshold mechanoreceptors in joints contribute to the coordination of muscular activity via the fusimotor system.[48]

The gamma motoneuron is not only influenced by joint receptors, however, but by joint, cutaneous, and muscle receptors in the ipsilateral as well as the contralateral extremity and by supraspinal nuclei (see Fig.

6.2).[7,41] In other words, afferent information from the joint, muscle, and skin receptors on the ipsilateral and contralateral side converges on the gamma motoneuron.[3,5,6,34,36,40,70] The gamma motoneuron functions as an integrator for input from different sources.[3,35,39,40] Based on this information, the "final common input" hypothesis was proposed.[4,6,34,35,39,41,63]

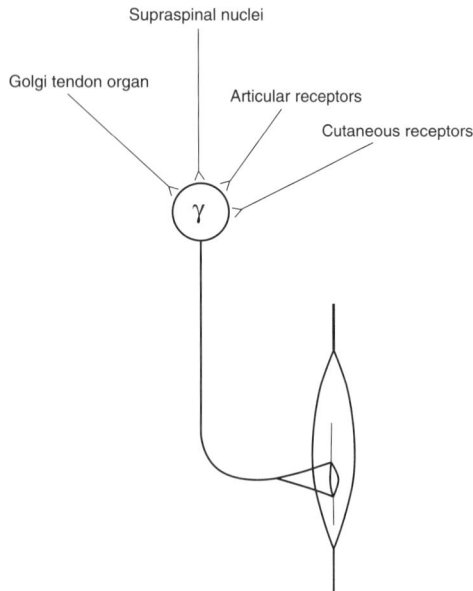

Fig. 6.2 Influence on the gamma motoneuron.

The final common input hypothesis implies that descending messages and peripheral receptor information are integrated in the gamma motoneuron and transmitted to the muscle spindle. In the muscle spindle, the integrated information undergoes a final adjustment according to the present length changes in the muscle. Thus, the information from the muscle spindle afferents is not only shaped by variations in muscle length, but also by signals from descending pathways and peripheral receptors. According to the final common input hypothesis, the gamma-spindle system is an integrative system of peripheral afference and descending input prior to the spindle cell delivering its afference to the alpha motoneuron.

Spindle Cell

The muscle spindle is considered the third most complex sensory organ, after the eye and the ear.[58] The receptor is arranged in parallel with the muscle fibers and is innervated by both afferent and efferent fibers.

The spindle cell contains intrafusal muscle fibers, whereas the true muscle fibers (i.e., the muscle fibers outside the spindle cell) are called the extrafusal fibers (see Fig. 6.3). Both the intrafusal and extrafusal muscle fibers are contractile and contain myofibrils. The intrafusal muscle fibers are innervated by gamma motoneurons, whereas the extrafusal fibers are innervated by alpha motoneurons.

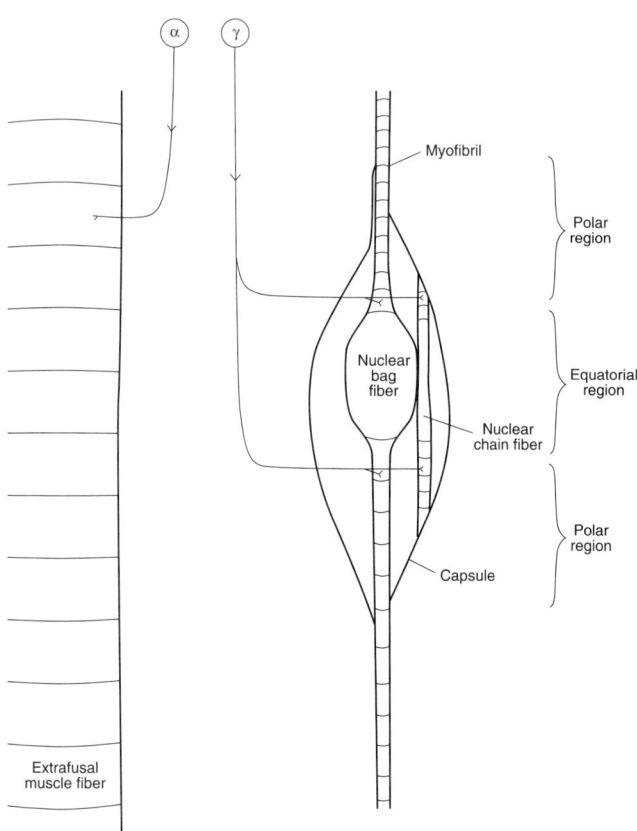

Fig. 6.3 The spindle cell and its efferent innervation.

The gamma motoneuron innervating the spindle cell was possibly first described in 1945 by Leksell.[45] He wrote that selective gamma fiber activity produced a contraction in the muscle of about 1 percent of a normal voluntary contraction.[45] Leksell observed that gamma fiber activity resulted in a discharge from the muscle proprioceptor (i.e., the spindle cell).

Within the muscle spindle, two different types of intrafusal muscle fibers are present.[13,14,71] One is the nuclear bag fiber, and the other is the nuclear chain fiber. Both types are largely contained within the capsule of the spindle. The nuclear chain fiber attaches to the inside of the capsule. The nuclear bag fiber extends beyond the capsule to attach to the endomysium of surrounding extrafusal muscle fibers. The intrafusal and extrafusal fibers run parallel to each other; thus, the length of the extrafusal fiber influences the length of the intrafusal fiber.

The two ends, or polar parts, of an intrafusal fiber attach to the muscle's connective tissue. The central part, or equatorial region, of the intrafusal fiber contains fewer myofibrils and has less resistance to lengthening than the two polar parts. Thus, active or passive lengthening of the muscle lengthens the equatorial part of the intrafusal fiber as well.

It is in the equatorial region that two different afferent nerve fibers (the primary and secondary afferents) contact the intrafusal fiber (see Fig. 6.4). The primary afferent fiber contacts the equatorial region of both the nuclear chain and the nuclear bag fibers through annulospiral endings. The secondary afferents have "flower spray" endings, which contact only the nuclear chain fibers. The endings of both nerve fibers are sensitive to lengthening of the equatorial part of the intrafusal muscle fibers. The annulospiral endings are rapidly adapting, whereas the flower spray endings are slowly adapting.

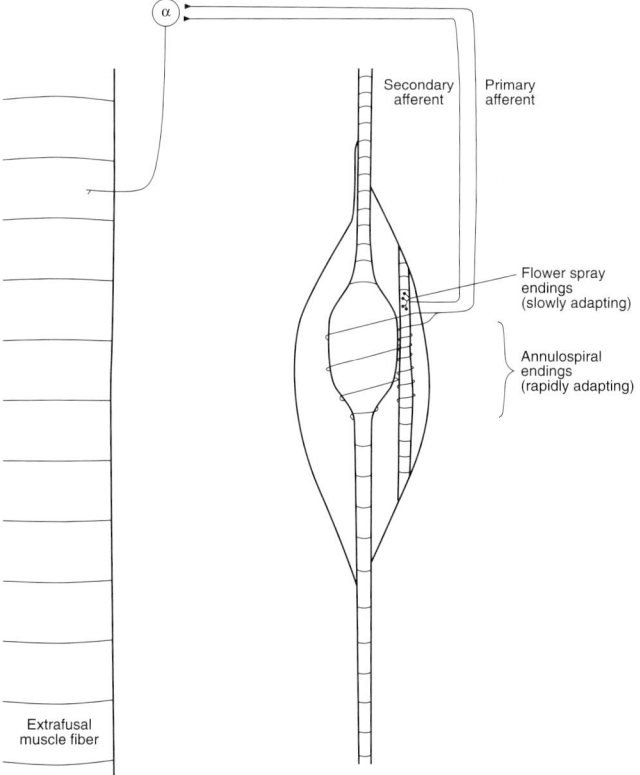

Fig. 6.4 The spindle cell and its afferent innervation.

Lengthening of the equatorial part of the intrafusal fiber occurs during active or passive lengthening of the muscle, as well as during a contraction of the myofibrils in the polar regions. The myofibrils in the polar regions receive their efferent innervation from the gamma motoneuron. A contraction of the polar regions becomes important when the muscle actively shortens. Without a contraction of the polar regions, an active shortening of the muscle would result in a shortening of the equatorial region of the intrafusal fibers, silencing the primary and secondary afferent nerve fibers. Further shortening would render the spindle afferents completely silent and incapable of providing feedback to the central nervous system. However, a contraction of the polar parts keeps the equatorial region "alert." Thus, gamma motoneuron activity keeps the spindle afferents alert or, depending on the amount of activity, could alter the spindle afferent discharge. Neuroscientists term this the *gamma bias* of the spindle.

The primary and secondary afferent nerve fibers conduct the spindle afference centrally to project at the alpha motoneuron. If the muscle lengthens due to an outside force, the receptors of the primary and secondary afferents at or near the equatorial region might register this stretch (depending on the gamma bias). Activity over the primary and secondary afferent nerve fibers will facilitate the alpha motoneuron and, when reaching the motoneuron's threshold, will cause depolarization of the alpha motoneuron, resulting in a contraction of extrafusal muscle fibers. The contraction counteracts the lengthening of the muscle. An example of this process is the myotatic reflex, better known as the deep tendon reflex. The quick muscle stretch by the reflex hammer causes the muscle to contract reflexly.

Activity of the primary and secondary spindle cell afferents facilitates the alpha motoneuron of the same muscle and its synergists. Synergistic muscles are those that control a similar action at the same joint.[27] At the ankle, for example, the gastrocnemius muscle is a synergist of the soleus muscle. Besides facilitating the alpha motoneuron of the same and the synergistic muscles, the alpha motoneurons of antagonistic muscles are inhibited (see Fig. 6.5). This configuration of facilitating one muscle group with concurrent inhibition of antagonistic muscles is termed *reciprocal innervation*.[17]

Thus, the myotatic reflex is a facilitation of the synergists with a simultaneous reciprocal inhibition of the antagonists. For example, the myotatic reflex of the quadriceps muscle (knee jerk reflex) results in a facilitation of the alpha motoneurons of the quadriceps with a simultaneous inhibition of the alpha motoneurons of the hamstrings. The net result may be a quick, temporary extension of the knee.

Activity of the gamma motoneurons and the muscle spindles largely controls the length and tone of a muscle.[16,17,33] Reflexes from peripheral afferents (e.g., joint afferents) to the gamma-spindle system obviously may also be important for length and tone of a muscle.[35,37,39,41,62,63,80] Muscle length and tone are therefore influenced by the amount of excitation and inhibition from peripheral afferents. Low muscle tone, for instance, may result from lack of gamma facilitation or from excessive inhibition. On the other hand, an abnormally short muscle, with reduced range of motion of the joint it spans, may result from an increased gamma bias. The gamma bias could be set too high in the presence of abnormal excitation or in the absence of proper inhibition.[50,51]

Static and Dynamic Gamma Motoneurons

Two types of gamma motoneurons exist: static and dynamic gamma motoneurons.[16,24,27,33] These gamma motoneurons selectively control the static and dynamic sensitivity of the spindle cell.[35]

Selective activation of the static gamma motoneuron causes the information from the spindles to reflect primarily the actual length of the muscle.[16,17,27,33,35] Thus, static gamma motoneuron activity causes the spindle cell to become more tonically sensitive (i.e., signal muscle length [see Table 6.1]).

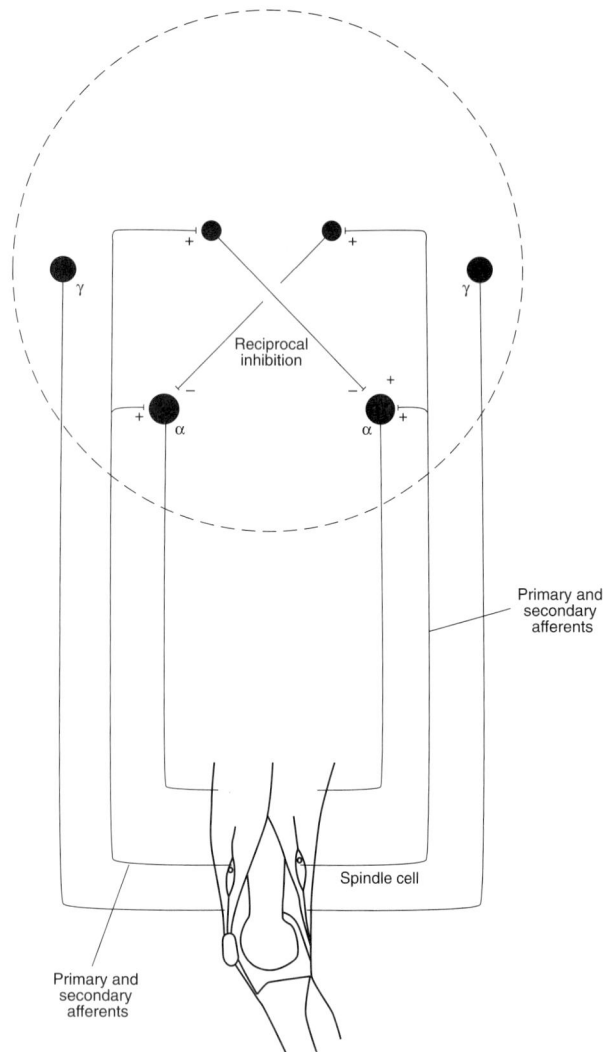

Fig. 6.5 Reciprocal innervation.
Adapted from Cranenburgh B van: Schema's Fysiologie. Lochem-Poperingen, Netherlands, De Tijdstroom, 1980, chapter 55, with permission from De Tijdstroom.

Table 6.1 Static and dynamic gamma motoneurons and spindle afferents.

Gamma Motoneuron	Controls the '...' Spindle Sensitivity	Spindle Cell Afferent	Afferent Receptor	Alpha Motoneurons
Dynamic	'Dynamic': Registers muscle length changes	Primary: Type Ia afferent Large diameter Fast conducting	Annulospiral ending Located on nuclear bag and nuclear chain fibers	Phasic (alpha-1)
Static	'Static': Registers muscle length	Secondary: Type II afferent Smaller diameter Slowly conducting	Flower spray ending Located on nuclear chain fibers	Static (alpha-2)

On the other hand, selective activation of the dynamic gamma motoneuron causes the spindle cell to be more dynamically responsive.[16,17,27,33,35] With dynamic gamma motoneuron activation, the spindle cell reports information about small and quick fluctuations in length (e.g., the myotatic reflex).

Static and Dynamic Spindle Responses

As mentioned earlier, two afferent nerve fibers derive from the spindle cell: the primary and the secondary.[9,24] The primary has a more dynamic function, whereas the secondary has a more static function.

The spindle cell's primary afferent originates from the annulospiral ending, which can be found on both the nuclear bag and the nuclear chain fibers. Of the two, the nuclear bag fiber is more involved with the dynamic responses to muscle length changes.[71] The annulospiral ending is rapidly adapting. It responds temporarily to lengthening of the muscle with an increased firing rate. The primary afferent itself is a large-diameter, fast-conducting Type Ia fiber. Thus, the primary afferent seems well equipped to notify the central nervous system about length changes in the muscle, such as occur during movement.[15]

The spindle cell's secondary afferent originates from the flower spray endings located predominantly at the nuclear chain fiber. The function of the secondary afferent may be a more static one, signaling the actual length of the muscle, for the flower spray endings are slowly adapting. The afferent nerve itself is a relatively slowly conducting Type II fiber and has a relatively small diameter. Thus, the secondary afferent, originating from the nuclear chain fiber, seems to function as a detector of static limb position.[15,71]

The dynamic activity of the primary spindle afferents monosynaptically influences "phasic" alpha motoneurons.[16,54] In contrast, the static activity of the secondary afferents seems to stimulate predominantly "tonic" alpha motoneurons.[16,54]

Tonic and Phasic Alpha Motoneurons and Muscle Fibers

Three types of alpha motoneurons exist: an alpha-1 motoneuron, an alpha-2 motoneuron, and an intermediate alpha motoneuron. The alpha-1 motoneuron has phasic characteristics. Its efferent axon has a large diameter and is fast conducting. In contrast, the alpha-2 motoneuron has more tonic characteristics and has a small, slowly conducting efferent axon (see Table 6.2).

Each alpha motoneuron innervates several extrafusal muscle fibers. All extrafusal muscle fibers that belong to the same alpha motoneuron share similar physiological characteristics. Muscle fibers innervated by an alpha motoneuron of another type have very different characteristics. The three groups of alpha motoneurons innervate three different groups of muscle fibers.[9,11,25,33,54] The phasic alpha-1 motoneuron innervates phasic muscle fibers, which have a large diameter and are quickly responding. The tonic alpha-2 motoneuron innervates tonic muscle fibers, which have a small diameter and are slowly responding. The intermediate alpha motoneuron innervates muscle fibers with intermediate characteristics.

The three different efferent nerve fibers innervate extrafusal muscle fibers with characteristics similar to their own. For instance, the large-diameter, fast-conducting efferent axon innervates likewise large-diame-

Table 6.2 Alpha motoneuron and muscle fiber types.

Alpha Motoneuron	Facilitation Threshold	Efferent Axon	Muscle Fiber
Alpha-1 (phasic)	High	Large diameter Fast conducting	Phasic: Large diameter Fast responding Rapidly fatigable
Alpha-2 (tonic)	Low	Small diameter Slowly conducting	Tonic: Small diameter Slowly responding Slowly fatigable
Intermediate		Medium diameter Intermediately conducting	Intermediate

ter, fast-responding muscle fibers. Small-diameter, slowly conducting nerve efferents serve small-diameter, slowly contracting muscle fibers. In other words, a phasic alpha motoneuron innervates phasic muscle fibers and a tonic alpha motoneuron innervates tonic muscle fibers. As a result, the speed of conduction along the axon is correlated with the muscle fiber's speed of contraction.

The phasic muscle fibers contract and relax rapidly. These muscle fibers can generate large amounts of force. Phasic muscle fibers can generate 100 times more force than tonic muscle fibers. One reason phasic muscle fibers can generate these large forces is that they have a larger diameter; however, the drawback to generating large amounts of force is that these muscle fibers are easily fatigued. The tonic muscle fibers contract slowly and generate less force; therefore, these tonic fibers are more fatigue resistant. The intermediate group of muscle fibers has properties that lie between the phasic and the tonic muscle fibers.

The tonic alpha-2 motoneuron has a relatively low facilitation threshold. In contrast, the phasic alpha-1 motoneuron has a higher facilitation threshold. The difference in facilitation threshold makes the low-threshold alpha-2 motoneuron discharge first. As a result, the tonic muscle fibers will contract before the phasic muscle fibers.[16,24,25,30,54] During a buildup of contraction force, the tonic motoneurons and muscle fibers are recruited first, and the phasic motoneurons and muscle fibers are recruited last. As a result, the tonic motoneurons and muscle fibers may be active in both moderate and strong muscle contractions. In contrast, the phasic motoneurons and muscle fibers are only active to generate a large contraction force.[16,24,25,30,54] The activity of phasic motoneurons and muscle fibers is added to the more constant activity of tonic motoneurons and muscle fibers.

Individual muscles contain a blend of muscle fibers of the three types. Some muscles, however, contain more tonic fibers, whereas other muscles contain more phasic fibers. When a muscle contains more phasic than tonic fibers, it is termed a *phasic muscle*. A muscle that contains a relatively large percentage of tonic fibers is termed a *tonic muscle*.

In summary, dynamic gamma motoneuron activity makes the spindle cell more responsive to dynamic, or changing, muscle lengths. The primary afferent from the spindle cell, which conveys predominantly dynamic information, monosynaptically contacts the dynamic alpha-1 motoneuron of the same muscle and synergistic muscles.[16,24,27,33,54] The dynamic, or phasic, alpha-1 motoneuron innervates phasic muscle fibers.

The static gamma motoneuron activity increases the spindle cell response to the actual length of the muscle. The secondary afferents from the spindle cell, which convey more static information about the muscle's length, may make more contact with the static alpha motoneurons.[16,24,27,33,54] These static, or tonic, alpha-2 motoneurons innervate tonic muscle fibers.

Static and Dynamic Articular Reflexes

Stimulation of articular receptors via reflexes onto static and dynamic gamma motoneurons causes changes in both primary and secondary spindle cell afferents.[2,35,36,37,38,41,62,63,70] The very modest load needed to evoke reflex effects on the gamma motoneuron indicates that the reflexes are not of a nociceptive character.[41] The receptors responsible for strong reflex effects on the gamma motoneurons are low-threshold articular receptors, which probably are active throughout the full range of the joint.[35,36,41,62] The Type I and II articular receptors fit the description of these receptors. The Type I and II joint receptors project polysynaptically to the gamma motoneuron.[22,74,78,79] Evidence suggests that the Type I and II afferents project to the gamma motoneuron via interneurons located in lamina IV of the spinal gray matter.[78]

Wyke suggested that the Type I afferent projects to the static gamma motoneuron.[74,78] Even in immobile joints, the Type I mechanoreceptor produces a discharge that influences the static gamma motoneuron.[55,78] Thus, the Type I joint receptor, via the static gamma motoneuron, can be expected to influence the tonic alpha motoneuron and the tonic muscle fibers.[54,55,76,78] Wyke wrote that the tonic discharge of the Type I receptor produces a constant background activity for the tonic alpha-2 motoneuron and the tonic muscle fibers.[76,78]

The normally present continuous discharge from the Type I receptor, together with information from the cutaneous and the muscle mechanoreceptors and descending input from the supraspinal nuclei, determine the degree of gamma bias during static positions.[3,5,6,7,22,34,36,40,41,70]

During movement, the alteration of cutaneous, muscle, and Type I receptor activity is augmented by the brief discharge from the Type II articular receptor. In contrast to the Type I articular receptor possibly influencing the tonic alpha-2 motoneuron, the Type II receptor may have a stronger influence on the phasic alpha-1 motoneuron and its phasic muscle fibers.[54,55,74,78] During movement, the altered activity of these peripheral receptors results in an adjustment of the gamma bias.[22]

Proximal extremity joints have a higher population density of Type I articular receptors, whereas distal extremity joints have a higher population density of Type II receptors. Therefore, movement of the proximal joints stimulates relatively more Type I articular receptors, whereas movement of the distal extremity joints stimulates relatively more Type II receptors.[18,73,78] Thus, movement of proximal joints may influence more tonic alpha-2 motoneurons, whereas distal joints influence more phasic alpha-1 motoneuron activity.[73] For instance, in contrast to the ankle joint, passive hip joint movements cause only a small amount of phasic muscle activity.[73] The main muscle activity over the hip joint is slowly adapting.[73]

Arthrogenic muscle wasting is the muscle atrophy caused by disease or trauma to the joint. Researchers have attempted to find out if arthrogenic muscle wasting preferred atrophy of tonic or phasic muscle fibers, but the limited research available has found no consistent preference of one fiber type atrophy over another.[11,65] Some researchers encountered atrophy of tonic fiber types, whereas others encountered atrophy of phasic muscle fibers. A possible explanation for the selective atrophy of tonic or phasic muscle fibers is selective inhibition of the tonic or phasic alpha motoneurons.[65,81]

Receptor Hysteresis and Residual Muscle Tension

When a joint moves into flexion, articular flexor receptors discharge. As the joint moves farther into flexion, the discharge frequency of the flexion receptors increases. However, when returning from a flexed position, these same receptors fire less than when the joint went into the flexed position. The reduced firing of the receptors is termed *hysteresis*. The word *hysteresis* is Greek for "lagging behind." In this case, the articular receptors are lagging behind, discharging less than when they were originally in this position. The cause of the receptors firing less might be the visco-elastic relaxation of the joint's ligamento-capsular tissue and/or the adaptation of the receptors themselves (see chapter 17).[28] Secondary to the hysteresis, every joint position has at least two firing frequencies of the articular receptors, depending on how the joint arrived at each position (e.g., from a flexed or an extended position).

The receptors' two different firing frequencies at the same joint angle may make accurate kinesthesia more difficult. Similarly, the reflex motor responses will also differ depending on whether the joint arrived at the position by flexion or by extension. Are the varying reflexes at the same joint angle an inadequacy of the body's design? Zill and Jepson-Innes came to an interesting conclusion and provided the following information.[82] An increase in motoneuron firing results in increased tone in the muscle it innervates. When the firing frequency of the motoneurons decreases to its previous level, the tone of the muscle decreases but a large residual increase in muscle tension remains. This phenomenon may allow maintenance of muscle tension without the necessity for high firing frequencies of the motoneurons. Zill and Jepson-Innes found that the hysteresis of the articular receptors counteracts the residual muscle tension.[82] For example, starting from an extended position, flexion of a joint will cause the flexion receptors to increase their firing rate, resulting in increased tone in the extensor muscle. On the joint's return to the extended position, the flexion receptors decrease their firing rate below the level they previously had. The decreased firing rate counteracts the residual tension still present in the extensor muscle. The net effect of the receptors' decreased firing rate and the residual tension in the extensor muscle is a tension in the extensor muscle identical to what it had when the joint arrived at its current position from an extended position.[82]

Summary

Articular receptors influence motoneurons polysynaptically. Type IV receptors polysynaptically influence alpha motoneurons and may be responsible for the distorted postures and movements associated with articular pain. Articular mechanoreceptor activity can inhibit Type IV nociception. Type III articular receptors have an influence on motoneurons, but it is not clear whether this influence is more directly on the gamma or on the alpha motoneurons.

The Type I and II articular receptors, together with other peripheral receptors, polysynaptically influence gamma motoneurons and subsequently the gamma bias of the spindle cell. Muscle tone and length are therefore influenced by articular afferents. Low tone and weakness of a muscle may result from a lack of facilitation of gamma motoneurons, whereas increased tone with muscle shortening may result from excessive facilitation of gamma motoneurons.

The Type I receptor may have a stronger influence on static gamma motoneurons and therefore on the static sensitivity of spindle cells, on tonic alpha-2 motoneurons, and on tonic muscle fibers. The Type II receptor may have a stronger influence on dynamic gamma motoneurons, dynamic sensitivity of the spindle cell, phasic alpha-1 motoneurons, and phasic muscle fibers.

Articular receptors show hysteresis (reduced firing) when a joint returns to its previous position from a movement that provoked the receptors into firing at higher frequencies. The hysteresis counters the residual muscle tension still present from the active movement. This mechanism seems to have the purpose of maintaining the same muscle tension relative to a joint position, regardless of how the position was achieved.

References

1. Appelberg B, Hulliger M, Johansson H, et al: Excitation of dynamic fusimotor neurones of the cat triceps surae by contralateral joint afferents. Brain Res 160:529-532, 1979
2. Appelberg B, Hulliger M, Johansson H, et al: Reflex activation of dynamic fusimotor neurons by natural stimulation of muscle and joint receptor afferent units. In Taylor A, Prochazka A (eds): Muscle Receptors and Movement. New York, Oxford University Press, 1981, pp 149-161
3. Appelberg B, Hulliger M, Johansson H, et al: Fusimotor reflexes in triceps surae elicited by natural stimulation of muscle afferents from the cat ipsilateral hind limb. J Physiol 329:211-229, 1982
4. Appelberg B, Hulliger M, Johansson H, et al: Actions on γ-motoneurones elicited by electrical stimulation of group III muscle afferent fibres in the hind limb of the cat. J Physiol 335:275-292, 1983
5. Appelberg B, Hulliger M, Johansson H, et al: Fusimotor reflexes in triceps surae muscle elicited by extension of the contralateral hind limb in the cat. J Physiol 355:99-117, 1984
6. Appelberg B, Johansson H, Sojka P: Fusimotor reflexes in triceps surae muscle elicited by stretch of muscles in the contralateral hind limb of the cat. J Physiol 373:419-441, 1986
7. Baxendale RH, Davey NJ, Ellaway PH, et al: The interaction between joint and cutaneous afferent input in the regulation of fusimotor neurone discharge. In Jami L, Pierrot-Deseilligne EP, Zytnicki D (eds): Muscle Afferents and Spinal Control of Movement. New York, Pergamon Press, 1991, pp 95-104
8. Baxendale R, Ferrell W, Wood L: Responses of quadriceps motor units mechanical stimulation of knee joint receptors in the decerebrate cat. Brain Res 453:150-156, 1988
9. Bernards JA, Bouman LN: Fysiologie van de Mens, ed 2. Utrecht, The Netherlands, Bohn, Scheltema & Holkema, 1977
10. Biedert RM, Stauffer E, Friederich NF: Occurrence of free nerve endings in the soft tissue of the knee joint. Am J Sports Med 20:430-433, 1992
11. Brooke M, Kaiser K: The use and abuse of muscle histochemistry. In Drachman DB (ed): Trophic Functions of the Neuron. Ann N Y Acad Sci 228: 121-144, 1974
12. Bush BM, Vedel JP, Clarac F: Intersegmental reflex actions from a joint sensory organ (CB) to a muscle receptor (MCO) in decapod crustacean limbs. J Exp Biol 73:47-63, 1978
13. Carew TJ: Spinal cord I: Muscles and muscle receptors. In Kandel ER, Schwartz JH (eds): Principles of Neural Science. New York, Elsevier, 1982, pp 284-292
14. Carew TJ: Spinal cord II: Reflex action. In Kandel ER, Schwartz JH (eds): Principles of Neural Science. New York, Elsevier, 1982, pp 293-304
15. Carli G, Fontani G, Meucci M: Static characteristics of muscle afferents from gluteus medius muscle: Comparison with joint afferents of hip in cats. J Neurophysiol 45:1085-1095, 1981
16. Cranenburg B van: Schema's Fysiologie. Lochem, The Netherlands, De Tijdstroom, 1980
17. Daube JR, Reagan TJ, Sandok BA, et al: Medical Neurosciences: An Approach to Anatomy, Pathology, and Physiology by Systems and Levels, ed 2. Boston, Little, Brown and Co, 1986
18. Dee R: Mechanoreceptors in hip joint capsule and ligamentum capitis femoris and their reflex contribution to posture. In Symposium on Osteoarthritis. St. Louis, C.V. Mosby Co, 1976, pp 52-65
19. Dee R: The innervation of joints. In Sokoloff L (ed): The Joints and Synovial Fluid. London, Academic Press, 1978, pp 177-204
20. Eldred E: Posture and locomotion. In Field J, Magoun HW, Hall VE (eds): Handbook of Physiology: A Critical, Comprehensive Presentation of Physiological Knowledge and Concepts: Section 1: Neurophysiology. Washington, DC, American Physiological Society, 1960, vol 2, pp 1067-1088

21. Feddina L, Hultborn H: Facilitation from ipsilateral primary afferents of interneuronal transmission in the Ia inhibitory pathway to motoneurones. Acta Physiol Scand 86:59-81, 1972

22. Freeman MAR, Wyke B: Articular reflexes at the ankle joint: An electromyographic study of normal and abnormal influences of ankle-joint mechanoreceptors upon reflex activity in the leg muscles. Br J Surg 54:990-1001, 1967

23. Freeman MAR, Wyke B: The innervation of the knee joint: An anatomical and histological study in the cat. J Anat 101(3):505-532, 1967

24. Geers A: Kinesiologie: Diagnostiek en Therapie van de posturale en fasische spieren. Eindhoven, The Netherlands, Stichting Manuele Geneeskunde, 1984

25. Ghez C: Muscles: Effectors of the motor systems. In Kandell ER, Schwartz JH, Jessell TM (eds): Principles of Neural Science, ed 3. New York, Elsevier, 1991, pp 548-563

26. Gordon JP: Spinal mechanisms of motor coordination. In Kandell ER, Schwartz JH, Jessell TM (eds): Principles of Neural Science, ed 3. New York, Elsevier, 1991, pp 581-595

27. Gordon J, Ghez C: Muscle receptors and spinal reflexes: The stretch reflex. In Kandell ER, Schwartz JH, Jessell TM (eds): Principles of Neural Science, ed 3. New York, Elsevier, 1991, pp 564-580

28. Grigg P, Greenspan BJ: Response of primate joint afferent neurons to mechanical stimulation of knee joint. J Neurophysiol 40:1-8, 1977

29. Grillner S, Hongo T, Lund S: Descending monosynaptic and reflex control of γ-motorneurones. Acta Physiol Scand 75:592-613, 1969

30. Hagenaars L: Tonische en fasische motore eenheden... Een nieuwe start. Nederlands Tijdschrift voor Fysiotherapie 93:55-61, 1983

31. Halata Z, Rettig T, Schulze W: The ultrastructure of sensory nerve endings in the human knee joint capsule. Anat Embryol 172:265-275, 1985

32. Harrison PJ, Jankowska E: Sources of input to interneurones mediating group I non-reciprocal inhibition of motoneurones in the cat. J Physiol 361:379-401, 1985

33. Henneman E: Peripheral mechanisms involved in the control of muscle. In Mountcastle VB (ed): Medical Physiology, ed 13. St. Louis, C.V. Mosby Co, 1974, vol 1, pp 617-635

34. Johansson H: Reflex integration in the γ-motor system. In Boyd IA, Gladden MH (eds): The Muscle Spindle. London, The MacMillan Press Ltd, 1985, pp 297-301

35. Johansson H: Role of knee ligaments in proprioception and regulation of muscle stiffness. Journal of Electromyography and Kinesiology 1:158-179, 1991

36. Johansson H, Sjölander P, Sojka P: Actions on γ-motoneurones elicited by electrical stimulation of joint afferent fibers in the hind limb of the cat. J Physiol 375:137-152, 1986

37. Johansson H, Sjölander P, Sojka P: Fusimotor reflexes in triceps surae muscle elicited by natural and electrical stimulation of joint afferents. Neuro-Orthopedics 6:67-80, 1988

38. Johansson H, Sjölander P, Sojka P: Activity in receptor afferents from the anterior cruciate ligament evokes reflex effects on fusimotor neurones. Neurosci Res 8:54-59, 1990

39. Johansson H, Sjölander P, Sojka P: A sensory role for the cruciate ligaments. Clin Orthop 268:161-178, 1991

40. Johansson H, Sjölander P, Sojka P: Fusimotor reflex profiles of individual triceps surae primary muscle spindle afferents assessed with multi-afferent recording technique. J Physiol Paris 85:6-19, 1991

41. Johansson H, Sjölander P, Sojka P: Receptors in the knee joint ligaments and their role in the biomechanics of the joint. Crit Rev Biomed Eng 18:341-368, 1991

42. Kawamura Y, Abe K: Role of sensory information from temporomandibular joint. Bull Tokyo Med Dent Univ 21(suppl):78-82, 1974

43. Klineberg I: Structure and function of temporomandibular joint innervation. Ann R Coll Surg Engl 49:268-288, 1971

44. Klineberg IJ, Greenfield BE, Wyke BD: Contributions to the reflex control of mastication from mechanoreceptors in the temporomandibular joint capsule. Dental Practitioner 21:73-83, 1970

45. Leksell L: The action potential and excitatory effects of the small ventral root fibres to skeletal muscle. Acta Physiol Scand 10(suppl 31):1-84, 1945

46. Marshall KW, Tatton WG: Joint receptors modulate short and long latency muscle responses in the awake cat. Exp Brain Res 83:137-150, 1990

47. McCough GP, Deering ID, Ling TH: Location of receptors for tonic neck reflexes. J Neurophysiol 14:191-195, 1951

48. McIntyre AK, Proske U, Tracey DJ: Fusimotor responses to volleys in joint and interosseous affer-

ents in the cat's hind limb. Neurosci Lett 10:287-292, 1978

49 McLain RF: Mechanoreceptor endings in human cervical facet joints. Spine 19:595-501, 1994

50 Mitchell FL jr: Elements of Muscle Energy technique. In Basmajian JV, Nyberg R (eds): Rational Manual Therapies. London, Williams & Wilkins, 1993, pp 285-321

51 Mitchell FL jr, Mitchell PKG: The Muscle Energy manual: Concepts & mechanisms, the musculoskeletal screen, cervical region evaluation and treatment. East Lansing, Michigan, MET Press, 1995, vol 1

52 Molina F, Ramcharan JE, Wyke BD: Structure and function of articular receptor systems in the cervical spine. J Bone Joint Surg Br 58:254-255, 1976

53 Neugebauer V, Schaible HG: Peripheral and spinal components of the sensitization of spinal neurons during an acute experimental arthritis. Agents Actions 25:234-236, 1988

54 Oostendorp RAB: De musculaire dysbalans bij de patiënt met lage rugklachten. Nederlands Tijdschrift voor Fysiotherapie 90:82-90, 1980

55 Oostendorp RAB, Sande JAW van de: Arthrokinetische reacties en musculaire stabiliteit. Nederlands Tijdschrift voor Fysiotherapie 93:63-72, 1983

56 Schaible HG, Neugebauer V, Schmidt RF: Osteoarthritis and pain. Semin Arthritis Rheum 18(suppl 2):30-34, 1989

57 Schutte MJ, Dabezies EJ, Zimny ML, et al: Neural anatomy of the human anterior cruciate ligament. J Bone Joint Surg Am 69:243-247, 1987

58 Schutte M, Happel L: Joint innervation in joint injury. Clin Sports Med 9(2):511-515, 1990

59 Scott DT, Ferrell WR, Baxendale RH: Excitation of soleus/gastrocnemius g-motoneurons by group II knee joint afferents is suppressed by group IV joint afferents in the decerebrate, spinalized cat. Exp Physiol 79:357-364, 1994

60 Sheer D: Manipulation/mobilization: Its effect on weight distribution. Scientific Physical Therapy 6:1-9, 1996

61 Shimamura M, Kogure I, Fuwa T: Role of joint afferents in relation to the initiation of forelimb stepping in thalamic cats. Brain Res 297:225-234, 1984

62 Sjölander P, Johansson H, Sojka P, et al: Sensory nerve endings in the cat cruciate ligaments: A morphological investigation. Neurosci Lett 102:33-38, 1989

63 Sojka P, Johansson H, Sjölander P, et al: Fusimotor neurones can be reflexly influenced by activity in receptor afferents from the posterior cruciate ligament. Brain Res 483:177-183, 1989

64 Solomonow M, D'Ambrosia R: Neural reflex arcs and muscle control of the knee stability and motion. In Scott WN (ed): Ligament and Extensor Mechanisms of the Knee. St. Louis, C.V. Mosby Co, 1991, pp 389-400

65 Stokes M, Young A: The contribution of reflex inhibition to arthrogenous muscle weakness. Clin Sci 67:7-14, 1984

66 Thoden U, Wenzel D: Tonic cervical influences on forelimb and hindlimb monosynaptic reflexes. In Granit R, Pompeiano O (eds): Reflex Control of Posture and Movement. Prog Brain Res 50: Amsterdam, Elsevier, 1979, pp 281-288

67 Tracey DJ: Joint receptors and the control of movement. Trends Neurosci 3:253-255, 1980

68 Voorhoeve PE, Kanten RW: Reflex behaviour of fusimotor neurones of the cat upon electrical stimulation of various afferent fibers. Acta Physiol Pharmacol Neerlandica 10:391-407, 1962

69 Vrettos XC, Wyke BD: Articular reflexogenic systems in the costovertebral joints. J Bone Joint Surg Br 5:382, 1974

70 Wadell I, Johansson H, Sjölander P, Sojka P, et al: Fusimotor reflexes influencing secondary muscle spindle afferents from flexor and extensor muscles in the hind limb of the cat. J Physiol Paris 85:223-234, 1991

71 Williams PL, Warwick R, Dyson M, et al (eds): Gray's Anatomy, ed 36. New York, Churchill Livingstone, 1989

72 Willer JC: Comparative study of perceived pain and nociceptive flexion reflex in man. Pain 3:69-80, 1977

73 Wyke B: Articular neurology - A review. Physiotherapy 58:94-99 1972

74 Wyke B: Morphological and functional features of the innervation of the costovertebral joints. Folia Morphol (Warsz) 23:296-305, 1975

75 Wyke B: Clinical significance of articular receptor systems. Ann R Coll Surg Engl 60:137, 1978

76 Wyke B: Conference on the ageing brain: Cervical articular contributions to posture and gait: Their relation to senile disequilibrium. Age Ageing 8:251-258, 1979

77 Wyke B: Neurology of the cervical spinal joints. Physiotherapy 65:73-76, 1979

78 Wyke B: The neurology of joints: A review of general principles. Clinics in Rheumatic Diseases 7:223-239, 1981

79 Wyke B: Receptor systems in lumbosacral tissues in relation to the production of low back pain. In White AA, Gordon SL (eds): The American Academy of Orthopaedic Surgeons: Symposium on Idiopathic Low Back Pain. St. Louis, C.V. Mosby Co, 1982, pp 97-107

80 Wyke B: Articular neurology and manipulative therapy. In Glasgow E, Twomey L (ed): Aspects of Manipulative Therapy. Edinburgh, Scotland, Churchill Livingstone, 1985, pp 72-77

81 Young A: Rehabilitation for wasted muscles. In Sarner, M (ed): Advanced Medicine. London, Pitman Medical, 1982, vol 18

82 Zill SN, Jepson-Innes K: Evolutionary adaptation of a reflex system: Sensory hysteresis counters muscle 'catch' tension. J Comp Physiol A 164:43-48, 1988

83 Zimny ML, Schutte M, Dabezies E: Mechanoreceptors in the human anterior cruciate ligament. Anat Rec 214:204-209, 1986

7. Articulo-Muscular Protective Reflex

During the past century, researchers have studied the reflexes initiated by articular receptor activity. Many researchers have reported how these reflexes altered muscle tone to improve joint stability and protect the joint from excessive movements. This chapter is dedicated to the protective reflex of the joint: the articulo-muscular protective reflex.

Ankle Mechanoreceptor Reflexes

Freeman and Wyke conducted a series of experiments to examine muscle tone alterations in response to stimulation of the articular receptors.[15,16,64,65] Their 1967 publication is very detailed and makes the reflex effects of the low-threshold articular receptors easy to understand.[16]

Freeman and Wyke conducted their experiments on cat ankle joints. Their objective was to examine the reflex effects of the ankle joint's capsular receptors on the gastrocnemius and anterior tibialis muscles. To isolate the capsular receptor reflexes from those arising from other sources during movement of the foot, the skin was resected and all tendons were tenotomized (cut). This procedure ensured that mechanoreceptors in the skin and muscles could not contribute to the reflex effects on the ankle muscles. All cats were lightly anesthetized. EMG recordings were obtained from the gastrocnemius and tibialis anterior muscles while the ankle joint was passively moved into plantar or dorsiflexion.

The study showed that passive ankle dorsiflexion produced increased activity of the gastrocnemius muscle with simultaneous reduction of tibialis anterior activity. The reverse occurred with plantar flexion; the gastrocnemius activity progressively diminished while tibialis anterior tone increased. The increased activity of the gastrocnemius during dorsiflexion and of the tibialis anterior during plantar flexion occurred in two phases. The first phase consisted of a brief initial, high-amplitude burst of motor unit activity. A motor unit consists of a single alpha motoneuron with the muscle fibers it innervates. The brief initial burst of motor unit activity was followed by the second phase, a prolonged, lower amplitude, lower frequency discharge that continued as long as the foot maintained the same position.[64,65] Freeman and Wyke suggested that the brief initial motor unit activity was the result of discharges of the rapidly adapting Type II receptors located in the stretched part of the joint capsule.[64,65] The sustained, slowly adapting response that followed was thought to be the result of the continuous discharge of slowly adapting Type I receptors in the same stretched area of the joint capsule.[65]

In the fully plantar flexed position, the gastrocnemius was relaxed. On passively dorsiflexing the ankle joint, the gastrocnemius initially remained relaxed. The gastrocnemius activity began when the ankle was about halfway dorsiflexed, the position where the posterior capsule started to be stretched. When the posterior capsule started to be stretched, the initial transient burst of motor unit activity was recorded. This brief initial burst lasted about 1 second and was followed by a sustained lower frequency discharge. The lower frequency discharge was maintained as long as the foot remained in the same position. When the ankle was moved back into plantar flexion, the sustained gastrocnemius activity was abruptly terminated. A comparable result was obtained in the tibialis anterior muscle during plantar flexion of the foot.

According to Freeman and Wyke, the articular receptors responsible for these reflexes were not Type III articular receptors. The Type III receptors only respond to high levels of stress, as could be reached at the end ranges of a joint. The Type IV receptor endings were likewise not involved in these articular reflexes. The muscular responses in these experiments were obtained in physiological ranges of the joint, without excessive stress to the capsule.

From the results of their study, Freeman and Wyke concluded that the articular reflexes on the gastrocnemius and tibialis anterior muscles serve to prevent (excessive) movement of the ankle joint. To further examine these results, they performed additional experiments. The ankle joint was immobilized in a plantar flexed position, where the gastrocnemius activity was minimal and the tibialis anterior activity was maximal. In this position, direct pressure on the posterior capsule of the ankle joint caused activity in the gastrocnemius, with a simultaneous reduction of tibialis anterior activity. The contraction of the gastrocnemius consisted of a brief initial burst of motor unit activity in the muscle, followed by a diminished sustained motor unit activity if the pressure on the capsule was maintained. Compression of the joint capsule after the capsule was electrocoagulated failed to produce this response.

Injection of the ankle joint tissues with a local anesthetic abolished the response of the muscles to the ankle movements. The muscular activity resumed after the anesthetic wore off. Electrocoagulation of the joint capsule produced effects similar to those of anesthetic injection except that they were not reversible. Evidently, destruction and local anesthesia of the nerve endings in the joint abolished the reflex responses to passive movement onto the gastrocnemius and anterior tibialis muscles.

When only the anterior capsule was coagulated, the gastrocnemius muscle's response to dorsiflexion was unaltered, but the response of the anterior tibialis to plantar flexion was substantially reduced. Additional coagulation of the posterior capsule and posterior fat pad, or coagulation of these posterior structures alone at the beginning of the experiment, caused a reduced response of the gastrocnemius to dorsiflexion. This observation suggests that the articular mechanoreceptors that cause facilitation of the gastrocnemius in response to dorsiflexion with simultaneous inhibition of the anterior tibialis are located dorsally in the ankle joint. Similarly, the articular mechanoreceptors responsible for anterior tibialis facilitation with simultaneous gastrocnemius inhibition during plantar flexion are located in the anterior part of the ankle joint.

Neurectomy (excision of part of a nerve) produced results similar to those obtained with coagulation. Neurectomy of the anterior tibial nerve, which innervates the anterior ankle joint, and sectioning of the lateral popliteal nerve, which innervates the lateral ankle joint, produced no change in the gastrocnemius response to dorsiflexion. Changes in the gastrocnemius muscle were observed after neurectomy of the posterior tibial nerve. The changes in response to neurectomy of the posterior tibial nerve were reduction or abolition of the gastrocnemius response. The fact that the gastrocnemius response was not completely abolished in all cases was explained by afferent articular nerves from the posterior ankle joint traveling proximally by other pathways than the posterior tibial nerve alone.

Freeman and Wyke found that the rapidly and slowly adapting responses in the gastrocnemius and anterior tibialis muscles disappeared with deep anesthesia, such as would be used for major surgical procedures. The responses were only present when very light anesthesia was used. At intermediate levels of anesthesia, the muscular response to joint movement was absent but the myotatic reflex (deep tendon reflex) was present. At deeper stages of anesthesia, the myotatic reflex also disappeared. When recovering from deep anesthesia, the myotatic reflex recovered first, before muscular responses to ankle joint movement were restored. Thus, the articular reflex system was more sensitive to anesthesia. Monosynaptic reflexes such as the myotatic reflex are less sensitive to barbiturate narcosis than polysynaptic reflexes. Freeman and Wyke's experiments suggest that the articular reflex is polysynaptic.

Freeman and Wyke wrote that the articular reflexes were likely mediated by gamma motoneurons influencing the muscle spindles. Gamma motoneurons need a certain amount of background stimulation for certain reflexes (e.g., articular reflexes) to function. Deep anesthesia deprives the gamma motoneurons of their background stimulation, and thus the articular reflex does not cause muscle tone changes.[56] Similarly, depriving gamma motoneurons of their background stimulation by decapitating the experimental animals, or spinalizing them, might also produce absence of the articular reflexes.[46,56] Another observation suggesting gamma motoneuron involvement was that after elimination of the joint receptor discharge (by electrocoagulation, for instance), the myotatic reflex had reduced excitability.

Freeman and Wyke's proposed mechanism for the low-threshold articular reflexes is shown in Fig. 7.1. Their hypothesis that the articular mechanoreceptors act polysynaptically on the gamma motoneurons has been confirmed by other researchers.[18,19,24,26,29,30,31,51,53,62] Researchers also took Freeman and Wyke's hypothesis a step further, adding that these reflexes may contribute to the "coordination of muscle tone and movement."[18,26,27,32,53] Thus, articular afferents appear to play a significant role in the regulation of posture and movement.[12]

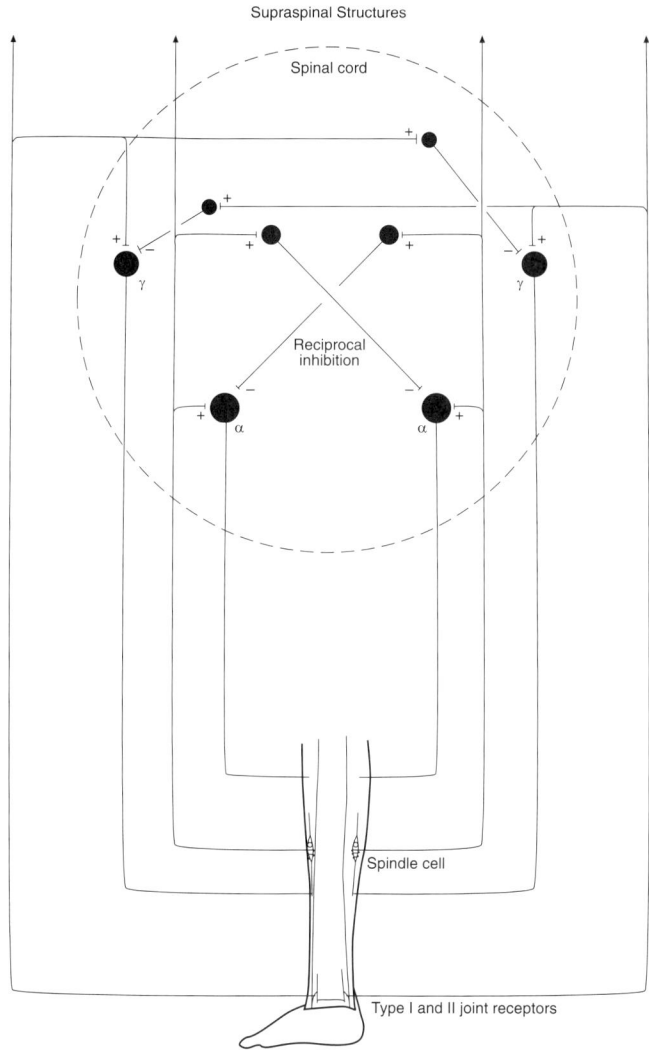

Fig. 7.1 Freeman and Wyke's proposed mechanism for the influence of the low-threshold articular afferents on the muscles.

Reciprocal Coordination of Articular Reflexes

The effect of spindle cell afferent activity is a facilitation of the agonists with a simultaneous reciprocal inhibition of the antagonists.[9] For example, the Achilles tendon reflex causes a facilitation of the gastrocnemius and soleus motor units with reciprocal inhibition of the anterior tibialis muscle. Similarly, the low-threshold Type I and II articular receptors cause reciprocally coordinated reflex changes in the motor units.[16,26,27,37,46,52,53,64,65,66]

The Type I and II articular receptors inhibit the muscles that stretch the part of the capsule in which these receptors are located and simultaneously facilitate the antagonists. An example is elbow flexion, in which the dorsal elbow capsule is stretched. Activation of the Type I and II articular receptors located in the stretched dorsal elbow capsule inhibits the biceps' motor units and facilitate the triceps' motor units. The changes in muscle tone of both the agonist (biceps) and antagonist (triceps) tend to prevent further active or passive movement into the range. Because the articular reflexes simultaneously inhibit agonists and facilitate antagonists, adequate functional stability can be created in the joint.[2,26,27,30] In other words, the reciprocal coordination of articular reflexes may serve to regulate joint stability.[26,27,53]

The reciprocal coordination of articular reflexes prevents excessive stretch of a joint's ligamento-capsular tissue and is an example of negative feedback. The negative feedback system tends to stop movements at the end of their range.[36] For example, as the wrist joint passively or actively moves into palmar flexion, the flexors are inhibited and the dorsiflexors are facilitated. This tends to stop the palmar flexion and protect the joint from moving too far. The negative feedback ensures that the muscle tone changes become more prominent, resisting the movement increasingly as the joint moves farther into the range. In contrast, a positive feedback system to the muscles would be potentially dangerous. For instance, as the wrist moves into palmar flexion, the palmar flexors are facilitated and the dorsiflexors are inhibited, which would allow the wrist to move farther into palmar flexion, increasing the tone of the palmar flexors even more. Positive feedback could result in damage to the joint. Negative feedback, however, mediated by the low-threshold articular mechanoreceptors, serves to protect the joint.[4,5]

Joint Stability

Two types of joint stability are recognized: passive and functional. The passive stability of a joint is determined by the joint geometry and the mechanical properties of the soft tissue within and around the joint.[30,32,56] Of the joint's soft tissues, the mechanical properties of its ligaments are considered most important for passive stability.[30,32,56] However, during strenuous activities such as downhill skiing, the passive stability of the joints seems insufficient to prevent injuries.[32] The joint, therefore, must rely on additional mechanisms to prevent damage to its ligaments.

In contrast to passive stability, functional stability of a joint is determined by the passive restraints, as well as the restraints offered by the muscles around the joint.[24,30,32] A contraction of the muscles around a joint significantly increases the joint's resistance to movement.[24,30,45]

The extensive innervation of ligaments makes it unlikely that their only function is passive restraint of movement.[24,30,33] More likely, ligaments, with their receptors, contribute to functional stiffness, which includes muscular contractions. Johansson and colleagues wrote that, when examining patients with knee ligament injuries, it may be useful to include tests aimed at the functional (neuronal) aspect of joint stability.[32] For proper execution of functional joint stability, the central nervous system, the sensory receptors, and the muscular tissue are essential.

Articulo-Muscular Protective Reflex

Functional stability is closely related to the articulo-muscular reflex. In a 1927 publication, Payr defined the articulo-muscular reflex as reflex contractures of muscles to protect damaged joints.[43,56]

In 1924, Partridge published a paper on the limitation of joint range of movement, seeking to answer the question, "Why does energetic muscular activity not normally endanger joints?"[42] He wrote that not only do the ligaments restrict range of movement in the joints, but the increased tone of the stretched muscles restricts range as well. Partridge named the latter a reflex contraction, which is probably what we now know as the myotatic (stretch) reflex of a muscle. Partridge concluded that these two restraints are insufficient to protect a joint from excessive range of movement.

Partridge referred to Hilton's law, which said, in part, that a joint is innervated by the same nerve trunk as the one that innervates the overlying muscle.[21,22] In other words, the flexor muscle of a joint shares the nerve supply with the flexor side of the joint, and the extensor muscle receives its innervation from the same source as the extensor side of the joint. Excessive movements are not stopped by passive restraints only, but by the capsule, on reaching a certain degree of tension, issuing a message to the overlying muscles to contract. For instance, extension of the knee is not mechanically stopped by the posterior capsular ligaments and the anterior cruciate ligament, but is stopped when the tension in them is sufficient to issue a reflex message to the hamstrings to contract. Partridge gave several other examples of this protective mechanism, as well as of diseases associated with the lack of it. He concluded that "muscular activity throughout the body is controlled by afferent stimuli from all the connective tissue structures . . . in any way affected by that activity; the connection between the afferent and efferent stimuli being close, and the reflex a quick one. . . . The constant nerve relationship suggests a simple reflex arc protecting joints by reflex muscular contraction, both from simple strain caused by excessive muscular activity, as well as from more violent injury. . . . Disease of the central nervous system leads to loss of the sensory side of the reflex arc for joint protection and this again to repeated assaults upon the joint by overstretching in various directions of the capsule and ligaments; and the subsequent synovitis, disorganization, and frequent dislocation of the joint is the inevitable result *(pp. 353–354)*."[42]

Some 14 years after Partridge's publication, in 1938, Palmer wrote that muscles act to protect ligaments.[40] In a 1958 publication, he reported that he observed contractions of the medial muscles of the knee (semimembranosus, sartorius, and vastus medialis) in response to stretching of the medial collateral ligament.[41]

In 1959, Stener co-authored three papers on the articulo-muscular protective reflex.[1,44,60] Similar to Palmer, Stener and colleagues observed the muscular response to passive valgus movement of the knee. Andersson and Stener's study did not produce evidence that the receptors which triggered muscular protection of the knee joint were located in the medial collateral ligament; however, the reflexes that affected muscle tone were found to originate from articular tissue adjacent to the medial collateral ligament.[1] A stretch of the medial collateral ligament (by passively abducting the knee) caused a reflex contraction of the vastus medialis, the sartorius, and the semimembranosus in patients with a mild injury to the medial collateral ligament.[44,61] Thus, Stener and Petersen considered the reflex to be a nociceptive reflex (Type IV receptors).[61]

Dee performed experiments on the hip joints of cats in which the hip muscles were sectioned from their attachment to the femur.[10] Passive abduction of the femur produced a progressive reduction in the activity of the abductors of the hip (inhibition of the abductors). When the passive movement was reversed, the previous inhibition of the abductors was abolished when the neutral position was achieved.

Jerosch and colleagues found decreased stability of the glenohumeral joint in healthy volunteers after an intraarticular injection of lidocaine.[23] The control group, which received an injection with saline and a contrast dye, had significantly more glenohumeral stability to passive translation ($p < .05$).[23]

The term *ligamento-muscular protective reflex* was used by Stener and colleagues.[1,44,60] Other authors used the terms *joint protective reflex* or the *joint's protective reflex movements*.[6,40] Modern researchers use the term *ligamento-muscular protective reflex*.[24,32,45,56] However, apparently not only joint ligaments initiate protective reflexes, but other articular tissues as well.[1,16,42,55,56] Kennedy and colleagues wrote that the ligamento-capsular structures may act to initiate joint protective reflexes by muscular splinting in situations of abnormal stress.[33] Because other structures seem capable of initiating the protective reflex, I prefer the term *articulo-muscular protective reflex*.

The result of stretching ligamento-capsular tissues often is observed as a contraction of the antagonistic muscles attempting to prevent the stretch of the ligamento-capsular tissue. The ligamento-capsular structures often are most significantly stretched near the end ranges of joint movement. Schaible and Schmidt suggested that the low-threshold articular receptors (i.e., Types I and II) are mainly occupied with signaling that the joint is about to leave its normal working range.[49] These warning signals increase in urgency as receptors with higher thresholds (possibly Type III) join the low-threshold articular receptors.[49]

The articulo-muscular protective reflex seems to serve the function of protecting joints from excessive movements. Some authors, however, expressed concern that the reflex was not fast enough to protect the joint in many circumstances.[7,45,55,56] Information from the articular receptors must travel centrally and then distally again to cause the reflex contraction of the muscles. The time needed from receptor discharge to contraction of the muscles is adequate in slow injuries but may be inadequate in faster injuries.[7,45,55,56]

The viewpoint that the articulo-muscular protective reflex is too slow for adequate protection of joints may be too narrow, since the articular mechanoreceptors provide continuous information to the gamma motoneuron throughout the range of a joint.[30,32] The articulo-muscular protective reflex is not only active when the joint exceeds its normal working range, but perhaps at all intermediate angles as well. The modulation of muscle tone is possibly accomplished over the entire range of joint movement.[32] Because simple protective reflexes would be too slow to protect the joint during fast movements, the main function of the articular afferents in control of functional joint stability may be the continuous preparatory adjustment of intrinsic muscle tone.[22,29,32]

Gamma Motoneurons

Activation of articular mechanoreceptors may contribute, via reflex effects on the gamma motoneurons and the spindle cells, to the continuous modulation of muscle stiffness and increase the joint's functional stability (see Fig. 7.2).[13,24,25,26,27,28,30,32,51,53,54] Johansson and colleagues wrote that adequate functional stability can be created in the joint because the articular reflexes are reciprocally coordinated.[2,26,27,30] Only very modest stretches of the knee ligaments are needed to elicit gamma motoneuron effects, suggesting that the reflexes are not of a nociceptive character.[25,28,32] The articular receptors affecting the spindle afferent responses had a low threshold to mechanical deformation and were slowly adapting (Type I recep-

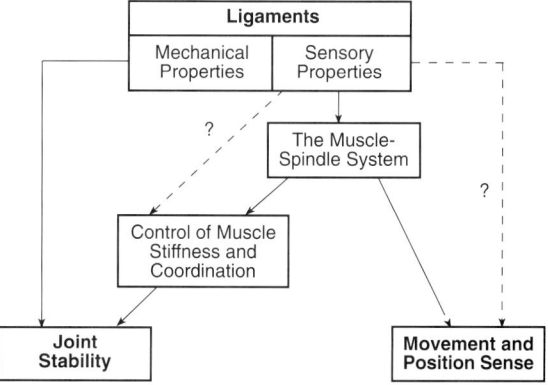

Fig. 7.2 Basic organizational plan of how reflexes from ligamentous joint afferents may contribute to the regulation of muscle stiffness around the joint, joint stability, and movement and position sense.

Adapted from Johansson H, Sjölander P, Sojka P: A sensory role for the cruciate ligaments. Clin Orthop Rel Res 268: 161-178, 1991, with permission from Lippincott-Raven.

tors).[24,25,26,28,51] Ferrell and colleagues found rapidly adapting (Type II) articular receptors also responsible for increased muscle tone.[13]

Although articular mechanoreceptor discharge affects the activity of gamma motoneurons first, and after that may influence alpha motoneurons, Ferrell and colleagues found strong reflex effects on alpha motoneurons.[13] They found quadriceps motor unit activity in response to firing of knee joint receptors. Stimulation of as few as one, but primarily two, articular receptors was capable of causing motor unit activity in the quadriceps. The joint receptors responsible for the quadriceps motor unit activation were both slowly and rapidly adapting (possibly Types I, II, and III) mechanoreceptors. Type IV articular receptors were not stimulated. Ferrell and colleagues wrote that these powerful reflexes might serve as the articulo-muscular protective reflex.

Anterior Cruciate Ligament

Schultz and colleagues suggested that the cruciate ligaments play a role in the kinesthetic reflex arcs that serve to protect the knee from deformation beyond its anatomical limits.[50] They wrote that, besides ligaments forming a passive restraint, normal ligaments may initiate a more active mechanism of protective joint restraint, a process that would cease with the interruption of afferent impulses from the ligaments.[50]

An example of the articulo-muscular protective reflex is the effect of the anterior cruciate ligament (ACL) on the knee muscles. Anterior displacement of the tibia on the femur stretches the ACL. Stretch of the ACL, resulting in simultaneous firing of its receptors, causes a reflex contraction of the biceps femoris and semimembranosus.[3,20,25,35,39,55,56,63] The contraction of the hamstrings assists the ACL in preventing the anterior displacement of the tibia by pulling the tibia posteriorly.[39,47,56] The hamstrings contraction produces this posterior translation force at all angles of the knee.[3,47] Thus, the hamstrings are synergists of the ACL.[11,56] The hamstrings could also be called the agonists of the ACL, with the quadriceps functioning as an antagonist.[63] A contraction of the quadriceps can cause an anterior displacement of the tibia.[47,56] Consequently, an inhibition of the quadriceps could also reduce stress on the ACL.

Draganich, Baratta, and colleagues found a coactivation of the hamstrings during active contraction of the quadriceps producing knee extension in healthy volunteers.[3,11] The amount of contraction varied with the hamstrings moment arm such that the hamstrings contracted less when the moment arm was longer and contracted more to compensate for a shorter moment arm.[3] Thus, the hamstrings produced a constant posterior translation throughout the knee range to reduce the strain on the knee ligaments.[3] The results of these studies by Draganich, Baratta, and colleagues support the hypothesis that the hamstrings function synergistically with the ACL to prevent anterior tibial displacement produced by an active quadriceps contraction.[3,11]

Wojtys and Huston found an alteration of the muscle recruitment time in response to an anterior tibia translation in subjects with ACL-deficient knees.[63] The recruitment sequence as a spinal reflex in normal individuals was first hamstrings, then quadriceps. In individuals with an acute ACL-deficient knee, recruitment of the hamstrings came after recruitment of the quadriceps. The lack of afference from the ACL may have caused a delay in the hamstrings response. Besides the change in recruitment sequence, the reaction time of both muscle groups in response to the ACL stretch was delayed in the group with ACL deficiencies. In patients with more chronic ACL-deficient knees, the normal sequence of muscle recruitment returned. It may be that secondary restraints to anterior tibia translation play a more important role in patients with a chronic ACL-deficient knee. Recruitment of both muscle groups, however, remained delayed. Similarly, in reviewing the literature, Johansson and colleagues wrote that there are clinical indications of a changed muscular activation pattern in patients with ACL-deficient knee joints.[32]

Comparable results were obtained by Solomonow and colleagues, who studied the ACL reflexes in cats and human subjects with and without damage to the ACL.[55] In the healthy human subjects, stretching the ACL caused facilitation of the hamstrings and inhibition of the quadriceps. In the ACL-deficient group, they observed a similar reflex, but the reflex took longer to elicit muscle tone changes. Joint receptors in the capsule, and possibly muscle receptors, were thought to be responsible for the reflex contraction.[55]

Other researchers confirmed that in patients with ACL-deficient knees, during maximal-effort extension causing an anterior subluxation of the tibia, the protective reflex still causes the hamstrings to contract.[55,56] However, the reflex activation of the hamstrings could not have come from receptors located in the torn ACL, but was elicited by mechanoreceptors located in other structures of the joint that were stretched during the anterior subluxation of the tibia.[56]

Osternig et al. studied the tone of the hamstrings during maximum quadriceps contraction.[39] The hamstrings exhibited considerable activity during maximum quadriceps contraction; however, the hamstrings in the ACL-injured extremity generated significantly less antagonistic activity than their normal contralateral extremities ($p < .05$). The results of the study suggest that an ACL dysfunction may result in reduced hamstrings activity during quadriceps contraction.

Grabiner and colleagues examined the tone in the hamstrings when stretching the ACL.[17] The anterior translation force delivered to the ACL was produced by a maximum isometric contraction of the quadriceps. Their study failed to show increased tone in the hamstrings, which contradicts the results obtained by other researchers.[11,20,55,56,63] Yack critiqued the study, believing that there may have been a problem with its methodology. He wrote that the basic assumption—that the ACL was stressed due to anterior tibial translation during isometric contraction of the quadriceps—may be incorrect.[67] Johansson and colleagues provided another reason the experiment was unable to demonstrate increased tone in the hamstrings, writing that the articulo-muscular protective reflex does not always cause contraction of muscles.[24,30,32] Contraction of muscles is caused by alpha motoneuron activity. The articulo-muscular protective reflex causes strong effects on gamma motoneurons, rather than causing reflex effects "directly" on the alpha motoneurons.[24]

Khalsa and Grigg studied the responses of slowly adapting mechanoreceptors in the posterior capsule of cat knee joints before and after the ACL was cut.[34] Despite the increased laxity of the knee joint (allowing more movement) after the ACL was transected, their study failed to show increased activity of the mechanoreceptors in the posterior capsule. If a different recruitment pattern of the knee muscles is present in patients with ACL-deficient knees, the mechanoreceptors in the posterior knee capsule may not be responsible.

Other studies that found no reflex activity from the ACL when using surgical techniques may have failed to do so because no effort was made to preserve the fine articular nerve twigs from the ACL.[56] Deep anesthesia also may suppress the articulo-muscular protective reflex.[55]

Considering the above information, rehabilitation of the hamstrings appears important for patients with injury or surgical repair of the ACL. Strong hamstrings contribute to the stability of the knee, preventing anterior subluxation of the tibia.[55]

Strengthening exercises for the hamstrings in patients with ACL-deficient knees reduces the possibility of anterior subluxation of the knee during, for instance, quadriceps contraction.[3,11,47,55] The hamstrings strengthening exercises can begin immediately after the ACL injury or repair, without placing undue stress on the damaged ligament.[3,11,47,55]

Ankle Sprains

Ligaments in other joints of the body likely serve the same purpose as the ACL in protecting the joint by increasing its functional stability.[24] Damage to ligaments may result in functional instability. For instance, in many cases, lateral ankle sprains result in functional instability of the foot with peroneal muscle weakness.[8,14,38,48,59] This functional instability produces complaints such as a "giving way" of the foot.[14,38] The cause of the functional instability is rarely an instability of the talus in the mortis.[14] Freeman and colleagues believed the usual cause of functional instability of the foot to be motor incoordination.[14] The motor incoordination is secondary to a loss of articular receptors in the ankle joint.[14] The lateral ankle sprain, or any traction injury to a ligament or joint capsule, may lead to rupture of the nerve and collagen fibers.[14,48] The likelihood of neural regeneration is poor; therefore, ligamentous and capsular trauma may lead to a partial joint deafferentiation that may be permanent.[14] The result of neural tissue damage is altered reflexes from the joint. For instance, after an ankle sprain, a kinesthetic deficit may be present, resulting in decreased balance when standing on the affected leg.[14] Functional instability of the ankle after a lateral ankle sprain is not always the result of nerve tissue damage. It also may be caused by other factors altering the function of the afferent or efferent nerves to the lower leg, such as swelling of the ankle joint or Type IV articular receptor activity.[38]

Elbow

In normal activity, the role of an agonist is to generate necessary and sufficient force in desired functions.[58] The amount of external force output by an agonist depends on the amount this muscle contracts and on the amount its antagonist relaxes.[56] Solomonow and colleagues studied the EMG activity of the elbow muscles during maximum voluntary flexor contraction and maximum voluntary extensor contraction throughout the joint range.[57,58] Surprisingly, they found that during slow concentric maximum contractions of the biceps, the triceps showed EMG activity. Thus, the triceps contracted and thereby reduced the flexor muscles' external output. Similarly, during slow concentric maximum voluntary contractions of the triceps, the biceps showed EMG activity. These researchers compared the activity of the antagonists with the joint angle and showed that the EMG activity of the antagonists was inversely related to the muscle's moment arm. They concluded that the agonist generates torques and forces that tend to destabilize the joint and separate its constituent bones. These torques and forces depend on the joint angle and its associated lever arm. Thus, the antagonist's role is to generate necessary and sufficient force to stabilize the joint.[57,58]

Summary

Stretching articular tissue stimulates Type I and II receptors in the stretched regions of the joint. Stimulation of Type I and II receptors evokes articular reflexes. These reflexes express themselves as increased tone in the antagonist to the movement with simultaneous decreased tone in the agonist. Therefore, the reciprocally coordinated muscle tone changes improve joint stability (i.e., the muscle tone changes attempt to return the joint toward a loose-packed position). Via the gamma-spindle loop, the articular reflexes continuously regulate muscle tone and functional joint stability. This is termed the *articulo-muscular protective reflex*.

The articulo-muscular protective reflex consists of two parts. The brief initial activity in the muscles may be predominantly in response to the phasic Type II receptor activity; the sustained muscle tone changes are the result of the more tonic activity of the Type I receptor.

References

1. Andersson S, Stener B: Experimental evaluation of the hypothesis of ligamento-muscular protective reflexes: II A study in cat using the medial collateral ligament of the knee joint. Acta Physiol Scand 48(suppl 166):27-48, 1959

2. Appelberg B, Hulliger M, Johansson H, et al: Reflex activation of dynamic fusimotor neurons by natural stimulation of muscle and joint receptor afferent units. In Taylor A, Prochazka A (eds): Muscle Receptors and Movement. New York, Oxford University Press, 1981, pp 149-161

3. Baratta R, Solomonow M, Zhou BH, et al: Muscular coactivation: The role of the antagonist musculature in maintaining knee stability. Am J Sports Med 16:113-122, 1988

4. Baxendale RH, Ferrell WR: Modulation of transmission in reflex pathways by knee joint afferent discharge in the decerebrate cat. Brain Res 202:497-500, 1980

5. Baxendale RH, Ferrell WR: The effect of elbow joint afferent discharge on the transmission in forelimb flexion reflex pathways to biceps and triceps brachii in decerebrate cats. Brain Res 247:57-63, 1982

6. Blockey NJ: An observation concerning the flexor muscles during recovery of function after dislocation of the elbow. J Bone Joint Surg Am 36:833-840, 1954

7. Biedert RM, Stauffer E, Friederich NF: Occurrence of free nerve endings in the soft tissue of the knee joint. Am J Sports Med 20:430-433, 1992

8. Bosien WR, Staples OS, Russell SW: Residual disability following acute ankle sprains. J Bone Joint Surg Am 37:1237-1243, 1955

9. Daube JR, Reagan TJ, Sandok BA, et al: Medical Neurosciences: An Approach to Anatomy, Pathology, and Physiology by Systems and Levels, ed 2. Boston, Little, Brown and Co, 1986

10. Dee R: Mechanoreceptors in hip joint capsule and ligamentum capitis femoris and their reflex contribution to posture. In Symposium on Osteoarthritis. St. Louis, C.V. Mosby Co, 1976, pp 52-65

11. Draganich LF, Jaeger RJ, Kralj AR: Coactivation of the hamstrings and quadriceps during extension of the knee. J Bone Joint Surg Am 71:1075-1081, 1989

12. Ferrell WR, Baxendale RH, Carnachan C, et al: The influence of joint afferent discharge on locomotion, proprioception and activity in conscious cats. Brain Res 347:41-48, 1985

13. Ferrell WR, Rosenberg JR, Baxendale RH, et al: Fourier analysis of the relation between the discharge of quadriceps motor units and periodic mechanical stimulation of cat knee joint receptors. Exp Physiol 75:739-750, 1990

14. Freeman MAR, Dean MRE, Hanham IWF: The etiology and prevention of functional instability of the foot. J Bone Joint Surg Br 47:678-685, 1965

15. Freeman MAR, Wyke B: Articular contributions to limb muscle-reflexes: An electromyographic study of the influence of ankle-joint mechanoreceptors upon reflex activity in the gastrocnemius muscle of the cat. J Physiol 171:20P-21P, 1964

16. Freeman MAR, Wyke B: Articular reflexes at the ankle joint: An electromyographic study of normal and abnormal influences of ankle-joint mechanoreceptors upon reflex activity in the leg muscles. Br J Surg 54:990-1001, 1967

17. Grabiner MD, Koh TJ, Miller GF: Further evidence against a direct automatic neuromotor link between the ACL and hamstrings. Med Sci Sports Exerc 24:1075-1079, 1992

18. Greenfield BE, Wyke B: Reflex innervation of the temporo-mandibular joint. Nature 211:940-941, 1966

19. Grillner S, Hongo T, Lund S: Descending monosynaptic and reflex control of γ-motorneurones. Acta Physiol Scand 75:592-613, 1969

20. Grüber J, Wolter D, Lierse W: Der vordere Kreuzbandreflex (LCA-Reflex). Unfallchirug 89:551-554, 1986

21. Hilton J: On the Influence of Mechanical and Physiological Rest in the Treatment of Accidents and Surgical Diseases, and the Diagnostic Value of Pain: A Course of Lectures, Delivered at the Royal College of Surgeons of England in the Years 1860, 1861, and 1862. London, Bell and Daldy, 1863, pp 166-167

22. Hilton J: Rest and Pain: A Course of Lectures on the Influence of Mechanical and Physiological Rest in the Treatment of Accidents and Surgical Diseases, and the Diagnostic Value of Pain. London, Bell and Sons, 1896

23. Jerosch J, Castro WHM, Halm H, et al: Does the glenohumeral joint capsule have proprioceptive capability? Knee Surg Sports Traumatol Arthrosc 1:80-84, 1993

24. Johansson H: Role of knee ligaments in proprioception and regulation of muscle stiffness. Journal of Electromyography and Kinesiology 1:158-179, 1991

25. Johansson H, Lorentzon R, Sjölander P, et al: The anterior cruciate ligament. Neuro-Orthopedics 9:1-23, 1990

26. Johansson H, Sjölander P, Sojka P: Actions on γ-motoneurones elicited by electrical stimulation of joint afferent fibers in the hind limb of the cat. J Physiol 375:137-152, 1986

27. Johansson H, Sjölander P, Sojka P: Fusimotor reflexes in triceps surae muscle elicited by natural and electrical stimulation of joint afferents. Neuro-Orthopedics 6:67-80, 1988

28. Johansson H, Sjölander P, Sojka P, et al: Reflex actions on the γ-muscle-spindle systems of muscles acting at the knee joint elicited by stretch of the posterior cruciate ligament. Neuro-Orthopedics 8:9-21, 1989

29. Johansson H, Sjölander P, Sojka P: Activity in receptor afferents from the anterior cruciate ligament evokes reflex effects on fusimotor neurones. Neurosci Res 8:54-59, 1990

30. Johansson H, Sjölander P, Sojka P: A sensory role for the cruciate ligaments. Clin Orthop 268:161-178, 1991

31. Johansson H, Sjölander P, Sojka P: Fusimotor reflex profiles of individual triceps surae primary muscle spindle afferents assessed with multi-afferent recording technique. J Physiol Paris 85:6-19, 1991

32. Johansson H, Sjölander P, Sojka P: Receptors in the knee joint ligaments and their role in the biomechanics of the joint. Crit Rev Biomed Eng 18:341-368, 1991

33. Kennedy JC, Alexander IJ, Hayes KC: Nerve supply of the human knee and its functional importance. Am J Sports Med 10:329-335, 1982

34. Khalsa PS, Grigg P: Responses of mechanoreceptor neurons in the cat knee joint capsule before and after anterior cruciate ligament transection. J Orthop Res 14:114-122, 1996

35. Krauspe R, Schmidt M, Schaible HG: Sensory innervation of the anterior cruciate ligament. J Bone Joint Surg Am 74:390-397, 1992

36. Lundberg A, Malmgren K, Schomburg ED: Role of joint afferents in motor control exemplified by effects on reflex pathways from Ib afferents. J Physiol 284:327-343, 1978

37. Molina F, Ramcharan JE, Wyke BD: Structure and function of articular receptor systems in the cervical spine. J Bone Joint Surg Br 5:254-255, 1976

38. Oostendorp RAB: Functionele instabiliteit na inversietrauma van enkel en voet: Een effectonderzoek pleisterbandage versus pleisterbandage gecombineerd met fysiotherapie. Geneeskunde en Sport 20:45-55, 1987

39. Osternig LR, Caster BL, James CR: Contralateral hamstring (biceps femoris) coactivation patterns and anterior cruciate ligament dysfunction. Med Sci Sports Exerc 27:805-808, 1995

40. Palmer I: On the Injuries to the ligaments of the knee joint: A Clinical Study. Stockholm, Acta Chirurgica Scandinavica 81(suppl 53):60-61, 1938

41. Palmer I: Pathophysiology of the medial ligament of the knee. Acta Chirurgica Scandinavica 115:312-312, 1958

42. Partridge EJ: Joints: The limitation of their range of movement, and an explanation of certain surgical conditions. J Anat 58:346-354, 1924

43. Payr E: Der Heutige Stand der Gelenkchirurgie. Arch f Klin Chirurgie 148:404-451, 1927

44. Petersén I, Stener B: Experimental evaluation of the hypothesis of ligamento-muscular protective reflexes: III A study in man using the medial collateral ligament of the knee joint. Acta Physiol Scand 48(suppl 166):51-61, 1959

45. Pope MH, Johnson RJ, Brown DW, et al: The role of the musculature in injuries to the medial collateral ligament. J Bone Joint Surg Am 61:398-402, 1979

46. Ramcharan JE, Wyke B: Articular reflexes at the knee joint: An electromyographic study. Am J Physiol 223:1276-1280, 1972

47. Renström P, Arms SW, Stanwyck TS, et al: Strain within the anterior cruciate ligament during hamstring and quadriceps activity. Am J Sports Med 14:83-87, 1986

48. Rottigni SA, Hopper D: Peroneal muscle weakness in female basketballers following chronic ankle sprain. The Australian Journal of Physiotherapy 37:211-217, 1991

49. Schaible HG, Schmidt RF: Mechanosensibility of joint receptors with fine afferent fibers. Exp Brain Res 9(suppl):284-297, 1984

50. Schultz RA, Miller DC, Kerr CS, et al: Mechanoreceptors in human cruciate ligaments. J Bone Joint Surg Am 66:1072-1076, 1984

51. Sjölander P, Johansson H, Sojka P, et al: Sensory nerve endings in the cat cruciate ligaments: A morphological investigation. Neurosci Lett 102:33-38, 1989

52. Skoglund S: Anatomical and physiological studies of knee joint innervation in the cat. Acta Physiol Scand 36(suppl. 124):1-101, 1956

53. Sojka P, Johansson H, Sjölander P, et al: Fusimotor neurones can be reflexly influenced by activity in receptor afferents from the posterior cruciate ligament. Brain Res 483:177-183, 1989

54. Sojka P, Sjölander P, Johansson H, et al: Influence from stretch-sensitive receptors in the collateral ligaments of the knee joint on the γ-muscle-spindle systems of flexor and extensor muscles. Neurosci Res 11:55-62, 1991

55. Solomonow M, Baratta R, Zhou BH, et al: The synergistic action of the anterior cruciate ligament and thigh muscles in maintaining joint stability. Am J Sports Med 15:207-213, 1987

56. Solomonow M, D'Ambrosia R: Neural reflex arcs and muscle control of the knee stability and motion. In Scott WN (ed): Ligament and Extensor Mechanisms of the Knee. St. Louis, C.V. Mosby Co, 1991, pp 389-400

57. Solomonow M, Guzzi A, Baratta R: EMG-force model of the elbow's antagonistic muscle pair. Am J Phys Med 65:223-244, 1986

58. Solomonow M, Guzzi A, Zhou BH, et al: The EMG of the elbow antagonist is inversely related to its moment arm. Proc IEEE Engr Med Biol Conf 8:620-622, 1986

59. Staples OS: Result study of ruptures of lateral ligaments of the ankle. Clin Orthop 85:50-58, 1972

60. Stener B: Experimental evaluation of the hypothesis of ligamento-muscular protective reflexes: IA method for adequate stimulation of tension receptors in the medial collateral ligament of the knee joint of the cat, and studies of the innervation of the ligament. Acta Physiol Scand 48(suppl 166):5-26, 1959

61. Stener B, Petersén I: Electromyographic investigation of reflex effects upon stretching the partially ruptured medial collateral ligament of the knee joint. Acta Chirurgica Scandinavica 124:396-415, 1962

62. Voorhoeve PE, Kanten RW: Reflex behaviour of fusimotor neurones of the cat upon electrical stimulation of various afferent fibers. Acta Physiol Pharmacol Neerlandica 10:391-407, 1962

63. Wojtys EM, Huston LJ: Neuromuscular performance in normal and anterior cruciate ligament-deficient lower extremities. Am J Sports Med 22:89-104, 1994

64. Wyke B: The neurology of joints. Ann R Coll Surg Engl 41:45-50, 1967

65. Wyke B: Articular neurology - A review. Physiotherapy 58:94-99 1972

66. Wyke BD: Articular neurology and manipulative therapy. In Glasgow EF, Twomey LT, Scull ER, et al (eds): Aspects of Manipulative Therapy, ed 2. New York, Churchill Livingstone, 1985, pp 72-77

67. Yack HJ: Further evidence against a direct automatic neuromotor link between the ACL and hamstrings (letter, comment). Med Sci Sports Exerc 25:407-408, 1993

8. Withdrawal Reflex

The articular receptors are not merely responsible for the articulo-muscular protective reflex. They also play a role in the withdrawal reflex. The withdrawal reflex may be better known as the flexion reflex but is also known as the nociceptive reflex and as the cutaneous reflex.[15,22,27] The withdrawal reflex consists of a number of polysynaptic reflexes that could result in movement of a body part. The receptors involved in the withdrawal reflex may be nociceptors but may also be low-threshold mechanoreceptors in the skin, muscle, or joint. Withdrawal reflexes have been described in all four extremities and in response to tactile stimulation of the head.[4,5,7,8,60] Gordon described the withdrawal reflex as creating flexion in the extremity, but under certain conditions, the opposite could occur: extension of the ipsilateral extremity.[22] This will be presented in more detail in this chapter.

As part of the withdrawal reflex, the extremity often moves away from the stimulus. An example of the withdrawal reflex is the motor response to stepping on a sharp object (see Fig. 8.1). The stimulation of the cutaneous receptors is conducted centrally to the posterior horn, where it alters the excitability of the interneurons activating the motoneurons. The flexor motoneurons of the extremity are facilitated while simultaneously the extensor motoneurons are inhibited.

The intensity of the stimulus modulates the muscle response in the withdrawal reflex. For instance, touching a warm stove might result in a moderately quick withdrawal; only the wrist and the elbow bend to move away from the stimulus. Touching a hot stove, however, might result in a quicker withdrawal with flexion of the entire extremity.

Willer studied the response to nociceptive electrical stimulation of the leg in normal volunteers, concluding that the threshold for the withdrawal reflex was identical to the threshold for pain.[57] In other words, the first observation of the withdrawal reflex occurred when the volunteers first perceived pain. Willer wrote that mechanoreceptor afferents can raise both the threshold for pain and the threshold for the withdrawal reflex.[56]

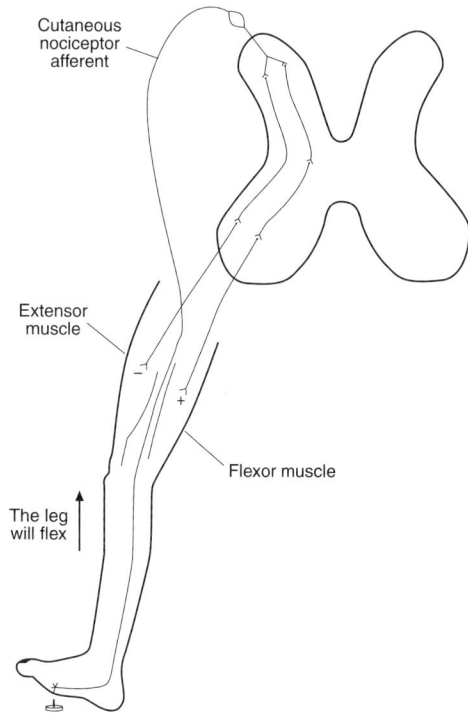

Fig. 8.1 Withdrawal reflex.

Reprinted from Morree JJ de: Dynamiek van het Menselijke Bindweefsel, ed 2. Houten, Netherlands, Bohn Stafleu Van Lochem, 1993, with permission.

Solomonow and D'Ambrosia considered the articulo-muscular protective reflex a special case of the withdrawal reflex. Whereas the withdrawal reflex serves the body by moving away from noxious stimuli applied to a wide range of tissues, the articulo-muscular protective reflex responds to stimulation of the articular tissues only.[45]

Interneurons

Interneurons have long been considered simple relay stations, located in spinal reflex arcs between afferent fibers and alpha motoneurons; this idea is too simplistic, however. Most, if not all, spinal interneurons receive a wide and diverse convergence of input from several different peripheral sources.[16,48] For instance, an interneuron might receive input from several synergistic muscles, the skin, and the joints. Fig. 8.2 shows a diagram that explains the simultaneous control of motoneurons by afferents from muscles, skin, and joints.[35] In addition, most interneurons receive extensive input from supraspinal sources (e.g., from the vestibular nuclei, the red nuclei, and the sensorimotor cortex).[16] Hence, the interneuron receives extensive convergence of afferent and descending input. This convergence has an important consequence. Depolarization of an interneuron depends not only on the amount of facilitation from afferents of a single source, but on the summation of all facilitory and inhibitory influences converging on this interneuron. For instance, the joint afferents can depolarize the interneuron, provided that the interneuron receives sufficient background facilitation from other afferents and supraspinal structures.

The joint afferents have access to the alpha motoneurons via these interneurons.[16] The afferent information from the joint influences the excitability of the interneurons.[16,36] By influencing the excitability of the interneurons, the articular afferents facilitate or inhibit transmission of the reflexes activated by other afferents (e.g., the myotatic reflex).[16,21,36,39]

Interneurons in the Withdrawal Reflex

The withdrawal reflex is not simply a stereotyped movement pattern but is modified by different afferent input.[9] For instance, the reflex is modified by the position of the extremity.[9] The withdrawal reflex is polysynaptic, with the interneurons receiving afferent information from different peripheral sources, not just nociceptive afferents, and from descending pathways.[1,12,22,25,28,34,35,36] The afferents that can evoke a withdrawal reflex include those from the skin, the joints, the Golgi tendon organ, and the secondary afferents of the spindle cell.[15,22,27,34,35,48,51] Thus, the afferents initiating the withdrawal reflex are not only nociceptors, but are also low-threshold mechanoreceptors.

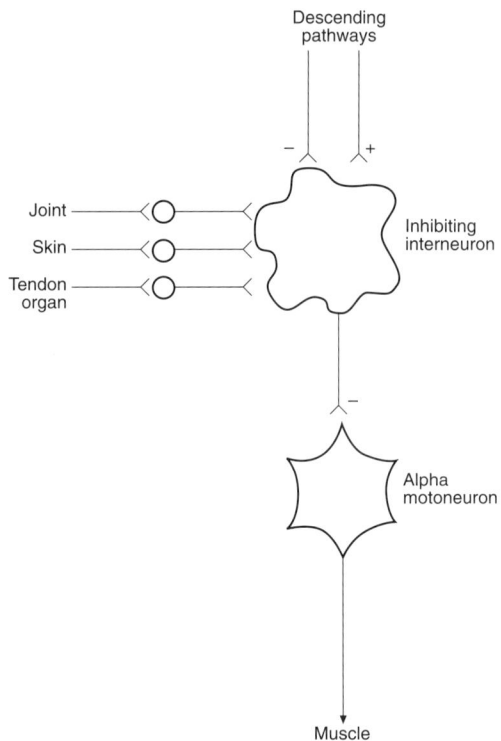

Fig. 8.2 Convergence of descending pathways and afferents from joint, skin, and muscle receptors onto an interneuron in the spinal cord.

The articular receptors responsible for influencing the inhibitory and facilitory interneurons in the withdrawal reflex are possibly the low-threshold Ruffini receptors (Type I articular receptors) and the Type IV nociceptors.[1,5,18,26,34,35] The articular receptors may project via interneurons to the motoneurons.[8] The influence from the joint afferents to the interneurons is possibly di- or trisynaptical.[18,25]

Cutaneous and tendon organ afferents most likely create a suitable background functioning of the interneurons against which articular afferents provide muscle tone changes.[6] This could explain why not all studies have shown muscle tone changes in response to articular afference. Obviously, studies in which the skin or muscles were removed may have altered the articular reflexes, failing to demonstrate any muscle tone changes.[6]

Reflex Reversal

The withdrawal reflex does not always lead to flexion of the involved extremity; it may also lead to extension of the involved extremity.[22] For instance, hitting the "funny" bone while the elbow is bent may result in a quick extension of the arm. Because the withdrawal reflex may lead to flexion as well as extension, it is helpful to describe the withdrawal reflex as having a flexion or an extension pattern. The pathways responsible for the flexion or the extension pattern are mutually inhibitory.[22] This means that activity of the withdrawal reflex with a flexion pattern inhibits the extension pattern and vice versa.

Baxendale and colleagues examined the effects of articular afferents on the withdrawal reflex.[2,3,4,5,7,8] Articular afferents project bilaterally within the cervical and lumbar regions of the spinal cord to modulate the excitability of the withdrawal reflex of the front or hind limb of cats.[5,7] The position of the extremity determines the reflex pattern of a stimulus.[3] If the extremity is extended, a flexion pattern is more likely, whereas if the extremity is flexed, a flexion pattern becomes unlikely. This may be partly due to stretch of the muscles, however, a strong influence from the articular receptors has been proven.[3,43]

In Baxendale and colleagues' experiments, a stimulus of sufficient intensity, applied to the upper or lower extremity of cats, created a withdrawal reflex.[2,3,4,5,7,8] As a result of the withdrawal reflex, the elbow or knee joint flexed when the joint was in an extended position. EMG recordings showed the flexion pattern as increased tone in the biceps brachii when stimulating the upper extremity, or as increased tone in the biceps femoris when stimulating the lower extremity. If the experiment was repeated with the joint in progressively more flexed positions, the flexion pattern weakened, and it became harder and harder to elicit a biceps response. Thus, the position of the knee or elbow joint modulated the flexion pattern of the withdrawal reflex. This modulation remained unaltered when a local anesthetic was injected into the skin or if the skin was removed. Tenotomizing the muscles and holding them at a constant length did not alter the modulation, either. However, an intraarticular injection of lignocaine (a local anesthetic) abolished the modulation of the withdrawal reflex. Anesthetizing the joint receptors by intraarticular injection removed the effect of joint position on the reflex. Baxendale and Ferrell concluded that the modulation of the withdrawal reflex by joint position, which is normally present, is mediated by the articular receptors.[8]

Grillner's research showed a negative feedback system similar to the one found by Baxendale and colleagues.[23,24] In his experiments, stimulation of the common peroneal nerve in spinal cats caused hip flexion if the hip had been positioned in extension. If the hip had been positioned in flexion, the same stimulus caused hip extension. Grillner thought that a gating process took the place of certain stimuli (e.g., peroneal nerve stimulation) to produce flexion when the limb was extended and extension when the limb was flexed. He termed this process reflex reversal.

Baxendale and Ferrell wrote that the articular receptors responsible for modulation of the withdrawal reflex are possibly the low-threshold midrange receptors (Type I and II articular receptors).[5] These findings seem to contradict Davidoff, who stated that the high-threshold myelinated articular afferents, in particular, influence the withdrawal reflex (the Type III and/or some Type IV afferents). Other researchers also found high-threshold joint afferents to provoke the withdrawal reflex.[18,20,51] It may be that the high-threshold articular receptors can evoke the withdrawal reflex, whereas the low-threshold articular receptors modulate this reflex.[20]

The presence of an inflammation in a joint increases the intensity of the withdrawal reflex.[20] An injection of lignocaine into the inflamed joint, deactivating the intraarticular receptors, markedly reduced the withdrawal reflex. However, whereas in a normal joint the withdrawal reflex is modulated by joint position, in an inflamed joint this modulation is absent.[20] For example, the flexion pattern of a withdrawal reflex is normally maximal when the knee joint is extended and minimal when the knee is flexed. In the inflamed knee joint, the flexion pattern of the withdrawal reflex is equally intense in all positions of the extremity. Ferrell and colleagues found that the withdrawal reflex with a flexion pattern was just as strong when the knee joint was flexed as when it was extended.[20] The absence of modulation by joint position suggests that the Type IV articular receptors can override the input from the low-threshold articular mechanoreceptors.[20]

In general, the articular receptors serve a negative feedback function in regulating limb position.[3,4,5,7,35] The negative feedback system could reflexly prevent excessive movements of a joint.[3,4,5,7] These reflexes that prevent excessive joint movement seem to follow a variation of von Uexküll's law, which stated: The response to a given input in the central nervous system, which can influence either muscle of an antagonistic pair, will appear as a contraction in the muscle that is most stretched.[26,37,38,43,47,49] According to von Uexküll's law, when a joint is flexed, and consequently the extensors are stretched, the response will appear in the extensors as a contraction. A contraction of the flexors will occur when the joint is extended.

The above information makes it clear that the position of a joint could be the single most important factor during articular reflex studies. The position of a joint could determine whether the result of a study will show a reflex contraction in the flexors or in the extensors of the joint. Obviously, the same stimulation with the joint in two opposing positions might lead to two different reflex patterns.

Prolonged Facilitation of the Withdrawal Reflex

It is well known that the withdrawal reflex lasts longer than the stimulus that caused it.[11,14,15,27,33] For instance, after touching a hot stove, the extremity may remain in a flexion pattern for several seconds after the hot stove is released. Depending on the tissue of origin of the withdrawal reflex (for instance, a nociceptive stimulus to the skin versus the muscle), different durations of the withdrawal reflex were observed. Wall and Woolf found that stimulation of a muscle resulted in excitability of the withdrawal reflex that lasted three times longer than did stimulation of the skin.[52] In a similar study, Woolf and Wall found that noxious stimuli applied to the skin, muscle, or joint generated prolonged facilitation of the withdrawal reflex of different durations.[59] Muscle afferents produced longer facilitation of the withdrawal reflex than did skin afferents. Joint afferents facilitated the withdrawal reflex longer than did the skin or muscle afferents. For instance, 5 mm of mustard oil applied to the skin, injected into muscle tissue, or injected into the joint produced a flexor reflex facilitation that lasted 0, 25 to 90, or more than 180 minutes, respectively.[59] The duration difference of withdrawal reflex excitability possibly relates to differences in the sensory consequences of injury to skin versus deep tissue.[26] The consequences of injury to a deep tissue, such as a joint, are possibly more serious than a superficial skin injury.

Since articular afferent input has more pronounced effects than cutaneous input, this may explain the more widespread sensory and motor disturbances that accompany joint injury (also see chapter 13, "Pseudoradiculopathy").[59] This may have therapeutic consequences as well, in that alteration of joint functioning possibly produces more changes than does treatment to more superficial structures such as the skin or muscles. Treatment of joints may have more impact on the body than treatment of skin or muscles.

Summary and Conclusion

Although Type IV receptor activation may produce a withdrawal reflex, Type I receptor activation modifies the reflex pattern. Interneurons mediate the modification of the withdrawal reflex. The low-threshold articular receptors serve as a negative feedback system, preventing excessive movements of a joint.

Research shows that the force developed by a muscle is largely determined by the number and discharge rate of the alpha motoneurons recruited.[16] Alpha motoneuron firing is partly determined by the descending fibers, the primary and secondary spindle cell afferents, and as discussed in this chapter, the facilitating and inhibiting interneurons from, among others, articular receptors.[16]

This concept, whereby facilitating and inhibiting influences determine alpha motoneuron firing, may explain why a maximum voluntary contraction is not an absolute amount of force. Rather, the alpha motoneuron discharge is determined by multiple alternating factors (e.g., position of the joint).[16] Muscle strength is largely determined by various neurophysiological factors, with the nervous system potentially altering its output. This could explain the dynamic changes in maximum voluntary muscle strength encountered in the clinic.[10,13,17,19,29,30,31,32,40,41,42,44,46,50,53,54,55,56,58,61]

References

1. Baldissera F, Hultborn H, Illert M: Integration in spinal neuronal systems. In Geiger et al (eds): Handbook of Physiology: A Critical, Comprehensive Presentation of Physiological Knowledge and Concepts: Section 1: The Nervous System. Bethesda, Maryland, American Physiological Society, 1981, vol II, pp 509-595

2. Baxendale RH, Conway BA, Ferrell WR: Conditioning of crossed reflex pathways by independent natural stimulation of labyrinth, neck and elbow joint afferents. Brain Res 377:41-46, 1986

3. Baxendale RH, Ferrell WR: Modulation of transmission in flexion reflex pathways by knee joint afferent discharge in the decerebrate cat. Brain Res 202:497-500, 1980

4. Baxendale RH, Ferrell WER: Modulation of transmission in forelimb flexion reflex pathways by elbow joint afferent discharge in decerebrate cats. Brain Res 221:393-396, 1981

5. Baxendale RH, Ferrell WR: The effect of knee joint afferent discharge on transmission in flexion reflex pathways in decerebrate cats. J Physiol 315:231-242, 1981

6. Baxendale RH, Ferrell WR: Facilitation of joint afferent mediated reflex effects by stretch-related muscle afferent discharge in decerebrate cats. J Physiol (Lond) 329:60P-61P, 1982

7. Baxendale RH, Ferrell WR: The effect of elbow afferent discharge on transmission in forelimb flexion reflex pathways to biceps and triceps brachii in decerebrate cats. Brain Res 247:57-63, 1982

8. Baxendale RH, Ferrell WR: Ascending and descending effects of joint afferent discharge on forelimb and hindlimb flexion reflex excitability in decerebrate cats. Brain Res 332:394-396, 1985

9. Belanger M, Patla AE: Corrective responses to perturbation applied during walking in humans. Neurosci Lett 49:291-295, 1984

10. Ben-Yishay A: Zuckerman JD, Gallagher M, et al: Pain inhibition of shoulder strength in patients with impingement syndrome. Orthopedics 17:685-688, 1994

11. Bernards JA, Bouman LN: Fysiologie van de Mens, ed 2. Utrecht, Netherlands, Bohn, Scheltema & Holkema, 1977

12. Carstens E, Campell IG: Responses of motor units during hind limb flexion withdrawal reflex evoked by noxious skin heating: Phasic and prolonged suppression by midbrain stimulation and comparison with simultaneous recorded dorsal horn units. Pain 48:215-226, 1992

13. Cibulka M, Rose S, Delitto A, et al: Hamstrings muscle strain treated by mobilizing the sacroiliac joint. Phys Ther 66:1220-1223, 1975

14. Costanzo LS: Physiology. London, Williams & Wilkins, 1995

15. Daube JR, Reagan TJ, Sandok BA, et al: Medical Neurosciences, ed 2. Boston, Little, Brown and Co, 1986

16. Davidoff RA: Skeletal muscle tone and the misunderstood stretch reflex. Neurology 42:951-963, 1992

17. DeAndrade JR, Grant C, Dixon AStJ: Joint distension and reflex inhibition in the knee. J Bone Joint Surg Am 47:313-322, 1965

18. Eccles RM, Lundberg A: Synaptic actions in motoneurones by afferents which may evoke the flexion reflex. Arch Ital Biol 97:199-221, 1959

19. Fahrer H, Rentsch HU, Gerber NJ, et al: Knee effusion and reflex inhibition of the quadriceps: A bar to effective retraining, J Bone Joint Surg Br 70:635-638, 1988

20. Ferrell WR, Wood L, Baxendale RH: The effect of acute joint inflammation on flexion reflex excitability in the decerebrate, low-spinal cat. Quarterly Journal of Experimental Physiology 73:95-102, 1988

21. Freeman MAR, Wyke B: Articular reflexes at the ankle joint: An electromyographic study of normal and abnormal influences of ankle-joint mechanoreceptors upon reflex activity in the leg muscles. Br J Surg 54:990-1001, 1967

22. Gordon J: Spinal mechanisms of motor coordination. In Kandell ER, Schwartz JH, Jessell TM (eds): Principles of Neural Science, ed 3. New York, Elsevier, 1991, pp 581-595

23. Grillner S: Locomotion in the spinal cat. In Stein RB, Pearson KG, Smith RS, et al (eds): Control of Posture and Locomotion. New York, Plenum Press, 1973, pp 515-535

24. Grillner S: Locomotion in vertebrates: Central mechanisms and reflex interaction. Physiol Rev 55:247-304, 1975

25. Harrison PJ, Jankowska E: Sources of input to interneurones mediating group I non-reciprocal inhibition of motoneurones in the cat. J Physiol 361:379-401, 1985

26. He X, Proske U, Schaible HG, et al: Acute inflammation of the knee joint in the cat alters responses of flexor motoneurons to leg movements. J Neurophysiol 59:326-340, 1988

27. Henneman E: Spinal reflexes and the control of movement. In Mountcastle VB (ed): Medical Physiology, ed 13. St. Louis, C.V. Mosby Co, 1974, vol 1, pp 651-667

28. Holmqvist B, Lundberg A: Differential supraspinal control of synaptic actions evoked by volleys in the flexion reflex afferents in alpha motoneurones. Acta Physiol Scand 54(suppl. 186):3-51, 1961

29. Iles J, Stokes M, Young A: Reflex actions of the knee-joint receptors on quadriceps in man. J Physiol 360:48P, 1984

30 Iles J, Stokes M, Young A: Reflex actions of knee afferents during contraction of the human quadriceps. Clin Physiol 10:489-500, 1990

31 Janda V: Muscle weakness and inhibition (pseudoparesis) in back pain syndromes. In Grieve GP (ed): Modern Manual Therapy of the Vertebral Column. New York, Churchill Livingstone, 1986, pp 197-201

32 Jayson MIV, Dixon AStJ: Intra-articular pressure in rheumatoid arthritis of the knee: III. Pressure changes during use. Ann Rheum Dis 29:401-408, 1970

33 Kandel ER, Schwartz JH, Jessell TM (eds): Principles of Neural Science, ed 3. Norwalk, Connecticut, Appleton & Lange, 1991

34 Lundberg A, Malmgren K, Schomberg ED: Convergence from Ib, cutaneous and joint afferents in reflex pathways to motoneurones. Brain Res 87:81-84, 1975

35 Lundberg A, Malmgren K, Schomburg ED: Role of joint afferents in motor control exemplified by effects on reflex pathways from Ib afferents. J Physiol 284:327-343, 1978

36 Lundberg A, Malmgren K, Schomburg ED: Reflex pathways from group II muscle afferents. Exp Brain Res 65:271-281, 1987

37 Magnus R: Zur regelung der Bewegungen durch das zentralnervensystem: II. Mitteilung. Pflügers Arch ges Physiol 134:253-269, 1909

38 Magnus R: Zur regelung der Bewegungen durch das zentralnervensystem: III. Mitteilung. Pflügers Arch ges Physiol 134:545-583, 1910

39 Marsden CD, Merton PA, Morton HB: Servo action and stretch reflex in human muscle and its apparent dependence on peripheral sensation. J Physiol 216:21-23P, 1971

40 Nicholas JA, Melvin M, Saraniti AJ: Neurophysiological inhibition of strength following tactile stimulation of the skin. Am J Sports Med 8:181-186, 1980

41 Nichols PJR: Rehabilitation Medicine: The Management of Physical Disabilities, ed 2. Boston, Butterworths, 1981, p 122

42 Pronsati M: Muscle dysfunction resolved with joint manipulation. Advance for Physical Therapists 3(22):16, 1992

43 Rossignol S, Gauthier L: An analysis of mechanisms of controlling the reversal of crossed spinal reflexes. Brain Res 182:31-45, 1980

44 Schmitt MA, Wijer A de, Schepman RWM, et al: Het psoasfenomeen op basis van een primaire sacroiliacale functiestoornis, een casus. Nederlands Tijdschrift voor Manuele Therapie 7:9-12, 1988

45 Solomonow M, D'Ambrosia R: Neural reflex arcs and muscle control of the knee stability and motion. In Scott WN (ed): Ligament and Extensor Mechanisms of the Knee. St. Louis, C.V. Mosby Co, 1991, pp 389-400

46 Stratford P: Electromyography of the quadriceps femoris muscles in subjects with normal knees and acutely effused knees. Phys Ther 62:279-283, 1981

47 Tatton WG, Bawa P: Input-output properties of motor unit responses in muscles stretched by imposed displacements of the monkey wrist. Exp Brain Res 37:439-457, 1979

48 Tracey DJ: Joint receptors and the control of movement. Trends Neurosci 3:253-255, 1980

49 Uexküll J v: Die ersten Ursachen des Rhythmus in der Tierreihe. Ergebn Physiol 3:1-11, 1904

50 Verreussel RLP: Een onbegrepen knieklacht. Nederlands Tijdschrift voor Manuele Therapie 8:88-90, 1989

51 Voorhoeve PE, Kanten RW: Reflex behaviour of fusimotor neurones of the cat upon electrical stimulation of various afferent fibers. Acta Physiol Pharmacol Neerlandica 10:391-407, 1962

52 Wall PD, Woolf CJ: Muscle but not cutaneous C-afferent input produces prolonged increases in the excitability of the flexion reflex in the rat. J Physiol 356:443-458, 1984

53 Warmerdam A: Arthrokinetische Therapie[SM]: Spierfunctie verbeteren door gewrichtsmobilisaties. Nederlands Tijdschrift voor Fysiotherapie 101:222-228, 1991

54 Warmerdam A: Arthrokinetic Therapy[SM]: Improving muscle performance through joint manipulation. Proceedings of the 1992 Conference of the International Federation of Orthopaedic Manipulative Therapists, Vail, Colorado, pp 204-207

55 Warmerdam A: Naschrift: Arthrokinetische Therapie[SM]. Nederlands Tijdschrift voor Fysiotherapic 102:58, 1992

56 Warmerdam ALA: Arthrokinetic Therapy[SM]: The relationship between lumbar spine and quadratus lumborum dysfunctions. Proceedings of the 1996 Conference of the International Federation of Orthopaedic Manipulative Therapists, Lillehammer, Norway, p 98

57 Willer JC: Comparative study of perceived pain and nociceptive flexion reflex in man. Pain 3:69-80, 1977

58 Wood L, Ferrell W, Baxendale R: Pressures in normal and acutely distended human knee joints and effects on the quadriceps maximal voluntary contractions. Quarterly Journal of Experimental Physiology 73:305-314, 1988

59 Woolf CJ, Wall PD: Relative effectiveness of C primary afferent fibers of different origins in evoking a prolonged facilitation of the flexor reflex in the rat. J Neurosci 6:1433-1442, 1986

60 Xin Y, Weiss KR, Kupfermann I: A pair of identified interneurons in Aplysia that are involved in multiple behaviors are necessary and sufficient for the arterial-shortening component of a local withdrawal reflex. J Neurosci 16:4518-4528, 1996

61 Young AS, Stokes M, Shakespeare DT, et al: The effect of intra-articular bupivacaine on quadriceps inhibition after meniscectomy. Med Sci Sports Exerc 15:154P, 1983

9. Dysfunctions and Pathology

Altered Articular Afference

Several conditions can result in altered articular afference, including trauma, aging, surgery, dysfunctions or diseases of the joints, and/or nervous system diseases. Altered input from articular mechanoreceptors results in three main clinical features: impaired position and movement sense, posture and movement disorders, and a decreased pain threshold.[26]

The altered or diminished articular afference causes impairment of posture and movement sense. For instance, Mallik, Hall, and colleagues found that patients with a hypermobility syndrome had impaired proprioception.[9,18] These patients perceived their joint to be less displaced toward the end of the range than it actually was. The altered kinesthesia may result in coordination and balance disturbances. This is especially evident in the elderly, where a progressive loss of articular afference is an inevitable consequence of aging.[25]

The altered articular reflex, whether due to the articulo-muscular protective reflex or modulation of the withdrawal reflex, produces disturbances of posture and movement. This could result in arthrogenous muscle dysfunction (i.e., muscle dysfunction resulting from a joint disorder).

Besides impaired kinesthesia and posture and movement disorders, loss of mechanoreceptor afferent input reduces the gating of nociception. This gives nociception a greater chance to reach the cortex and produce the sensation of pain. These conditions, which are due to loss of mechanoreceptor afference, may be caused by aging.

Aging

Position and movement sense declines with age.[2,3,6,25,26,27] Deterioration of kinesthetic sense with advancing age is a generalized phenomenon that probably affects all joints.[2,6,26,27] The decline of kinesthetic sense is partly due to the progressive degenerative loss of the mechanoreceptor afferents.[25]

Disorders such as arthritis or arthrosis accumulate over the years and further disturb the information received from the remaining receptors.[25] These disorders may contribute to the subjective and objective disturbance of kinesthetic sense and also further disturb the articular reflexes.[25]

The receptors in the cervical spine may be of greatest importance in the process of senile loss of balance, because these receptors have widespread influence on muscle tone.[14,25] Inflammatory, degenerative, or traumatic affections of the cervical joints have been found to aggravate the loss of balance.[14,25]

Trauma

A joint sprain, or any traction injury to a joint's ligamento-capsular tissues, can lead to a rupture of the collagen and the nerve fibers.[7,22] The results of neural tissue damage are diminished kinesthesia, altered articular reflexes, and possibly loss of nociceptive gating.[17]

The ligaments in the joints of the body provide passive and functional stability to the joint.[10] Therefore, damage to the ligaments can result in passive and functional instability. The functional instability is created by delayed activation of the muscles that protect the joint and by an altered recruitment pattern of these muscles.

Not only does trauma lead to a loss of articular receptors, but it could also lead to swelling of the joint or to Type IV articular receptor activity.[7] These conditions may also cause afference disturbances.[21]

Surgery

Due to the invasive nature of surgery, nerve tissue damage is inevitable. The decision for surgery should therefore be made with caution.[4,11,19] Surgical denervation of parts of a joint produces serious and lasting changes in the articular reflexes from the joint.[24]

Surgical alteration of the joint mechanics may also disturb the articular reflexes. For instance, Clark, Klineberg, and colleagues reported that extraction of the teeth and their prosthetic replacement may modify the static and dynamic activity of the articular mechanoreceptors of the temporomandibular joint.[5,15] These researchers reported a case in which malocclusion of the dentures caused abnormal afference from the temporomandibular joint. The abnormal afference resulted reflexly in an almost complete inhibition of the temporal muscles during active occlusion by the patient.[15] Restoration of normal joint mechanics by remodeling the dentures restored normal muscular activity.

Many authors point to the importance of minimizing sensory damage to a joint during surgery to reduce disturbance of the articular reflexes.[8,10,11,12,13,19,20] To reduce sensory loss, an arthroscopy is preferable to an open surgery and surgical repair is preferable to substitution with a graft.[4,8,10,12,13,23] Even surgical incisions themselves should be designed to preserve the nerve receptors.[1] Krauspe makes a case for preserving the nerve supply of the anterior cruciate ligament as completely as possible during ACL surgery, writing that loss of innervation may be responsible for the high rate of unsatisfactory results following this type of surgery.[15]

Summary

Altered afference from a joint could result in sensory and motor disorders. The altered afference may be the result of conditions such as aging, trauma, or surgery. Besides changes in the tissue hosting the articular receptors, these conditions may directly affect articular receptors and their afferents. The result of the altered articular afference could be reduced pain threshold, impaired posture and movement sense, and posture and movement disorders. The posture and movement disorders may result from altered articulo-muscular protective reflexes and from altered modulation of the withdrawal reflex.

References

1 Andrews JR: Commentary to Kennedy JC, Alexander IJ, Hayes KC: Nerve supply of the human knee and its functional importance. Am J Sports Med 10:329-335, 1982

2 Barrack RL, Skinner HR, Cook SD, et al: Effect of articular disease and total knee arthroplasty on knee joint-position sense. J Neurophysiol 50:684-687, 1983

3 Barrett DS, Cobb AG, Bentley G: Joint proprioception in normal, osteoarthritic and replaced knees. J Bone Joint Surg Br 73:53-56, 1991

4 Biedert RM, Stauffer E, Friederich NF: Occurrence of free nerve endings in the soft tissue of the knee joint. Am J Sports Med 20:430-433, 1992

5 Clark RKF, Wyke BD: Contributions of temporomandibular articular mechanoreceptors to the control of mandibular posture: An experimental study. J Dent 2:121-129, 1974

6 Ferrell WR, Crighton A, Sturrock RD: Age-dependent changes in position sense in human proximal interphalangeal joints. Neuroreport 3:259-261, 1992

7 Freeman MAR, Dean MRE, Hanham IWF: The etiology and prevention of functional instability of the foot. J Bone Joint Surg Br 47:678-685, 1965

8 Halata Z, Haus J: The ultrastructure of sensory nerve endings in the human anterior cruciate ligament. Anat Embryol 179:415-421, 1989

9 Hall MG, Ferrell WR, Sturrock RD, et al: The effect of the hypermobility syndrome on knee joint proprioception. Br J Rheumatol 34:121-125, 1995

10 Johansson H: Role of knee ligaments in proprioception and regulation of muscle stiffness. Journal of Electromyography and Kinesiology 1:158-179, 1991

11 Johansson H, Lorentzon R, Sjölander P, et al: The anterior cruciate ligament. Neuro-Orthopedics 9:1-23, 1990

12 Johansson H, Sjölander P, Sojka P: A sensory role for the cruciate ligaments. Clin Orthop 268:161-178, 1991

13 Johansson H, Sjölander P, Sojka P: Receptors in the knee joint ligaments and their role in the biomechanics of the joint. Crit Rev Biomed Eng 18:341-368, 1991

14 Jong PTVM de, Vianney de Jong JMB, Cohen B, et al: Ataxia and nystagmus induced by injection of local anaesthetic in the neck. Ann Neurol 1:240-246, 1977

15 Klineberg IJ, Greenfield BE, Wyke BD: Contribution to the reflex control of mastication from mechanoreceptors in the temporomandibular joint capsule. The Dental Practitioner 21:73-83, 1970

16 Krauspe R, Schmidt M, Schaible HG: Sensory innervation of the anterior cruciate ligament. J Bone Joint Surg Am 74:390-397, 1992

17 Lephart SM, Kocher MS, Fu FH, et al: Proprioception following anterior cruciate ligament reconstruction. Journal of Sports Rehabilitation 1:188-196, 1992

18 Mallik AK, Ferrell WR, McDonald AG, et al: Impaired proprioceptive acuity at the proximal interphalangeal joint in patients with the hypermobility syndrome. Br J Rheumatol 33:631-637, 1994

19 Mclain RF: Mechanoreceptor endings in human cervical facet joints. Spine 19:495-501, 1994

20 Nyland J, Brosky T, Currier D, et al: Review of the afferent neural system of the knee and its contribution to motor learning. J Orthop Sports Phys Ther 19:2-11, 1994

21 Oostendorp RAB: Functionele instabiliteit na inversietrauma van enkel en voet: Een effectonderzoek pleisterbandage versus pleisterbandage gecombineerd met fysiotherapie. Geneeskunde en Sport 20:45-55, 1987

22 Rottigni SA, Hopper D: Peroneal muscle weakness in female basketballers following chronic ankle sprain. The Australian Journal of Physiotherapy 37:211-217, 1991

23 Schutte MJ, Dabezies EJ, Zimny ML, et al: Neural anatomy of the human anterior cruciate ligament. J Bone Joint Surg Am -A:243-247, 1987

24 Wyke B: The neurology of joints. Ann R Coll Surg Engl 41:25-50, 1967

25 Wyke B: Conference on the ageing brain: Cervical articular contributions to posture and gait: Their relation to senile disequilibrium. Age Ageing (8):251-258, 1979

26 Wyke BD: Articular neurology and manipulative therapy. In Glasgow EF, Twomey LT, Scull ER, et al (eds): Aspects of Manipulative Therapy, ed 2. New York, Churchill Livingstone, 1985, pp 72-77

27 Wyke B: The neurology of low back pain. In Jayson MIV (ed): The Lumbar Spine and Back Pain, ed 3. New York, Churchill Livingstone, 1987, pp 56-99

10. Arthritis

Terminology

Dorland's Medical Dictionary defines *arthritis* as "rheumatism in which the inflammatory lesions are confined to the joint."[13] Cyriax objected to such a definition of the word *arthritis* and suggested using the word *rheumatism* only for diseases accompanied by rheumatic fever.[10] He suggested restricting use of the word *arthritis* to conditions with inflammation of the joint. The meaning of the Latin word *arthritis* is inflammation of the "arthron," or joint. The cause of arthritis may be a rheumatoid disease but may also be trauma, bacteria, deposition of crystals (e.g., gout), or allergy.[10]

A further differentiation of the word *arthritis* needs to be made. Osteoarthritis, in spite of its *-itis* ending, is considered noninflammatory. A better name for this degenerative condition is arthrosis.[10]

Arthrosis and all the forms of arthritis cause changes in muscle function, such as weakness and reduced extensibility. These muscle function changes are partly responsible for the movement restrictions that accompany arthritis and arthrosis.

Every acute arthritis is accompanied by effusion of the joint. Just as every inflammation consists of "rubor, calor, dolor, and tumor" (redness, increased temperature, pain, and swelling), these conditions also occur in inflamed joints.[49] The swelling of the joint is known as *effusion*.

Effusion

The synovial membrane both secretes and absorbs the synovial fluid. The force of the synovial absorption is stronger than the secretion force, resulting in a negative or subatmospheric pressure inside healthy joints. The relative vacuum inside the joints is a common finding in most research on intraarticular pressure of normal joints.[22,35,41,43,51,64,81,94] In healthy knees, the negative pressure is present throughout most of the range, whether the joint moves passively or actively.[93,88] The negative pressure may increase the stability of the joints; the suction draws the articulating surfaces together.[35]

In contrast, patients with arthritis often have positive pressure inside their arthritic joints.[42] The positive pressure is probably due to the increased amount of synovia within the joint (effusion) and to a stiffer joint capsule allowing less expansion.[42] The amount of positive pressure inside the arthritic joint is not static but varies with, for instance, the position of the joint.

Eyring and Murray examined the intraarticular pressure in human cadavers within two hours after death.[15] These researchers reported that the joints examined were normal at the time the patients were admitted to the hospital and normal at the time of the study. In postmortem examinations, they studied the intraarticular pressure of artificially effused shoulder, elbow, wrist, hip, knee, and ankle joints.[15] They found the midrange position of these effused synovial joints to have the least amount of pressure.[15] Any movement away from this midrange position increased the intraarticular pressure, which was confirmed in similar studies by other researchers.[11,42,43,94] The midrange position, Eyring and Murray reported, is often spontaneously assumed by patients with symptomatic effused joints.[15]

The increased intraarticular pressure in joints with acute arthritis has significant consequences for muscle function. In decerebrate cats, stimulation of the knee joint receptors by increasing the intraarticular pressure caused inhibition of the quadriceps.[14] In humans, researchers observed a similar weakening of the quadriceps in the presence of a knee joint effusion.[16,40,41,43,67,83,94,98] Most knee joint experiments were performed with a cannula inserted into the joint through which the distending fluid or an anesthetic could be injected and the pressure inside the joint measured (see Fig. 10.1).[11,20,40,41,42,46,81,92,94] When the knee joints of healthy subjects were injected with a saline fluid or with human plasma, the quadriceps lost strength as measured by voluntary contraction, electromyographic (EMG) recording, or Hoffmann reflex recording.[11,40,41,43,46,81,94]

The Hoffmann reflex is the electrical equivalent of a myotatic reflex (stretch reflex of the muscle).[58] The Hoffmann reflex in these experiments consisted of a low-voltage stimulus, selectively stimulating the primary spindle cell afferents in the femoral nerve, which in turn excited the alpha motoneurons in the anterior horn, causing a quadriceps contraction.[46,58] Articular receptors can inhibit or facilitate the Hoffmann reflex.[40,46]

The amount of quadriceps inhibition was strongly related to the amount of intraarticular volume and to the amount of pressure inside the joint.[41,43,46,81] Any degree of joint distention appeared to have some inhibitory effect on the quadriceps.[41] With increasing amounts of intraarticular fluid, the quadriceps became more inhibited[11,43,46,81]; however, removal of the fluid reversed the inhibitory process.

A quadriceps lag is a dysfunction many patients with inflamed knees display. A quadriceps lag is the inability to complete active full extension of the knee, although passive knee extension is full.[83,94] In the effused knee, the quadriceps lag is the result of progressive diminished quadriceps activity as the knee extends more.[83] During extension, the intraarticular pressure of the effused knee increases, causing increased reflex inhibition of the quadriceps.[83]

Wood and colleagues studied the quadriceps lag after saline fluid was injected into the knee joint.[94] The subjects with saline fluid in their knee were asked to extend the knee. The subjects were blindfolded so they could not see the amount of effusion or the angle of their knee joint. Although the subjects' knee joints were lacking about 20 degrees of extension, they reported that their knees were fully extended and that they could not extend them any more. Shakespeare and col-

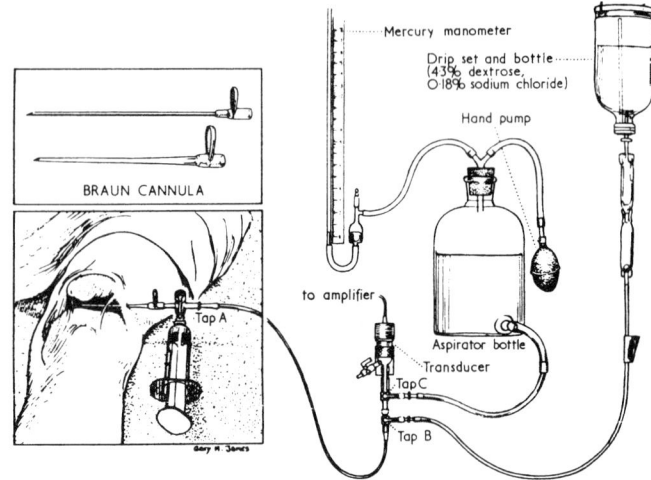

Fig. 10.1 Experimental design of knee effusion research.

Reprinted from Jayson MIV, Dixon AStJ: Intra-articular pressure in rheumatoic arthritis of the knee. Ann Rheum Dis 29: 261-265, 1970, with permission from BMJ Publishing Group.

leagues observed a similar weakening in patients who were postmeniscectomy.[80] The quadriceps in these patients tested weaker in extension as compared with flexion.

The increased amount of fluid in the knee joint does not mechanically obstruct the quadriceps from extending the knee. Two observations suggest that the distending fluid does not mechanically obstruct knee extension. First, full passive knee extension was comfortably available.[83,94] Second, when a local anesthetic was injected into the knee joint with the distending fluid, the quadriceps extended the knee with no sign of inhibition.[11,46,81,94] Young and colleagues studied the effects of injecting a local anesthetic into the knee joint immediately following a meniscectomy.[98] They found diminished inhibition of the quadriceps (a stronger quadriceps) in the group that had received the anesthetic as compared with the control group.[96,98] The local anesthetic reduced the activity of the intracapsular receptors and/or their afferents, and therefore inhibition of the quadriceps did not occur.[11,81]

A third observation suggesting that the distending fluid does not mechanically obstruct active knee extension comes from electromyographic research. EMG readings showed that the quadriceps was less active when the joint was distended.[11,16,83] More specifically, these EMG readings showed action potentials of both lesser amplitude and number.[11] These findings also showed that the loss of active knee extension resulted from quadriceps weakness.[11,16,81]

A fourth indication that the distending fluid does not mechanically obstruct active knee extension comes from observation of a patient with a neuropathic disease of the knee (i.e., a joint disease resulting from a functional disturbance in the peripheral nervous system). DeAndrade described a patient with a neuropathic disease of the knee joint who did not develop muscle weakness, even when the joint was distended at high pressure.[11] This suggests that, in the presence of an effusion, an intact nervous system is essential for inhibition of the quadriceps.[11]

In 1965, DeAndrade and colleagues examined inhibition of the quadriceps by joint distention.[11] Their healthy human subjects were positioned supine with a rolled sheet beneath a 10-degree flexed knee. Human plasma was injected into the knee joint in small increments. After each injection, the subjects were asked to raise their heel off the table. With progressive distention, all subjects remarked that attempting to straighten their knee required more concentration and effort. Two subjects indicated their surprise that a painless knee could be so helpless.[11] Other researchers confirmed that the progressive quadriceps inhibition due to increasing intraarticular pressure was not caused by pain.[30,40,41,43,82,83,94] Neither was the quadriceps inhibition after a meniscectomy found to be due to pain.[98]

Because pain is not an essential component of quadriceps inhibition by effusion, Iles and colleagues assumed that the inhibition was mediated by the mechanoreceptors in the joint capsule and not by the Type IV receptors.[41] Various researchers found different articular mechanoreceptor discharges in response to joint distention. Ferrell and colleagues found increased activity of the afferents from Type I and II articular receptors after the knee joint was injected with a distending fluid.[20] Several others found that the slowly adapting mechanoreceptors, possibly the Type I receptors, responded to increased intraarticular volume.[1,17,22,26,46,92] When more fluid was injected, more receptor discharges were observed.[17] Researchers found an almost linear relationship between the pressure inside the joint and the discharge rate of the articular mechanoreceptors.[41,81] When lignocaine was injected into the synovial cavity, there was an initial increase in the discharge rate of the receptors, secondary to the increased volume.[26] This response was followed by a silencing of the receptors.[26] The fact that anesthetic injections abolished the inhibition suggests that the receptors are located intracapsularly (i.e., in the synovial membrane).[26,81,94]

Researchers have found receptors (possibly Type I) sensitive to intraarticular pressure.[1,11,16,17,27,43,59,81,92] In the effused knee joint, these pressure-sensitive receptors were responsible for inhibition of the quadriceps' alpha motoneuron.[11,16,43,81] These receptors were insensitive to capsular tension, suggesting that they do not reside in the fibrous capsule. Grigg and colleagues demonstrated that these receptors were located in the synovial membrane, close to the fibrous capsule.[27]

Wood and colleagues found that active positioning of the distended human knee joint caused higher intraarticular pressures than passive positioning of the joint.[94] During active positioning, contraction of the muscles caused increased tension of the joint's ligamento-capsular structures.[94] The increased tension was possibly due to slips of the muscles attaching to the joint's ligaments and capsule.[94] The muscle contraction also caused increased intraarticular pressure due to compression of the joint itself.[94]

Jayson and Dixon found the intraarticular pressure to increase (become positive) in people with rheumatoid knees during the stance phase of walking, whereas in normal subjects the intraarticular pressure remained negative ($p < .001$).[43] Contractions of the quadriceps also increased the intraarticular pressure in these effused knee joints, whereas in normal subjects, the intraarticular pressure remained subatmospheric.[43,94] Jayson and Dixon's findings suggested an increased inhibition of the quadriceps during walking and activities when the quadriceps contracts. The disturbance of the normal articular reflexes by acute effusion may explain the tendency for acutely effused joints to "give way."[20] Since knee flexion and extension also raise intraarticular pressure, the effused knee tends to assume a loose-packed position, where the intraarticular pressure is lowest.[11,15,42,43,94]

An effusion of the knee joint also alters kinesthetic performance at the joint.[2,18] Baxendale and Ferrell found that an impaired kinesthetic sense was present at the extremes of motion.[2,18] Their subjects had the sense that the effused knee had moved farther into the range than it actually had. Subsequent injection with local anesthetic reduced the erroneous perception. The erroneous sensation that the effused knee had moved farther into the range may be related to increased activity of the articular receptors. Ferrell and colleagues found that an effusion enhanced the response of the articular receptors.[20]

Fahrer and colleagues aspirated synovial fluid from chronically effused knees in patients with various forms of arthritis.[16] The aspiration caused an average 13.6 percent increase in quadriceps strength as measured by EMG and isometric strength testing. The quadriceps strength increased by another 8 percent after an additional anesthetic injection of lignocaine. Fahrer and colleagues concluded that two mechanisms might be responsible for the quadriceps inhibition. The first mechanism is the raised intraarticular pressure, causing pressure-sensitive receptors to inhibit the muscle. Aspiration of the joint eliminated this mechanism of inhibition. Injection of the anesthetic further improved the strength, proving that another unknown mechanism was also responsible for inhibition.[16] This "anesthetic-sensitive" form of quadriceps inhibition is possibly mediated by receptors on the inside of the joint capsule.[94] The anesthetic-sensitive mechanism of quadriceps inhibition may be more important in patients with chronic effusions.

Jones and colleagues found that quadriceps strength in patients with chronically effused knees did not improve after aspiration.[44] The inhibition of the quadriceps may have been caused by this anesthetic-sensitive mechanism. Unfortunately, Jones and colleagues did not inject an anesthetic to observe if this would increase quadriceps strength. Ferrell provided another explanation as to why aspiration did not improve quadriceps strength in patients with chronically effused knees, writing that the effusion may have stretched out the joint capsule and thereby reduced articular mechanoreceptor firing.[17] Thus, aspiration did not reduce articular mechanoreceptor activity, because the stretched-out joint capsule had already done so.

Inhibition of muscles could lead to muscle atrophy.[60,63] More specifically, in the effused knee joint, inhibition of the quadriceps could be responsible for quadriceps atrophy.[11,43,60,81] Inhibition and atrophy of the vastus medialis seems to get more attention than inhibition and atrophy of other parts of the quadriceps; however, in the presence of a knee joint effusion, reflex inhibition of the quadriceps seems to affect the entire quadriceps, not just the vastus medialis.[82] Similar to inhibition, atrophy also seems to take place uniformly in the quadriceps.[55,82,97] Spencer and colleagues stated the following:

> It must be recognized that reflex inhibition due to effusion affects the entire quadriceps muscle, not just the vastus medialis. Normally when reporting a weakened quadriceps muscle, atrophy of the vastus medialis muscle is singled out as being responsible for the entire muscle's dysfunction. Although this muscle appears to be the most sensitive to inhibitory influences, and because its bulk and orientation of its fibers may be the most important in aligning the patella in the intercondylar groove, the results clearly indicate an inhibition of other quadriceps components as well.[81, p. 176]

Morrissey found that a knee injury (not specifically a knee "effusion") caused reflex quadriceps inhibition, but the part least affected was the rectus femoris, probably because of its additional function as a hip flexor. The other parts of the quadriceps were perhaps more inhibited because they are active only over the knee.[63]

Performing quadriceps exercises such as quad sets is considered an ineffective way to strengthen the muscle if an effusion is present.[16,43,81,83] The inhibited quadriceps will only contract submaximally, and the cause of the inhibition and atrophy is not addressed. In addition, the active and passive use of the effused knee joint produces high intraarticular pressures instead of the normal subatmospheric pressures.[43] These high intraarticular pressures place nonphysiological stress on the articular surfaces and the joint capsule.[43] If the articular cortex is weakened by osteoporosis or by rheumatoid granulation tissue, these pressures may be sufficient to disrupt the surface and burst through to the bone marrow, producing rheumatoid bone cysts.[43] The high intraarticular pressures may also be sufficient to produce an acute joint rupture or "blow out" the capsule and produce synovial cysts.[43]

Muscle Weakness

More than a century ago, in a translation of his clinical lectures, Jean Martin Charcot, a French neurologist, reported on older publications describing muscle weakness resulting from joint afflictions.[8] In his publication, Charcot added his personal observations to what he referred to as "paralysis having an articular origin." He described the paralysis as "a frequent complication of different idiopathic or traumatic lesions affecting the corresponding articulation."[8, p. 24]

> Thus, one not infrequently sees atrophy of the deltoid muscle after different lesions of the scapulo-humeral articulation; or as a consequence of an arthritis, sprain, or other injury of the hip-joint, the buttock is sometimes affected in the same manner; or if it be the knee-joint which is attacked, the nutrition and mobility of the quadriceps extensor femoris is affected.[8, p. 24]

Charcot described the primary joint afflictions causing the paralysis as often painful and very severe, but not always so. A minor sprain, a small effusion without pain, can cause the paralysis. There is not necessarily a relationship between the intensity of the joint affliction and the paralysis. In explaining the paralysis, some of Charcot's contemporaries hypothesized that the articular inflammation spread little by little into the neighboring muscles.[8] Charcot objected to this theory because his observations showed that the atrophy existed equally along the entire length of the muscle. He also noted that the muscle atrophy had no trace of inflammation. Charcot also objected to the hypothesis that the paralysis was the result of disuse atrophy due to prolonged rest. One of his arguments against this hypothesis was that the articular affliction is often so slight that it does not require more than a brief rest period.

Charcot's explanation of "paralysis having an articular origin" was as follows. The articular affliction causes afferent articular nerves to transmit "irritant impulses" to the spinal cord. The impulses modify the motor and trophic nerves cells of origin innervating the muscles that move the joint. The altered function of the motor and trophic nerves causes paralysis and atrophy of the muscle.[8]

In 1890, Fulgence Raymond, a French neurologist, published a paper on muscle atrophy resulting from traumatic arthritis.[71] Working on dogs, Raymond cut the posterior nerve roots of the lower three lumbar and upper three sacral nerves on one side of the dog's body. He created an arthritis in both knee joints by burning the joint surface with a hot iron bar and injecting an irritant solution (turpentine or silver nitrate) into the joint. Raymond found the quadriceps on the side of the intact posterior nerves greatly wasted. The quadriceps on the side of the sectioned nerves showed only minimal atrophy. Raymond concluded that joint lesions reflexly produced changes in the spinal cord, which caused muscle atrophy.

These publications from the previous century provide early descriptions of arthritic joints causing muscle weakness. Both Charcot and Raymond agreed on the nervous system relaying information from the arthritic joints to the muscles moving these joints. These findings were not unlike those four years later of Beryl Harding, who created arthritis in the knee joints of rabbits and found a subsequent quadriceps atrophy.[29]

Harding studied rabbits and cats who had the left posterior nerve roots cut at the level of the lower three lumbar and upper three sacral nerves.[30] The right side was left intact to serve as a control. All animals were very ataxic for the first 10 to 14 days, then gradually learned to control their deafferentiated left limb. After one month, injections were given into both knee joints with varying organisms or chemicals. All animals used their arthritic limbs freely and appeared to have very little pain. Another group of animals, used as controls, showed that only sectioning of the left posterior roots produced no wasting of the muscles. In other control animals, the knee injections produced a wasting of the quadriceps muscles equal to that in the right knee (with intact nerves) of the experimental group. However, the left quadriceps of the experimental group with the nerve lesions on the same side showed hardly any wasting. Harding concluded that arthritis provides an abnormal stimulus to the afferent nerve endings, causing changes in the spinal cord. This creates an abnormal discharge of impulses to the muscles and produces atrophy. Harding termed the atrophy a *reflex atrophy,* which differs from disuse atrophy in that the latter has no reflex basis.

Most studies have been done on the knee joint; however, Meyer made a case for extrapolating the findings on distended knee joints to the spinal facet joints.[60] The inhibition of the spinal muscles would depend on spinal positions where the intraarticular pressure had increased.[60] Inhibition would be absent in the loose-packed position of the facet joint. Additionally, Nichols suspected that the principle of an arthritic knee inhibiting the quadriceps could be applied to arthritic hip and ankle joints as well.[67] The arthritic hip and ankle joints would inhibit their local muscles.[67]

In 1954, Blockey recorded a series of experiments on patients who had undergone reduction of an elbow joint dislocation.[4] Following the procedure, the patients always had limited range of motion, which Blockey believed was caused by two factors: adhesions in the joint and inhibition of the elbow muscles. In testing the strength of the elbow flexors, Blockey found that the flexors could contract quite strongly if they were tested in midrange positions of the elbow. At maximal flexion, however, the flexors remained soft, although the patients were contracting with all the strength at their disposal. Blockey stated that pain was not the reason the elbow flexors could not contract. Only a few of his subjects experienced pain, and if they did, it did not prevent them from attempting to contract strongly. Using needle electrodes, Blockey recorded the activity of the biceps. With the elbow in maximal flexion, the electromyograph recorded only a few action potentials, and these were lower in amplitude and frequency than those obtained when the muscle contracted in midrange. Blockey suggested "that some reflex mechanism exists which inhibits the motor muscle at a point just short of that which would be required to break down adhesions . . . This joint-protection reflex seems to be initiated by receptors other than pain receptors; it does not reach a conscious level."[4, p. 839]

Besides inhibition of muscles by articular mechanoreceptors, Scott et al. found inhibition by articular nociceptors.[79] They reported that the muscle weakness accompanying inflamed joints may be caused by nociception. Articular nociception inhibited the facilitating influences on muscles by articular mechanoreceptors.

If a joint disorder such as an elbow dislocation is not entirely responsible for a disability, the resulting muscle weakness and atrophy may be a contributing factor.[10,63,82] In addition, the muscle weakness and atrophy may render the joint vulnerable to further damage.[82]

Substance P

The dorsal root ganglion of the nociceptive fibers synthesizes neurotransmitters, or neuropeptides.[33,52,56,68] Axoplasmic flow within the nerve transports these neurotransmitters centrally and peripherally to the distant terminals of the nerve. Both the central and the peripheral terminals of the nociceptive fibers secrete these neurotransmitters following depolarization.

Substance P is one neurotransmitter secreted by the nociceptive fibers.[33,52,54,56,68,99] The synapses of the nociceptive fibers secrete substance P. Centrally, the nociceptive fibers release substance P in the dorsal horn to transmit the nociceptive signal to other neurons.

The nociceptor afferents do not release substance P only in the dorsal horn. It is also released in the periphery through the *axon reflex*. Branches of the Type IV afferent fibers curve back to run in a distal direction, ending in the neighborhood of their Type IV receptor (see Fig. 10.2). Thus, Type IV nociceptor activity not only results in conduction of impulses to the spinal cord, but also in impulse conduction in branches of the axon ending near blood vessels close to the receptor.

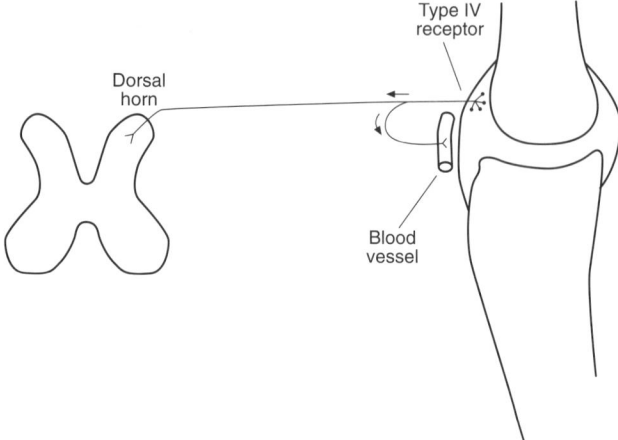

Fig. 10.2 Axon reflex.

The release of sufficient substance P and possibly other neurotransmitters results in an acute local inflammation.[21,48,50,56,99] More specifically, substance P controls a variety of inflammatory activities, such as vasodilation, plasma extravasation, and release of histamine from mast cells.[7,21,56] Thus, substance P influences the major features of inflammation.[7,21,56] The process of nociceptor activity creating an inflammation is termed *neurogenic inflammation*.[56,57,91]

Matucci-Cerinic and Partsch wrote that the literature strongly supports the role of substance P in the development of human arthritis.[56,57] Unfortunately, substance P not only contributes to the inflammation; it also specifically exacerbates joint destruction in adjuvant-induced arthritis.[54] Thus, in theory, an inflammation of a joint might be caused solely by Type IV articular receptor activity.

The Nervous System and Arthritis

Rheumatoid arthritis tends to be bilateral and symmetrical.[3,52] The involvement of the nervous system could explain the symmetry of the joints affected by rheumatoid arthritis. Levine and colleagues found evidence suggesting connections across the spinal cord to mediate the contralateral response.[54] They referred to this contralateral response as a neurogenic inflammation. The nervous system seems responsible for the symmetrical character of joint inflammation in rheumatoid arthritis.[54]

The bilateral and symmetrical nature of rheumatoid arthritis seems to take exception in patients with hemiplegia. In these patients, the development of rheumatoid arthritis seems confined to the uninvolved half of the body, sparing the paralyzed side.[24,85] Lower motoneuron paralysis resulted in similar sparing of the paralyzed limbs.[24] Glick studied 12 cases of rheumatoid arthritis that developed in patients with preexisting poliomyelitis whose paralyzed limbs were spared totally or partially by the arthritic process.[24] Glick reported similar findings in a patient with hemiplegia and in several patients with arthrosis after poliomyelitis. The greater the severity of the paralysis, the more pronounced the sparing of the extremity. Glynn and Clayton reported a similar sparing effect on the development of gout.[25] Patients with hemiplegia developed less gout in the partially paralyzed extremities.

Besides the tendency for rheumatoid arthritis to be bilateral and symmetrical, distal joints tend to be more severely involved than proximal joints.[52,54] Working with rats, Levine and colleagues found a greater nociceptive innervation density in the ankle joint compared with the knee joint.[54] In addition, the ankle joints examined had a lower nociceptive threshold and contained more substance P than the knee (a low-risk joint for rheumatoid arthritis). Levine et al. wrote that the presence of higher levels of substance P and the increased density of nociceptive afferents suggest a relationship with risk for more severe joint injury in adjuvant-induced arthritis.

Some observations and studies provide additional proof for the involvement of the nervous system in arthritis. For instance, although nerve lesions cause a denervation of their target tissue, injury to the target tissue caused less inflammation compared with tissue that was still innervated.[32] Administering morphine into the central nervous system produced "significantly less severe arthritis,"[54, p. 846s] as shown by radiographs.

The beneficial effect of gold therapy in patients with rheumatoid arthritis may have its basis in the nervous system as well. Gold salts slowed the progression of joint erosion in rheumatoid and psoriatic arthritis; however, a side effect of the treatment may be a peripheral neuropathy.[19,56] Levine and colleagues wrote that gold selectively destroys unmyelinated axons in the peripheral nervous system.[52] The gold salts may inhibit the release of substance P from the nociceptive terminals, preventing neurogenic inflammation and reducing nociception.[56] Thus, the nervous system is likely to contribute to the severity of joint affliction in arthritic patients.

Type IV Nociceptors

INFLAMMATION

Tissue damage stimulates the release of chemical mediators such as histamine, serotonin, bradykinin, and prostaglandin E2. This has two major effects: The chemical mediators create an inflammation, and they stimulate the nociceptors.[86]

The chemical mediators initiate an inflammatory response that, via the process of vasodilation and increased permeability, causes localized swelling. The localized swelling may lead to increased pressure around the Type IV receptors, or possibly swelling of the receptors themselves.[66,75] With joint capsule damage, the swelling is an intraarticular edema (effusion). The increased circulation that accompanies inflammation raises the local temperature. The elevated temperature in the inflamed joint may contribute to the increased responsiveness of the Type IV receptors.[66,75]

In summary, the inflammatory mediators indirectly stimulate the Type IV receptors through the inflammatory process; however, direct stimulation is also possible.[45,66,75,76,77] The neurotransmitter substance P can also directly increase Type IV receptor activity.[7,21,56,91] Thus, an inflammation of the joint through various means stimulates the Type IV articular receptors.[9,28,34,39,65,66,74,75] It does so by altering the behavior of these receptors in three ways[9,28,34,75,77]:

1. Normally silent Type IV receptors become active.
2. The Type IV receptors increase their discharge frequency.
3. The threshold of the Type IV receptors to movement is lowered.

A healthy joint is painful only when it receives intense mechanical stimuli.[74,91] Joints with acute arthritis, however, can be painful throughout their range.[74] Even movements in the joint's midrange can facilitate Type IV activity.[9,28,74,75,91]

SPINAL PERCEPTUAL CHANGES

In addition to changes in Type IV nociceptor functioning, acute arthritis causes changes in segmentally related regions of the spinal cord.[31,65,73,74,76,95] These spinal cord changes consist of alterations in the responses of "nociceptive-specific" neurons and "polymodal" neurons. The nociceptive-specific neurons normally do not respond to brushing of the skin or to deep pressure of the tissues outside the joint, but respond only to nociception from the joint. In contrast, the polymodal neurons respond to a variety of stimuli, such as pressure to a joint, stimulation of the ipsilateral skin, or sometimes even stimulation of skin in the contralateral extremity. After the development of acute arthritis, both types of neurons become more sensitive to other inputs as well and respond, or respond more strongly, to brushing of the skin and to deep pressure in regions remote from the joint. Previously subthreshold afferent inputs from remote regions are now able to excite these spinal neurons.[31,65,73,74] It may be that arthritis causes these spinal neurons to become more sensitive. Practically, this means that arthritis may cause a decreased threshold for pain from regions other than the joint. Pain on palpation of the skin or muscles (trigger points?) may be the result of Type IV nociceptor activity.

ALPHA/GAMMA MOTONEURON CHANGES

Activation of the Type IV articular receptor by an acute inflammation not only causes changes in polymodal and nociceptive-specific spinal neurons, but changes are also observed in the activity of the alpha and gamma motoneurons.[31] Some motoneurons respond with increased activity, whereas others respond with decreased activity. According to He and colleagues, the altered motoneuron activity is consistent with a withdrawal reflex.[31]

Ferrell and colleagues also found that the presence of an inflammation in the knee joint increased the intensity of the withdrawal reflex.[23] However, the acute inflammation abolished the normal pattern of modulation of the withdrawal reflex. Different joint positions did not alter the withdrawal reflex, and the reflex intensities were equal in all positions. Thus, instead of being modulated by joint position, the magnitude of the reflex was increased in all positions.[23] The absence of modulation by position suggests that the Type IV articular receptors can override the input from the low-threshold articular mechanoreceptors.[23]

The enhanced excitability of the withdrawal reflex by acute joint inflammation could have implications for the arthritic diseases. The deformities observed in patients with chronic inflammatory arthritis such as rheumatoid arthritis may be the result of chronic facilitation of the motoneurons.[23]

CONCLUSION

In summary, nociceptive afferents, through substance P and other neurotransmitters, may contribute to the inflammation in rheumatoid arthritis.[53] This has been termed *reflex neurogenic inflammation*. Nociception can thus produce or contribute to the inflammation of joints. In chronic arthritis, the release of neurotransmitters could cause destruction of the joint. The available research suggests that the reduction of pain through physical or other means might reduce the progression of rheumatoid arthritis. Hogeweg wrote that, considering present knowledge about the relation between the nervous system and arthritis, the reduction of nociception is more than just symptomatic treatment.[38] The reduction of nociception seems to affect joint inflammation directly.

Capsular Pattern

Nociception is not the only consequence of arthritis. Neither are weakness and atrophy the only results of an articular lesion, according to Charcot.[8] An articular lesion may produce a reflex contracture of the muscles of the joint. Charcot called these contractures "reflex contractures of articular origin."[8, p. 52] The contractures of the muscles are not voluntary, nor is their purpose to relieve joint pain; they are a spinal reflex.[8] Charcot believed that the deformities of joints with chronic rheumatoid arthritis are the result of these reflex muscle contractions.

The contracture and the weakness are mainly limited to the muscles moving the joint but can spread to other muscles in the extremity. Charcot concluded that the contractures and the muscular atrophy are two extreme phases of the same morbid process.[8]

In lectures published in 1863 and 1896, the English surgeon John Hilton spoke about arthritic joints and the stiffness that accompanies them. He believed that a reflex muscle contraction was responsible for the stiffness in arthritis. The nervous system mediated between the arthritic joint and the contracted muscles. Hilton's description of stiffness in response to arthritis could be interpreted as a capsular pattern of movement restrictions.[10]

Late in the 19th century, Hilton wrote about his observations of the innervation of muscles, joints, and skin. He stated that "the same trunks of nerves, whose branches supply the groups of muscles moving a joint, furnish also a distribution of nerves to the skin over the insertions of the same muscles; and—what at this moment more especially merits our attention—the interior of the joint receives its nerves from the same source."[37, p. 168]

This statement that the nerve trunk which innervates a muscle also innervates the joint underneath the muscle and the skin overlying the muscle later became known as Hilton's law.[13,69] Hilton wrote that due to the mutual innervation of muscle and joint, a harmonious cooperation exists between them.[36] For instance, the nerves are capable of "telling" the muscles that the articular structures are overexerted. Overexertion of a joint, pain, inflammation, or an irritation can cause the articular afferent nerves to influence the function of the motor nerves in the spinal cord. The motor nerves cause a contraction of the muscles that oppose movement of the joint. Thus, he wrote, the joint is at once rendered involuntarily "rigid and stiff" for the purpose of keeping it at rest.[37, p. 169] Hilton described the inflamed joint as "never straight, but always flexed, the degree of that flexed condition depending upon the intensity or the long duration of the mischief."[37, p. 174] He also wrote that chloroform can "destroy" the muscular contraction, and "the fixed and flexed conditions of the joint are for the time entirely lost."[37, p. 172]

Four years later, in Germany, Payr wrote that joint damage produced a reflex contraction of the muscles.[70] Although the joint damage produced a contraction of all muscles acting over the joint, certain muscles responded with more tone than other muscles, causing the typical antalgic position. The muscle hypertonus was reversible when the irritation disappeared from the joint, or when anesthesia interrupted the reflex. However, if the hypertonus of the muscle had led to structural changes of the muscle, the improvement in muscle length would be quite slow.

In 1932, Brunschwig and Jung described six cases in which injured joints where accompanied by reflex muscle "rigidity."[6] In these subjects, injections of Novocain around the nerve endings in the traumatized joints abolished the pain and the reflex muscle rigidity.

Swearingen and Dehne found the acute limitation of motion and the decrease of voluntary control to be associated with muscle tone changes.[84] These researchers recorded an inhibition of agonist muscles and an increased activity in the muscle groups opposing stress in the injured joint (antagonists). They described these muscular reactions as resulting from a decreased stress tolerance of the injured joint, hypothesizing that the muscle tone changes were mediated by the articular receptors, the gamma system, and the muscle spindles.

Although arthritis and arthrosis affect the entire joint capsule, the movement restrictions are not equal in all directions, but in capsular proportions. For example, a capsular pattern of movement restrictions in the glenohumeral joint displays itself as a major limitation of external rotation, a somewhat lesser restriction of abduction, and a slight restriction of internal rotation; adduction might be completely free of any restriction. Although the amount of restriction may vary, all shoulder joints with arthritis or arthrosis will be most limited in external rotation, followed by abduction. Every joint in the body, when afflicted with arthritis or arthrosis, has its own unique set of movement restrictions that form the capsular pattern of that joint (see Table 10.1).

Only joints controlled by muscles can develop movement restrictions in a capsular pattern.[10] The muscles act to prevent tension on the joint capsule; therefore, a joint that relies only on ligaments for its stability does not have a capsular pattern.[10] The joints that rely on ligaments for their stability are the sternoclavicular, acromioclavicular, sacroiliac, and sacrococcygeal joints, the symphysis pubis, and the tibiofibular joint. These joints do not have a protective reflex to restrict their movement excursion.[10] For instance, all muscles crossing the sacroiliac joint, such as the psoas major and the piriformis, are muscles that move the hip or lumbar joints. We cannot voluntarily move the sacroiliac joint through a range using these or any other muscles. If muscle tone changes were to occur in response to sacroiliac arthritis, these changes would restrict lumbar or hip joint movements but not sacroiliac joint movement. Thus, when afflicted by arthritis, joints without an active voluntary range do not display limited range, but only hurt when their capsule is sufficiently stretched. The severity of arthritis is evidenced by the amount of pain a patient feels when the capsule is stretched.[10]

Except at those joints without a voluntary active range, arthritis and arthrosis present themselves clinically as a limitation of movements in the capsular pattern.[10] The muscle contractions (spasms) occur always in the muscles opposing the joint movement.[10] In a loose-packed position all the muscles might be relaxed, but the contractions occur as soon as a movement passes a certain point in the range where the capsule starts to become stretched. With repeated movements,

Table 10.1 Capsular patterns of movement restriction.[10,61,87,88,89,90]

Joint	Movement Restrictions
Temporormandibular	Depression limited. Occlusion free.
Cervical spine	Facet extension limited (extension, ipsilateral rotation, and ipsilateral side bending). Flexion full but often painful.
Thoracic spine	Bilateral rotation limited.
Lumbar spine	Extension and side bending limited. Flexion full but often painful.
Sacroiliac, sacrococcygeal, and symphysis pubis	Pain when the joints are stressed.
Sternoclavicular	Pain at extremes of range.
Acromioclavicular	Pain at extremes of range.
Shoulder	External rotation most limited. Abduction less limited. Internal rotation less limited than abduction. Adduction free.
Elbow	Flexion most limited. Extension less limited. In severe cases, rotation painful.
Proximal radioulnar and distal radioulnar	Pain at end of pronation and end of supination.
Wrist	Palmar flexion and dosiflexion equally limited. Ulnar and radial deviation free.
Carpometacarpal I	Extension and abduction equally limited. Flexion and adduction free.
Other thumb joints and the finger joints	Flexion most restricted. Extension less restricted.
Hip	Internal rotation most limited. Flexion, abduction, and extension less limited. External rotation and adduction least limited.
Knee	Flexion grossly limited. Extension much less limited. Rotations affected only at end stage of arthritis.
Tibiofibular	Pain on stretching.
Ankle	Plantar flexion most limited. Dorsiflexion less limited.
Subtalar	Inversion grossly limited. Eversion less limited.
Midtarsal	Equal limitation of supination, plantar flexion, and adduction. Extension less limited. Abduction and pronation free.
Metatarsophalangeal I	Extension markedly limited. Flexion slightly limited.
Metatarsophalangeal II–IV	Flexion and extension variably limited, but in the end stage, flexion more limited.
Interphalangeal	Extension often more limited than flexion.

this point in the range does not change. During movements in other restricted ranges of the joint, different muscles, again opposing the movement, contract to restrict the range.[10] In addition to the muscles opposing the movement (the antagonists) contracting, the muscles moving the joint farther into the range (the agonists) become inhibited. Both muscle changes restrict the movement excursion.

The presence of a capsular pattern indicates that arthritis or arthrosis is present with lesions of the joint capsule or its synovial membrane.[10] Why is it that lesions of the joint capsule or its synovial membrane cause different movements to be restricted in varying degrees? Is it the strength of the antagonist muscles that determines the degree of restriction? For example, at the shoulder joint, the internal rotators are about twice as strong as the external rotators, and a greater restriction of external rotation seems understandable. This does not, however, explain the movement restrictions found in arthritic elbow and finger joints. These joints have stronger flexor muscles than extensor muscles.[78] However, these stronger flexor muscles cannot be responsible for flexion being more restricted than extension. The capsular pattern of these joints is a greater restriction of flexion than extension. Meanwhile, the reason that certain movements are more restricted than others remains unknown. Future research will undoubtedly reveal the precise mechanism.

Cyriax believed that it is not necessarily pain that produces limited movement in arthritis.[10] Using the example of a hip joint with considerable restriction of movement but without any pain, even when the capsule was strongly stretched, he suggested that the nociceptor system (Type IV afferents) is responsible for the restricted movement, but that it does not necessarily produce pain.

Zimmermann, however, does not think that Type IV articular input directly causes increased muscle tone of the antagonistic muscles.[99] He wrote that Type IV afference may increase the *excitability* of the antagonistic muscles, calling the resulting immobility and stiffness a "reflex neurogenic immobilization of that joint."[99]

Causes for a capsular pattern of movement restrictions can be classified into two categories.[47] The first category consists of conditions with joint effusion and/or synovial inflammation; the second category consists of joints with capsular fibrosis (scar formation).[47,72] Similarly, the inflammation of a joint consists of two responses.[72] The first is the inflammation itself; the second is the repair.[72] The repair starts during the inflammation phase and consists of replacement of the destroyed cells and tissues.[72] The destroyed tissue is often replaced by scar tissue, and this may lead to reduced mobility of the joint.[72] The proportion of lost extensibility is likely related to the lost mobility during the acute inflammation phase, since during this period scar formation is taking place. Thus, a capsular pattern can exist both during the inflammatory phase and during and after the repair phase.

Acute Arthritis

Although the phrase "movement restrictions in a capsular pattern" suggests a contracture of the capsule, this is not the case in acute arthritis.[10] Cyriax wrote that gross movement restrictions in a capsular pattern might be present only a few hours after a severe joint sprain.[10] Obviously, a contracture of the capsule cannot develop in only a few hours.[10] The movement restrictions in a capsular pattern are often believed to be caused by the effusion that accompanies the joint's synovitis.[10] Cyriax believed that nociceptive afference from the joint, or from its neighborhood, provokes an involuntary muscle spasm, which restricts mobility in capsular proportions.[10] According to Cyriax, "arthritis can thus be seen to behave clinically as if capsular contracture had supervened, even though it has not."[10, p. 55]

During the acute stage of arthritis, the inflammation of the capsule produces increased amounts of synovium. This increases the intraarticular volume and pressure.[43] Portions of the capsule that are normally lax to allow movement become taut because of the effusion.[43] When moving toward an end range, parts of the capsule stretch even more, because the increasing intraarticular pressure pushes the capsule outward.[43] To prevent these high intraarticular pressures, the joint with acute arthritis assumes a loose-packed position.[47]

Irritation of the synovial membrane causes a muscle spasm to prevent stretching of the capsule.[10] Thus, the movement restrictions in acute arthritis are caused by a reflex contraction of the muscles around the joint. Additionally, effusion enhances the response of the articular receptors.[20] The result is increased facilitation of the muscles preventing the movement (antagonists), with a simultaneous increased inhibition of the agonists.[84] For instance, a patient with acute arthritis of the knee, while attempting to extend the knee, is expected to show an inhibition of the quadriceps and a facilitation of the hamstrings. These articular reflexes, restricting range in the acutely arthritic joints, are expressions of the enhanced ligamento-muscular protective reflex.[46] Clinically, the ligamento-muscular protective reflex is often called "guarding" or "splinting" of the joint. Because the limited range in these joints is reflex based, the limitation of movement might disappear under general anesthesia.[5,37,62,70]

Research suggests a relationship between the amount of fluid in a joint (effusion) and the stiffness of the joint.[42] In other words, a normal knee joint that contains very little synovial fluid has a capsule with normal extensibility, whereas an arthritic knee with an effusion is a stiff joint that resists extension of the capsule.[42] Possibly, the increased vascularization of the synovial membrane, causing an effusion, also produces gradual fibrosis of subsynovial tissue. In time, the fibrosis might reduce the excursion of the joint.

Chronic Arthritis

Hilton reported that chloroform removes the involuntary reflex contraction, and "the fixed and flexed conditions of the joint are for the time entirely lost."[37, p. 172] Hilton's statement applied to recently inflamed joints.

"In cases of long continued or chronic inflammation of joints, the contraction and fixed condition depend partly, no doubt, on the encumbrance of the joint by the new material."[37, p. 173] Hilton probably spoke of what we now call newly formed collagen fibers in the chronically inflamed joint reducing the joint's excursion. "Hence the deformity of the joint at that period cannot be entirely removed even under the influence of chloroform."[37, p. 172]

Similar to acute arthritis, chronic arthritis is accompanied by restricted mobility in capsular proportions.[10] However, chronic arthritis may lack signs of inflammation. Restricted range in the chronic forms of arthritis seems to be caused less by enhanced articulo-muscular protective reflexes and more by fibrosis of the ligamento-capsular tissue. The fibrosis causes limited extensibility of the collagen fibers within the ligamento-capsular tissue.[12,47]

An example of arthritis in which effusion is absent and the movement restrictions are secondary to stiffness of the ligamento-capsular tissue is immobilization. Immobilization causes reduced extensibility of the joint's soft tissues.[47] Another example with similar features is arthrosis.

Acute arthritis is often accompanied by muscle tone changes that restrict range. Such changes may be beneficial in acute arthritis because it prevents movement at the joint.[10] In acute arthritis, the muscle guarding may initially help reduce nociception from the joint.[99] However, in chronic arthritis, or other conditions with intraarticular adhesions, the reflex muscle contraction might prevent recovery.[10,99] With an intraarticular adhesion, a forceful movement into the range might break the adhesion and provide permanent improvement of the condition.

Acute arthritis with inflammation and chronic arthrosis with a stiff ligamento-capsular tissue are two extreme conditions that cause movement restrictions in a capsular pattern. Most patients with a capsular pattern of movement restriction display a combination of both conditions. They have some restricted movement secondary to acute inflammation and some restricted movement secondary to capsular stiffness.

Summary

Normal joints have negative intraarticular pressure. Joints with acute arthritis are inflamed, possibly with effusion, and therefore have positive intraarticular pressure. Effused joints have the least amount of intraarticular pressure in loose-packed positions. Movement toward an end-range position increases intraarticular pressure and increases articular receptor activity. In response to acute arthritis, the articular reflexes become enhanced articulo-muscular protective reflexes; agonists are strongly inhibited, and antagonists are strongly facilitated. The inhibition of agonists displays itself as muscle weakness and possibly muscle atrophy, whereas the facilitation of antagonist displays itself as movement restriction in the capsular pattern. Joints without active range do not develop a capsular pattern of movement restrictions; they only exhibit pain when the capsular tissues are stretched.

As part of a joint inflammation, fibrosis of the ligamento-capsular tissue develops. In case of arthrosis or chronic arthritis, the movement restrictions in a capsular pattern are caused by the fibrosis.

Through substance P and other neurotransmitters, articular nociception can produce or contribute to inflammation of a joint. The release of neurotransmitters could cause destruction of the joint. Thus, this mechanism in the nervous system is likely to contribute to the severity of joint afflictions in arthritic patients. Therefore, reducing articular nociception may be more than symptomatic treatment for patients with arthritis.

References

1. Andrew BL, Dodt E: The deployment of sensory nerve endings at the knee joint of the cat. Acta Physiol Scand 28:287-296, 1953

2. Baxendale RH, Ferrell WR: Disturbance of proprioception at the human knee resulting from acute joint distension. J Physiol 392:60P, 1987

3. Bennett JC: Rheumatoid arthritis: Clinical features. In Schumacher HR (ed): Primer on Rheumatic Diseases, ed 9. Atlanta, Arthritis Foundation, 1988, pp 87-92

4. Blockey NJ: An observation concerning the flexor muscles during recovery of function after dislocation of the elbow. J Bone Joint Surg Am 36:833-840, 1954

5. Brügger A: Der Nozizeptive-somatomotorische Blockierungseffekt als wichtiges Element der pseudoradiculären Syndrome. Presented at Stichting Manuele Geneeskunde, Zeist, The Netherlands, 1983

6. Brunschwig A, Jung A: The importance of the periarticular innervation in the pathological physiology of sprained joints. J Bone Joint Surg Am 14:273-276, 1932

7. Cavanaugh JM: Neural mechanisms of lumbar pain. Spine 20:1804-1809, 1995

8. Charcot JM: Clinical Lectures on Diseases of the Nervous System Delivered at the Infirmary of la Salpetriere (p 52-60). London, The New Sydenham Society, 1889, vol 3

9. Coggeshall RE, Hong KAP, Langford LA, et al: Discharge characteristics of fine medial articular afferents at rest and during passive movements of inflamed knee joints. Brain Res 272:185-188, 1983

10. Cyriax J: Textbook of Orthopaedic Medicine: Diagnosis of Soft Tissue Lesions, ed 8. London, Baillière Tindall, 1982, vol 2

11. deAndrade JR, Grant C, Dixon AStJ, et al: Joint distension and reflex inhibition in the knee. J Bone Joint Surg Am 47:313-322, 1965

12. Donatelli R, Owens-Burkhart H: Effects of immobilization on the extensibility of periarticular connective tissue. J Orthop Sports Phys Ther 3:67-72, 1981

13. Dorland's Illustrated Medical Dictionary, ed 25. Philadelphia, W.B. Saunders Co, 1974

14. Ekholm J, Eklund G, Skoglund S: On the reflex effects from the knee joint of the cat. Acta Physiol Scand 50:167-174, 1960

15. Eyring EJ, Murray WR: The effect of joint position on the pressure of intra-articular effusion. J Bone Joint Surg Am 46:1235-1241, 1964

16. Fahrer H, Rentsch HU, Gerber NJ, et al: Knee effusion and reflex inhibition of the quadriceps: A bar to effective retraining. J Bone Joint Surg Br 70:635-638, 1988

17. Ferrell WR: The effect of acute joint distension on mechanoreceptor discharge in the knee of the cat. Quarterly Journal of Experimental Physiology 72:493-499, 1987

18. Ferrell WR: Discharge characteristics of joint receptors in relation to their proprioceptive role. In Hnik P, Soukop T, Vejsada R, et al (eds): Mechanoreceptors, Development, Structure and Function. New York, Plenum Press, 1988, pp 383-388

19. Ferrell WR, Crighton A, Sturrock RD: Position sense at the proximal interphalangeal joint is distorted in patients with rheumatoid arthritis of finger joints. Exp Physiol 77:675-680, 1992

20. Ferrell WR, Nade S, Newbold PJ: The interrelation of neural discharge, intra-articular pressure, and joint angle in the knee of the dog. J Physiol 373:353-365, 1986

21. Ferrell WR, Russell NJW: Extravasation in the knee induced by antidromic stimulation of articular C fibre afferents of the anaesthetized cat. J Physiol 379:407-416, 1986

22. Ferrell WR, Wood L: The effect of increased intra-articular volume on the discharge of stretch receptors in the cat knee joint. J Physiol (Lond) 329:59P-60P, 1982

23. Ferrell WR, Wood L, Baxendale RH: The effect of acute joint inflammation on flexion reflex excitability in the decerebrate, low-spinal cat. Quarterly Journal of Experimental Physiology 73:95-102, 1988

24. Glick EN: Asymmetrical rheumatoid arthritis after poliomyelitis. BMJ 3:26-28, 1967

25. Glynn JJ, Clayton ML: Sparing effect of hemiplegia on tophaceous gout. Ann Rheum Dis 35:534-535, 1976

26. Godwin-Austen RB: The mechanoreceptors of the costo-vertebral joints. J Physiol 202:737-753, 1969

27. Grigg P, Hoffman AH, Fogarty KE: Properties of Golgi-Mazzoni afferents in cat knee joint capsule, as revealed by mechanical studies of isolated joint capsule. J Neurophysiol 47:31-40, 1982

28. Guilbaud G, Iggo A, Tegnér R: Sensory changes in joint-capsule receptors of arthritic rats: Effect of aspirin. In Fields HL, et al (eds): Advances in Pain Research and Therapy. New York, Raven Press, 1985, vol 9, pp 81-89

29. Harding AEB: Arthritic muscular atrophy. Journal Pathology Bact XXVIII:179-187, 1925

30. Harding AEB: An investigation into the cause of arthritic muscular atrophy. The Lancet 28:433-434, 1929

31. He X, Proske U, Schaible HG, et al: Acute inflammation of the knee joint in the cat alters responses of flexor motoneurons to leg movements. J Neurophysiol 59:326-340, 1988

32. Helme RD, Andrews PV: The effect of nerve lesions on the inflammatory response to injury. J Neurosci Res 13:53-459, 1985

33. Henry JL: Relation of substance P to pain transmission: Neurophysiological evidence. In Porter R, O'Conner M (eds): Substance P in the Nervous System (Ciba Foundation Symposium 91). London, Pitman Co, 1982, pp 206-224

34. Heppelmann B, Schaible HG, Schmidt RF: Effects of prostaglandins E1 and E2 on the mechanosensitivity of group III afferents from normal and inflamed cat knee joints. In Fields HL, et al (eds): Advances in Pain Research and Therapy. New York, Raven Press, 1985, vol 9, pp 91-101

35 Hettinga DL: Inflammatory response of synovial joint structure. In Gould JA, Davies GJ (eds): Orthopaedics and Physical Therapy. Toronto, C.V. Mosby Co, 1985, vol 2, pp 87-117

36 Hilton J: On the Influence of Mechanical and Physiological Rest in the Treatment of Accidents and Surgical Diseases, and the Diagnostic Value of Pain: A Course of Lectures, Delivered at the Royal College of Surgeons of England in the Years 1860, 1861, and 1862. London, Bell and Dalby, 1863, pp 166-167

37 Hilton J: Rest and Pain: A Course of Lectures on the Influence of Mechanical and Physiological Rest in the Treatment of Accidents and Surgical Diseases and the Value of Pain. London, George Bell & Sons, 1896, pp 166-189

38 Hogeweg J: Pain Threshold and Tissue Compliance in Juvenile Chronic Arthritis. Thesis. Utrecht, The Netherlands, Universiteit Utrecht, 1995

39 Iggo A, Guilbaud G, Tégner R: Sensory mechanisms in arthritic rat joints. In Kruger L, Liebeskind JC (eds): Advances in Pain Research and Therapy. New York, Raven Press, 1984, vol 6, pp 83-93

40 Iles JF, Stokes M, Young A: Reflex actions of knee-joint receptors on quadriceps in man. J Physiol 360:48P, 1984

41 Iles JF, Stokes M, Young A: Reflex actions of knee-joint afferents during contraction of the human quadriceps. Clin Physiol 10:489-500, 1990

42 Jayson MIV, Dixon AStJ: Intra-articular pressure in rheumatoid arthritis of the knee: I. Pressure changes during passive joint distension. Ann Rheum Dis 29:261-265, 1970

43 Jayson MIV, Dixon AStJ: Intra-articular pressure in rheumatoid arthritis of the knee: III. Pressure changes during joint use. Ann Rheum Dis 29:401-408, 1970

44 Jones DW, Jones DA, Newham DJ: Chronic knee effusion and aspiration: The effect on quadriceps inhibition. Br J Rheumatol 26:370-374, 1987

45 Kanaka R, Schaible HG, Schmidt RF: Activation of fine articular afferent units by bradykinin. Brain Res 327:81-90, 1985

46 Kennedy JC, Alexander IJ, Hayes KC: Nerve supply of the human knee and its functional importance. Am J Sports Med 10:329-335, 1982

47 Kessler RM, Hertling D: Management of Common Musculoskeletal Disorders: Physical Therapy Principles and Methods. New York, Harper & Row, 1983

48 Konttinen YT, Grönblad M, Hukkanen M, et al: Pain fibers in osteoarthritis: A review. Semin Arthritis Rheum 18(suppl):35-40, 1989

49 Korst JK van der: Gewrichtsziekten. Utrecht, The Netherlands, Bohn, Scheltema & Holkema, 1980

50 Lam FY, Ferrell WR: CGRP modulates nerve-mediated vasoconstriction of rat knee joint blood vessels. Ann N Y Acad Sci 30:519-521, 1992

51 Levick JR: Joint pressure-volume studies: Their importance, design and interpretation. J Rheumatol 10:353-357, 1983

52 Levine JD, Coderre TJ, Basbaum AI: The peripheral nervous system and the inflammatory process. In Dubner R, Gebhart GF, Bond MR (eds): Proceedings of the Vth World Congress on Pain. New York, Elsevier, 1988, pp 33-42

53 Levine JD, Collier DH, Basbaum AI, et al: Hypothesis: The nervous system may contribute to the pathophysiology of rheumatoid arthritis. J Rheumatol 12:406-411, 1985

54 Levine JD, Moskowitz MA, Basbaum AI: The contribution of neurogenic inflammation in experimental arthritis. J Immunol 135:843s-847s, 1985

55 Lieb FJ, Perry J: Quadriceps function: An anatomical and mechanical study using amputated limbs. J Bone Joint Surg Am 50:1535-1548, 1968

56 Matucci-Cerinic M: Sensory neuropeptides and rheumatic diseases. Rheum Dis Clin North Am 19:975-988, 1993

57 Matucci-Cerinic M, Partsch G: The contribution of the peripheral nervous system and the neuropeptide network to the development of synovial inflammation. Clin Exp Rheumatol 10:211-215, 1992

58 McIlwain JS, Hayes KC: Dynamic properties of human motor units in the Hoffmann-reflex and M response. Am J Phys Med 56:122-135, 1977

59 McLain RF: Mechanoreceptor endings in the human cervical facet joints. Spine 19:495-501, 1994

60 Meyer P: Strategy: Application of knee research to facet joint reflex inhibition. Orthopaedic Practice 2:28-29, 1990

61 Mink AJF, Veer HJ ter, Vorselaar JAC Th: Extremiteiten: Functie-onderzoek en Manuele Therapie, ed 6. Houten, The Netherlands, Bohn Stafleu Van Loghum, 1993

62 Morree JJ de: Heeft bindweefsel wel een reflextonus nodig? Nederlands Tijdschrift voor Fysiotherapie 101:126-130, 1991

63 Morrissey MC: Reflex inhibition of thigh muscles in knee injury: Causes and treatment. Sports Med 7:263-276, 1989

64 Nade S, Newbold PJ: Factors determining the level and changes in intra-articular pressure in the knee joint of the dog. J Physiol 338:21-36, 1983

65 Neugebauer V, Schaible HG: Peripheral and spinal components of the sensitization of spinal neurons during an acute experimental arthritis. Agents Actions 25:234-236, 1988

66 Neugebauer V, Schaible HG, Schmidt RF: Sensitization of articular afferents to mechanical stimuli by bradykinin. Pflügers Arch ges Physiol 415:330-335, 1989

67 Nichols PJR (ed): Rehabilitation Medicine: The Management of Physical Disabilities, ed 2. Boston, Butterworths & Co. 1981, p 122

68 Otsuka M, Yanagisawa M: Does substance P act as a pain transmitter? Trends Pharmacol Sci 8:506-510, 1987

69 Partridge EJ: Joints: The limitation of their range of movement, and an explanation of certain surgical conditions. J Anat 58:346-354, 1924

70 Payr E: Der heutige Stand der Gelenkchirurgie. Arch f Klin Chirurgie 148:404-451, 1900

71 Raymond F: Recherches expérimentale sur la pathogénie des atrohies musculaires consecutives aux arthritis traumatiques. Revue de Médicine 10:374-392, 1890

72 Robbins SL, Cotran RS: Pathologic Basis of Disease, ed 2. Philadelphia, W.B. Saunders Co, 1979, pp 55-106

73 Schaible HG, Neugebauer V, Cervedo F, et al: Changes in tonic descending inhibition of spinal neurons with articular input during the development of acute arthritis in the cat. J Neurophysiol 66:1021-1032, 1991

74 Schaible HG, Neugebauer V, Schmidt RF: Osteoarthritis and pain. Semin Arthritis Rheum 18(suppl 2):30-34, 1989

75 Schaible HG, Schmidt RF: Effects of an experimental arthritis on the sensory properties of fine articular afferent units. J Neurophysiol 54:1109-1122, 1985

76 Schaible HG, Schmidt RF: Discharge characteristics of receptors with fine afferents from normal and inflamed joints: Influence of analgesics and prostaglandins. Agents Actions (suppl 19):99-117, 1986

77 Schaible HG, Schmidt RF: Time course of mechanosensitivity changes in articular afferents during a developing experimental arthritis. J Neurophysiol 60:2180-2195, 1988

78 Schreuder TOR, Brandsma JW: Het onderzoek van de spierkracht van de hand. FysioPraxis 5(8):14-19, 1996

79 Scott DT, Ferrell WR, Baxendale RH: Excitation of soleus/gastrocnemius γ-motoneurones by group II knee joint afferents is suppressed by group IV joint afferents in the decerebrate, spinalized cat. Exp Physiol 79:357-364, 1994

80 Shakespeare D, Stokes M, Sherman KP, et al: The effect of knee flexion on quadriceps inhibition after meniscectomy. Clin Sci 65:64P-65P, 19??

81 Spencer JD, Hayes KC, Alexander IJ: Knee joint effusion and quadriceps reflex inhibition in man. Arch Phys Med Rehabil 65:171-177, 1984

82 Stokes M, Young A: The contribution of reflex inhibition to arthrogenous muscle weakness. Clin Sci 67:7-14, 1984

83 Stratford P: Electromyography of the quadriceps femoris muscles in subjects with normal and acutely effused knees. Phys Ther 62:279-283, 1981

84 Swearingen RL, Dehne E: A study of pathological muscle function following injury to a joint. J Bone Joint Surg Am 46:1364, 1964

85 Thompson M, Bywaters EGL: Unilateral rheumatoid arthritis following hemiplegia. Ann Rheum Dis 21:370-377, 1962

86 Will TE: The biochemical basis of manipulative therapeutics: Hypothetical considerations. J Manipulative Physiol Ther 1:153-156, 1978

87 Winkel D, Aufdemkampe G, Meijer OG, et al: Orthopedische Geneeskunde en Manuele Therapy: Diagnostiek Extremiteiten. Houten, The Netherlands, Bohn Stafleu Van Loghum bv, 1993, vol 2b

88 Winkel D, Aufdemkampe G, Meijer OG, et al: Orthopedische Geneeskunde en Manuele Therapy: Diagnostiek Extremiteiten. Houten, The Netherlands, Bohn Stafleu Van Loghum bv, 1992, vol 2c

89 Winkel D, Aufdemkampe G, Meijer OG, et al: Orthopedische Geneeskunde en Manuele Therapy: Wervelkolom - Diagnostiek en Therapy. Houten, The Netherlands, Bohn Stafleu Van Loghum bv, 1991, vol 4b

90 Winkel D, Aufdemkampe G, Meijer OG, et al: Orthopedische Geneeskunde en Manuele Therapy: Wervelkolom - Diagnostiek en Therapy. Houten, The Netherlands, Bohn Stafleu Van Loghum bv, 1991, vol 4c

91. Wong HY: Neural mechanisms of joint pain. Ann Acad Med 22:646-650, 1993

92. Wood L, Ferrell WR: Response of slowly adapting articular mechanoreceptors in the cat knee joint to alterations in intra-articular volume. Ann Rheum Dis 43:327-332, 1984

93. Wood L, Ferrell WR: Fluid compartmentation and articular mechanoreceptor discharge in the cat knee joint. Quarterly Journal of Experimental Physiology 70:329-335, 1985

94. Wood L, Ferrell WR, Baxendale RH: Pressures in normal and acutely distended human knee joints and effects on quadriceps maximal voluntary contractions. Quarterly Journal of Experimental Physiology 73:305-314, 1988

95. Wyke B: The neurology of joints. Ann R Coll Surg Engl 41:25-50, 1967

96. Young A: Rehabilitation for wasted muscles. In Sarner M (ed): Advanced Medicine. London, Pitman Medical, 1982, vol 18, pp 138-142

97. Young A, Hughes I, Round JM, et al: The effect of knee injury on the number of muscle fibers in the human quadriceps femoris. Clin Sci 62:227-234, 1982

98. Young A, Stokes M, Shakespeare DT, et al: The effect of intra-articular bupivacaine on quadriceps inhibition after meniscectomy. Med Sci Sports Exerc 15:154P, 1983

99. Zimmermann M: Pain mechanisms and mediators in osteoarthritis. Semin Arthritis Rheum 18(suppl 2):22-29, 1989

11. Neurogenic Arthropathy

The previous chapter discussed how articular nociception is capable of causing joint inflammation and possibly joint destruction through the release of neurotransmitters. However, joint degeneration as a result of altered joint afference may not be caused by Type IV receptor activity alone.

Loss of articular sensation could also lead to joint degeneration. The chronic progressive degeneration of a joint caused by sensory abnormalities goes by various names: neurogenic arthropathy, neuropathic arthropathy, neuropathic arthritis, or Charcot's joint.[4] A variety of neurological diseases, such as diabetes mellitus, leprosy, and tabes dorsalis, can result in neurogenic arthropathy.[11] These neurological disorders have in common a loss of sensation. For instance, tabes dorsalis, the late stage of syphilis, is accompanied by, among other conditions, degeneration of the sensory nerve trunks. Besides the above neurological diseases, other reasons for denervation of a joint are a lesion of the articular afferent nerves or damage to the articular receptors, for example, as in a ligamento-capsular lesion.[6,12,14] The reason sensory loss leads to progressive joint degeneration may be twofold.

In studying rabbit knee joints that had been sensory denervated by cutting the posterior nerve roots, Finsterbush and Friedman observed a progressive cell atrophy in all the knee structures.[5] The degenerative changes were present in the denervated joints even if protected by a plaster cast. Finsterbush and Friedman's microscope research suggested that the degenerative changes were due to a nutritional deficiency and that joint trauma did not play a primary role.[5] Thus, sensory denervation of a joint may impair nutrition (see chapter 12).

O'Connor and colleagues studied dog knee joints after posterior nerve root ganglionectomies of L4 through S1.[11] These researchers found no degenerative changes in the deafferentiated knees for 16 months. Apparently, deafferentation alone was not the cause of rapid damage to the joint; another factor needed to be present. If the anterior cruciate ligament was cut in these dogs, only three weeks later, striking lesions were found in the articular cartilage. O'Connor and colleagues concluded that, in the presence of sensory denervation, the protective reflexes are inadequate to protect unstable joints from becoming rapidly and severely damaged.[11] Other researchers affirmed these conclusions.[1,3,7,8,9,10,11,13]

Articular mechanoreceptor afference warns against joint and ligament deformations that could result in joint damage.[1,2] Loss of articular afference probably results in absence or diminishing of the articulo-muscular protective reflex.[3,7,8,9,11,13] Absence of the protective reflexes permits excessive joint range of motion and allows trauma to the articular and periarticular tissues.[11] Severe forms of joint degeneration can result.[11] Even altered afference from stretched or damaged ligaments might lead to unpredictable "giving way," functional instability, and progressive joint damage.[6,9]

In summary, the rapid degeneration of joints following denervation may result from lack of nutrition and/or lack of articulo-muscular protective reflexes. The protective reflexes may not be important in ordinary activities, but they seem critical in protecting damaged or unstable joints from early degeneration.[10]

References

1. Barrack RL, Skinner HB, Buckley SL: Proprioception in the anterior cruciate deficient knee. Am J Sports Med 17:1-6, 1989

2. Biedert RM, Stauffer E, Friederich NF: Occurrence of free nerve endings in the soft tissue of the knee joint. Am J Sports Med 20:430-433, 1992

3. DeAndrade JR, Grant C, Dixon AStJ: Joint Distension and reflex muscle inhibition in the knee. J Bone Joint Surg Am 47:313-322, 1965

4. Dorland's Illustrated Medical Dictionary, ed 25. Philadelphia, W.B. Saunders Co, 1974

5. Finsterbush A, Friedman B: The effect of sensory denervation on rabbits' knee joints. J Bone Joint Surg Am 57:949-956, 1975

6. Freeman MAR, Dean MRE, Hanham IWF: The etiology and prevention of functional instability of the foot. J Bone Joint Surg Br 47:678-685, 1965

7. Freeman MAR, Wyke B: Articular contributions to limb muscle reflexes: The effects of partial neurectomy of the knee-joint on postural reflexes. Br J Surg 53:61-69, 1966

8. Freeman MAR, Wyke B: Articular reflexes at the ankle joint: An electromyographic study of normal and abnormal influences of ankle-joint mechanoreceptors upon reflex activity in the leg muscles. Br J Surg 54:990-1001, 1967

9. Kennedy JC, Alexander IJ, Hayes KC: Nerve supply of the human knee and its functional importance. Am J Sports Med 10:329-335, 1982

10. McLain RF: Mechanoreceptor endings in human cervical facet joints. Spine 19: 595-501, 1994

11. O'Connor BL, Palmoski MJ, Brandt KD: Neurogenic acceleration of degenerative joint lesions. J Bone Joint Surg Am 67:562-572, 1985

12. Schutte MJ, Dabezies EJ, Zimny ML, et al: Neural anatomy of the human anterior cruciate ligament. J Bone Joint Surg Am 69:243-247, 1987

13. Wyke B: The neurology of joints. Ann R Coll Surg Engl 41:25-50, 1967

14. Zimny ML, Schutte M, Dabezies E: Mechanoreceptors in the human anterior cruciate ligament. Anat Rec 214:204-209, 1986

12. Radiculopathy

A well-functioning nervous system is required for generation of adequate muscle contraction. Therefore, impaired conduction of impulses by the nervous system makes adequate contraction impossible. Not only is conduction of motor impulses essential, but also conduction of sensory impulses. Sensory impulses are important, first, because the sensory input can initiate muscle tone changes (e.g., as in reflexes), and second, the sensory input provides feedback to a contraction. The proper execution of a skilled movement, such as an isometric contraction against varying resistance, is hardly possible if correct feedback is unavailable. Conduction over dysfunctional sensory and/or motor nerves may yield weakness on muscle tests. This type of weakness will likely be present throughout the range of the muscle, contrary to arthrogenic inhibition, where weakness is likely to be found only in an end-range position of the joint (the position in which the muscle is shortened).

Compression of a nerve may lead to impaired conduction. A possible location of nerve tissue compression is within the spinal foramen. The nerve root within the foramen is especially sensitive to compression, since it lacks an epineurium (i.e., the connective tissue that covers a peripheral nerve).[3] The lack of an epineurium makes the nerve root more susceptible to compression than a peripheral nerve.[5,12] In addition, the blood vessels inside the nerve root and ganglion are more permeable to plasma protein than blood vessels within the peripheral nerves.[8] Due to this increased permeability of the capillaries, compression can easily create edema within the nerve root or ganglion.

The foramen, through which the spinal nerve runs, is like a tunnel. The anterior part of the foramen is formed in the lumbar spine by the disc and in the cervical spine by the disc and the uncinate process (see Fig. 12.1). Superiorly and inferiorly, the tunnel is formed by the pedicles. The posterior boundary is formed by the ligamentum flavum covering the facet joints. A reduced diameter of the foramen may result from any combination of the following: posterior disc expansion with or without osteophytes, osteophytes projecting anteriorly from the facet joint, and effusion of the facet joint.

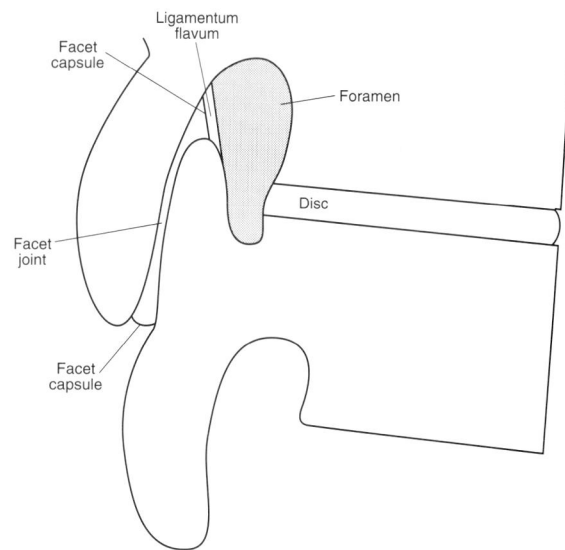

Fig. 12.1 Foramen.

Adapted from Paris SV: Anatomy as related to function and pain. Orthop Clin North Am 14: 475-489, 1983, with permission from W.B. Saunders Co.

Natural movements also alter the diameter of the foramen. Its diameter can be increased by flexion of the superior vertebra, whereas extension reduces its diameter (see Fig. 12.2). Pathological conditions are more likely to compress the nerve root if they occur with extension of the superior vertebra. In addition, the presence of inflammatory agents, released by a herniated or degenerated disc, may contribute to pressure on the nerve root.[10]

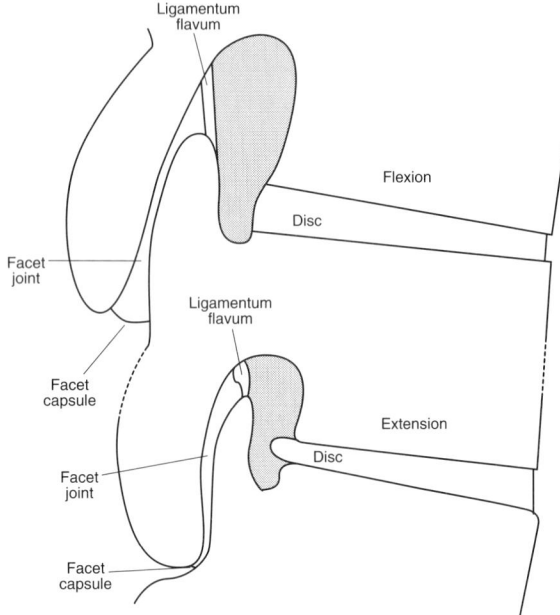

Fig. 12.2 Foramen size after flexion and after extension of its superior vertebra.

Adapted from Paris SV: Anatomy as related to function and pain. Orthop Clin North Am 14: 475-489, 1983, with permission from W.B. Saunders Co.

For any structure in the human body to receive circulation, pressure within the arterioles must be higher than that within the capillaries and pressure within the capillaries must be higher than in the venules. The drop in pressure gradient allows the blood to flow from arterioles (with a higher pressure) to venules (with a lower pressure gradient). If the pressure in the foramen is higher than the pressure in the nerve's venules, or worse, higher than the pressure in the capillaries, circulation in the nerve root is impeded.[12] Lymph circulation is also susceptible to pressure. Although structurally intact, the nerve could stop functioning for lack of blood and possibly lack of lymph circulation. This ischemic condition of the nerve has been termed *neuropraxia*.[6,7]

In addition, plasma flow within the nerve can be impeded.[1,2,9] Various elements, such as neurotransmitters (for instance, substance P), neuropeptides, and so on, are transported within the axon.[1] The transport is from the cell body to the periphery and back.[1] Peripheral nerves not only conduct electrical impulses to or from their target organs (the tissues they innervate), but they also exchange proteins and macromolecules with their target tissue.[4] These trophic or neurotrophic influences can be essential for the development, growth, and survival of the target organ.[4] Any factor that impedes the transport mechanism within the axon, such as compression of the nerve, could adversely affect the nutrition of the nerve itself and its target organ.[4] The nerve, although still structurally intact, could stop functioning for lack of internal plasma circulation. This condition is also called neuropraxia.[6,7] The blood circulation and plasma flow might be restored in a neuropraxic nerve if the pressure is removed from it.

Besides impeding blood, lymph, and axon flow, compression of a nerve root may deform the nerve itself. The myelin sheath around the nerve fiber is sensitive to compression and might deform, or demyelination of the nerve might occur.[1]

Not all nerve fibers are equally sensitive to compression. The spinal nerve root consists of nerve fibers of different sizes. The smaller-diameter fibers are more resistant to compression; thus, compression mainly affects large-diameter myelinated fibers.[8,11]

A radiculopathy creates weakness of all the muscles innervated by the nerve root. The amount of weakness depends on the percentage of the muscle innervated by the nerve root. Muscle strength may vary with different positions of the foramen. Opening of the foramen (flexion of the superior vertebra) may reduce pressure on the nerve and restore muscle strength. Extension of the superior vertebra may decrease muscle strength.

Summary

The nerve root is more susceptible to compression than the peripheral nerve. Increasing pressure within the foramen reduces blood and lymph flow of the nerve root before it compresses the nerve fibers. The reduced blood flow within the foramen can cause ischemia of the nerve root. Besides causing nerve root ischemia, compression of the nerve root could impede the nerve's axon flow or deform the nerve itself. Large myelinated nerve fibers are more susceptible to compression than small unmyelinated nerve fibers.

Closing the foramen may increase nerve root pressure, whereas opening the foramen may decrease nerve root pressure. Closing the foramen may reduce nerve root circulation and decrease the strength of the muscles innervated by the nerve root, whereas opening the foramen may restore nerve root circulation and restore muscle strength.

References

1. Butler DS: Mobilisation of the Nervous System. New York, Churchill Livingstone, 1991
2. Dahlin LB, McLean WG: Effects of graded experimental compression on slow and fast axonal transport in rabbit vagus nerve. J Neurol Sci 72:19-30, 1986
3. Dorland's Illustrated Medical Dictionary, ed 25. Philadelphia, W.B. Saunders Co, 1974
4. Korr IM: Neurochemical and neurotrophic consequences of nerve deformation. In Glasgow EF, Twomey LT, Scull ER, et al: Aspects of Manipulative Therapy, ed 2. New York, Churchill Livingstone, 1985, pp 64-71
5. Matsui H, Kitagawa H, Kawaguchi Y, et al: Physiologic changes of nerve root during posterior lumbar discectomy. Spine 20:654-659, 1995
6. Morree JJ de: Dymaniek van het Menselijke Bindweefsel: Functie, Beschadiging, en Herstel, ed 2. Houten, Netherlands, Bohn Stafleu Van Loghum, 1993
7. Mumenthaler M: Neurology, ed 5. Stuttgart, George Thieme Verlag, 1976
8. Rydevik B, Brown MD, Lundborg G: Pathoanatomy and pathophysiology of nerve root compression. Spine 9:7-15, 1984
9. Rydevik B, McLean WG, Sjostrand J, et al: Blockage of axonal transport induced by acute graded compression of the rabbit vagus nerve. J Neurol Neurosurg Psychiatry 43:690-698, 1980
10. Saal JS: The role of inflammation in lumbar pain. Spine 20:1821-1827, 1995
11. Seddon H: Surgical Disorders of the Peripheral Nerves. Baltimore, Williams & Wilkins Co, 1972, pp 32-56
12. Sunderland S: Traumatized nerves, roots and ganglia: Musculoskeletal factors and neuropathic consequences. In Korr IM (ed): The Neurobiologic Mechanisms in Manipulative Therapy. New York, Plenum Press, 1977, pp 137-153

13. Pseudoradiculopathy

This chapter discusses the phenomenon of pseudoradiculopathy. As the name implies, this condition clinically resembles a radiculopathy; however, its pathomechanics are quite different. To understand pseudoradiculopathy, it is useful to understand the concept of referred pain.

Referred Pain

Referred pain is the phenomenon whereby a nociceptive stimulus is perceived, not at the origin, but elsewhere in the body. This means that pain is perceived distant from the area of stimulation. An example is a patient having a heart attack who might perceive some of the pain in the left arm, although this is not the location of the heart.[1,7,10] A second example is pain originating from the sacroiliac joint that is felt in the heel.[9]

A few phenomena have been observed regarding the perception of referred pain.[8,9,24] These phenomena are as follows:

1. Referred pain never crosses the midline. Pain originating from one side of the body will be felt in the same side of the body.

2. Referred pain tends to be felt distal to its origin. Sacroiliac joint irritation may be felt distally as far down as the heel. An irritation of the toes, however, will not refer pain proximally into the leg.

3. Pain felt in a certain dermatome must arise from a structure with the same segmental innervation.

4. The size of the area of pain reference depends on the intensity of the nociceptive stimulus. The stronger the stimulus, the larger the area of pain reference. For example, pain originating from a mild sacroiliac joint lesion might produce referred pain in the buttock.[14,15] However, more intense pain may be felt in the leg or heel as well.[9] Thus, as the intensity of the nociceptive stimulus increases, referred pain might be experienced more distally in the extremity. As the stimulus intensity decreases, the referred pain area becomes smaller and the patient is better able to localize the pain correctly. This principle could be compared with McKenzie's peripheralization and centralization of pain.[33,34] Pain becoming shorter, more localized, and more proximal in the extremity would be termed *centralization* of pain, whereas *peripheralization* is the experience of pain more distally in the extremity.

5. Irritation of structures located deep in the body refer pain to a larger area than superficial irritations. Woolf and Wall found that nociception from deep structures, such as joints, produced more widespread sensory disturbances than nociception from more superficial tissue, such as skin or muscle.[49]

The reason an irritation might be experienced as pain distant from its source is due to how the parts of the body are innervated. The human spine consists of a row of segments, each of which contains two spinal nerves that innervate a part of the skin (the dermatome), a part of the muscles (the myotome), and a part of the skeleton (the sclerotome) (see Fig. 13.1).[30] In addition, the spinal nerves might innervate a part of the blood vessels (the vasotome) and a part of the visceral organs (viscerotome). Thus, all these tissues, possibly with widespread locations in the body, are innervated by the same spinal nerve pair.

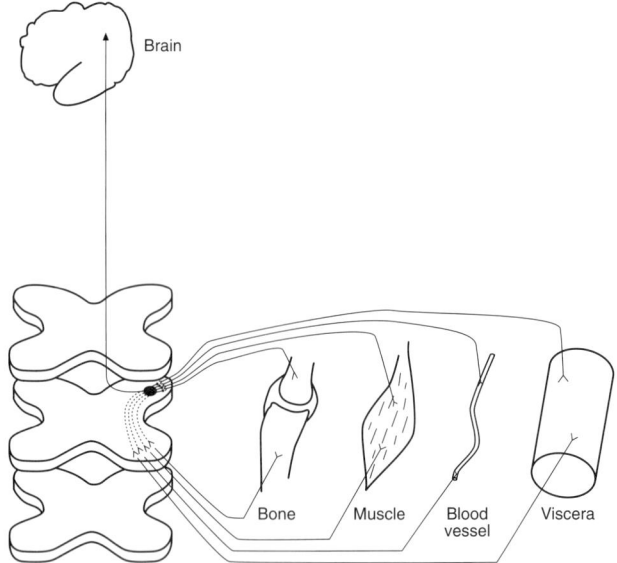

Fig. 13.1 The segmental innervation of the body and the convergence of afferents.

The afferent nerves from all tissues related to the same segment converge to the spinal cord segment. The most accepted explanation for referred pain is the convergence of the afferents from all these tissues onto the same dorsal horn neurons (see Fig. 13.1).[6,7,10,19,26,30,48] These dorsal horn neurons project via the spinothalamic tract to the thalamus, and from there to the cerebral cortex. The brain, in this case, cannot distinguish from which tissue the information derived. The pain could be experienced in the dermatome, but also in the myotome or sclerotome.[7]

Facilitated Segment

Increased activity of the nociceptive afferents causes elevated activity of the related spinal cord segment(s).[7] A barrage of nociceptive input to the spinal cord produces increased activity of the dorsal horn neurons, as well as changes in the neurons and interneurons located in the lateral and anterior horn.[3,21,43] This nociceptive barrage can sensitize these neurons and interneurons so that the segment becomes "facilitated."[3] In the lateral horn, the nociceptive input stimulates the sympathetic preganglionic neurons.[3,5,7,22,25,31,43] In the anterior horn, nociceptive input might alter the function of the motoneurons with a low threshold.[5,21,25,45] This altered motoneuron function results in altered muscle function.[3,7,31] Two examples of muscle tone changes in response to irritation of another tissue with related innervation are (1) abdominal muscle spasm in response to an appendicitis and (2) cervical extensor muscle spasm with limited cervical flexion in response to meningitis.[6,9]

The facilitated segment can express itself in every body part it innervates.[7] In addition, the muscle tone changes in the myotome, the stimulation of the sympathetic preganglionic neurons, and the pain experienced in the dermatome, myotome, or sclerotome might occur simultaneously.

Paris described the concept of the facilitated segment as a spinal cord segment receiving nociceptive input.[39] Because this spinal cord segment is at an elevated level of irritation, the segmental facilitation will cause changes in the dermatome, myotome, and sclerotome.[39] "Consequently further input, particularly noxious input, may trigger a painful response of muscle twitch either locally or distally but definitely segmentally related."[39, p. 153] The cause of the changes in the dermatome, myotome, and sclerotome is not in these tissues themselves, but may be in a spinal facet joint or a costovertebral joint.[39]

Nociceptive input is not the only phenomenon that can facilitate a spinal cord segment. Large-diameter afferent fibers from the joints (Type I, II, and III afferents) also can provoke polysynaptic reflex discharges in segmentally related efferent nerves.[16,27]

MYOTOME CHANGES

Intense afferent input can facilitate a spinal cord segment, and this facilitation may cause referred pain and muscle function changes. Since the afferent input can alter neuron function in both the anterior and posterior horns, it seems logical that some phenomena which apply to referred pain (mentioned earlier in this chapter) also apply to muscular changes. The following are phenomena that apply to muscular changes:

1. The muscular changes can be expected on the same side of the body as the stimulus causing the facilitation of the segment.[40]

2. The muscular changes share the same segmental innervation as the stimulus (i.e., the muscle changes are within the myotomes of the facilitated segment).[2,12,40]

3. The stronger the stimulus, the more pronounced the muscle (myotome) responds with altered function.

4. Deeper lesions express themselves more prominently in muscular reactions.[46,49]

Although nociception of a variety of tissues can facilitate a segment, deeper tissues (e.g., joints) can create more pronounced spinal cord changes.[49] Woolf and Wall found that joints create more prominent spinal cord changes than muscle or skin tissue.[49] This may explain the more widespread muscle tone disturbances that accompany joint injury.[49]

TENDERNESS

Woolf and Wall wrote that tenderness in response to an injury may be partly due to spinal cord changes produced by the afferent nociceptive barrage.[49] This means that the nociceptive input entering the spine may cause spinal cord changes such that tenderness may be felt from tissues other than the one originally producing the nociception.[6,49] Normally, much of the information entering the spinal cord remains suppressed; however, nociceptive input can reduce the suppression.[26] The central nervous system thus becomes more sensitive to other afferent input.[26] The enhanced spinal cord sensitivity may explain the hyperalgesia of periarticular skin and muscles that often accompanies diseases such as arthritis.

Some clinicians have found specific tender points in relation to facet joint dysfunctions.[11,18,23,28,42] A facet joint dysfunction may create tenderness to palpation of the spinous or transverse processes. In addition, lesions of the thoracic vertebrae may cause tender rib angles, whereas lesions of the lower lumbar segments and the sacroiliac joint may produce tenderness in the posterior superior iliac spine. The tenderness of these specific points depends on the position of the spinal segments.[23,28] The vertebra produces more or less afferent input to the spinal cord, depending on its position, and consequently alters the sensitivity of the tender points.

The areas that are abnormally sensitive in response to tissue irritation may be present locally, but also may be far removed in the extremity.[18] Regardless of whether the tender points are located locally in the trunk or distally in the extremity, they are always in the "innervation territory" of the facilitated segment(s).[18]

MUSCLE FUNCTION CHANGES

Bogduk and Munro examined the reflex effects from the spine to the semitendinosus muscle in cats.[2] The synovial joints and other local structures in the spine were chemically and mechanically stimulated, and individual dorsal rami were electrically stimulated. The activity of the semitendinosus was electromyographically recorded. The study showed that the reflex effects on the semitendinosus were segmental and polysynaptic. The only areas of the spine that evoked a reflex contraction of the semitendinosus were those that shared the same segmental nerve supply.

In the same publication, Bogduk and Munro described an experiment in which five volunteers received saline fluid injections in the L5-S1 interspinous ligament.[2] These injections caused back pain and evoked activity in the gluteus medius and tensor fasciae latae muscles. Both muscles may share their L5 innervation with the interspinous ligament.[38,47]

Pedersen and colleagues recorded reflex contractions of the lumbar extensor muscles in response to mechanical stimulation of the lumbar facet joints.[40] A contraction was recorded from the hamstrings as well, especially when the lower lumbar or sacral region was stimulated. The muscle contraction appeared predominantly ipsilateral. Other muscles, mainly in the thigh and the abdominal wall, also responded reflexly, but no recordings were made from them.[40] Similar findings were obtained by Nade and colleagues on cats.[36] Electrical stimulation of the articular nerves of the spine caused responses in paraspinal, thigh, and abdominal muscles.

England and Deubert described an experiment whereby manipulation of the T9 area resulted in decreased motor activity of the erector spinae at the same level.[12] The decreased motor activity was evidenced by palpation and electromyographic recordings.

Sweeting and colleagues studied patients with anterior knee pain as well as dysfunctions in the lumbar spine.[44] The patients were tested for strength before and after a lumbar manipulation. The muscles tested were the quadriceps, hip abductors, hamstrings, and foot evertors. The muscles that tested weak before the manipulation tested 15 to 20 percent stronger immediately following the manipulation. Sweeting and colleagues concluded that the central nervous system had inhibited the muscles. The manipulation corrected the lumbar dysfunction, altered the afferent impulses to the central nervous system, and removed the inhibition from the muscles. Bourdillon reported a similar experience.[4] He repeatedly observed the cause of anterior knee pain (chondromalacia patella) to be in the spine.

Pseudoradiculopathy

In summary, intense afferent input, especially nociception from deep structures such as joints, can alter spinal cord function. The spinal cord with its function altered has been termed a *facilitated segment*. The facilitated segment can create sensitivity or pain in the dermatome, myotome, or sclerotome, and it can alter the activity of the sympathetic nervous system, as well as the tone of the muscles. These alterations in the dermatome, myotome, and so on, are segmentally related to the structure producing the intense afferent input. This resembles a radiculopathy, hence the term *pseudoradicular syndrome*.

Besides perceived changes in the afference from the dermatome and myotome, and the actual changes in these structures (muscle tone changes, for instance), the nerve tissue itself may be affected. A facilitated segment can cause intense sustained activity of its peripheral nerves.[27] Presumably, the barrage of afferent impulses from the proprioceptors and other nerve endings in the stressed tissue maintains the intense sustained activity.[27] The continuous intense activity of the neurons places increased energy demands on the nerve tissue, affecting the nerves' metabolism and probably their synthesis and turnover of protein, neuropeptides, and neurotransmitters.[27] The hyperactivity could impair the axoplasma transport and probably the exchange of nutrients with the target tissue.[27] This means that the nerves related to the facilitated spinal segment become diseased and demonstrate, for instance, decreased conductivity. Nerve provocation tests, such as a straight leg raise test, might become positive. Myotatic reflexes, for example, the ankle jerk, may become diminished, and segmentally related muscles may test weak. Again, this resembles a true radiculopathy.

Thus, in addition to anatomical compression of nerve tissue, a diminished reflex might also be caused by central nervous system disturbances of the related segment.[17] Epstein described four patients with bilaterally diminished ankle jerks.[13] Even though each of the patients had a herniated disc, the herniated disc was only present unilaterally and did not approach the opposite nerve root.

Injection of an irritant fluid into the facet joint has been found to cause referred pain patterns identical to those associated with a "disc syndrome."[35] Mooney and Robertson injected hypertonic saline into the lower lumbar facet joints of healthy subjects and patients with chronic back pain and sciatica.[35] They found both increased pain down the leg and a diminished straight leg raise.[35] An injection of a local anesthetic, Xylocaine, into the facet joint eliminated the pain and the limited straight leg raise. In the patients with a straight leg raise of 70 degrees or less before injection, the intra-articular injection of Xylocaine restored the straight leg raise to normal. Three patients in the experimental group had depressed tendon reflexes before the lumbar injection. Following injection, the reflexes were normal compared with the noninjected side. Mooney and Robertson assumed that nociceptive stimuli from the facet joint inhibited the segmentally related anterior horn motoneurons and that the local anesthetic injection into the facet joint abolished the inhibition. These authors concluded that limited straight leg raising or reflex changes do not necessarily implicate nerve root pressure by disc protrusion.

Lewit described how a successful lumbar manipulation immediately increased the strength of a paretic extremity muscle and at times even restored reflexes to some degree.[29] Because of the success of the manipulation, Lewit doubted that these positive results were obtained on patients with true radiculopathies, but considered the muscle weakness due to inhibition.

Any structure producing sufficient afferent input to the spinal cord could facilitate its segment(s). If the structure responsible for the facilitation is part of the movement apparatus, movement alters the amount of afferent input and could change the level of facilitation. For instance, a spinal facet joint with restricted flexion produces higher levels of afferent input, possibly nociception, when the segment moves into the restriction (i.e., moves into flexion). Moving away from the restriction (i.e., out of the flexed position) reduces the afferent input to the spinal cord and possibly reduces the facilitation. Function of the segmentally related muscles is also expected to alter with the spinal movements.

Radiculopathy or Pseudoradiculopathy?

Despite obvious differences in pathology, radiculopathy and pseudoradiculopathy have common elements. Dermatomal pain, lowering of the reflexes, muscle weakness, and positive nerve provocation tests can be present in both conditions. Although strong similarities are present between radiculopathy and pseudoradiculopathy, some differences exist. Radiculopathy and pseudoradiculopathy may distinguish themselves in the pattern of pain. According to Cyriax, radiculopathy causes "pins and needles" felt distally in the dermatome.[9] In contrast, when pressure is applied to the dural sleeve surrounding the nerve root (causing referred pain without radiculopathy), pain can be felt in any part of the related dermatome.

McCullogh et al. studied the effects of electrical stimulation of different spinal structures.[32] They wrote that there was usually a clear distinction between root pain and referred pain.[24,32] Stimulation of the spinal ligaments, muscles, facet joints, the annulus, and the nucleus pulposus produced pain, referred to the buttock and the upper leg but rarely below the upper calf. This pain was dull and poorly localized and was rarely described as paresthesia. In contrast, stimulation of the nerve roots produced a sharper, well-localized pain, often with some paresthesia. Stimulation of the L5 and S1 nerve roots nearly always radiated to or below the ankle. Two studies by Fortin and colleagues on subjects with pain from the sacroiliac joint confirm that referred pain is predominantly felt proximally in the extremity.[14,15]

Grieve stated that it is difficult to distinguish between radicular pain and referred pain. Each can have elements of the other's pattern. Two short examples illustrate his more extensive publication on this topic.[20] The first example involves referred pain from the sacroiliac joint. Irritation of this joint may elicit referred pain into the heel, thus well below the upper calf.[20] The second example entails cases of severe radiculopathy without extremity pain. O'Laoire found an absence of sciatic pain in some patients with discogenic cauda equina compression, producing severe disturbance of micturition with perineal pain.[37]

In summary, although exceptions exist, the literature describes radicular pain as tending to be more distally located, whereas referred pain tends to be more centrally located (see Table 13.1). Radicular symptoms may be sharper and well localized and may include paresthesia. Referred pain tends to be dull and poorly localized.

Radiculopathy and pseudoradiculopathy distinguish themselves more readily in the evaluation of muscle function. Both conditions can cause weakness of segmentally related extremity muscles and central muscles. The muscles test weak throughout their range in both a shortened and a lengthened position. Both radiculopathy and pseudoradiculopathy can cause muscle strength to vary with the position of the cervical or lumbar spine.

The difference between a radiculopathy and a pseudoradiculopathy is the specific position that weakens the muscles. A radiculopathy tends to cause muscles to test weaker with extension of the related spinal segment (reduced foraminal diameter), whereas flexion improves muscle strength. An exception might be radiculopathy caused by a lumbar disc herniation; flexion could increase pressure on the nerve root.[41] A pseudoradiculopathy could cause muscles to be weaker in vertebral extension, but could also cause weakness in flexion, depending on what segmental movement is dysfunctional. Therefore, an arm muscle that weakens when the cervical spine is flexed always suggests a pseudoradiculopathy.

A second distinction can be made using duration. A muscle affected by radiculopathy gets weak in lumbar extension and tends to get weaker as the spine remains in extension. With sustained lumbar extension, the nerve root becomes more ischemic, and consequently, the muscle gets weaker. In contrast, segmental stiffness, responsible for pseudoradiculopathy, gradually stretches out and/or the articular receptors adapt,

Table 13.1 Possible differences between radiculopathy and pseudoradiculopathy.

Factor	Radiculopathy	Pseudoadiculopathy (originating from the spinal movement segment)
Dermatomal pain	Distally	Proximally
	Sharp	Dull
	Well localized	Poorly localized
Paresthesia	May be present	Absent (?)
Nerve provocation test	Present	Possibly present
Muscle weakness worse in cervical	Extension	Extension or flexion
Muscle weakness worse in lumbar	Extension (foramen) or flexion (disc herniation)	Extension or flexion
Maintaining the weakening spinal position will make the muscle	Weaker	Stronger
Movement restriction of a related segment	May be present	Always present

causing less inhibition of the muscle over time. Thus, if the spinal position that caused muscle weakness is maintained and the inhibited muscle slowly becomes stronger, the cause of the weakness is a pseudoradiculopathy.

The third way to distinguish between radiculopathy and pseudoradiculopathy is through evaluation of the spine itself. If no movement restriction of the involved spinal segment can be detected, pseudoradiculopathy from the spinal movement segment cannot be responsible.

Summary

Intense afferent input, especially nociception from deep structures such as joints, can alter function of the related spinal cord segment. This spinal cord segment with altered function has been termed a *facilitated segment*. The facilitated segment may create sensitivity or pain in the dermatome, myotome, or sclerotome, and it may alter the activity of the sympathetic nervous system and the tone of segmentally related muscles. These alterations in function occur in multiple structures that are segmentally related to the facilitated segment and to the structure producing the intense afferent input. The intense sustained activity of the segment's nerves may cause these nerves to become diseased. All these changes make this condition resemble a radiculopathy, hence the term *pseudoradicular syndrome*.

Differences exist between a radiculopathy and a pseudoradiculopathy. Radicular pain may be more distally located, whereas referred pain may be more centrally located. Radicular symptoms may be sharper in sensation and well localized and may include paresthesia. Referred pain tends to be dull and poorly localized.

Extremity muscle weakness in a radiculopathy differs from that in a pseudoradiculopathy. Muscle weakness in both a radiculopathy and a pseudoradiculopathy may alter with various spinal positions; however, maintaining the spinal position that brought on or worsened the weakness will aggravate the condition in a radiculopathy, whereas in a pseudoradiculopathy the muscle function will improve.

References

1. Bernards JA: Over de fysiologie van de pijn. Nederlands Tijdschrift voor Fysiotherapie 12:384-388, 1975
2. Bogduk N, Munro RR: Dorsal ramus-ventral ramus reflexes in the cat and man. J Anat 118:394 (abstract), 1974
3. Bonica JJ: Clinical importance of hyperalgesia. In Willis WD (ed): Hyperalgesia and Allodynia. New York, Raven Press, 1992, pp 17-43
4. Bourdillon JF: The importance of the thoracic spine. Proceedings of the 1984 International Conference of the International Federation of Orthopaedic Manipulative Therapists, Vancouver, Canada, pp 188-191
5. Brügger A: Die Erkrankungen des Bewegungsapparates und seines Nervensystemes. New York, Gustav Fischer, 1977
6. Cranenburg B van: Inleiding in de Toegepaste Neurowetenschappen: Opvattingen over Zenuwstelsel en Hersenen. Lochem, Netherlands, De Tijdstroom, 1983, vol 1
7. Cranenburg B van: Inleiding in de Toegepaste Neurowetenschappen: Pijn. Lochem, Netherlands, De Tijdstroom, 1987, vol 3
8. Cyriax J: Cyriax on Orthopaedic Medicine: Study Guide: The diagnosis and treatment of the soft tissue lesions. London, OM Publications, 1981
9. Cyriax J: Textbook of Orthopaedic Medicine, ed 8. London, Ballière Tindall, 1982, vol 1
10. Daube JR, Reagan TJ, Sandok BA, et al: Medical Neurosciences, ed 2. Toronto, Little, Brown and Co, 1986
11. Dvorák J, Dvorák V: Manuelle Medizin: Diagnostik. New York, Thieme, 1983
12. England RW, Deubert PW: Electromyographic studies: I. Consideration in the evaluation of osteopathic therapy. J Am Osteopath Assoc 72:211-223, 1972
13. Epstein BS: The Spine: A Radiological Text and Atlas, ed 3. Philadelphia, Lea & Febiger, 1969
14. Fortin JD, Aprill CN, Ponthieux B, et al: Sacroiliac joint: Pain referral maps upon applying a new injection/arthrographic technique: II. Clinical evaluation. Spine 19:1483-1489, 1994
15. Fortin JD, Dwyer AP, West S, et al: Sacroiliac joint: Pain referral maps upon applying a new injection/arthrography technique: I. Asymptomatic volunteers. Spine 19:1475-1482, 1994
16. Freeman MAR, Wyke B: Articular reflexes at the ankle joint: An electromyographic study of normal and abnormal influences of ankle-joint mechanoreceptors upon reflex activity in the leg muscles. Br J Surg 54:990-1001, 1967
17. Grieve GP: Common Vertebral Joint Problems. New York, Churchill Livingstone, 1981, p 252

18. Grieve GP: Referred pain and other clinical features. In Grieve GP (ed): Modern Manual Therapy. New York, Churchill Livingstone, 1986, pp 233-248

19. Grieve GP: Thoracic joint problems and simulated visceral disease. In Grieve (ed): Modern Manual Therapy. New York, Churchill Livingstone, 1986, pp 377-394

20. Grieve GP: Referred pain and other clinical features. In Boyling JD, Palastanga N (eds): Grieve's Modern Manual Therapy, ed 2. New York, Churchill Livingstone, 1994, pp 271-292

21. Gunn CC: "Prespondylosis" and some pain syndromes following denervation supersensitivity. Spine 5:185-192, 1980

22. Jinkins JR, Whittemore AR, Bradley WG: The anatomic basis of vertebrogenic pain and the autonomic syndrome associated with lumbar disk extrusion. Am J Roentgenol 152:1277-1289, 1989

23. Jones LH: Strain and Counterstrain. Indianapolis, American Academy of Osteopathy, 1981

24. Kellgren JH: Observations on referred pain arising from muscle. Clin Sci 3:175-190, 1938

25. Keyser A: Neurologische aspecten van afwijkingen aan de cervicale wervelkolom. In Sneep R (ed): De Cervicale Wervelkolom. Alphen aan de Rijn, Netherlands, Stafleu, 1983

26. Kidd BL, Mapp PI, Blake DR, et al: Neurogenic influences in arthritis. Ann Rheum Dis 49:649-652, 1990

27. Korr IM: Neurochemical and neurotrophic consequences of nerve deformation. In Glasgow EF, Twomey LTT, Scull ER, et al: Aspects of Manipulative Therapy, ed 2. New York, Churchill Livingstone, 1985

28. Kusunose RS, Wendorff R: Strain and Counterstrain Syllabus. Encinitas, California, Jones Institute

29. Lewit K: The contribution of clinical observation to neurobiological mechanisms in manipulative therapy. In Korr IM (ed): The Neurobiologic Mechanisms in Manipulative Therapy. New York, Plenum Press, 1977

30. Lohman AHM: Segmentale innervatie, somatisch en visceraal. Nederlands Tijdschrift voor Fysiotherapie 12:389-392, 1975

31. Lowenstein MB: Osteopathic theories and practice. Virginia Medical Monthly 102:25-28, 1975

32. McCullogh JA, Waddell G: Variation of the lumbosacral myotomes with bony segmental anomalies. J Bone Joint Surg Br 62:475 480, 1980

33. McKenzie RA: The Lumbar Spine: Mechanical Diagnosis and Therapy. Lower Hutt, New Zealand, Spinal Publications, 1981

34. McKenzie RA: The Cervical and Thoracic Spine: Mechanical Diagnosis and Therapy. Waikanae, New Zealand, Spinal Publications, 1990

35. Mooney V, Robertson J: The facet syndrome. Clin Orthop 115:149-156, 1976

36. Nade S, Bell E, Wyke B: Articular neurology of the feline lumbar spine. J Bone Joint Surg Br 60:292 (abstract), 1978

37. O'Laoire SA, Crockard HA, Thomas DG: Prognosis for sphincter recovery after operation for cauda equina compression owing to lumbar disc prolapse. BMJ 282:1852-1854, 1981

38. Paris SV: Anatomy as related to function and pain. Orthop Clin North Am 14:475-489, 1983

39. Paris SV: S3: Course Notes. St. Augustine, Florida, Institute Press, Division of Patris Inc, 1988, p 153

40. Pedersen HE, Blunck CFJ, Gardner E: The anatomy of lumbosacral posterior rami and meningeal branches of spinal nerves (sinu-vertebral nerves). J Bone Joint Surg Am 38:377-391, 1956

41. Schnebel BE, Watkins RG, Dillin W: The role of spinal flexion and extension in changing nerve root compression in disc herniations. Spine 14:835-837, 1989

42. Schneider W, Dvorák J, Dvorák V, et al: Manuelle Medicin: Therapie. New York, George Thieme Verlag Stuttgart, 1986

43. Sutter M: Versuch einer wesenbestimmung pseudoradikulären syndrome. Schweiz Rundschau Med (Praxis) 63:842-845, 1974

44. Sweeting RC, Fowler C, Crocker B: Anterior knee pain and spinal dysfunction in adolescence. Manual Medicine 4:65-68, 1989

45. Vujnovich AL: Neural plasticity, muscle spasm and tissue manipulation: A review of the literature. Journal of Manual Manipulative Therapy 3:152-156, 1995

46. Wall PD: The gate control theory of pain mechanisms: A re-examination and re-statement. Brain 101:1-18, 1978

47 Williams PL, Warwick R (eds): Gray's Anatomy, ed 36. New York, Churchill Livingstone, 1980

48 Winkel D, Aufdemkampe G, Meijer OG, et al: Orthopedische Geneeskunde en Manuele Therapie: Diagnostiek Extemiteiten, Algemeen Gedeelte. Houten, The Netherlands, Bohn Stafleu Van Loghum, 1992, vol 2A

49 Woolf CJ, Wall PD: Relative effectiveness of C primary afferent fibers of different origins in evoking a prolonged facilitation of the flexor reflex in the cat. J Neurosci 6:1433-1442, 1986

14. Arthrogenic Muscle Dysfunction

Joint dysfunction may cause muscle dysfunction. The mediators between these two are the articular receptors and the nervous system; therefore, altered stimulation of articular receptors may result in altered muscle function. The appropriate stimulation for articular mechanoreceptors is tension in the ligamento-capsular structures.[7,10,12,13] The active receptors will discharge with a higher frequency as the tension in the ligamento-capsular structures increases.[8,9,26] In addition, with increased ligamento-capsular tension, more receptors become active.[8,9,23,26] A joint with decreased extensibility (stiff ligamento-capsular tissue) reaches its end range sooner and has more soft tissue tension when the joint moves toward an end-range position, which subsequently may cause more depolarization of its mechanoreceptors. Any increased activity of the articular mechanoreceptors is likely to provide stronger articulo-muscular protective reflexes and stronger modulation of the withdrawal reflex. Therefore, when an end-range position strongly stretches a tight ligamento-capsular structure, increased inhibition of the agonist muscle is expected, with simultaneous increased reciprocal facilitation of its antagonist.[10,19,20,21,28] When stretching a tight ligamento-capsular tissue, the increased reflex activity is present whether the taut structure is part of a joint with restricted range in multiple directions (as in an arthritic joint with a capsular pattern of movement restrictions) or whether it is the only portion of the joint that has decreased extensibility.

Weakness of the muscles as a result of altered articular afferent input has been called by different names throughout the years, including:

- arthritic atrophy,[15]
- arthrogenic atrophy,[15]
- arthrogenic inhibition,[30]
- arthrogenous muscle wasting,[33]
- arthrogenous muscle weakness,[31]
- atrophic articular paralysis,[3]
- joint-protection reflex,[2]
- paralysis having an articular origin,[3]
- reflex atrophy,[15] and
- reflex inhibition.[5,25,29,30,31]

Over the years, clinicians have written about the influence of restricted joint mobility on muscle function. Although most authors believe the altered muscle function to be mediated by the nervous system, they have differing viewpoints.

Van Bachum and colleagues considered a limited extension mobility of specific carpal bones in the wrist possibly responsible for lateral elbow pain (lateral epicondylitis).[1] Lateral epicondylitis is an irritation of part of the extensor carpi radialis longus and/or brevis. The extensor carpi radialis longus and brevis run directly over the scaphoid, capitate, and lunate at the wrist and produce wrist extension on contraction. A limited extension mobility of these carpal bones may be responsible for the lateral epicondylitis. Van Bachum and colleagues reported a remarkable improvement in about half of their patients in response to mobilization of the scaphoid, capitate, and lunate. They considered their findings an example of a dysfunction of the passive movement apparatus (carpals) being the cause of complaints in the active movement apparatus (extensor carpi radialis longus or brevis).[1]

In 1986, Vladimir Janda wrote that certain muscles respond to conditions such as pain or altered joint afference with tightness and shortness, whereas other muscles respond with inhibition and weakness.[16] Janda studied the hip muscle activity in 51 adults and 22 children. In subjects with a nonpainful sacroiliac joint subluxation, he found a striking hypoactivity in the gluteus maximus or medius during hip extension and abduction. Janda considered the altered afference from the sacroiliac joint to be the cause of the muscle hypoactivity.[16] He reported a personal communication with Lewit, who in some cases observed an immediate improvement of the tone of the gluteal muscles after an anesthetic injection of the sacroiliac joint.[16] Supporting Janda's publication, Dorman and colleagues reported that "the abductors of the hip, the gluteus medius muscle mainly, are facilitated or inhibited by the position of the sacroiliac articulation. We presume this action is facilitated through a reflex arc modulated by joint position sensors in proximity to the sacroiliac articulations (page 89)."[6]

Janda considered inhibition of muscles to be clinically demonstrated by three main syndromes: hypotonia, weakness, and delayed activation in specific activities.[16] During specific activities, Janda observed normal muscles to contract in a certain sequence; however, inhibited muscles were often noticed to have delayed muscle activation, which changed the order in which the muscles contracted. The noninhibited synergists and stabilizers usually became activated before the inhibited muscle. At the completion of the movement, the inhibited muscle relaxed prematurely.[11] In extreme cases, an electromyograph recorded complete silence of the inhibited muscle when the patient performed the specific activity for that muscle.[16] The inhibited muscle would simulate a peripheral paresis, and thus Janda described the condition as "pseudoparesis."

Janda wrote that the quality of muscle function depends directly on the central nervous system activity and any inhibitory phenomena occurring there.[16,18] An inhibition of the muscle may, through the central nervous system, originate from a joint dysfunction.[16,17,22] "Altered joint function (movement restriction or decrease of joint play) is another factor which principally affects the quality of muscles which cross the particular joint. Even though it may seem that joint pain quickens the development of inhibition, pain does not necessarily precede inhibition. Evidently, the altered proprioception from the joint plays the more important role, influencing the muscle either in an inhibitory or facilitory way *(p. 199)*."[16] Besides joints, other sources of inhibition are upper or lower motor neuron lesions, pain, and reciprocal inhibition.[15,16]

In 1988, Schmitt and colleagues described the psoas phenomenon in a 45-year-old female patient as a painful and weak psoas major when tested manually.[27] Besides the painful and weak psoas, the therapist's main finding was limited mobility of the right sacroiliac joint. Among other techniques, the therapist performed mobilizations and manipulations of both sacroiliac joints. The patient was completely pain free after the treatments. Thus, Schmitt and colleagues proposed that nociception from a sacroiliac joint dysfunction can cause weakness of the psoas major muscle.[27]

Verreussel described a treatment of joints that resulted in muscle tone changes.[32] In a patient with knee complaints, the main findings were a shortening of the right iliopsoas, a possible compression of the femoral nerve by the psoas, causing knee pain, and limited mobility of the right sacroiliac joint, L5-S1, and T12-L1. The treatment consisted of manipulation of the right sacroiliac joint into counternutation and T12 and L5 into flexion, left side bending, and left rotation. Immediately after the treatment, the tone in the iliopsoas muscle had reduced and the patient had 70 percent fewer complaints in the week that followed. During the second treatment, the same manipulation of the right sacroiliac joint and T12 was repeated. In the following week, all complaints disappeared.

In 1992, Mennell wrote about the necessity of built-in mechanical play for moving parts in general and for human joints specifically.[24] The play in joints is logically called "joint play." According to Mennell, joint play is a prerequisite for the performance of normal voluntary movements and function. Mennell believed that joint play is essential to *active* movement.

The above examples, as well as other information presented in this book, make a case for considering the muscle and its joint as a single unit. The muscle moves the joint and the joint influences muscle function. A separation between joint and muscle function is artificial, since the action of one influences the other. Studies that presuppose a strict functional separation between the muscles and joints are flawed. Cummings, for instance, found an increased elbow extension range in normal adults after general anesthesia with a muscle-paralyzing agent.[4] However, he may have been incorrect in concluding that, since the muscle-paralyzing agent brought about an increase in range, it must have been the muscle that limited extension. The muscle-paralyzing agent also blocks the expression of the articular reflexes to the muscles crossing the joint. In other words, the muscle-paralyzing agent inhibited the articulo-muscular protective reflex.

In response to a chronic inhibition of muscles, muscle atrophy might ensue. DeAndrade and colleagues summarized arthrogenous muscle dysfunction and muscle atrophy by stating, "There is some association between reflex muscle inhibition and the muscle atrophy that is seen in patients with joint disease. However, neurogenic muscle atrophy results only from a lower motoneuron lesion. On the other hand, disuse atrophy is due to a lack of functional excitation of the motoneuron supplying the muscles. A depression of excitation has been demonstrated to result from reflex inhibition due to joint distention and other stimuli. Thus, the muscle atrophy seen in association with joint disease may be functional in origin and due to reflex inhibition."[5, p. 320]

Summary

Ligamento-capsular tissues with decreased extensibility are likely to create stronger articular reflexes. Stretching the tissues with limited extensibility causes even more activity of the articular receptors, with stronger inhibition of agonists and stronger facilitation of antagonists. Chronic inhibition of muscles may lead to muscle atrophy. Improving the extensibility of tissues with decreased extensibility may normalize the articular reflexes and consequently normalize muscle tone.

References

1 Bachum A van, Elkhuizen JW, Tilstra S: Epicondyalgie lateralis. Nederlands Tijdschrift voor Manuele Therapy 2:2-20, 1983

2 Blockey NJ: An observation concerning the flexor muscles during recovery of function after dislocation of the elbow. J Bone Joint Surg Am 36:833-840, 1954

3 Charcot M: On the muscular atrophy that follows certain joint lesions. In Clinical Lectures on the Diseases of the Nervous System. London, The New Sydenham Society, 1889, vol 2, pp 20-31

4 Cummings GS: Comparison of muscle to other soft tissue in limiting elbow extension. J Orthop Sports Phys Ther 5:170-174, 1984

5 deAndrade JR, Grant C, Dixon AStJ, et al: Joint distension and reflex inhibition in the knee. J Bone Joint Surg Am 47:313-322, 1965

6 Dorman TA, Brierly S, Fray J, et al: Muscles & pelvic clutch: Hip abductor inhibition in anterior rotation of the ilium. Journal of Manual Manipulative Therapy 3:85-90, 1995

7 Eklund G, Skoglund S: On the specificity of the Ruffini like joint receptors. Acta Physiol Scand 49:184-191, 1960

8 Ferrell WR: The adequacy of stretch receptors in the cat knee joint for signalling joint angle throughout a full range. J Physiol 299:85-99, 1980

9 Ferrell WR, Nade S, Newbold PJ: The interrelation of neural discharge, intra-articular pressure, and joint angle in the knee of the dog. J Physiol 373:353-365, 1986

10 Freeman MAR, Wyke B: Articular reflexes at the ankle joint: An electromyographic study of normal and abnormal influences of ankle-joint mechanoreceptors upon reflex activity in the leg muscles. Br J Surg 54:990-1001, 1967

11 Geers A: Kinesiology: Diagnostiek en Therapie van de Posturale en Fasische Spieren. Eindhoven, The Netherlands, Stichting Manuele Geneeskunde, 1984

12 Grigg P, Hoffman AH, Fogarty KE: Properties of Golgi-Mazzoni afferents in cat knee joint capsule, as revealed by mechanical studies of isolated joint capsule. J Neurophysiol 47:31-40, 1982

13 Halata Z, Rettig T, Schulze W: The ultrastructure of sensory nerve endings in the human knee joint capsule. Anat Embryol 172:265-275, 1985

14 Harding B: An investigation into the cause of arthritic muscular atrophy. The Lancet 1:433-434, 1929

15 Janda V: Muscles, central nervous motor regulation and back problems. In Korr M (ed): The Neurobiologic Mechanisms in Manipulative Therapy. New York, Plenum Press, 1978, pp 27-41

16 Janda V: Muscle Weakness and inhibition (pseudoparesis) in back pain syndromes. In Grieve GP (ed): Modern Manual Therapy of the Vertebral Column. New York, Churchill Livingstone. 1986, pp 197-201

17 Janda V: Muscle spasm - a proposed procedure for differential diagnosis. Manual Medicine 6:136-139, 1991

18 Janda V: Treatment of chronic back pain. Manual Medicine 6:166-168, 1992

19 Johansson H, Sjölander P, Sojka P: Actions on γ-motoneurons elicited by electrical stimulation of joint afferent fibers in the hind limb of the cat. J Physiol 375:137-152, 1986

20 Johansson H, Sjölander P, Sojka P: Fusimotor reflexes in triceps surae muscle elicited by natural and electrical stimulation of joint afferents. Neuro-Orthopedics 6:67-80, 1988

21 Johansson H, Sjölander P, Sojka P: A sensory role for the cruciate ligaments. Clin Orthop 268:161-178, 1991

22 Jull GA, Janda V: Muscles and motor control in low back pain: Assessment and management. In Twomey LT, Taylor JR (eds): Physical Therapy of the Low Back. New York, Churchill Livingstone, 1987, pp 253-278

23 McLain RF: Mechanoreceptor endings in the human cervical facet joints. Spine 19:495-501, 1994

24 Mennell JMcM: The Musculoskeletal System: Differential Diagnosis from Symptoms and Physical Signs. Gaithersburg, Maryland, Aspen Publishers Inc, 1992

25 Morrisey MC, Reflex inhibition of thigh muscles in knee injury: Causes and treatment. Sports Med 7:263-276, 1989

26 Schaible HG, Schmidt RF: Responses of fine medial articular nerve afferents to passive movements of knee joint. J Neurophysiol 49:1118-1126, 1983

27 Schmitt MA, Wijer A de, Schepman RWM, et al: Het psoasfenomeen op basis van een primaire sacro-iliacale functiestoornis, een casus. Nederlands Tijdschrift voor Manuele Therapie 7:9-12, 1988

28 Sojka P, Johansson H, Sjölander P, et al: Fusimotor neurones can be reflexly influenced by activity in receptor afferents from the posterior cruciate ligament. Brain Res 483:177-183, 1989

29 Spencer JD, Hayes KC, Alexander IJ: Knee effusion and quadriceps reflex inhibition in man. Arch Phys Med Rehabil 65:171-177, 1984

30 Stener B: Reflex inhibition of the quadriceps elicited from subperiosteal tumour of the femur. Acta Orthop Scandinav 40:86-91, 1969

31 Stokes M, Young A: The contribution of reflex inhibition to arthrogenous muscle weakness. Clin Sci 67:7-14, 1984

32 Verreussel RLP: Een onbegrepen knieklacht. Nederlands Tijdschrift voor Manuele Therapie 8:88-90, 1989

33 Young A: Rehabilitation for wasted muscles. In Sarner M (ed): Advanced Medicine. London, Pitman Medical, 1982, vol 18, pp 138-142

15. Manual Muscle Testing

Various methods exist for testing muscle strength. For instance, muscles could be tested manually, with a hand-held dynamometer, or with an isokinetic station. Although an isokinetic station may provide the greatest degree of objectivity, it has several disadvantages: it may be expensive; it has limited ability to be used in several places; and not all muscles or muscle groups can be tested.[10] In addition, the large loads incurred during isokinetic testing could jeopardize the integrity of the muscle or the joint it crosses.[14,17]

Unlike isokinetic stations, manual muscle tests are financially affordable for every clinic and can be included in every physical examination.[12,13,24] Despite being less objective and limited with regard to data collection, manual muscle tests have the advantages that they are easy to perform and that small muscles or muscle groups can be tested on almost every patient.[10,11,15,18] In addition, manual muscle tests can be performed more gently than muscle tests on machines. This will be discussed later in this chapter.

Testing Movements

Some authors have advocated the use of testing movements, often entailing muscle groups, whereas others have advocated testing individual muscles. Both systems have their merits.

Testing movements is done by testing muscles in the cardinal planes of the body.[4,5,11,13] For instance, when testing the shoulder, the following movements could be tested (in sequence of capsular pattern restrictions): external rotation, abduction, internal rotation, and adduction. When testing movement directions, we may be testing a single muscle or multiple muscles. For instance, in testing elbow extension, we test a single muscle, the triceps. In testing elbow flexion, we test three muscles simultaneously: the biceps, the brachialis, and the brachioradialis.

Although multiple muscles are often tested when testing movements, it may still be possible to identify an individual muscle. A lesion of an individual muscle can be identified by combining the results of testing the various movements.[5] For example, if resisted internal rotation of the shoulder is painful, the muscle responsible for the pain could be the pectoralis major, latissimus dorsi, teres major, or subscapularis. The first three are also adductors of the humerus. Thus, if resisted adduction is pain free, the muscle responsible for the patient's pain is probably the subscapularis.[4] However, if the painful resisted movement involves more than one muscle, as when resisted adduction is painful, testing of the individual muscles may be necessary to determine the muscle at fault.

An advantage to testing movement directions is that motions are generally the result of the action of multiple muscles.[11] Although motions may be the result of the contraction of predominately one muscle, the importance of secondary or synergistic muscles should not be discounted.[11] The testing of movements allows for simultaneous testing of these synergists, as well as examining the adaptation of the synergists. This could be important if a muscle is congenitally absent. In this case, a strength test of this muscle should yield weakness. Weakness caused by a congenitally absent muscle, however, may be unimportant when synergistic muscles are substituting. For example, the palmaris longus muscle is absent in 12.8 percent of the population.[23] A carefully conducted muscle test of the palmaris longus should yield weakness when this muscle is absent; however, this weakness is insignificant for the patient's rehabilitation. Muscle testing the joint movements, such as wrist flexion, is preferable here. In cases of possibly reversible muscle weakness, however, testing individual muscles may be preferable.

Testing movement directions is valuable for evaluation of function and can even be used to find the level of spinal dysfunction (see chapter 16).[22] The muscle testing of movements, however, does not allow proper evaluation of peripheral nerve lesions.

When evaluating for possible inhibition from the ligamento-capsular structures, strength testing in the cardinal planes is often preferable. Testing in the cardinal planes tests the function of the ligamento-capsular structures in these same planes. For instance, testing wrist flexion strength evaluates for the inhibiting influence of the dorsal ligaments and capsule of the wrist. Strength testing ulnar deviation tests the radial ligaments and capsule.

Testing Individual Muscles

Every muscle is a prime mover for some specific action.[15] No two muscles in the body have the same function.[15] Because of their unique function, most muscles can be individually tested.

The greatest advantage to testing individual muscles might be in the assessment of (peripheral) neurological conditions. These conditions are often characterized by weakness. The pattern of weak and strong muscles points to the type of condition and suggests the level at which the nerve is affected. For instance, all muscles distal to a peripheral nerve injury may show weakness or paralysis.

Testing individual muscles is more differentiated than testing muscles in the cardinal planes of movement, as the latter often entails testing muscle groups. Some muscles allow a further differentiation. Kendall described how different parts of fan-shaped muscles can be tested individually.[15] For example, the upper, middle, and lower parts of the pectoralis major and the trapezius muscle can be tested separately.[15] Even further differentiation is possible, since a dysfunction of a single thoracic or lumbar vertebra or a single rib can inhibit the fibers attaching to the vertebra or rib. In these cases, a slightly different angle of testing could make the difference between designating the muscle (fibers) strong or weak.

Multijoint Muscles

A muscle spanning a single joint normally has sufficient contractility to shorten throughout the full range allowed by the joint. Thus, the test position for single-joint muscles could be in a shortened position at the end of the range. Normal two-joint muscles, however, may not possess sufficient contractility to shorten throughout the full range that both joints allow. For instance, the hamstrings cannot shorten to such a degree that the hip is fully extended and the knee is fully flexed. When strength testing the multijoint muscles, the therapist could lengthen the muscle over one or more joints while shortening and testing the muscle over the remaining joint(s). For example, when testing the strength of the finger flexors over the interphalangeal joints, the wrist and the metacarpophalangeal joints are in a loose-packed position or dorsiflexed/extended to elongate the flexors, and the interphalangeal joints are flexed. The examiner tests the patient's ability to maintain interphalangeal flexion.

Some multijoint muscles can be tested while shortened over all the joints they cross. Most intrinsic spinal muscles can be tested this way; however, if the multijoint muscle is inhibited by any of the joints it crosses (arthrogenous inhibition), testing the muscle will not reveal which of the joints is responsible. For instance, the right upper trapezius muscle could be tested by bringing the origin and insertion together before assessing its strength.[15] The patient moves the neck into lateroflexion to the right and rotation to the left while the shoulder girdle is elevated. Between the origin, the occiput, the insertion, and the lateral clavicle are several joints that the upper trapezius moves: all the bilateral cervical facet joints and the acromioclavicular and sternoclavicular joints. Even the upper thoracic facet joints are involved in this movement of the head and neck. Any of these joints could be responsible for a weakness of the upper trapezius.

Testing a multijoint muscle in its maximally shortened position can give an overview of function of all its joints. If it is possible to alter the position of one or more of the individual joints a muscle crosses, the muscle test can become more specific. For example, the hamstrings can be used to evaluate the knee, the hip, and the sacroiliac joint. The hamstrings flex the knee, extend the hip, and posteriorly rotate the innominate.[2] To differentiate among these three joints, only one joint is allowed to be in the position that shortens the hamstrings. The other two joints are preferably positioned in a loose-packed position, or in a position that lengthens the hamstrings. The examiner tests the hamstrings over the joint that shortened the muscle. A more detailed description of the three different hamstrings muscle tests is included in chapter 19. The multijoint spinal muscles can be tested in a similar manner, evaluating one facet joint at a time.

Isometric Versus Concentric Manual Muscle Testing

Muscles can be manually tested isometrically and concentrically. During isometric muscle tests, the patient is requested to keep the body part in a certain position while the examiner gradually pushes against this part, attempting to move it in a specific direction. During a concentric muscle test, the patient shortens the muscle, moving the body part against the resistance offered by the examiner or against gravity.

The easiest way to determine muscle strength manually may be to test muscles isometrically. The task is circumscribed for the patient ("keep the body part in place"), and grading muscle performance by the examiner is circumscribed: any deviation from the task of keeping the body part in place (i.e., an eccentric contraction) could be termed weakness.

Concentric methods of manual muscle testing contain more variables than isometric muscle tests. During concentric muscle testing, the examiner has to control speed of movement and provide resistance in various positions and directions as the patient's body part moves through an arc around the joint. To examine strength manually throughout the entire range, several isometric muscle tests can be done, each in a different part of the range.

Grading Muscle Strength

Two methods of grading muscle strength can be distinguished. The first method is to determine the amount of force a muscle can produce (a quantitative approach). The second method is to determine the quality of the contraction.

QUANTITATIVE GRADING

Most methods of quantitative muscle testing use gravity to test the muscles.[11,13,15] In quantitative grading, the zero through five grading scale is most commonly used (see Table 15.1).[11,13] Grade 3, or "fair," is the ability to hold the antigravity test position without any resistance from the examiner. The antigravity position for a muscle is a shortened position whereby the body part has moved up against the force of gravity. For grades above "fair" (grades 4 and 5), the examiner applies pressure in the antigravity position of the muscle. The examiner could test the strength of the weight-bearing muscles by having the muscles lift the body weight against gravity. For instance, the ankle plantar flexors could be tested in the standing patient by having the patient raise the heels off the floor (standing on the toes). Muscles that move parts which are less affected by gravity, such as the distal extremity muscles moving, for instance, the fingers and toes, take exception to the use of gravity. Because these muscles are less affected by gravity, muscle testing can be done in any direction.

Table 15.1 Zero through five scale for quantitative grading.

Grade	Symbols	Description
5	Normal	Complete movement through a full range against gravity and full pressure.
4	Good	Complete movement through a full range against gravity and some pressure.
3	Fair	Complete movement through a full range against gravity.
2	Poor	Complete range of motion in a gravity-eliminated position.
1	Trace	A contraction is visible or palpable. No movement occurs.
0	Zero	No contraction is palpable.

Gravity is not used as a grading factor in muscles testing less than "fair." A "poor" grade (grade 2) denotes the patient's ability to move the body part through a full range in the horizontal plane. Again, the tests for distal extremity muscles might take exception, since the weight of these body parts is minimal.[11] The grades "trace" (grade 1) and "zero" (grade 0) are given to minimal or no contraction of the muscle, respectively. The examiner determines these grades by palpating or observing the muscle or its tendons during attempts by the patient to move the body part.

Kendall and colleagues wrote that the extent of involvement may be determined by three simple gradings: "zero," "weak," and "normal."[15] Detailed grading of muscle strength is more important in determining the rate and degree of muscle strength return. For instance, a muscle might be weak for weeks but during this time improve from grade "poor" to "fair."

Kendall and colleagues' textbook is a good source for quantitative testing of individual muscles.[18] A good source for quantitative grading of movements with grades "fair" and lower is the book by Hislop and Montgomery.[11]

QUALITATIVE GRADING

In this book, muscle tests are a determination of the quality of the contraction rather than the maximum output (a quantitative approach). Analogous to muscle testing are tests for range of motion. A quantitative approach attempts to determine the excursion, or *quantity*, of the range. A goniometer might be used to get a reading of the range. In contrast, a qualitative approach is concerned with the *quality* of the range, such as the dysfunctional, physiological, and anatomical barriers. The examiner concerned about the quality of the range might experience the end-feel (e.g., muscular, bony, empty, or guarding), and this end-feel provides more information about what treatment approach should be used. For instance, a therapist interested in quantity of range might decide to stretch the elbow if flexion is limited to 30 degrees. In performing a qualitative examination, however, the therapist would not stretch the elbow, despite its limited range, if the examination revealed a bony end-feel. In this case, the end-feel is more important than the range in determining the therapy.

Qualitative Muscle Testing Technique

The examiner might be any health care provider who is trained to evaluate the musculoskeletal system of patients. The test for quality of the contraction starts by positioning the patient and asking that they hold the body part in a certain position. The examiner often grasps the body part distally, where the pressure of the examiner's hand might provide better stimulus to the muscle being tested and less force is required to test the muscle. In addition, minute excursions might be better sensed distally on the body part. If available, contacting the patient at bony prominences will also provide better judgment of minute movements of the body part. The simple instruction to the patient could be "resist," after which the examiner gradually applies pressure against the body part to attempt to push it in a specific direction.

The patient must understand the instruction to keep the body part in the same position; therefore, the test may not be appropriate when the patient's comprehension of the task is limited (e.g., patients under the age of six). The patient does not necessarily need to know beforehand in which direction the body part will be pushed. The gradual application of pressure by the examiner gives the patient enough time to respond with an isometric contraction, regardless of the direction of the examiner's force.

Another restriction on the patient population is that the patient should not have a contraindication to the specific movement being tested or to contracting the muscle. An example of a contraindication to a specific movement is horizontal adduction of the hip in a patient with a total hip replacement. Testing abduction of the hip or testing other joints may not be contraindicated. An example of a contraindication to muscle contraction might be recent surgical repair of a muscle.

In the absence of the above restriction, manual muscle testing may be indicated in patients with musculoskeletal problems and patients with neurological problems affecting their musculoskeletal system.

During the muscle test, while the examiner gradually increases force to move the patient's body part, the examiner should focus fully on the patient's response to the pressure. Does the patient hold the body part in place, or does the body part yield to the pressure? In other words, does the patient's muscle contract isometrically, or does it contract eccentrically? In this book, a "strong" muscle is defined as a muscle *isometrically* contracting against the therapist's gradually applied force. A "weak" muscle is defined as a muscle *eccentrically* yielding to the therapist's gradually applied force.

The examiner can both feel and see the patient's response to the gradually increasing pressure. A weak muscle can be felt by the examiner as a "softness" of the patient's body part. The softness may be felt before the body part is seen yielding to the pressure. In contrast, a strong muscle keeps the body part solidly in place.

To be able to pay full attention to the response of the patient's muscle during the test, it is important that the examiner be positioned optimally. Uncomfortable examiner positions, or positions in which exerting pressure is an effort for the examiner, reduce the examiner's ability to grade the muscle. Optimal testing positions are often those wherein the examiner tests the patient in front of and close to the examiner, thus reducing the moment arm.

All muscle tests should be performed very gently. If the patient has significant pain, it is important not to aggravate the condition and to test even more gently, if possible. If, despite the gentleness, pain is provoked during the muscle test, it is important to stop the test immediately. By this time, the examiner may already have felt the muscle's response to the pressure: eccentric or concentric. If the test was stopped before the therapist sensed the muscle's response to pressure, the test could be repeated with less pressure than previously produced pain. The gradual application of pressure should be stopped before any pain is provoked.

Assessment of Muscle Strength

A healthy joint–nervous system–muscle complex will have no problem keeping the body part steady during the muscle test. In the healthy joint–nervous system–muscle complex, the gradually increasing pressure by the examiner is adequately perceived by the patient, and the central nervous system adjusts its output accordingly. The patient counters the examiner's pressure with an equal force in the opposite direction (see Fig. 15.1). The uninhibited muscle will not yield in response to the force from the therapist, regardless of the amount of force used, provided the therapist does not "break" the muscle (i.e., push with more force than the muscle's maximum output).

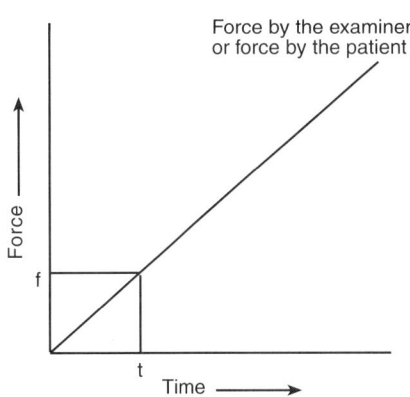

Fig. 15.1 Force–time diagram of the therapist's force and the counterforce by the patient's healthy muscle. In response to pressure by the therapist, the patient's body part behaves rigidly (does not comply with the therapist's pressure.

The graph represents the gradually increasing force produced by the therapist and the gradually increasing force produced by the patient. The graph representing the therapist's force is identical to that representing the patient's counterforce. At any time during the muscle test, the patient counters the examiner's force with an equal counterforce. For instance, at time t, the examiner presses with f amount of force. At the same time, the patient counters with the same f amount of force.

In contrast, an inhibited muscle always produces less counterpressure than requested by the examiner (see Fig. 15.2). An inhibited muscle will yield (an eccentric contraction) at the first application of pressure. With the gradual application of more pressure, the muscle will yield further. The muscle will continue yielding to the increased application of pressure as long as it is inhibited.

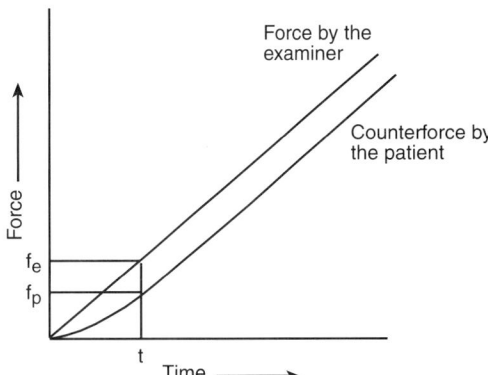

Fig. 15.2 Force–time diagram of the therapist's force and the counterforce by the patient's muscle when inhibited by radiculopathy or pseudoradiculopathy. In response to pressure by the therapist, the patient's body part behaves in a compliant manner.

At time t, the examiner pushes with f_e amount of force; however, the inhibition causes the patient to push back with f_p. The difference between f_e and f_p represents the amount of force reduction secondary to inhibition. Since the patient's counterforce is less than the force by the examiner, the patient's muscle will contract eccentrically. The inhibition (and the eccentric yielding) is present in every part of the range where the muscle can be tested.

When testing an inhibited muscle, the amount of inhibition may remain constant during an eccentric contraction or may decrease. The inhibition may remain constant if a radiculopathy or a pseudoradiculopathy is the cause of the weakness; any eccentric yielding of the extremity muscle during the test does not alter the inhibition from the spine (see Fig. 15.2). In contrast, in the case of arthrogenic inhibition, during the eccentric contraction, inhibition will decrease (see Fig. 15.3). For instance, an arthrogenically inhibited flexor muscle might test weak in full flexion and strong in a loose-packed position of the joint it crosses. During the muscle test, the eccentric contraction of the muscle slowly brings the joint out of its flexed position into the loose-packed position. In the loose-packed position, the inhibition is absent and the muscle tests strong. Thus, as the muscle yields to pressure, the inhibition decreases and the muscle tests stronger. Therefore, arthrogenic inhibition of a muscle is most easily found at the extreme of the range, where the muscle is maximally shortened and arthrogenic inhibition may be maximal.

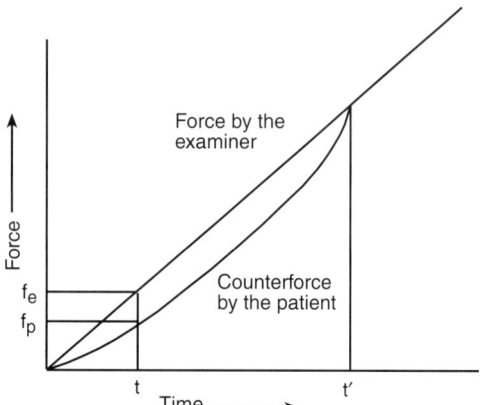

Fig. 15.3 Force–time diagram of the therapist's force and the counterforce by the patient's muscle when arthrogenically inhibited.

The examiner's increasing force and the reduced counterforce by the inhibited muscle cause the patient's muscle to contract eccentrically. At time *t*, the joint moves out of its dysfunctional range, and the inhibition disappears. Beyond *t*, the patient's muscle provides a counterpressure equal to the examiner's.

Despite the possibility of continued yielding of an inhibited muscle, this condition need not be demonstrated; the first application of pressure will reveal to the examiner if the muscle is inhibited. If the examiner wishes to assess muscle strength in other parts of the range, other isometric muscle tests could be performed in these positions.

Evaluation of Muscle Strength

If the examiner found weakness during the assessment of muscle strength, the test could be repeated in a different position of the joint it crosses, or in a different position of the part of the spine that innervates the muscle, to find the source of the inhibition. Four different outcomes are possible as a result of testing a muscle in multiple positions.

1. The muscle tests weak in its shortened position and strong in a lengthened position. The muscle is arthrogenically inhibited by the joint it crosses. The joint causing the inhibition may be an extremity joint or a spinal joint but is always mechanically related to the muscle. For example, the quadratus lumborum might test weak after the patient exhaled and strong after the patient inhaled. Since the quadratus lumborum attaches to the twelfth rib and is an accessory exhalation muscle, the inhibition of the quadratus lumborum after exhalation suggests a limited mobility of the twelfth rib for exhalation.

   Muscle weakness caused by inhibition from a joint without active range, such as the sacroiliac joint, is a special case of arthrogenic inhibition. Inhibition from these joints will be discussed later in this chapter under the heading "Joints Without Active Range."

2. The strength of the extremity muscle is altered by spinal positions, which signifies a radiculopathy or a pseudoradiculopathy influencing segmentally related extremity muscles. For example, the dorsiflexors of the foot are strong in lumbar flexion but weak in lumbar extension. An L4 or an L5 radiculopathy or pseudoradiculopathy may be present.

3. Nothing alters the muscle's weakness. The muscle has poor contraction quality in every position of the joint it crosses and in every position of the spine. Possible causes are inactivity, immobilization, aging, and myopathy. Circulatory disorders, nociception, and disorders of nerve conduction may, at times, also produce weakness throughout the working range of the muscle. With the exception of inactivity, immobilization, aging, and possibly peripheral nerve lesion, traditional strengthening exercises are only symptomatic treatment, for the cause is not addressed.[25]

4. Nothing alters the muscle's strength. The examiner can find no position in which the muscle tests weak. This is the goal of therapy. The result of successful treatment is that a previously weak muscle tests strong and remains so in any position of the joint it crosses and in any position of the spine. For instance, the quadratus lumborum is strong when tested in maximal ipsilateral side bending combined with exhalation.

Often, muscle weakness has multiple causes. For instance, a multijoint muscle could be arthrogenically inhibited by two or more joints. Combinations of arthrogenic and spinal causes of weakness are common.

Loose-Packed Position

Muscle tests can be performed in various positions of a joint. When a muscle is tested in its maximally shortened position, more than just the muscle is involved. The joint the muscle spans is in an end-range position as well, and parts of the ligamento-capsular tissue are stretched. Thus, a muscle test at end range stresses multiple structures. If a muscle being tested at end range produces pain, the cause of the pain might be located in the muscle that contracted or in the structures that are stretched, such as the capsule, the ligaments, or the antagonistic muscles.

To prevent stretching multiple structures, the muscle could be tested in a loose-packed position of the joint. In this position, all the ligamento-capsular tissues are maximally relaxed, the antagonistic muscles are more relaxed, and contraction of the agonistic muscle will only stress the structures related to this muscle (the contractile tissue, fascia, and tendon).

Cyriax advocated manually testing muscles by requesting a strong contraction in a loose-packed position of the joint.[5] A strong contraction in a loose-packed position provides information about the muscle's strength and whether the contraction affects pain. Cyriax described possible outcomes of muscle testing, which are presented here with modifications.[4,5]

1. All muscles are strong and painless. If all the muscles acting over a joint are strong and do not produce any pain on strong contraction, these muscles can be eliminated as a source of pain. The cause of the pain is located outside the muscles. To determine whether any nonmuscular structures are responsible for the patient's complaints, the examiner could perform passive movements.

   A strong and painless contraction does not necessarily mean that the muscle is without dysfunction. For example, a shortened and hypertonic muscle will probably test strong and pain free. However, a passive movement into the opposite direction will stretch the muscle and may indicate this muscle's shortness.

2. The muscle is weak and painful. When the muscle tests weak and a contraction of the muscle causes pain, a minor or major lesion of the muscle, the tendon, or its attachment to the bone could be the cause.

   If the muscle lesion is minor, testing the muscle in a midrange position does not always reproduce the patient's pain. Adding passive tension to the muscle test may reveal such minor lesions (see

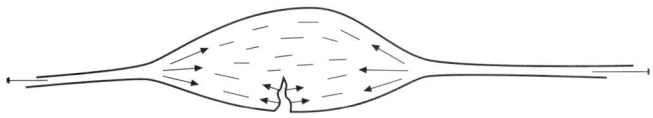

Fig. 15.4 The effect of contraction and lengthening on a muscle tear. Contraction, especially when a muscle is in a lengthened position, opens the tear.

Fig. 15.4). Passive tension is added by testing the muscle in a lengthened position.

A strong contraction is not needed to determine the quality of the contraction but may be necessary to determine if the contraction influences pain. If a muscular lesion produces nociception on contraction, the nociception will inhibit the muscle to reduce pain. Therefore, a lesion in a muscle producing pain can always be picked up as a weakness, even in gentle muscle tests. A carefully executed muscle test is more sensitive than the patient's experience of pain. An examiner can pick up weakness in the muscle with careful and gentle testing *before the contraction produces pain*.

Requesting a forceful contraction from the patient to find out if it reproduces the patient's pain may be indicated to *diagnose* the patient's disorder; however, it may also jeopardize the patient's internal healing process. For instance, in case of a muscular lesion with interruption of the muscle fibers (i.e., a muscle tear), a strong contraction might pull the two ends of the muscle fibers apart, further severing the lesion. Second, strong nociceptor input might engrave itself in the posterior horn of the spinal cord, causing nociceptive memory.[3] Once nociceptive memory is present, peripheral nociceptive stimuli are no longer needed to produce pain; the pain can be brought on "spontaneously." It is therefore important for the patient's healing process that minimal pain be provoked. The choice of whether to attempt to reproduce pain with muscle tests or to avoid pain becomes a tradeoff. An examiner might decide to request strong contractions on the initial evaluation to determine if the patient's pain is reproduced by the muscle's contraction. On subsequent visits, the therapist might decide to judge the patient's progress using submaximal pain-free contractions.

3. The muscle is weak and painless. Constant pain may be present, but the (attempted) contraction does not increase the pain. This finding suggests

a complete rupture of the muscle or tendon or an impairment of the nervous system. For instance, decreased conductivity of the nerve supplying the muscle may cause weakness without any pain.

Cyriax reported two more possible outcomes of muscle tests[4,5]:

1. All resisted movements are painful. Most people with well-functioning muscles, joints, and nervous system will perceive the increased tension, or a sense of effort, when contracting a muscle. Normally, this sensation is not uncomfortable; however, fatigue, emotional stress, or a very low pain threshold could alter the patient's perception of the contraction and cause a sensation of pain. Cyriax wrote that muscles, even when far removed from the source of pain, can hurt when tested. For instance, testing a patient's shoulder muscles could be painful in a patient with low back pain.[5]

2. The muscle is painful on repeated testing. This may be the result of insufficient blood supply to the muscle or the nerve innervating the muscle. Examples are intermittent claudication (insufficient blood supply to the lower extremity muscles), spinal claudication (insufficient blood supply to the cauda equina), or thoracic outlet compression syndrome (insufficient blood supply to the arm muscles and/or nerves in the brachial plexus).

Testing in Multiple Positions

If a muscle tests strong and pain free in the maximally loose-packed position, it is likely healthy and the nervous system is intact; however, this does not mean that the muscle, the joint, and the nervous system are functioning adequately. In other words, the hardware may be all right but the software may be malfunctioning. By testing the muscle in multiple positions, the examiner obtains more information about the functioning of muscle, joint, and nervous system.

The best quality of contraction is often obtained in the muscle's lengthened position. As the muscle shortens and antagonistic structures are stretched, inhibition from the stretched tissues may decrease the muscle's output. In other words, the maximally shortened position of the muscle may be the most difficult position for obtaining a quality contraction.

Before performing the actual muscle test in the shortened position, the examiner positions the joint at its end range. While doing so, the examiner should note the available excursion at the joint.[13] This gives information about the extensibility of the antagonistic structures. The available range may be reduced in conjunction with weakness of the agonistic muscle, especially if the muscle is arthrogenically inhibited.

If the muscle tests weak in a shortened or in a midrange position, the test could be repeated in a more lengthened position to determine if this position yields a strong muscle. When a muscle is weak in a shortened position and strong in a lengthened position, it is likely to be inhibited in its shortened position.

Somewhere between the lengthened position, where the muscle tested strong, and the shortened position, where the muscle tested weak, there is a crossover point. On one side of this point, the muscle tests weak, and on the other side, the muscle tests strong. This point in the range is called the "dysfunctional barrier." The dysfunctional barrier is located where the inhibiting antagonistic structure just begins to become tight. The exact location of the dysfunctional barrier can be found by performing a few muscle tests in various parts of the range.

If the muscle tests weak in a shortened position, and in no other point in the range does it test strong, the cause of this weakness is likely nonarthrogenic. The cause for weakness throughout the range could be neurogenic, for instance, a radiculopathy, or could be a lesion of the muscle itself.

Muscle testing in the lengthened position should not be done at the absolute end of the range, but short of it. The danger of testing a weak muscle in its maximally lengthened position is overstretching of the joint or the muscle. When testing in a *submaximally* lengthened position, an eccentric contraction could still be within the physiological range of the joint. Testing in a lengthened position, wherein the stress of passive lengthening is added to the muscle's contraction, might reproduce pain if a minor muscle lesion is present. The results of muscle tests in various positions are summarized in Table 15.2.

Barrier to Movement

Most joints can move in multiple planes. In a single plane, a joint has an excursion in one direction and an excursion in the opposite direction. Somewhere between the two excursions lies a midline neutral point.[9] In healthy joints, this point often coincides with the joint's maximally loose-packed position in this plane (see Fig. 15.5). For instance, in the frontal plane,

Table 15.2 The results of muscle testing in different positions.

	Testing in a shortened position	Testing in a midrange position	Testing in a lengthened position
Major muscle lesion: strong contraction reproduces pain?	Yes?	Yes	Yes
Minor muscle lesion: strong contraction reproduces pain?	Possibly	Less likely	More likely
Arthrogenic inhibition yields a muscle that is	Weak	Weak or strong	Strong
Nerve dysfunction or muscle tissue lesion yields a muscle that is	Weak	Weak	Weak

the glenohumeral joint has an abduction excursion of 90 degrees and an adduction excursion of 0 degrees. The midline neutral point and the maximally loose-packed position in the frontal plane might be located at 45 degrees of abduction. In addition to a maximally loose-packed position in the frontal plane, there is also a maximally loose-packed position in the sagittal plane for flexion and extension and a maximally loose-packed position in the transverse plane for internal and external rotation. The position that accommodates the maximally loose-packed positions of all movement planes of a joint is the position where the entire joint is maximally loose packed. This is the position where the least tension exists on the joint's capsule and ligaments.

When dealing with restricted range of motion, it is important to understand the barrier concept. Osteopaths distinguish several barriers to movement.[1,9,20] The *physiological barrier* is an accumulation of tension that limits active movement in the joint (see Fig. 15.5). Further movement is only passively available. The elastic quality of the physiological barrier may be felt by passive range-of-motion testing or end-feel testing. The *anatomical barrier* is the final limitation to movement. Tissue damage will occur if movement exceeds the anatomical barrier.

Besides these normal barriers (physiological and anatomical barriers), *pathological barriers* may exist within the range of movement.[8,19] The barrier that "restricts" movement is called the "restrictive" or "dysfunctional" barrier (Fig. 15.6).[19] The examiner can detect the dysfunctional barrier with very subtle passive movements; it is the earliest beginning of a sudden increase in tension or a sudden resistance to movement. The dysfunctional barrier differs from the physiological barrier in that the latter has a more gradual resistance to movement. Goodridge compared the sensing by the examiner of the initial increase in tension

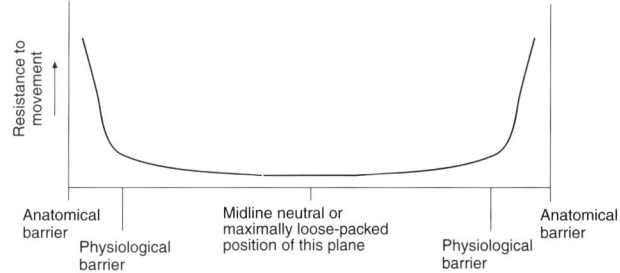

Fig. 15.5 Resistance to movement of a healthy joint.

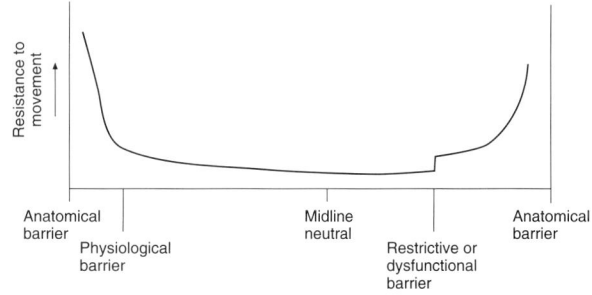

Fig. 15.6 Resistance to movement of a joint with a minimal restriction.

with sensing of the initial illumination by a rheostat-controlled light switch.[8]

The dysfunctional barrier can be located anywhere within the total range of a joint, and it is possible to have considerable range available beyond the "feather edge" of the dysfunctional barrier. The dysfunctional barrier can be located in the beginning of the range, before the normal midline neutral point would be reached. It can also be located farther into the range, beyond the normal midline neutral point. Accordingly, osteopathy distinguishes two types of movement restrictions.[9] The first type is a minimal restriction (see Fig. 15.6); the second is a major restriction (see Fig. 15.7). A minimal restriction is a restricted movement excursion wherein the dysfunctional barrier lies

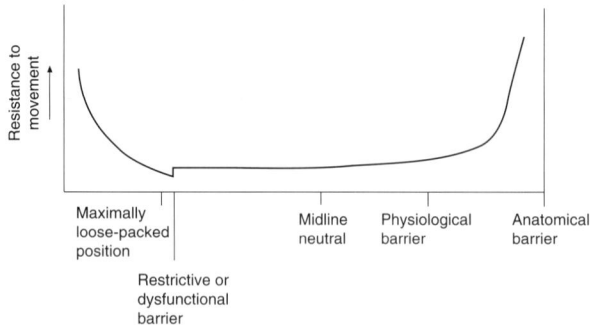

Fig. 15.7 Resistance to movement of a joint with a major restriction.

beyond the normal midline neutral point. In contrast, a major restriction is a reduced excursion range wherein the dysfunctional barrier lies before the midline neutral position.

For example, a major restriction with a dysfunctional barrier at 20 degrees of shoulder abduction does not mean that no movement is available beyond 20 degrees. It simply means that the examiner can first notice the dysfunctional barrier at 20 degrees. Movement beyond 20 degrees is possible but with increasing (pathological) resistance to movement.

When a joint's available excursion in one direction reduces, the maximally loose-packed position moves away from the restriction (see Fig. 15.7). A healthy shoulder's maximally loose-packed position in the frontal plane might be located at 45 degrees of abduction. In a shoulder with restricted abduction, the pathologically maximally loose-packed position might be at 15 degrees of abduction.

If an antagonistic ligamento-capsular structure has decreased extensibility and is responsible for a limited excursion, the examiner can sense the dysfunctional barrier by performing careful passive movements. Most disorders of the connective tissue, such as gelosis of the ground substance or scars in the connective tissue, can by themselves reduce the movement excursion; however, muscle tone changes often accompany these disorders.[19]

The dysfunctional barrier is the point in the range where the tight antagonistic ligamento-capsular structure starts to become stretched. It is also the point in the range where the agonistic muscle becomes inhibited. The barrier characterizes itself as a sharp delineation in this range between where the joint functions normally and where it is dysfunctional. The agonist tests strong before the restrictive barrier is engaged but tests weak at any point in the range beyond the dysfunctional barrier.

The resistance to movement, starting at the dysfunctional barrier, consists of three elements. First, the antagonistic ligamento-capsular structure with reduced extensibility provides resistance to movement. Second, as soon as this structure with limited extensibility is stretched, reflexes cause an inhibition of the agonistic muscle. Third, simultaneous with the inhibition of the agonists, a facilitation of the antagonists occurs. An example will clarify this process.

The patient has a restricted extensibility of the inferior part of the hip joint capsule, limiting abduction. Moving passively, at 10 degrees of abduction, the examiner senses the dysfunctional barrier. Beyond 10 degrees of abduction, the inferior capsule and ligaments of the hip joint are stretched and provide resistance to abduction. Passive abduction with increasing resistance is available to 30 degrees. Beyond the dysfunctional barrier at 10 degrees of abduction, the abductors become weak and the adductors increase in tone. The changes in muscle tone are produced by the articular reflex originating at the tight inferior capsule.[26] The three changes (i.e., capsular tightness, inhibition of the agonist, and facilitation of the antagonist) create a resistance to abduction. As the leg abducts from 10 to 30 degrees, the three changes in the muscle–joint complex become more evident, and it becomes harder to passively or actively abduct the leg.

The examiner can find the exact location of the dysfunctional barrier by carefully moving the joint passively, but also by performing multiple muscle tests. In the above example, the abductor muscles, if tested 1 degree before the dysfunctional barrier, would be fully strong, whereas a repeated test slightly beyond the barrier would yield weak abductors. The fact that the muscle tested strong before the dysfunctional barrier means that the muscle itself is healthy; that is, it has enough adenosine triphosphate (ATP) and creatine phosphate (CP) to produce a strong contraction. The weakness found in the shortened position of the muscle means that the position caused the weakness.

In the case of a major restriction, the midline neutral point is located beyond the dysfunctional barrier. When testing the agonist at or near the midline neutral point, the examiner will find the muscle weak. A joint with a major restriction arthrogenically inhibits its muscle over more than 50 percent of the joint range. The muscle will test strong in only a small part of the range, and in this small part, the muscle is always lengthened. A joint with a major restriction can thus be defined as a joint with arthrogenic inhibition (weakness) of a muscle when tested at a midline neutral point. The clinical significance of a major restriction is that the joint's resting position is dysfunctional.

A joint might have two or more restrictions in one or more planes of movement. For instance, both internal and external rotation might be restricted. Also, these restricted barriers might overlap. If dysfunctional barriers of two antagonistic movements overlap, no point in the range can be found where the patient has not engaged a barrier. In any position of the joint, a barrier is engaged and a muscle will test weak. Moving the joint to one extreme end of the range will cause one muscle to be weak, whereas moving the joint to the opposite end of the range will cause its antagonist to be weak. Testing the muscles in a neutral position yields a weakness of both muscles. The significance of overlapping inhibitions of antagonistic muscles is that the joint is unable to function well in any position. Often, the patient presenting with this condition has a painful joint and can find no position to ease the pain.

Joint dysfunction will result in muscle dysfunctions by changing the gamma bias of the spindle cells. Besides muscle testing and passively testing for the dysfunctional barrier, many other methods exist to evaluate the muscle–joint unit. The dysfunctional joint also can reveal itself by a reduced excursion, end-feel testing, translation testing, or palpation of the stiff structure itself. The weak agonist might be atrophic, or feel soft to palpation, whereas the antagonist might test short on length tests, show increased tone on palpation, feel "crampy" to the patient, or have trigger points.

If a muscle tests weak in its shortened position, how can one confirm that the muscle is arthrogenically inhibited? Using muscle tests, three methods exist for confirming that the cause of the muscle weakness is arthrogenic inhibition. First, the muscle tests strong in a lengthened position. Second, a multijoint muscle can be tested over another joint while its length is maintained by lengthening it over the possibly dysfunctional joint. In the case of arthrogenic inhibition, the muscle will test strong. The final confirmation that the limited extensibility of the joint's ligamento-capsular tissue was inhibiting the muscle comes from treatment. After a mobilization of the joint's tightness, the muscle regains its strength and will remain strong when tested on subsequent visits.

Joints Without Active Range

Most joints in the body can be moved actively through a range. Some joints, however, do not possess an active range through which they can be moved. Examples of joints that cannot be voluntarily moved actively through a range are the pelvic joints (the sacroiliac joints and the symphysis pubis), the proximal and distal tibiofibular joints, and the tarsal joints. Joints without active range are the same joints that produce no capsular pattern of movement restriction when affected with arthritis. Again, the reason is that these joints have no active range that muscles could limit. Joints without active range can only move in response to:

1. Passive movements of the bones that form the joint. An outside force might place a stress on a joint without active movement. The outside force might directly stress the joint, such as during a passive mobilization of the joint, or it might indirectly move the joint. An example of an indirect movement is posterior rotation of the innominate as a result of strong passive hip flexion. After the slack at the hip joint is removed, the innominate follows the femur into a posterior rotation. A prerequisite for indirect passive movement is that the neighboring joint is brought past its physiological barrier when the slack is removed.

2. Contractions of muscles acting over the neighboring joints. Many muscles may attach to the bones that form the joint without active range. These muscles move the neighboring joints through their active range, but they influence the joints without active range as well. For instance, the iliacus, a hip flexor, tends to rotate the innominate anteriorly, stressing the sacroiliac joint. Another example is the hip adductors, which, when contracting in the anatomical position, pull the ipsilateral pubis inferiorly.

Arthrogenic inhibition from joints without active range is assessed by testing the muscles acting over the neighboring joints with an active range. The weakness exists throughout a large part of the neighboring joint's range and only gradually changes to strength. For instance, the anterior rotation mobility of the innominate could be evaluated by strength testing the iliacus. In the case of limited anterior rotation mobility of the innominate, the iliacus will test weak in maximal hip flexion, less weak in 90 degrees of hip flexion, and may test only slightly weak in 70 degrees of flexion. The dysfunction at the sacroiliac joint or the symphysis pubis inhibiting the iliacus does this throughout a large range of the hip joint. No crossover point where a weak iliacus suddenly turns strong can be found in the range of the hip joint. A second example is limited mobility of the pubis in an inferior direction causing inhibition of the adductors in a large range of the hip joint. The adductors will pull the pubis inferiorly whether the hip joint is adducted, in a loose-packed position in the frontal plane, or in some degree of abduction. In contrast, an adduction stiffness at the hip joint might only produce adductor weakness when tested in adduction. Arthrogenic inhibition from the lumbar spine or the hip often causes muscle dysfunction in only a small part of the range.

Not only does the difference in range distinguish between inhibition from a joint without active range and a joint with active range. In addition, muscle weakness from arthrogenic inhibition of a joint with active range occurs abruptly when the restriction (dysfunctional barrier) is engaged.

Goodheart associated subluxations of the sacroiliac joints with weakness of the muscles attaching to the pelvis.[7,8] He stated that the subluxation was always opposite the pull of the weak muscle. Goodheart's "subluxation away from the pull of a weak muscle" could be interpreted as stiffness of the sacroiliac joint toward the pull of the muscle.

To evaluate the pelvic joints with muscle strength tests, the examiner tests the "pelvic" muscles over their active joints (hip or lumbar spine) neighboring the pelvis. The position of these active joints in testing for pelvic joint dysfunction is not critical.

Other joints without active range are the tibiofibular joints. The biceps muscle attaches to the proximal fibula. Depending on the knee position, the biceps can stress the proximal tibiofibular joint capsule by pulling the fibular head in a proximal direction (at 0 degrees of flexion), in a posterior direction (at 90 degrees of flexion), or inferiorly (in full knee flexion). Even though the biceps attaches to the fibular head, the proximal tibiofibular joint does not have an active range that can be voluntarily moved. Stiffness in the joint may inhibit the biceps.

The tendons of the invertor muscles (tibialis posterior, flexor digitorum longus, and flexor hallucis longus) run posterior to the medial malleolus. A contraction of the invertors will push the medial malleolus anteriorly at the distal tibiofibular joint. Limited anterior movement of the medial malleolus at the distal tibiofibular joint will inhibit the invertors. The inhibition of the invertors occurs in a large part of the inversion–eversion range at the subtalar joint. A contraction of the invertors will push the medial malleolus anteriorly, whether the subtalar joint is inverted, in a loose-packed position, or everted. Similarly, the evertor tendons push the lateral malleolus anteriorly, irrespective of the ankle position. A limited mobility at the distal tibiofibular joint for anterior translation of the distal fibula will inhibit the evertors through a large part of the range in the subtalar joint.

The tarsals and metatarsals act as joints without active range for plantar flexion, dorsiflexion, inversion, and eversion of the ankle. These ankle movements can be inhibited throughout most of their range by limited tarsal or metatarsal mobility.

Besides the pelvic joints, the tibiofibular joints, and the tarsals, other joints may at times act as joints without active range. For instance, the knee joint has an active range into flexion and extension. Abduction and adduction (varus and valgus) movements of the knee joint range only a few degrees, and the knee cannot be actively abducted or adducted. Therefore, the knee joint functions as a joint without active range for abduction and adduction. The knee passively abducts and adducts during varus-valgus testing, for instance, and in response to hip abduction and adduction resisted at the ankle.

During varus and valgus movements of the knee joint, the posteromedial and posterolateral parts of the knee capsule are stressed. To evaluate for inhibition from these parts of the knee, the hip abductors and adductors can be tested. The degree of hip abduction or adduction during the muscle test is hardly relevant, though, for the knee will inhibit the hip muscles throughout most of their range.

The spine and ribs can function as joints without active range when extremity muscles that attach to them are tested. Examples of extremity muscles that attach to the spine or ribs are the pectoralis major, serratus anterior, middle and lower trapezius, rhomboids, latissimus dorsi, and psoas major. One spinal segment or a single rib can arthrogenically inhibit specific fibers of these muscles.

From personal experience, I have concluded that these specific muscle fibers can be singled out by a muscle test if the muscle is fan-shaped. Non-fan-shaped muscles such as the psoas major will test weak if the inhibition of the muscle fibers is intense enough. The muscles attaching to the spine or the ribs are tested over the shoulder or the hip joint using the humerus and the femur as levers. Whether the muscle or its specific fibers are shortened or lengthened over the shoulder girdle joints or the hip joint, the inhibition from the spine or ribs is not altered. The spine and the rib joints function as joints without active movement for these shoulder girdle and hip muscles.

Another joint that may at times act as a joint without active range is the first metatarsophalangeal joint. This base joint of the big toe is not a passive joint, for we can flex and extend it at will; however, for most people, abduction and adduction of this joint is not voluntary, making it function as a joint without active range for these movements. Stiff abduction of the hallux can inhibit muscles in the foot. Wearing deforming "fashionable" shoes over a life span, or even a much shorter time, could cause this inhibition. Limited abduction mobility of the hallux inhibits the invertors throughout the inversion–eversion range at the ankle. The invertors are inhibited throughout their ankle range, since it is not the subtalar joint that causes the inhibition of the invertors, but tension in part of the first metatarsophalangeal joint capsule. The abductor hallucis muscle, with abduction of the big toe, may take part in the inversion pattern of the entire foot.

Generalized Muscle Weakness

Generalized muscle weakness, or weakness of multiple or all muscles in the body, could result from causes other than inhibition by tight antagonistic structures. When weakness is not caused by tight antagonistic structures, the muscle often tests weak throughout the range of the joint it crosses and in all positions of the spine. Weakness caused by inactivity or immobilization of muscles, circulatory disorders, aging, myopathy, nociception, and disorders of nerve conduction are often present throughout the range of the muscle, and no position can be found where the muscle tests completely strong. Exceptions may be muscle weakness caused by either nociception or compression of blood vessels or nerve fibers; here, the muscle may be strong in specific parts of the range, provided the nociception or compression is eliminated in this position.

Often a patient with weak muscles caused by a condition other than tightness of antagonistic structures has arthrogenic inhibition as well. The chronic weakness caused by, for instance, a myopathy or nerve compression may have prevented the joint from moving fully throughout its range, resulting in stiffness of the ligamento-capsular structures. This stiffness created a second cause of muscle weakness: arthrogenic inhibition.

A particular case may provide more insight to the reader. A patient recently diagnosed with multiple sclerosis was referred for physical therapy to strengthen her weak leg muscles. During the evaluation, I found that all her lower extremity muscles tested weak in a shortened position. I told her that I could do nothing about her multiple sclerosis, but if the muscles were weak because of inhibition, there was a good chance we could remove the inhibition and make the muscles stronger. During subsequent treatments, all the muscles in her lower extremities turned out to be arthrogenically inhibited by local joints or lumbar facet joints (pseudoradiculopathy). The patient was discharged with full strength in all her lower extremity muscles. This case is interesting because the weakness in all the lower extremity muscles could easily have been diagnosed as resulting from the multiple sclerosis. In the past, these muscles had likely become weak because of the multiple sclerosis, but at the time of her visits, the patient was in remission and the weakness was directly due only to arthrogenic inhibition and pseudoradiculopathy.

This case points out the importance of carefully evaluating patients with muscle weakness. The condition the patient presents with, such as myopathy or nerve compression, may not be the reason a specific muscle is weak. Only after careful evaluation can the examiner be reasonably sure of the cause of the weakness. Most patients with the diagnosis of "radiculopathy" have muscle weakness due to a variety of causes, of which the radiculopathy is just one.

In another interesting case, a patient was referred to me for low back pain with sciatica producing pain in the foot. During treatment, the patient's foot pain and the weakness of the ankle and foot muscles turned out to be caused by a minor L5 *pseudoradiculopathy* and by major dysfunctions in the ankle and tarsal joints. Thus, the patient's complaint in the foot were mainly caused by local changes and only minimally by his back problem. The patient had no radiculopathy at all.

Active Exercises

Contractions of a muscle normally result in acute facilitation of that muscle. This means that following a contraction, the subsequent contraction is easier and/or stronger. Increased use of a muscle results in immediate adaptations in the spindle cell and the Golgi tendon organ. In the spindle cell, the intrafusal muscle fibers reset to a higher gain after the contraction.[12] The Golgi tendon afferent input, which could inhibit the muscle, undergoes a brief inhibition.[12] These two short-term adaptations to contraction increase the excitability of the alpha motoneurons; therefore, exercising a muscle facilitates subsequent contractions.

The principle that exercising a muscle facilitates subsequent contractions can be helpful during evaluation. When a muscle tests borderline strong or borderline weak, exercises that involve contractions of the muscle tend to facilitate the muscle (make it immediately stronger). These exercises might consist of jumping in place or simulated throwing of a ball. If the muscle tests fully strong after the exercises, it could be categorized as strong.

To the contrary, after doing the exercises, the muscle may test weaker. Obviously, in this case, the exercises inhibited the muscle. They repeatedly exposed the muscle to a source of inhibition. Following the exercises, the examiner can evaluate for the cause of inhibition (e.g., is the weakness throughout the range or only in part of the range).

The plantar flexors might serve as an example. The examiner finds that the plantar flexors are not fully strong, but they do not test weak either. The patient does some jumping in place to contract the plantar flexors repeatedly. Next, on retesting, the muscles test weak in a shortened position but strong in a midrange position. Further evaluation reveals limited plantar flexion at the ankle. During the jumping, the repeated plantar flexion into the restriction brought out the dysfunction.

This principle of repeated contractions causing facilitation or inhibition of the muscle can be important when a patient has progressed well and may be ready for discharge. When a patient is ready to return to a physically demanding job or sport activity, the evaluation should be done immediately following the activity. For instance, the patient might come 20 minutes early for treatment and go for a 20-minute walk or run. Only if the exercise feels comfortable to the patient will he or she do it for the full 20 minutes. By evaluating the patient immediately after the exercise, the examiner can determine more precisely if the patient has reached the goal of treatment.

Reliability and Validity

The muscle tests described in this book and the interpretation of the test results have been developed in a clinical setting. The inferences from the muscle tests may not be based on a valid measurement, since research data are not yet available to substantiate the judgments and decisions based on the muscle tests. However, until research data is available, our critical thinking may be the best way to deal with qualitative muscle tests and their interpretation. To stimulate critical thinking and scientific research, the following paragraphs will discuss the reliability and validity of qualitative muscle testing.

RELIABILITY

Lamb summarized the research regarding *quantitative* manual muscle tests and concluded that these tests may have low intertester reliability.[16] Since the reliability of *qualitative* manual muscle tests is not known, and since qualitative and quantitative forms of manual muscle tests have many elements in common, qualitative manual muscle tests may also have low intertester reliability.

For both forms of muscle testing, possible errors that reduce reliability may fall into the following categories: patient's position, patient's attention, examiner's instructions, examiner's position, application of pressure by the examiner, and grading of muscle performance.

Patient's Position—Muscle strength may vary with different positions of the patient. For this reason, it is important to either standardize the patient's position or to record the position(s) in which the muscle was tested. For example, if the patient is allowed to position her head freely while an upper extremity muscle is tested, subsequent repeated tests by the same or different examiners may yield different results. For example, in the case of a patient with cervical pseudoradiculopathy, turning her head might make a difference between a strong and a weak test result. This decreases both intratester and intertester reliability.

Reliability should improve with close attention to the patient's position. Qualitative muscle testing attempts to control all related aspects of the patient's position. The effect of various positions on muscle performance is systematically assessed.

Patient's Attention—Clearly, the attentiveness, or mental cooperation, of the patient is important in muscle testing. For instance, testing in a gym area where other events are occurring, possibly with radio or television sets playing, may reduce the patient's level of attention. In addition to outside distractions, the reduced attentiveness may be due to factors related to the patient. The patient's mental status may not allow optimal attentiveness (e.g., young patients or patients under the influence of medications at the time of the visit).

Examiner's Instructions—The examiner's instructions to the patient should always be consistent (e.g., "resist" in a neutral tone of voice). Differences in tone or in the instructions themselves could make the patient more or less motivated and thus affect the test results.

Examiner's Position—To improve reliability, the examiner should always try to perform muscle tests in the same manner. This is not always possible, because the examiner must adapt to each patient's size and shape. Besides adapting to circumstances as required to perform a muscle test, the examiner should attempt to standardize his method of manual muscle testing.

The examiner's position is very important to obtaining reliable test results. The examiner should be in an optimal position and should use minimum effort to test the muscle. For instance, testing with straight arms, as opposed to having the load close to the examiner's body, often requires more effort on the examiner's part, resulting in less sensitivity and less attention to the patient's response to the test.

Application of Pressure by the Examiner—An inhibited muscle will yield in response to the examiner's force regardless of the amount of force used. A strong muscle will test strong regardless of the amount of testing force, provided the examiner does not break the muscle's contraction. The fact that qualitative muscle tests are not dependent on the amount of force the examiner uses may improve the reliability of this method of testing. If an examiner were to test the maximum force a muscle could deliver (a quantitative test), the amount of force delivered by the examiner, dynamometer, or isokinetic station would be critical. However, pressure must be applied gradually while the examiner assesses the patient's muscle response.

Grading of Muscle Performance—Occasionally, it is not clear to the examiner whether a muscle should be graded weak or strong. Retesting the muscle may provide more clarity. If repeated testing does not make it easier to categorize a muscle's performance, the patient needs to be repositioned. In particular, the joint that is suspected of being responsible for any possible inhibition should be positioned to maximize the inhibition. Qualitative muscle testing is gentle enough that a muscle could be tested repeatedly without overexerting it.

Based on clinical observations, repeated tests do not seem to result in increased muscle strength, as do active exercises. It may that the contraction forces during the muscle tests are too low to create adaptations that increase the excitability of the alpha motoneuron.

INTRATESTER AND INTERTESTER RELIABILITY

Because the tester is able to control many of the above factors, skill and experience are required to test muscles manually with optimal reliability.[11,21] In the classes I teach, I am often asked by a therapist to grade the strength of a person she has just tested, seeking feedback on her muscle testing skills. If the therapist and I disagree on the muscle's performance, I always ask the therapist to show me how she tested the muscle. If I duplicate the test (i.e., position myself and the patient the same way the therapist did), I often get the same result she did. Does this signify poor intertester (between the therapist's and my first measurement) and poor intratester (between my first and my second test) reliability? I do not think so.

In the past, intratester and intertester differences have mainly been caused by variables that normally are easy to control. Often, the patient's position was responsible for the difference in test results. Finding differences in muscle test responses to various patient positions is exactly the goal of qualitative manual muscle testing. The differences in strength between one position and another are what directs the examiner's attention to a specific structure that may be responsible for the inhibition. The skill and experience of some examiners is probably what helps them instantly recognize the cause of a difference in muscle performance and use it to the patient's benefit, whereas the novice examiner is puzzled by this difference in muscle performance.

CONSTRUCT VALIDITY

The preceding chapters provide an account of muscle inhibition by joint dysfunction based predominantly on research publications. The muscle tests described later in this book attempt to assess muscle inhibition. Decreased contraction performance of a muscle (eccentric contraction) during qualitative isometric manual muscle testing is interpreted as resulting from inhibition. This interpretation is based on the references in the preceding chapters and on clinical observations. No direct research evidence exists that joint dysfunctions causing muscle inhibition can be assessed by manual muscle tests; however, we have aids that help us make this assumption. If inhibition from a joint dysfunction can be altered, the results of qualitative manual muscle testing should alter with it. Therefore, testing in a different position such that a specific structure is stressed (e.g., as in a tight joint capsule) or compressed (e.g., as in a radiculopathy), or the reverse, such that stress or compression is removed, is expected to result in a predictable change in muscle performance.

CONTENT VALIDITY

If a muscle is inhibited by a structure, testing the muscle seems an appropriate way to assess the inhibition. Is testing a single muscle adequate for inhibition by a structure, or should more muscles be tested? Perhaps more important than the number of muscles to be tested is finding a position where a single weak muscle becomes strong. The difference in position between where a muscle tested weak and where it tested strong indicates the source of the inhibition. Since it is impossible to determine the source of inhibition after a single muscle test, a minimum of two tests is necessary for weak muscles, provided the second test yields a strong muscle. If the second test in a less inhibiting position does not reveal normal strength in the muscle, another test of the same muscle may be indicated if another position can be found that removes more inhibition from the muscle. Thus, errors in content validity could result from testing a single weak muscle in only one position. If the first muscle test yielded a strong result with the muscle in a maximally inhibited position, a single test would be sufficient for this muscle.

Depending on the relationship between the muscle and the inhibiting structure, other muscles with a similar function over the joint it crosses, or with similar innervation, should demonstrate the same pattern of inhibition. For instance, in the case of limited flexion mobility of a joint, all muscles that flex the joint are expected to demonstrate the same pattern of weakness. Likewise, in the case of a radiculopathy, all muscles receiving sufficient innervation from the nerve root in question should demonstrate the same pattern of weakness.

If a muscle is inhibited by a specific structure, performing multiple tests of the same or different muscles should help the examiner make rational decisions about the source of inhibition and the number of muscles to be tested. Subsequent chapters in this book will provide more detailed information about what muscles to test and in what positions.

CONCURRENT VALIDITY

Do the results of qualitative manual muscle tests have concurrent validity to infer whether a muscle is inhibited? To answer this question, the results of manual muscle tests would have to be compared with another test that has been shown to measure arthrogenic inhibition, but no direct methods exist for doing this. For instance, afference from a joint could be measured with invasive procedures, but this would not mean that the afference measured is representative for inhibition, or for inhibition of a specific muscle.

The phenomenon that muscle test results alter in response to positions of specific joints suggests that muscle performance is influenced by joint position. With the exception of postmobilization weakness, which will be described in chapter 17, my clinical observations have been that, in the case of arthrogenic inhibition, muscle strength alters immediately when joint position is changed, and in the case of inhibition from radiculopathy, a change in muscle strength may be delayed by a few minutes.

PRESCRIPTIVE VALIDITY

The results of the various muscle tests direct the examiner to the cause of inhibition and guide the therapist in providing treatment. Treatment aims to correct the dysfunction and remove inhibition from the muscle. Thus, inferences toward treatment are made based on the results of qualitative muscle tests.

Chapter 17 contains a summary of literature presenting limited evidence that a joint manipulation can result in muscle tone changes. These muscle tone changes were observed with an isokinetic station and with EMG.

Although these publications indicate that manual therapy could cause muscle tone changes, it has not yet been established that the results of muscle tests validly predict appropriate treatment. For fifteen years, I have assessed the correctness of the prediction that joint mobilization will correct inhibition, testing my patients after each treatment. Based on my clinical observations, I have found qualitative manual muscle test measurements to have prescriptive validity for removing inhibition through joint mobilization. To date, no research has been conducted to support this.

Every clinician using muscle tests is strongly encouraged, for the sake of providing optimal patient care, to evaluate muscle function after therapeutic intervention. In this way, every clinician can test the prescriptive validity of his conclusions from the evaluation.

Summary

Various methods of manual muscle testing exist. Depending on the force of the contraction requested, or the point in the range where the muscle is tested, different information is obtained from the test. To find the cause of a muscle's inhibition, it must be tested at various points in the range. A muscle arthrogenically inhibited by the joint it crosses will cause weakness of the muscle in its shortened position only. The muscle will demonstrate a sharp delineation in the range of the joint it crosses between where it tests strong and where it tests weak. Hypomobility in a joint without active range causes muscle weakness over a large part of the range of a neighboring joint. This weakness only gradually changes to strength in other positions of the joint. Weakness of an extremity muscle throughout its range may result from disease, a neurological disorder, or a spinal dysfunction.

References

1. Bourdillon JF, Day EA, Bookhout MR: Spinal Manipulation, ed 5. London, Butterworth Heinemann, 1992
2. Cibulka MT, Rose SJ, Delitto A, et al: Hamstring muscle strain treated by mobilizing the sacroiliac joint. Phys Ther 66:1220-1223
3. Cranenburg B van: Inleiding in de Toegepaste Neurowetenschappen: Pijn. Lochem, The Netherlands, Tijdstroom, 1987
4. Cyriax J: Cyriax on Orthopaedic Medicine: Study Guide: The Diagnosis and Treatment of the Soft Tissue Lesions. London, OM Publications, 1981
5. Cyriax J: Textbook of Orthopaedic Medicine, ed 8. London, Baillière Tindall, 1982, vol 1
6. Goodheart GJ: Sacroiliac and ilio sacral problems. Digest of Chiropractic Economics January/February: 1972
7. Goodheart GJ: Sacroiliac and ilio sacral problems: Part 2. Digest of Chiropractic Economics July/August:42-45, 1972
8. Goodridge JP: Muscle Energy technique: Definition, explanation, methods of procedure. J Am Osteopath Assoc 81:249-254, 1981
9. Greenman PE: Principles of Manual Medicine, ed 2. London, Williams & Wilkins, 1996
10. Guffey JS, Burton BJ: A critical look at muscle testing. Clinical Management 11(2):15-19, 1991
11. Hislop HJ, Montgomery J: Daniels and Worthingham's Muscle Testing: Techniques of Manual Examination, ed 6. London, W.B. Saunders Co, 1995
12. Hutton RS, Atwater SW: Acute and chronic adaptations of muscle proprioceptors in response to increased use. Sports Med 14:406-421, 1992
13. Janda V: Muskelfunctionsdiagnostik: Muskeltest, Untersuchung Verkurzter Muskeln, Untersuchung der Hypermobilität. Leuven, Belgium, ACCO, 1979
14. Kannus P: Isokinetic evaluation of muscular performance: Implications for muscle testing and rehabilitation. Int J Sports Med 15:S11-S16, 1994
15. Kendall FP, McCreary EK, Provance PG: Muscles: Testing and Function, ed 4. London, Williams & Wilkins, 1993
16. Lamb RL: Manual muscle testing. In Rothstein JL (ed): Measurement in Physical Therapy. New York, Churchill Livingstone, 1985, pp 47-55
17. Maletius W, Gillquist J, Messner K: Acute patellar dislocation during eccentric muscle testing on the biodex dynamometer. Arthroscopy 10:473-474, 1994
18. Marino M, Nicholas JA, Gleim GW, et al: The efficacy of manual assessment of muscle strength using a new device. Am J Sports Med 10:360-364, 1982
19. Mitchell FL: Elements of Muscle Energy technique. In Basmajian JV, Nyberg R (eds): Rational Manual Therapies. London, Williams & Wilkins, 1993
20. Mitchell FL Jr, Mitchell PKG: The Muscle Energy manual: Concepts & mechanisms, the musculoskeletal screen, cervical region evaluation and treatment. East Lansing, Michigan, MET Press, 1995, vol 1
21. Nicholas JA, Strizak AM, Veras G: A study of thigh muscle weakness in different pathological states of the lower extremity. Am J Sports Med 4:241-248, 1976
22. Peck D, Brower TD: Algorithms for the segmental motor innervation of the extremities. Am Surg 53:270-273, 1987
23. Reimann AF, Daseler EH, Anson BJ, et al: The palmaris longus muscle and tendon: A study of 1600 extremities. Anat Rec 89:495-505, 1944
24. Schwartz S, Cohen ME, Herbison GJ, et al: Relationship between two measures of upper extremity strength: Manual muscle test compared to hand-held myometry. Arch Phys Med Rehabil 73:1063-1068, 1992
25. Van Meeteren N: Modulation of peripheral nerve repair by exercise training and chronic stress in the rat. Thesis. Utrecht, The Netherlands, Universiteit Utrecht, 1994
26. Williams PL, Warwick R (eds): Gray's Anatomy, ed 36. New York, Churchill Livingstone, 1989

16. Evaluation of Spinal and Multiple Dysfunctions

Evaluation

Evaluation for spinal dysfunctions is important for spinal conditions as well as for extremity problems, which may have a spinal component to their dysfunction. An example will clarify this. Limited arm elevation could be caused by reduced thoracic extension. The limited thoracic extension mechanically restricts arm elevation but also inhibits the lower trapezius muscle. Both thoracic extension and uninhibited lower trapezius functioning are needed for elevation of the arm. Another example is a patient with de Quervain's syndrome whose wrist flexors were weak. Mobilization of the carpals restored most of the strength during treatment, but on subsequent visits, the flexors were weak again. This repeated itself until the patient's lower cervical facet dysfunctions were corrected.

Evaluation of the spine begins with ruling out potentially serious spinal conditions.[2] These conditions include fracture of the spine, a tumor, an infection, or cauda equina syndrome. Having ruled out these conditions, the evaluation for a spinal dysfunction is often simple. One way to evaluate the spine is to answer the following four questions:

1. Is it an extension or a flexion dysfunction?

   A patient who perceives pain on forward bending may have a flexion dysfunction at the segment producing pain or at a functionally related segment. An example of a flexion dysfunction at a segment that is functionally related to the painful segment is one that occurs near a painful and hypermobile segment. A mobilization of this flexion restriction might reduce pain on flexion, even though the mobility of the hypermobile segment has not changed. Likewise, pain on extension could be related to an extension dysfunction at the painful segment or at a functionally related segment. If the patient has constant pain, the examiner needs to know what influences this pain. For instance, flexion may make it worse, whereas extension produces a different pain. The pain on extension could be caused by another structure, and the practitioner may want to deal with that after the primary pain has been addressed. If the patient's constant pain is not influenced by movement, it may not derive from the structures that moved during the test.

   A good method of evaluating for flexion and extension dysfunctions is to test the unilateral spinal flexors and extensors (four muscle tests in total). These muscle tests reveal flexion and extension dysfunctions, but also answer the next question.

2. Is it a right- or a left-sided dysfunction?

   Often, this question is easily answered, for example, right-sided muscles becoming strong or weak depending on the position of the spine. The patient's response to movements can also guide the therapist to the side of the dysfunction. Asymmetrical movements such as unilateral facet flexion or extension can be used to find the (most) dysfunctional side. The asymmetrical movements stress one side at a time.

   Asymmetrical movements are also useful if the patient has bilateral or central pain. For instance, on extension, a patient feels pain over the spinous process of L5, perceiving the pain exactly in the middle of her back. To find the side of her extension dysfunction, the examiner has the patient extend and side bend her lumbar spine to the right. This will extend the right lumbar facet joints while leaving the left facet joints in a loose-packed position. Next, the patient extends and side bends to her left. Most likely, one asymmetrical movement will reproduce her symptoms, whereas the other will not. Using side bending, three-dimensional unilateral facet movements, or other asymmetrical tests, the examiner can selectively flex or extend the facet joints on either side of the patient's spine. After finding the movement that reproduces the pain, the examiner can evaluate for dysfunctions within this movement.

3. What is the level of the dysfunction?

Using extremity and spinal muscle tests, it is easy to find the level of a spinal dysfunction. First, the cervical or lumbar spine is brought into the dysfunctional position. The dysfunctional position is the position that was found in answering the previous two questions. It is a combination of flexion or extension (question 1) with an asymmetrical movement (question 2).

Next, the examiner tests the spinal or the extremity muscles. Using the spinal muscles, the level of spinal dysfunction is found by bringing one spinal segment into the dysfunctional position and testing the muscles that maintain this segmental position. This is repeated with other segments. When testing the extremity muscles to find the level of segmental involvement, all the cervical or all the lumbar segments are brought into the dysfunctional position. The pattern of extremity muscle weakness will suggest the spinal segment by virtue of its segmental innervation. The spinal evaluation is more precise, since it tests only one spinal segment at a time.

It may be advantageous to test the extremity muscles in a loose-packed position of the extremity joint they cross. Testing in a loose-packed position of the extremity joint prevents picking up arthrogenic inhibition originating from this joint. It also makes the test more specific for influences of the spine. If the extremity muscles are strong, the spinal dysfunction lies outside the area of the spine that innervates these extremity muscles.

For instance, if the right cervical spine has an extension dysfunction but the right arm muscles test strong, the extension dysfunction lies outside the C5 through T1 region.

A test is positive for a spinal dysfunction when a weak extremity or spinal muscle, tested in the position of spinal dysfunction, becomes strong in a loose-packed position of the spine or in an opposite position of the spine. For example, the right wrist flexors test weak in right cervical facet extension (cervical extension, right rotation, and right side bending) but test strong in right cervical facet flexion (cervical flexion, left rotation, and left side bending).

In addition to extremity or spinal muscle testing, other methods exist to aid in finding the level of a spinal dysfunction. These include:

a. Intervertebral motion palpation. Palpation of vertebral position, excursion, and end-feel to movement are included here. Intervertebral motion palpation forms an integral part of testing the segmental spinal muscles.

b. Nerve provocation tests.[3] Nerve provocation tests may be indicated for a possible radiculopathy. Different nerves derive from different spinal levels, and a positive nerve provocation test suggests involvement of a spinal level. For instance, the tibial nerve often derives from the L4,5 and the S1 nerve roots. The femoral nerve often derives from the L3 and L4 nerve roots. A combination of positive provocation tests of both nerves may be caused by their common nerve root, L4. A positive provocation test of the femoral nerve with a negative provocation test of the tibial nerve may be caused by an L3 radiculopathy.

c. Tender points at the spinous or the transverse processes.[6,12,16]

d. The distribution of the area of pain or paresthesia.[1,8,9,18] The area of pain reference or paresthesia in response to spinal dysfunctions may be helpful in discerning the spinal level or origin. The patient may feel pain or paresthesia in the dermatome of a specific segment. Some authors have linked mapped areas of pain reference or paresthesia to spinal joints.[1,2,8,9,18]

4. What technique will do best here?

Certain techniques may be contraindicated. For instance, extension techniques may be contraindicated in patients with a radiculopathy caused by a lateral spinal stenosis, since an increased extension mobility could cause more nerve impingement in the foramen.[22,29] Besides choosing a specific technique that will improve the condition, the therapist must make a decision about the intensity of the technique. During the evaluation, the examiner will get a sense of the irritability of the spinal dysfunction and the general tolerance of the patient. Certain treatment techniques may be poorly tolerated, whereas other techniques may be well accepted. Gentler techniques are safer and are more appropriate for new patients, until the examiner has a better idea of the patient's reactivity.

Spinal Origin of Peripheral Dysfunctions

Radiculopathy and pseudoradiculopathy can cause peripheral muscle weakness. Upper extremity muscle weakness may derive from dysfunctions in the cervical or the upper thoracic spine, and lower extremity weakness may be caused by lumbar and/or sacral dysfunctions. A spinal origin of peripheral muscle weakness is suggested by any of the following observations:

1. The extremity muscle is weak throughout its range. Unlike weakness caused by a peripheral joint (arthrogenic inhibition), the centrally inhibited muscle is weak in its shortened and lengthened positions and any midposition.

2. The weakness in the extremity muscle is related to the position of the spine. The weakness depends on the position of the facet joint in the case of a pseudoradiculopathy, on the size of the foramen in the case of a radiculopathy caused by foraminal stenosis, or on the vertebral position in the case of a radiculopathy caused by central stenosis.[26,29] The position of the facet joint and the size of the foramen are strongly related.

Facet joint movement could be described as flexion or extension. Flexion of the spine is accomplished by bilateral facet flexion (see Fig. 16.1). After maximal spinal flexion, each facet joint may allow more flexion to accommodate asymmetrical movements. Adding contralateral side bending (see Fig. 16.2) and contralateral rotation to the spinal flexion may bring the facet joint farther into flexion. For instance, unilateral facet flexion on the right side is accomplished by flexion, side bending to the left, and rotation to the left. The position of unilateral facet flexion is identical to that of foraminal opening. Maximal unilateral flexion might flex the individual facet joint more than bilateral facet flexion (symmetrical spinal flexion).

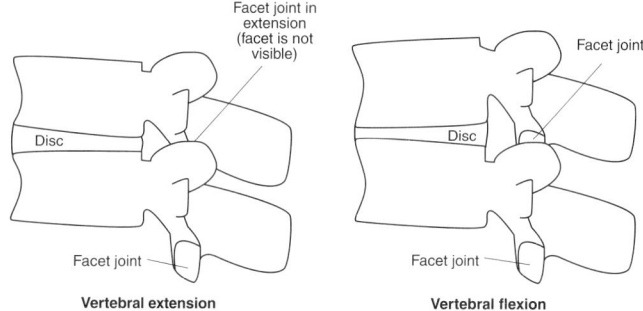

Fig. 16.1 Facet flexion and extension.

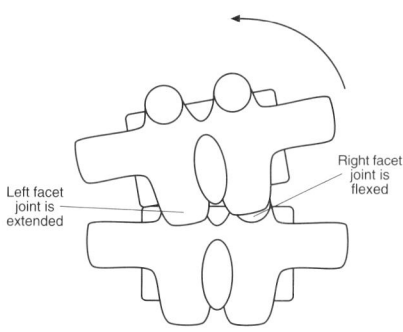

Fig. 16.2 Left side bending. The left facet joint is extended. The right facet joint is flexed.

Extension of the spine entails extension of both the left and right facet joints. Unilateral facet extension is accomplished by extension, ipsilateral side bending, and ipsilateral rotation. Unilateral facet extension might extend the individual facet joint more than bilateral facet extension (spinal extension). The position of unilateral facet joint extension is identical to the position of foraminal closing. Of course, the entire spine need not be flexed or extended, but only one segment on the ipsilateral side.

3. Other muscles innervated by the same spinal level may have the same pattern of weakness. The weak peripheral muscle is segmentally related to the dysfunctional spinal segment. For

instance, wrist flexor muscles (innervated by C6,7,8) may be weak because of a dysfunction of one lower cervical segment.[24] Other segmentally related muscles may have the same pattern of weakness. This means that other muscles might also test weak throughout the range of the extremity joint they cross and test strong in a different spinal position. For example, the triceps, which has the same innervation as the wrist flexors, might test weak throughout the range of the elbow and test stronger in a different position of the lower cervical spine. The possibility of other segmentally related muscles joining the pattern of weakness, and the degree to which they do so, depends on the amount of innervation by the involved spinal segment.

4. The spinal position that inhibited the extremity muscle also inhibits the muscles acting over the involved spinal segment. The muscles in the extremity and in the spine may both test weak in certain positions of the spine. The weak spinal muscles are those that moved the involved spinal segment into its dysfunctional range. These spinal muscles become inhibited the moment the spinal segment moves past its restrictive barrier. This means that if right lumbar extension causes weakness of a certain right leg muscle, the right lumbar extensors also will show weakness when tested in right lumbar extension. The spinal muscles as well as the extremity muscle may test fully strong in other spinal positions.

Despite the similarities between the extremity and the spinal muscles, these muscles do not share the same pattern. The extremity muscle is inhibited throughout the full range of the extremity joint it spans, whereas the spinal muscles act as arthrogenically inhibited muscles. The spinal muscles are inhibited through part of the range of the facet joint they cross.

To evaluate for cervicothoracic or lumbar dysfunctions, we need to position these spinal regions before testing the extremity or spinal muscles. For radiculopathy, we need to test the spine in foraminal opening, in foraminal closing, and in midpositions. For pseudoradiculopathy, we need to test in unilateral facet flexion, in unilateral facet extension, and possibly in a midposition. The position of foraminal opening is identical to facet flexion, and the position of foraminal closing is identical to facet extension. Thus, in each part of the spine we have one position that tests both right foraminal opening and right facet flexion and one position that does the same thing on the left. Similarly, we have one position that both closes the right foramen and extends the right facet joint and one position that does the same thing on the left. Following are the test positions for the right side of the spine:

1. Right cervical foraminal opening/right cervical facet flexion (see Fig. 16.3). The patient is seated with her lumbar and thoracic spine in a neutral position and her neck flexed, side bent to the left, and rotated to the left. This position is identical to one in which she looks toward her left axilla. While the patient maintains this position, the examiner tests the patient's right arm muscles or the flexors of her right cervical facet joints.

2. Right cervical foraminal closing/right cervical facet extension (see Fig. 16.4). The patient is seated with her lumbar and thoracic spine in a neutral position and her neck extended, side bent to the right, and rotated to the right. This position can be achieved by asking her to bring her right ear to her right shoulder blade. In this position, the patient's right arm muscles or right cervical facet extensors can be tested.

3. Right lumbar foraminal opening/right lumbar facet flexion.

    a. Lumbar flexion and side bending (see Fig. 16.5). The patient is supine with her legs to the left (lumbar spine side bent to the left). The patient holds her left knee with her hands and pulls the knee to her chest. In this position, the examiner tests the patient's right leg muscles.

    b. Lumbar flexion and rotation (see Fig. 16.6). The patient brings both knees to her chest (lumbar flexion). Next, she brings her knees to her right side on the table (left rotation of the superior vertebra). The examiner supports the patient's upper leg (left leg), and the muscles of the leg closest to the table (right leg) are tested for strength.

4. Right lumbar foraminal closing/right lumbar facet extension (see Fig. 16.7). The supine position naturally provides some lumbar extension and foraminal closing. More foraminal closing can be achieved by placing a pillow in the hollow of the patient's back to increase the lordosis, or the patient's legs can be positioned to her right (right

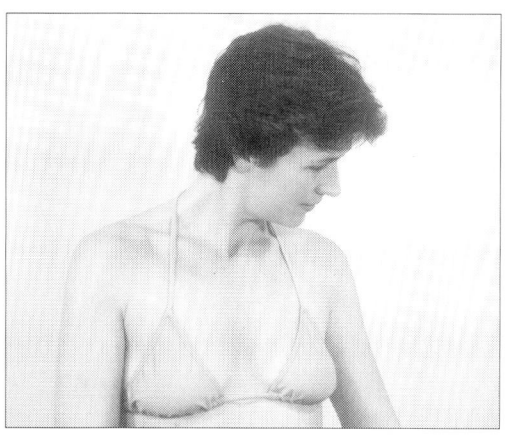

Fig. 16.3 Right cervical foraminal opening/right cervical facet flexion.

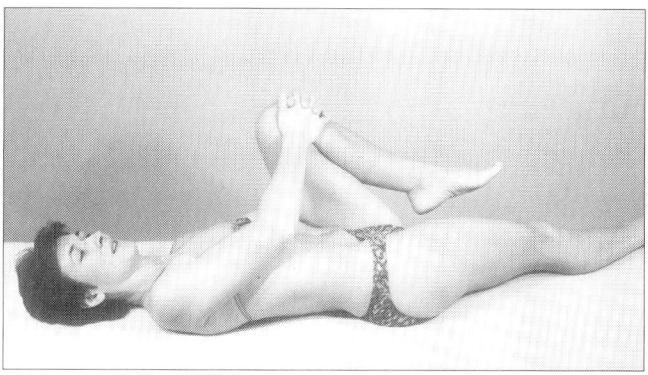

Fig. 16.5 Right lumbar foraminal opening/right lumbar facet flexion (lumbar flexion and left side bending).

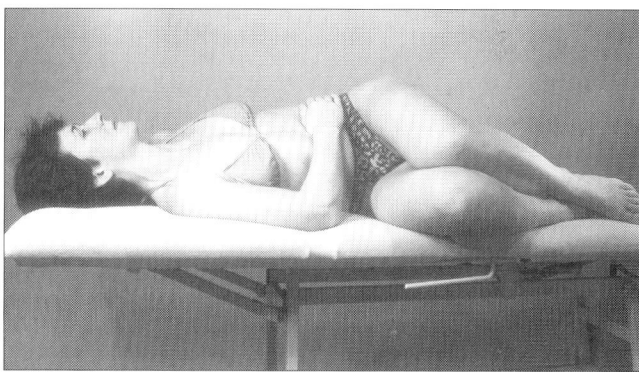

Fig. 16.6 Right lumbar foraminal opening/right lumbar facet flexion (lumbar flexion and right rotation).

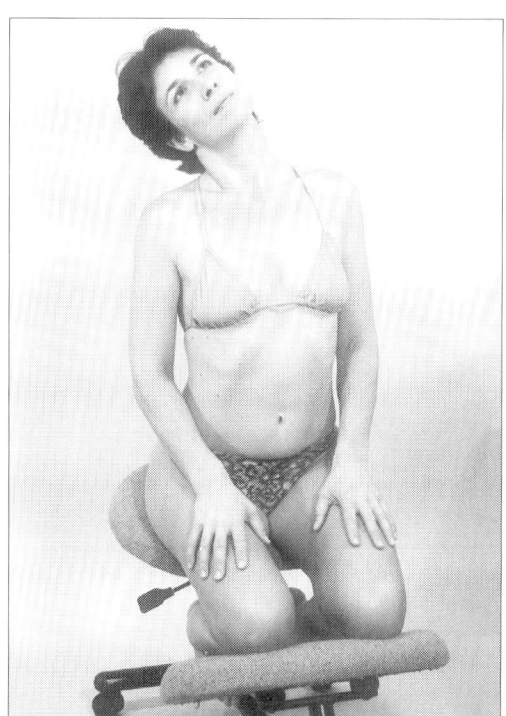

Fig. 16.4 Right cervical foraminal closing/right cervical facet extension.

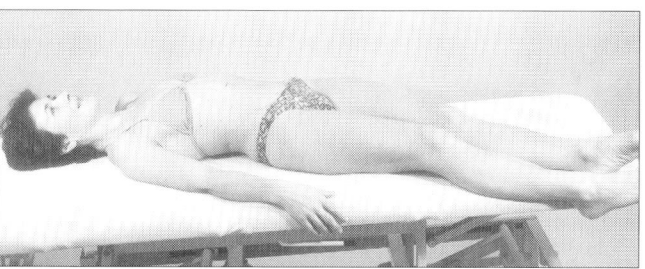

Fig. 16.7 Right lumbar foraminal closing/right lumbar facet extension.

lumbar side bending). The right leg muscles are tested in this position.

Radiculopathy

Besides the patient's history and regular orthopedic exam, the muscle tests presented in this book provide additional information about the possibility of a radiculopathy. Muscle strength may be dependent on the spinal position. For instance, a radiculopathy caused by foraminal stenosis might show itself by muscle strength declining in a position where the foramen is closed, whereas opening the foramen might improve muscle strength. In contrast, a lumbar disc herniation compromising functioning of a nerve root might cause decreased muscle strength in response to spinal flexion.[26] Both possibilities will be discussed in more detail below.

1a. The weakness is worse in a position where the foramen is closed. Opening the foramen improves muscle strength.

   Closing the foramen reduces its diameter and may (further) compress a jeopardized nerve root. Foraminal opening removes pressure from the nerve root and improves its function. More advanced stages of neural inflammation, possibly with scar tissue interrupting the continuity of the axons, may not readily respond with increased strength of the segmentally related muscles.

   The spinal nerve consists of fibers of various sizes. The fibers with a smaller diameter are more resistant to compression, which mainly affects large-diameter myelinated fibers.[4,25,27] Restoration of circulation inside the nerve might revive small-diameter (nociceptive) nerve fiber conduction, but it may not improve large myelinated axon functioning. Restoration of nociception in a previously numb area is a sign of improvement, which may not always be appreciated by the patient, who now experiences pain where before there was numbness. The patient needs to understand that experiencing pain rather than numbness is a sign of improvement. Gradually, if larger diameter nerve fibers are restored to function, other extremity functions may improve as well.

1b. The weakness is worse in spinal flexion. Spinal extension improves muscle strength. Obviously, foraminal stenosis cannot be responsible for the increased weakness in spinal flexion. Schnebel and colleagues demonstrated that an L4-5 disc herniation could compress the L5 nerve root in spinal flexion.[26] Spinal extension reduces the nerve root compression.

2. Maintaining the foraminal position exacerbates the muscle's response. Although foraminal opening improves muscle strength, prolonged foraminal opening could improve muscle strength even more. While the foraminal opening is maintained, the restored nerve root circulation gradually reduces the nerve root ischemia and slowly improves muscle strength.

   Similarly, although foraminal closure may immediately increase nerve root pressure, impede nerve root circulation, and cause muscle weakness, maintaining foraminal closure will increase nerve root ischemia and further weaken the muscle.

   To evaluate for the possibility of a radiculopathy caused by central or foraminal stenosis, the muscle must be tested in different spinal positions. If this does not provide a clear difference in muscle strength, the procedure should be repeated, this time maintaining each position for, say, 30 seconds.

If the evaluation suggests compression of a specific nerve root, the site of compression could be the foramen through which the nerve exits, or it could be one level higher caused by a disc protrusion. Although nerve root compressions in most regions of the spine occur at the level of exit, in the lower lumbar spine, the nerve may also be compressed one level above the level of exit. For instance, an L5 radiculopathy could be caused by compression at the L5 foramen (lateral foraminal stenosis with, for example, encroachment from the facet joint), or it could be caused at the level of L4 by an L4 disc protrusion.[26] If the cause of an L5 nerve root compression is an L4 disc protrusion, the compression might be less in spinal extension and greater in spinal flexion.[26] Flexion of the lumbar spine could increase the pressure of a nerve root crossing a herniated nucleus pulposus. Lumbar extension may reduce the compression of the herniation on the nerve.[26]

Pseudoradiculopathy

Bombardment of the spinal cord with afferent stimuli could create a facilitated segment. If the spinal cord bombardment originates in afferent stimuli from a taut facet joint capsule, simple movements of the spine can alter the afferent stimuli. For instance, a facet joint with limited flexion mobility will produce significant stimulation of the Type I and II articular receptors when the segment is flexed. If the spinal segment moves aggressively into the restriction, the high-threshold Type III and IV receptors may discharge as well. When the facet joint moves out of the restricted position into a loose-packed position or into the opposite movement, the spinal cord bombardment from the facet joint afferents ceases. The facilitation of the segment can thus be turned on and off.

The strength of the segment's myotome alters with the presence or absence of an afferent bombardment of the spinal cord. In the case of facet joint stiffness, the strength of the muscles in the myotome alters with the position of the facet joint. Facet joints with reduced extensibility cause increased afferent impulse traffic when the facet capsule is stretched. The increased afference entering the spinal cord facilitates the segment, causing referred pain or muscular changes. Just as certain spinal movements might cause pain while other movements ease the pain, these same movements might cause muscle dysfunction while other movements improve muscle function. For example, a patient with low back pain that worsened in flexion had weakness of the right quadriceps femoris and the right ankle dorsiflexors. The weak muscles suggested an L4 segment involvement. Positioning the right lumbar facets in extension restored strength in the quadriceps and the dorsiflexors. Flexion of the right lumbar spine weakened the muscles more than any other position of the spine. After restoring the right L4 facet flexion mobility, the quadriceps and dorsiflexors remained strong in every position of the spine. The pain had disappeared as well.

Three observations are pathognomonic for a pseudoradiculopathy. The first is noted when spinal flexion and/or extension weakens the extremity muscles. An extremity muscle changing strength in response to different spinal positions alerts the examiner to a spinal dysfunction. However, even if different spinal positions do not alter muscle function, the spine may still be responsible for an extremity muscle weakness. If a facet joint has a stiffness for both flexion and extension, both movements will create peripheral muscle weakness. This could confuse the examiner, who believes that no position of the spine alters the muscle's strength. However, performing one muscle test in a loose-packed position of the facet joint could show that the muscle weakness is position dependent, provided the muscle tests strong in that position. In addition, unilateral facet flexion and extension often do not inhibit the extremity muscle to the same degree. Carefully testing the muscle in unilateral spinal flexion and extension could reveal that one position creates more weakness than the other.

The second observation is that sustained positioning of the spine in the inhibiting position slowly reduces the inhibition. Unlike a radiculopathy, where sustained positioning in the inhibiting position slowly increases muscle weakness, the opposite happens in the case of a pseudoradiculopathy. Sustained positioning of the spine in the pathogenic position slowly diminishes the weakness. While the joint is in the inhibiting position, the stiffness is slowly stretched and/or the articular receptors slowly adapt. Either way, the result is diminished firing of the articular receptors and a lessening of the inhibition. Thus, the muscle will immediately test weak when the spine is brought into the inhibiting position but will test less weak when the position is maintained.

The final observation is that a spinal dysfunction is always present. If the pseudoradiculopathy is based on increased afference from the facet joint, limited facet joint mobility is always present. The cervical or lumbar extensor or flexor muscles, acting over the vertebra with limited mobility, will always test weak in the case of a pseudoradiculopathy.

Weakness associated with pseudoradiculopathy from a spinal facet joint is a special form of arthrogenic weakness. In both arthrogenic inhibition and pseudoradiculopathy, the joint with limited extensibility inhibits muscles when the restrictive barrier is engaged. In the case of a pseudoradiculopathy, however, the muscular changes are more generalized. Not only is tone altered in the muscles acting over the facet joint, but also in the segmentally related muscles in the extremity.

In the case of extremity dysfunction, an evaluation for contributing factors from the spine is often essential to a good treatment result. If a spinal component is left untreated, the patient's peripheral problem may not respond to the treatments or may recur. In many cases, patients who test positive for a spinal component to their extremity dysfunction are unaware of their spinal dysfunction.

The following example may clarify the need for a spinal evaluation in the case of an extremity problem. A patient with right shoulder pain had weakness of her shoulder internal rotators. All other shoulder muscles were strong. The internal rotators were weak in internal rotation only (arthrogenically inhibited). An internal rotation mobilization restored the strength of the muscles; however, the weakness returned on a subsequent visit. The cervical spine was evaluated for a spinal component to the weakness. The internal rotators were tested in right cervical flexion and in right cervical extension. In right cervical extension, the internal rotators lost their strength. Mobilization of a stiff right C5 facet into extension restored internal rotator strength.

This example also points to the variability in muscle innervation. The external rotators and abductors of the glenohumeral joint normally share a common segmental innervation with the internal rotators. All three muscle groups receive innervation from the C5,6 spinal segment. In this patient, however, with the cervical spine in extension, only the internal rotators became weak; the external rotators and abductors remained strong. An individual variation in innervation probably was responsible.

Innervation of Extremity Muscles

To find the segmental level of a radiculopathy or a pseudoradiculopathy, peripheral muscle tests can be performed. Some authors have linked single muscles to a spinal segment.[7,17] For instance, the deltoid muscle is used to gain information about the C5 level, the biceps about C6, the triceps about C7, and so on. This method has clinical usefulness, but it may be an oversimplification of the innervation of muscles.

Muscles are innervated by multiple segments. The deltoid muscle is innervated by C5 and C6, the biceps by C5 and C6, and the triceps by (C6), C7, C8, and (T1).[14,28] Most muscles are innervated by multiple spinal segments, and most segments innervate multiple muscles. For instance, the L4 segment innervates more than 29 muscles in the lower body.[14]

In the patient mentioned earlier in this chapter, stiffness of the right C5 facet joint for extension was causing weakness of the internal rotators of the shoulder; however, the other shoulder muscles innervated by this segment did not exhibit weakness. This clinical example shows that the clinician can be confused by too strictly assigning segmental levels to the innervation of muscles without allowing for individual variation. Many patients display individual variation in the innervation of their muscles.

Several authors have described anomalies in innervation.[10,19] Kadish and Simmons found anomalies of the lumbosacral nerve roots in 14 percent of their specimens.[13] Among the anomalies found were intradural anastomoses between rootlets of different levels and extradural anastomoses between nerve roots. Both types of anastomoses cause nerve fibers originating from one spinal cord level to exit the spinal canal at an adjacent level. D'Avella and Mingrino confirmed the presence of anastomoses between lumbosacral roots, finding them in 100 percent of the 30 cadavers they examined.[5] Parke and Watanabe found intersegmental anastomoses as well. These researchers found axons of the motoneurons (rootlets) joining spinal nerves one or more segments inferior to their original spinal cord level.[23] Occasionally, the (anterior) rootlets joined a dorsal nerve root.[23] Pallie found anastomoses in the cervical and lumbar spine between the posterior spinal rootlets of adjacent segments.[21] These interconnections between adjacent dorsal rootlets may contribute to variations in dermatome distribution between individuals.

Kadish and Simmons described the presence of extradural (but intraspinal) divisions of nerve roots that caused a single nerve root to split and exit the spinal canal at two different levels.[13] Kikuchi and colleagues described the presence of an independent nerve root (the furcal nerve) in 100 percent of 69 cadavers dissected.[15] In 93 percent of the specimens, the furcal nerve originated from the L4 level within the spinal cord. It often exited the spine at the L4 level as well, running independently beside the L4 nerve root. However, part of the furcal nerve was also seen exiting the spinal canal at L3 or L5, or the entire nerve exited at the L5 level. The authors presented a case in which the atypical neurological changes were caused by a compression of the furcal nerve and the L5 nerve root. The compression occurred at the level of L5. Infiltration of the fifth lumbar nerve root was performed, and muscle weakness occurred in the iliopsoas and the quadriceps muscles, which were both innervated by L4 (the furcal nerve). Sensory impairment, however, was localized to the fifth lumbar dermatome, not the fourth. An L5 nerve root decompression relieved all the patient's symptoms.

Hasue, Neidre, and colleagues described two classifications of nerve root anomalies that produce variations in segmental innervation of the target organs.[11,20] The first type of anomaly is the presence of two separate nerve roots exiting the same foramen. The second type is formed by two adjacent nerve roots linked by a connecting root (an anastomosis). Combinations of both types also occurred.

Ten percent of patients may have anomalies such as a variable number of vertebrae or a transitional vertebra.[19] Two other anomalies that occur are a prefixed or a postfixed lumbar plexus.[19] A prefixed lumbar plexus is one in which every nerve root exits the lumbar spine one level higher. Thus, the nerve root normally exiting at L5 will exit at the L4 foramen. In a post-fixed lumbar plexus, all nerve roots exit one level lower (i.e., the L5 nerve root would exit at the S1 foramen).

Considering these variations in extremity innervation, it seems inappropriate to adhere too strictly to predetermined innervation tables. Clinicians would be best served by a system that correlates segmental innervation levels with muscles but allows individual flexibility. The system should be extensive enough to include the muscle's multiple innervation levels, but should be easy to memorize. Peck's and Brower's algorithm may fulfill these requirements.

Peck and Brower created a clinically useful algorithm with which the clinician can quickly determine the segmental level of muscle innervation (see Fig. 16.8).[24] In the extremities, the proximal joints have more degrees of freedom than the distal joints. For instance, the scapulothoracic, glenohumeral, and hip joints all have three degrees of freedom, whereas the finger and toe joints have only one degree of freedom. Several muscles are needed to control a joint with three degrees of freedom, whereas only two muscles are needed to control a joint with one degree of freedom. Thus, the multiple muscles controlling the proximal joints derive their innervation from multiple spinal segments. In contrast, the distal joints, being controlled by fewer muscles, receive motor innervation from fewer segments. Peck's and Brower's algorithm is predicated on the patient being in the piano-playing position. Two segments innervate the muscles that move the upper and lower extremity joints against gravity, and three levels innervate the muscles that move a joint downward with gravity. Exceptions are the flexors of the fingers and toes and the plantar flexors at the ankle, which are distal muscles that are controlled by only two segments. Most muscles that move the upper extremity joints downward with gravity (in the piano-playing position) share a thoracic innervation, with the exception of the triceps, whose innervation is only cervical. In the lower extremity, all muscles moving the joints downward with gravity share a

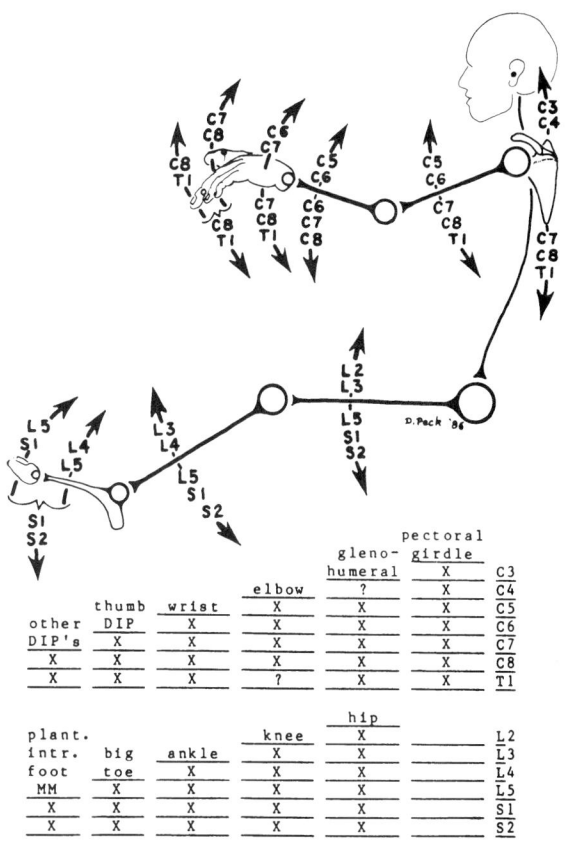

Fig. 16.8 Algorithms for segmental motor innervation of the extremities.[24]

With the patient placed in the position illustrated by the stick figure, the most cranial two spinal cord levels of a group controlling a joint are responsible for upward movement around that joint. The most caudal three (two in the case of the digits) spinal cord levels of that group control downward movement.

Reprinted with permission from Peck D, Brower TD: Algorithms for the segmental innervation of the extremities. Am Surg 53: 270-273, 1987

sacral innervation. An example of the ease with which extremity innervation can be memorized using Peck's and Brower's algorithm is the innervation of the anterior lower extremity muscles. The muscles that move the lower extremity joints against gravity, from proximal to distal, always jump one segment. The hip flexors are innervated by L2 and L3, the knee extensor by L3 and L4, the ankle dorsiflexors by L4 and L5, and the toe extensors by L5 and S1.

In addition to using Peck's algorithm to determine muscle innervation, two more muscles can be used. The sternocleidomastoid muscle can be used to gain information about the C2 and C3 segments, and the diaphragm can be used for the C4 and C5 segments.

Evaluation of Weakness from Multiple Sources

A muscle can be tested in various positions of the joint it crosses and in various positions of its related spinal segment (see Table 16.1). If a muscle tests weak in any one position, the examiner should attempt to find the position in which the muscle tests strong. The difference between the position where the muscle tests strong and where it tests weak suggests the origin of inhibition. Before discussing the evaluation for inhibition from multiple sources, a review of the various forms of inhibition may be in order (see Table 16.1).

ARTHROGENIC INHIBITION

Extremity and spinal joints can arthrogenically inhibit the muscles that cross them. The inhibited muscle is the muscle that, by its contraction, stretches (the part of) the joint capsule that has limited extensibility. The arthrogenically inhibited muscle tests weak when the taut antagonistic ligamento-capsular tissue is being stretched. In other parts of the range, where this tissue is not under tension, the muscle tests strong. The point in the range where the muscle strength changes is abrupt.

Muscles that are arthrogenically inhibited by neighboring joints without active range test weak throughout their full range or a large part of their range. If a neighboring joint with an active range functions as a joint without active range (e.g., spine or knee for hip abduction and adduction), movements of this neighboring joint can increase or abolish the inhibition (e.g., knee or spinal flexion or extension).

Pelvic joint dysfunctions affect muscles over large parts of their range. The delineation between where the muscle tests strong and where it tests weak is gradual.

RADICULOPATHY

Radiculopathy can cause weakness throughout the range of the segmentally related peripheral muscle. If the radiculopathy is caused by lateral foraminal stenosis, the weakness worsens with ipsilateral facet extension and deteriorates even more if ipsilateral extension is maintained. Facet flexion improves strength, and strength improves even more if flexion is maintained.

Radiculopathy caused by a herniated lumbar nucleus pulposus also causes weakness throughout the range of the segmentally related extremity muscle. The weakness may worsen with lumbar flexion and may

Table 16.1 The result of muscle tests in different positions linked with causes of weakness.

	Testing the extremity muscle in a:		Testing in spinal:	
	Shortened position	Lengthened position	Extension	Flexion
Arthrogenic inhibition				
• Extremity joint	Weak	Strong	No change	No change
• Spinal joint	--	--	Weak or strong	Strong or weak
• Pelvic joint	Weak	Stronger	No change	No change
• Other joints without active range				
– Extremity	Weak	Weak	No change	No change
– Spine	No change	No change	Weak or strong	Strong or weak
Radiculopathy				
• Lateral foraminal stenosis	No change	No change	Weaker	Stronger
• Central lumbar stenosis	No change	No change	Stronger	Weaker
Other peripheral nerve dysfunction or muscle tissue disorders	Weak	Weak	Weak	Weak
Pseudoradiculopathy	No change	No change	Weak or strong	Strong or weak
Disease causing weakness	Weak	Weak	Weak	Weak

deteriorate even more if flexion is maintained. Lumbar extension may improve strength, and if extension is maintained, strength improves even more.

PSEUDORADICULOPATHY

Pseudoradiculopathy can also cause weakness of the segmentally related peripheral muscle, brought on by ipsilateral flexion or extension of the facet joint. The opposite movement completely restores the muscle's strength, unless facet flexion and extension are both restricted. Some of the muscle strength is restored if the spinal position responsible for the weakness is maintained.

DISEASES, MUSCLE TISSUE DISORDERS, AND PERIPHERAL NERVE LESIONS

The muscle is weak throughout its range and in every position of the spine. No position can be found where the muscle is strong.

Pain experienced during strong contraction of a damaged muscle is often a combination of nociception from the lesion and stiffness. The input to the central nervous system about the stiffness also could be misinterpreted by the cerebral cortex as being nociception. A mobilization of the joint to restore its mobility may improve strength and reduce pain as well.

Combinations of different sources of inhibition are common. Weakness throughout the range of a muscle is often seen in combination with arthrogenic inhibition. It may be that one caused the other. The weakness throughout the range created a lack of active excursion, followed in time by stiffness of the joint and arthrogenic inhibition. The muscle was thus inhibited throughout its range by the original inhibition and became arthrogenically inhibited as well.

Forms of inhibition that cause weakness throughout the range tend to obscure arthrogenic inhibition with weakness in part of the range. Weakness in part of the range might show up after the weakness throughout the full range is corrected. For instance, after a successful spinal treatment, an extremity muscle weakness will improve from weakness throughout the range of the extremity joint to weakness in only part of the range. The next step would be to remove the arthrogenic inhibition as well.

Besides a combination of peripheral arthrogenic and spinal causes of inhibition, other combinations are also possible. For instance, I once treated a patient who had a weak right quadratus lumborum throughout the lumbar range. There was nothing to indicate, however, that the patient had a dysfunction of her twelfth rib or her sacroiliac joint, either of which can cause quadratus lumborum weakness throughout its range. It turned out that the right quadratus lumborum was arthrogenically inhibited by a right L5 facet joint with limited extension and a right L2 facet joint with limited flexion, causing a pseudoradicular inhibition of the quadratus lumborum. The right L5 facet joint with stiffness for extension inhibited the quadratus lumborum when the muscle was in a shortened position, whereas the right L2 facet joint with stiffness for flexion inhibited the quadratus lumborum when the muscle was lengthened. The quadratus lumborum also tested weak in a midposition, possibly due to one or both facet dysfunctions being a major restriction, inhibiting the quadratus lumborum over more than 50 percent of its range. Other combinations of dysfunctions occur that include radiculopathy, inhibition from a joint without active range, diseases, muscle tissue disorders, and peripheral nerve lesions.

Summary

Muscles are often inhibited by spinal conditions such as radiculopathy and pseudoradiculopathy. These conditions may cause weakness of spinal muscles in part of their range and weakness of extremity muscles in their entire range. The muscle weakness is often dependent on the position of the spine. Both spinal flexion and extension could be responsible for muscle inhibition. One of the differences between radiculopathy and pseudoradiculopathy is that maintaining the inhibiting spinal position tends to increase inhibition in the former and to diminish inhibition in the latter.

Weakness from a combination of inhibitions is common. In the extremities, dysfunctions causing full-range weakness tend to obscure arthrogenic inhibition causing weakness in part of the range. A careful evaluation, possibly including mini-trials of treatment, could reveal the causes of muscle weakness.

References

1. April C, Dwyer A, Bogduk N: Cervical zygapophyseal joint pain patterns II: A clinical evaluation. Spine 15:458-461, 1990
2. Bigos S, Bowyer O, Braen G, et al: Acute Low Back Problems in Adults: Clinical Practice Guideline No. 14. AHCPR Publication No. 95-0642. Rockville, Maryland, Agency for Health Care Policy and

Research, Public Health Service, US Department of Health and Human Services, 1994

3. Butler DS: Mobilisation of the Nervous System. New York, Churchill Livingstone, 1991

4. Cranenburgh B van: Inleiding in de Toegepaste Neurowetenschappen: Opvattingen over zenuwstelsel en hersenen. Lochem, The Netherlands, De Tijdstroom, 1983, vol 1

5. D'Avella D, Mingrino S: Microsurgical anatomy of lumbosacral spinal roots. J Neurosurg 51:819-823, 1979

6. Dvorák J, Dvorák V: Manuelle Medicin: Diagnostik. New York, George Thieme Verlag Stuttgart, 1983

7. El A van der, Lunacek P, Wagenaar A: Manuele Therapie: Wervelkolom Onderzoek. Rotterdam, Manuwel, 1983, vol 1

8. Fortin JD, April CN, Ponthieux B, et al: Sacroiliac joint: Pain referral maps upon applying a new injection/arthrography technique: II. Clinical evaluation. Spine 19:1483-1489, 1994

9. Fortin JD, Dwyer AP, West S, et al: Sacroiliac joint: Pain referral maps upon applying a new injection/arthrography technique: I. Asymptomatic volunteers. Spine, 19:1475-1482, 1994

10. Grieve GP: Referred pain and other clinical features. In Boyling JD, Palastanga N (eds): Grieve's Modern Manual Therapy, ed 2. New York, Churchill Livingstone, 1994, pp 271-292

11. Hasue M, Kikuchi S, Sakuyama Y, et al: Anatomic study of the interrelation between lumbosacral nerve roots and their surrounding tissue. Spine 8:50-58, 1983

12. Jones LH: Strain and Counterstrain. Indianapolis, The American Academy of Osteopathy, 1981

13. Kadish LJ, Simmons EH: Anomalies of the lumbosacral nerve roots: An anatomical investigation and myelographic study. J Bone Joint Surg Br 66:411-416, 1984

14. Kendall FP, Kendall McCreary E, Provance PG: Muscles: Testing and Function, ed 4. London, Williams & Wilkins, 1993

15. Kikuchi S, Hasue M, Nishiyama K, et al: Anatomical features of the furcal nerve and its clinical significance. Spine 11:1002-1007, 1986

16. Kusunose RS, Wendorff R: Strain and Counterstrain Syllabus. Encinitas, California, Jones Institute, 1990

17. Maitland GD: Musculo-Skeletal Examination and Recording Guide, ed 4. Glen Osmond, South Australia, Lauderdale Press, 1986

18. McCall IW, Park WM, O'Brien JP: Induced pain referral from posterior lumbar elements in normal subjects. Spine 4:441-446, 1979

19. McCullogh JA, Waddell G: Variation of the lumbosacral myotomes with bony segmental anomalies. J Bone Joint Surg Br 62:475-480, 1980

20. Neidre A, Macnab I: Anomalies of the lumbosacral nerve roots: Review of 16 cases and classification. Spine 8:294-299, 1983

21. Pallie W: The intersegmental anastomoses of posterior rootlets and their significance. J Neurosurg 16:188-196, 1959

22. Paris SV, Nyberg R, Irwin M: S2 Course Notes. St. Augustine, Florida, Institute of Graduate Physical Therapy, 1991

23. Parke WW, Watanabe R: Lumbosacral intersegmental epispinal axons and ectopic ventral nerve rootlets. J Neurosurg 67:269-277, 1987

24. Peck D, Brower TD: Algorithms for the segmental motor innervation of the extremities. Am Surg 53:270-273, 1987

25. Rydevik B, Brown MD, Lundborg G: Pathoanatomy and pathophysiology of nerve root compression. Spine 9:7-15, 1984

26. Schnebel BE, Watkins RG, Dillin W: The role of spinal flexion and extension in changing nerve root compression in disc herniations. Spine 14:835-837, 1989

27. Seddon H: Surgical Disorders of the Peripheral Nerves. Baltimore, Williams & Wilkins, 1972, pp 32-56

28. Williams PL, Warwick R (eds): Gray's Anatomy, ed 36. New York, Churchill Livingstone, 1980

29. Yoo JU, Zou D, Edwards T, et al: Effect of cervical spine motion on the neuroforaminal dimensions of human cervical spine. Spine 17:1131-1136, 1992

17. Therapy

Traditional Strengthening Exercises

Therapy that addresses the cause of muscle weakness should not simply be a program of traditional strengthening exercises. Muscle weakness is not limited to the muscle itself; the central nervous system plays a major role in its etiology and resolution. To clarify the role of the central nervous system in muscle weakness, I will discuss muscle weakness caused by diminished use (atrophy), a form of muscle weakness that may be looked upon as clearly located in the muscle itself.

Although it is widely accepted that immobilization causes muscle atrophy and weakness, the fact that the nervous system contributes to the weakness is less well known.[55] Immobilization of an extremity produces both stiffness of the connective tissue and weakness of the muscles, the latter of which is often attributed to atrophy. In a study of patients who had had one leg immobilized by a cast, Fuglsang and Scheel measured quadriceps strength and electrical activity immediately after removal of the cast, eight days later, and again 30 days later.[11] These measurements were compared with the contralateral side. Quadriceps strength on the day after the cast was removed was 40 percent of that on the contralateral side. Without training, eight days later, quadriceps strength improved to 80 percent of the healthy contralateral side. This quick recovery rate does not seem compatible with the time required for recovery of the cross-sectional area of the muscle fibers. Initial EMG recordings from the quadriceps after the cast had been removed revealed a reduction in the number of active motor units. Eight days later, the number of active motor units had increased to normal (i.e., to the same number as on the contralateral side).

Mayer and colleagues also found changes in the nervous system related to immobilization of muscles, concluding that immobilization of a limb altered the input from the primary spindle cell afferents, which synapse in the spinal cord.[35] At these synapses, the input from the primary afferents is transmitted onto other neurons. Immobilization reduced the efficiency of input transmission from the spindle cell's primary afferents, resulting in decreased facilitation of the muscle (i.e., decreased strength).

With few exceptions, traditional strengthening exercises for weak muscles rarely address the cause of weakness. Such exercises might, for instance, increase muscle bulk and energy storage in the muscle while leaving the source of inhibition as is. Even while doing the strengthening exercises, the muscle could be inhibited and contract submaximally. The presence of inhibition frustrates the purpose of a strengthening program, as any gain in improved muscle function (e.g., muscle bulk and energy storage) is quickly lost after the exercises are discontinued. In most cases, traditional strengthening exercises for inhibited muscles can be likened to mopping a wet floor while the faucet (the inhibition) keeps running in an overflowing sink. At best, strengthening exercises might address the source of inhibition in cases of disuse (e.g., after cast removal) or arthrogenic inhibition, provided they mobilize the stiff joint.

A strengthening exercise can stretch a stiffness if it meets one of two conditions. The first condition is that the muscle must contract and position the joint beyond its dysfunctional barrier (positional mobilization). Maintaining the joint at or near its limited end range stretches the taut ligamento-capsular structures. The exercise addresses the cause of the weakness by positioning the joint into its restricted range, where mobilization can take place. Once the joint is correctly positioned, the muscle can relax, provided the joint remains in this position. For instance, in the supine patient, the hip abductors might position the hip joint in abduction, stretching the medial-inferior hip joint capsule. Once the hip of the supine patient is abducted, the abductors can relax, provided the hip remains in this abducted position. This form of mobilization, in which the inhibited muscle positions the joint into its restricted range, may be the least effective.

The second condition is that the muscle contraction must translate the bones that make up the joint in such a way that the restricted range is mobilized. This could be termed an *active mobilization* of the joint. Active mobilizations are very effective means of improving joint range. For example, in the bent knee, contraction of the hamstrings translates the tibia posteriorly on the femur, mobilizing flexion (see Figure 17.1). In contrast to positional mobilization, an active mobilization only mobilizes while the muscle contracts.

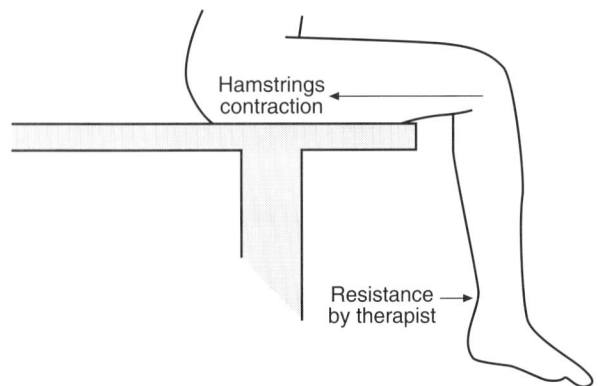

Fig. 17.1 An active flexion mobilization of the knee joint by an isometric contraction of the hamstrings.

Mobilization

Normal extensibility of the joint's ligamento-capsular tissue is essential for range of joint motion and related muscle function. During mobilization, the joint's ligamento-capsular tissue is stretched with a combination of a specific position and a translatory movement. The direct goal of mobilization is not to restore the translatory movement of the joint, but to improve the extensibility of the ligamento-capsular tissue.[38,52]

While a joint is held in an end-range position, a slow relaxation of antagonistic tissues occurs, making it easier to maintain this position.[17,46] The relaxation presumably results from viscoelastic elongation of the joint's ligamento-capsular tissue.[17] This relaxation is termed *hysteresis*.[46]

Mechanically, the collagen fibers within the capsule and ligaments have viscoelastic properties and could be compared with a spring–dashpot arrangement (see Figure 17.2).[9,60] When the ligamento-capsular tissues are stretched, tension develops in the "springs" initially; however, maintained stretch results in a viscous "creep" of the tissues. This creep is a product of time, and with longer duration of the stretch, the tissue's extensibility further improves. The increased extensibility of the ligamento-capsular tissue may be permanent, since the joint capsule proved to have some plasticity (i.e., permanent deformation).[31]

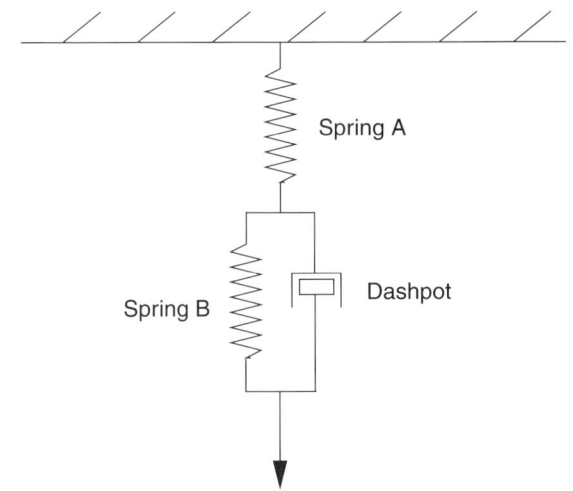

Fig. 17.2 The spring–dashpot model of collagen fiber lengthening. During initial lengthening of the construction, spring A lengthens (elastic deformation). Maintaining the lengthened position allows the dashpot (with spring B) to slowly move (plastic deformation). The movement by the dashpot represents viscous or plastic deformation.

Several mechanisms explain why mobilizations reduce articular receptor activity. The Type I, II, III, and possibly Type IV articular receptors are attached to collagen fibers.[20,21] In line with the relaxation (hysteresis) of the ligamento-capsular tissues, the discharge frequency of the articular receptors reduces.[17] Viscoelastic relaxation of ligamento-capsular tissue leads to "relaxation" of articular receptor discharge, because the receptor is coupled to the capsule and the ligaments.[19] A second reason for reduction of articular receptor discharge is a possible electrochemical adaptation of the receptors themselves.[1,19,37] Hysteresis of articular receptors could thus be due to relaxation of the ligamento-capsular tissues, as well as to adaptation of the receptors themselves.[32,33,42]

Mobilization of a joint results in a temporary reduction of intraarticular pressure. During the mobilization, the end-range position initially increases intraarticular pressure, and the increased pressure increases the absorption rate of the intraarticular fluid.[46,61,62] In addition, maintaining the end-range position relaxes the joint capsule, further decreasing the intraarticular pressure.[22,46] Therefore, sustained end-range positions, as in mobilizations, decrease intraarticular pressure due to hysteresis of the joint capsule and resorption of the intraarticular fluid.[46] Even when the joint returns from its end-range position to a loose-packed position, the intraarticular pressure is lower than before the end-range position was assumed.[46] The reduced intraarticular pressure may contribute to the decreased firing of the articular receptors.

Articular receptors display hysteresis not only to position, but also to movement.[24,32,33,37,42] Schaible and Schmidt wrote that repetitive movements desensitize receptors.[53] The hysteresis to movement occurs in the articular mechanoreceptors, as well as the articular Type IV receptors.[24] Therefore, both static and dynamic mobilizations (e.g., oscillations) can reduce articular receptor activity.

In conclusion, the decreased firing rate of articular receptors following a joint mobilization could be explained in three ways:

1. An indirect reduction of receptor activity through a relaxation of the ligamento-capsular tissue,

2. An indirect reduction of receptor activity through reduced intraarticular pressure, or

3. An adaptation of the articular receptors themselves.

Any reduction in articular receptor activity following a mobilization reduces the impulse traffic to the gamma motoneurons. Both active and passive mobilization techniques alter gamma bias and thus alter muscle function. A therapeutic influence on gamma bias or the spindle cell is not unique to active and passive mobilization techniques. Other treatment techniques also affect gamma bias. For instance, Muscle Energy's isometric contractions may alter the gamma bias as well.[43,44]

The decreased articular receptor activity in the stretched part of the joint capsule reduces tone in the muscle opposing stretch (antagonistic muscle) and increases tone in the agonist muscles. These changes make it easier for the patient to keep the joint in the end-range position. The successful normalization of muscle function indicates that the joint dysfunction that caused the muscle function changes has also been corrected.

Immediate and Aftereffects of a Mobilization

Mobilizations have an immediate effect (i.e., while the mobilization is taking place), as well as an effect after the mobilization is discontinued. During the mobilization, the articular mechanoreceptors in the stretched regions of the ligamento-capsular tissues initially discharge at high rates. With discharges at a sufficient rate, the tone of the muscle opposing stretch (antagonist) increases and the tone of the agonist muscle decreases. The intensity of the mobilization needed to produce these muscle tone changes may be very low. Johansson found that only very modest stretches of the ligaments were needed to influence the spindle afferents (also see chapter 7).[25] During the mobilization, the articular mechanoreceptors gradually decrease their firing rate, secondary to adaptation of the receptors and the ligamento-capsular tissue and to decreased intraarticular pressure. Due to the receptors' decreasing firing rate, the stretch-induced muscle tone alterations gradually diminish.

Therefore, although the goal of the mobilization is to decrease tension in the ligamento-capsular tissues and decrease the discharge rate of the articular receptors, the mobilization itself temporarily has the opposite effect. For example, during an evaluation, the therapist finds a weakness of the flexors of the base joint of the index finger (MCP II joint) when tested over a maximally flexed MCP joint; however, the flexors test strong when the MCP joint is in a loose-packed position. Thus, the therapist realizes that the flexor muscles may be inhibited by a limited flexion mobility of the MCP joint. A reduced flexion range at the MCP joint with stiff end-feel and limited joint mobility confirm a limited mobility of the dorsal joint structures, so the therapist decides to mobilize MCP flexion. The patient's MCP joint is flexed beyond its dysfunctional barrier, and the proximal phalanx is translated in a volar direction on the metacarpal (see Figure 17.3). The stiff capsule and ligaments in the dorsal portion of the MCP joint are stretched by the joint position and by the volar translation of the phalanx. Stretching of the capsule and ligaments causes a high discharge rate of the articular mechanoreceptors. During the mobilization, the articular receptor activity inhibits the MCP flexors and increases tone in the extensors. The tone changes in these muscles resist further flexion but do not prevent the therapist from translating the phalanx (see Figure 17.3). In addition, the translation of the articular bones is nearly friction free, even with increased pressure of one articular surface on the other, such as when the muscles contract.

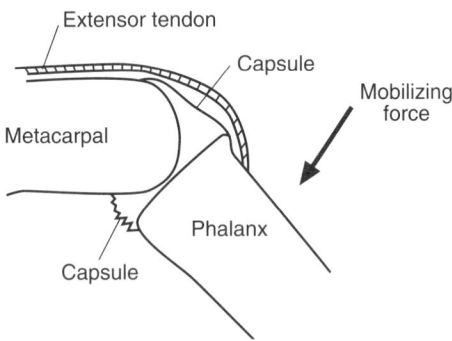

Fig. 17.3 A mobilization at the second metacarpophalangeal joint. The increased tone of the extensors during the mobilization prevents further flexion but affects the mobilization only slightly. The clinician can still translate the phalanx in a volar direction on the metacarpal.

During the mobilization, the articular mechanoreceptors gradually decrease their firing rate, and the stretch-induced muscle tone changes subside. In our example, the tone of the extensors is gradually reduced during the mobilization, and the tone of the flexors is gradually increased.

The aftereffects of a mobilization start when the therapist stops the mobilization. Ideally, the joint mobilization is discontinued when normal extensibility of the ligamento-capsular tissues is obtained. In practice, the therapist may have interrupted the mobilization a few times to assess whether this goal has been obtained. When the mobilization is discontinued, the stretch of the capsule and ligaments ceases and a further quick drop in the firing rate of the Type I and III articular mechanoreceptors occurs. The Type II articular receptors (responding to any tension change in the host tissue) temporarily increase their firing rate during the release. Following the release, the Type II receptors quickly (within about 0.5 second) return to a low discharge rate. Therefore, after the mobilization has ceased, all three mechanoreceptors have a low discharge rate. The mobilization is completed, and the previously inhibited muscles (the MCP flexors) have regained normal strength, whereas the facilitated muscles (the MCP extensors) have reduced in strength.

Sustained Versus Oscillatory Mobilizations

Mobilizations can be performed in a sustained (mobilizing force remains constant) or in an oscillating manner. Both methods have merit. Sustained mobilizations will cause a discharge of mainly the low-threshold Type I receptors. The quickly adapting Type II receptors may be completely inactive during a sustained mobilization. Thus, steady mobilizations may provide little articular receptor activity and consequently small muscle tone changes. If the therapist first mobilizes using a strong force, then follows with a sustained mobilization using less force, the Type I articular receptors will discharge at even lower rates because of the adaptation that occurred (see "Receptor Hysteresis and Residual Muscle Tension" in chapter 6).

In contrast, oscillatory mobilizations will produce a barrage of impulses to the central nervous system from the tonically active Type I receptors and the dynamically active Type II receptors. The fluctuating tension in the ligamento-capsular tissues provides constant stimulus for the Type II receptors. The afferent barrage produces increased reflex effects and increased gating of nociception.

Active Mobilization

Besides passive manual therapy techniques, active mobilization is another technique for improving ligamento-capsular extensibility. In contrast to a passive mobilization, in which the patient relaxes and the clinician provides the mobilizing force, an active mobilization is performed by the patient. In other words, the mobilizing force is delivered by a contraction of the patient's own muscles. During an active mobilization, the patient contracts a specific muscle or muscle group against resistance and translates the articular bones. Not every movement or every joint can be mobilized actively. A prerequisite for an active mobilization is that the line of pull of the muscle(s) is such that it can translate the articular bones. An example is a hamstrings contraction to mobilize a limited flexion mobility of the knee joint (see Figure 17.1). If the patient has knee flexion limited to about 90 degrees, at that point, or just short of it, the therapist can hold the patient's ankle and provide resistance to a biceps contraction. The hamstrings contraction translates the tibia posteriorly on the femur.

The translating force created by a muscle contraction does not need to be parallel with the joint surface, as in the above example. A translation with compression may be equally effective. An example might be a cranial glide of the fibula on the tibia by a biceps femoris contraction (see Figure 17.4). If the patient contracts the biceps over a straight knee, the fibula is pulled cranially and may compress the proximal tibiofibular joint. Since most joint surfaces provide almost no friction to translation, even while bearing weight, a mobilization with compression still serves as an effective mobilization technique. The near-frictionless joint surfaces alter the muscle pull into a translation parallel with the joint surfaces.

Another possibility is that a mobilization translates the articular bones while separating the joint surfaces as well. This can also be effective, as distraction of the joint surfaces with or without translation is a useful mobilization technique. For example, in an active mobilization for an elbow with a 10-degree flexion limitation (see Figure 17.5), a biceps brachii contraction in maximal flexion mobilizes the radiohumeral joint by distracting and translating the radius on the humerus.

Active mobilizations stretch ligamento-capsular tissue, regardless of any accompanying compression or distraction. This also applies to passive mobilization techniques. Passive mobilizations need not necessarily be done perfectly in line with the joint surface. Any mobilization (distraction with or without translation, translation, or translation with compression) that stretches the tissue with limited extensibility can be an effective mobilization. To improve the effectiveness of a mobilization, factors such as hand placement, comfort, or dosage may be more important than the precise direction of the mobilizing force.

To mobilize a joint actively, either the agonistic or the antagonistic muscle to the restricted movement may be used. If an agonist can be used effectively, its attachment closest to the joint is often onto the concave joint partner, as in the previous examples. When using the agonist, the inhibited muscle is always used for the mobilization (i.e., the inhibited hamstrings can be used to mobilize restricted knee flexion).

If an antagonist to the restricted movement can be used to translate the articulating surfaces, the antagonist's insertion closest to the joint is often on the con-

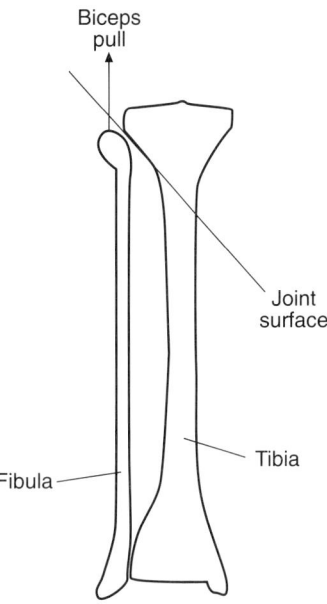

Fig. 17.4 An active cranial mobilization of the fibula on the tibia. The hamstrings contraction produces a cranial translation of the fibula on the tibia, but also causes a compression of the articulating surfaces at the proximal tibiofibular joint. At the distal tibiofibular joint, the same hamstrings contraction causes a cranial translation of the fibula without compression.

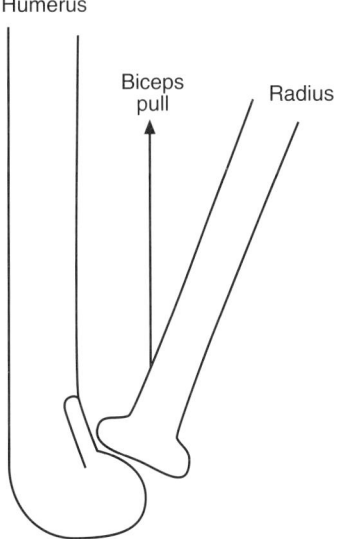

Fig. 17.5 An active flexion mobilization of the radiohumeral joint. The pull by the biceps contraction translates the joint surfaces, but also separates them.

vex joint partner. An example is a contraction of the glenohumeral adductor muscles to improve abduction mobility (see Figure 17.6). The adductors' insertion closest to the glenohumeral joint is on the convex humeral head. On normal abduction, the elbow moves cranially while the humeral head translates caudally on the glenoid (see Figure 17.7). For adequate abduction mobility, sufficient inferior translation of the humeral head on the glenoid is necessary. If inferior translation of the humeral head is insufficient, the adductors can be used to translate the humeral head inferiorly. Thus, the antagonists (adductors) are used to improve abduction mobility. The agonists (abductors) are completely useless here, since they do not provide a caudal translation of the humeral head. A deltoid contraction can even produce a cranial translation!

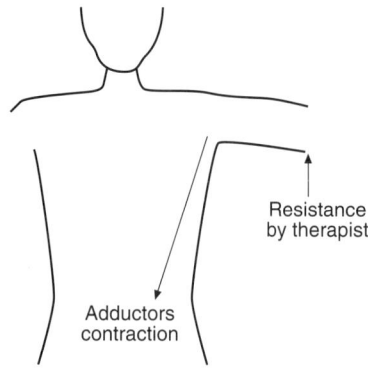

Fig. 17.6 An active abduction mobilization at the glenohumeral joint by an isometric contraction.

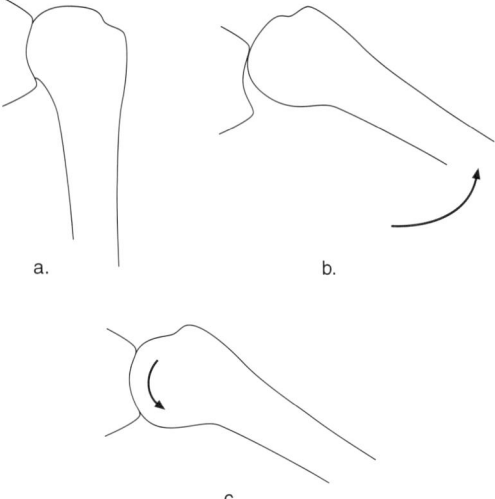

Fig. 17.7 The arthrokinematic movement of glenohumeral abduction. Fig. 16a represents 0 degree abduction, Fig. 16b, 40 degrees abduction without inferior translation, and Fig. 16c, 40 degrees abduction with inferior translation.

Abduction is a combination of the distal humerus moving cranially (b) while the humeral head translates inferiorly on the glenoid (c). The translation ensures that the convex joint partner (humeral head) does not "roll cranially out of its socket" (b).

Most muscles do not lend themselves to active mobilization of joints. Their line of pull is such that no effective translation can occur. For instance, no muscle is available to translate the femoral head inferiorly; therefore, no active mobilization exists for hip abduction. Most hip muscles center the femoral head firmly into the acetabulum or, like the adductors, may produce a cranial translation.

Clinically, active mobilizations are more effective than passive mobilizations. Often, less time is required to mobilize a joint when using active mobilizations. The effectiveness of active mobilizations may be due to the following two factors:

1. The active mobilizing force often far exceeds any passive mobilizing force. Thus, the clinician can opt to make the active mobilization very comfortable (using a gentle contraction) and still provide an effective mobilization force.

2. Active mobilizations address muscle imbalance better than passive mobilizations. Limited mobility is associated with weak agonists (stretching the taut tissue) and hypertonic antagonists (opposing the stretch) beyond the dysfunctional barrier. If a weak agonist can be used for the active mobilization, the muscle is invited to contract in its shortened, inhibited range. The contraction of the muscle trains it to function in this shortened range and reciprocally inhibits the hypertonic antagonist.

If an antagonistic muscle is available for the mobilization, this tight hypertonic antagonist is invited to contract in its lengthened position (where the dysfunctional barrier is engaged). The antagonist, with its increased tone, is stretched, and the contractions in this lengthened position may serve as a self-mobilization of the muscle. Therefore, the active mobilizations, whether by agonist or antagonist, may address the muscle imbalance by requesting the muscle to function in an unfamiliar part of the range.

In the next section, the above-described advantages of active mobilizations will be discussed using two examples: one for a mobilization by an agonist and one for a mobilization by an antagonist.

ACTIVE MOBILIZATION BY AN AGONIST

To explain active mobilization by an agonist, we will use the example of a patient with limited knee flexion mobility. Knee flexion is limited to 95 degrees, and the hamstring muscle is weak when tested in more than 70 degrees of knee flexion. The therapist brings the knee to about 90 degrees of flexion and invites the patient to contract the hamstrings comfortably. The therapist holds the patient's ankle and provides unyielding resistance to the hamstrings contraction (see Figure 17.8). The contraction of the hamstrings produces a posterior glide of the tibia on the femur (i.e., a flexion mobilization).

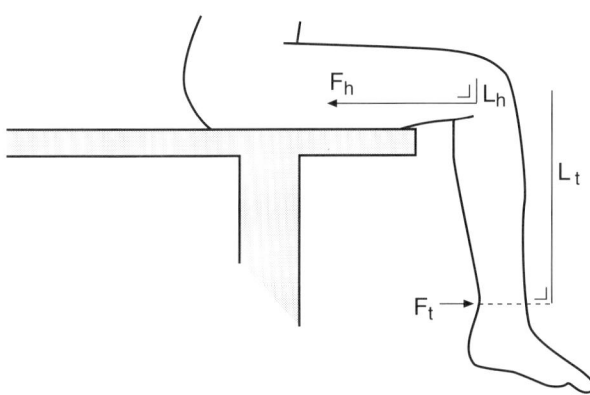

Fig. 17.8 An active flexion mobilization of the knee joint by an isometric contraction of the hamstrings.

F_h, or the force of the hamstrings, is the magnitude of the force produced by the hamstrings.

L_h is the length of the line perpendicular to the force F_h, running from force F_h to the knee's instantaneous axis of rotation.

F_t, or the force of the therapist, is the magnitude of the force produced by the therapist pulling the distal tibia anteriorly.

L_t is the length of the line perpendicular to the force F_t, running from force F_t to the knee's instantaneous center of rotation.

$$L_h \times F_h = L_t \times F_t$$

The magnitude of the force produced by the hamstring muscle multiplied by its lever arm equals the magnitude of the force exerted by the therapist at the ankle multiplied by that lever arm. Since the value of L_t is about six times greater than L_h, we can substitute these values in the equation:

$$1 \times F_h - 6 \times F_t \text{ or } F_h = 6 \times F_t$$

This means that the mobilizing force at the knee is six times greater than the magnitude of the therapist's force pulling at the ankle.

In calculating the mobilizing force at the knee joint, we will assume that the hamstrings contract isometrically. This means that no movement is taking place at the knee joint. The lever arm over which the hamstrings work is one-sixth the length of the lever arm of the counterforce produced by the therapist holding the ankle. Thus, the force produced by the hamstrings is six times greater than that applied by the therapist, bringing to bear a large translation force on the knee joint.

Even though the mobilizing force at the knee joint is six times greater than the magnitude of the therapist's force pulling at the ankle, since the therapist asked the patient to contract the hamstrings comfortably, the patient often experiences this strong mobilization as comfortable.

Whereas before the mobilization, the hamstrings tested weak beyond 70 degrees of flexion, due to the sustained contraction of the hamstrings at 90 degrees, this muscle "learns" to contract at this knee angle. In other words, the contraction of an arthrogenically inhibited muscle within its inhibited range trains the muscle (and the nervous system) to function in this range.

In the previous example, the mobilizing force was six times greater than the therapist's force. If instead we had used the example of specific trapezius fibers mobilizing a thoracic segment, the mobilizing force would have been even greater. To mobilize the thoracic segment using the trapezius, the therapist holds the patient's wrist and provides resistance to the trapezius contraction. The therapist's lever arm, from the thoracic vertebra to the wrist, is about 80 cm. The trapezius, attaching to the spinous process, provides the mobilizing force to the thoracic vertebra. Since the moment arm of the mobilizing force is about 5 cm (the distance between the vertebra's rotational axis and the tip of the spinous process), the mobilizing force is 16 (80 divided by 5) times the therapist's force at the patient's wrist!

ACTIVE MOBILIZATION BY AN ANTAGONIST

An example of an active mobilization by an antagonistic muscle is a contraction of the glenohumeral adductors to improve abduction mobility. A taut inferior glenohumeral capsule limits abduction range, inhibits the abductor muscles, and reciprocally increases the tone of the adductors. When passively abducting the patient's shoulder, the therapist might describe the increased adductor tone as guarding or as a limited abduction range with restrictive barriers. Increasing the mobility of the inferior glenohumeral capsule improves abduction mobility, improves abductor strength, and reduces adductor tone.

To improve limited abduction of the glenohumeral joint, no agonist (deltoid or supraspinatus) is available to produce caudal translation of the humeral head. Any contraction of the supraspinatus would only center the humeral head in the glenoid, and a contraction of the deltoid might even pull the humeral head cranially in the glenoid; however, the antagonists of these muscles, the adductors (latissimus dorsi, teres major, and pectoralis major), produce a caudal glide of the humerus in the glenoid.

In this example, the patient has only 75 degrees of abduction available at the glenohumeral joint. The therapist supports the patient's elbow at 70 degrees of glenohumeral abduction, then asks the patient to press the elbow gently toward the floor (see Figure 17.9). The therapist provides unyielding resistance to the adductors' contraction.

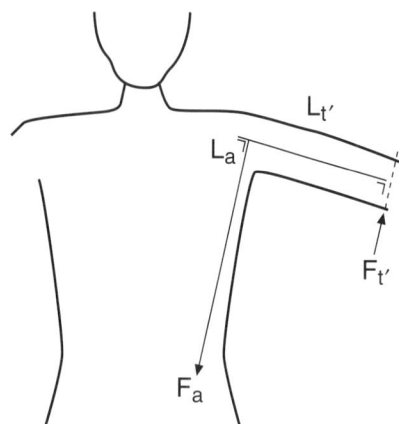

Fig. 17.9 An active abduction mobilization at the glenohumeral joint by an isometric contraction of the adductors.

F_a, or the force of the adductors, is the magnitude of the force produced by the adductors.

L_a is the length of the line perpendicular to the force F_a, running from force F_a to the glenohumeral joint's instantaneous axis of rotation.

$F_{t'}$, or the force of the therapist, is the magnitude of the force produced by the therapist pushing the distal humerus into abduction.

$L_{t'}$ is the length of the line perpendicular to force $F_{t'}$, running from force $F_{t'}$ to the glenohumeral joint's instantaneous center of rotation.

$$L_a \times F_a = L_{t'} \times F_{t'}$$

The value of $L_{t'}$ is about five times that of L_a.

$$1 \times F_a = 5 \times F_{t'} \text{ or } F_a = 5 \times F_{t'}$$

Thus, the mobilizing force at the shoulder joint (F_a) is about five times greater than the force produced by the therapist ($L_{t'}$).

A voluntary contraction of the adductor muscles produces two direct therapeutic effects. First, the contraction produces a caudal glide of the humerus on the glenoid, and second, by contracting in a lengthened position, the adductors might mobilize themselves.[13]

Besides these two direct mechanical effects, the increased mobility of the inferior capsule of the glenohumeral joint normalizes the muscle imbalance. Neurophysiologically, the tone in the hypertonic adductors decreases and the tone in the inhibited abductors increases. Ironically, it appears that the adductors are trained to make the abductors stronger.

Postmobilization Weakness

Occasionally, immediately following a mobilization, the muscle intended to be strengthened fails to test stronger and may even test weaker than before the mobilization. This can happen despite a correctly chosen and correctly performed mobilization technique. In fact, it only happens when the mobilization technique effectively stretches the target tissue. The weakness, however, is temporary, and after a few minutes the muscle tests strong. Such cases of temporary weakness following a mobilization have five elements in common:

1. The structure being stretched is quite stiff. Tissues with minimal stiffness that require only a minimal mobilization do not cause temporary weakness.

2. The therapist uses a strong mobilization technique for a long duration.

3. Following the mobilization, the patient experiences stiffness and often only carefully moves out of the mobilization position. The sensation of stiffness usually lasts a minute or less.

4. During the patient's subjective experience of stiffness, the muscle (intended to be made stronger) tests weak.

5. The subjective experience of stiffness and the weakness of the muscle disappear simultaneously. After the patient reports that the stiffness has disappeared and that the joint feels normal again, the muscle tests stronger.

Why does a correctly performed but intense and enduring mobilization cause temporary weakness when it was intended to improve the muscle's strength? For the answer, we must look at the articular reflexes.

Increased weakness of the agonist is perfectly normal during a mobilization. Not only is the joint brought into the position where the muscle tested weak, but in addition, the mobilization took place. Both the position and the mobilization stretch the tight ligamento-capsular tissue and cause inhibition of the agonist; however, this type of muscle weakness is normally limited to the duration of the mobilization.

The occasional temporary weakness after the mobilization is linked directly to the severity of stiffness and the intensity of the mobilization. A strong mobilization of stiff tissue causes greater than usual stimulation of Type I receptors as well as stimulation of higher threshold articular receptors (possibly the Type III and IV receptors). The afference from these high-threshold receptors superimposes on the low-threshold afference and causes strong inhibition of the muscle. The relatively long duration of the mobilization may ingrain the inhibitory pathways in the central nervous system, creating a self-perpetuating circuitry within the spinal cord that endures temporarily following the mobilization. This can be likened to a spinal memory of the inhibition. The spinal circuitry may be responsible for both the temporary sensation of stiffness and the post-mobilization weakness.

Clinically, it is important to know that, following a mobilization, a muscle may test weak due to temporary perseverance of inhibition. If this occurs, the therapist should ask the patient if the body part feels stiff. If it does, the therapist should ask the patient to move the joint until it feels loose, after which the muscle should be tested again for increased strength. Often, the transient postmobilization weakness and the feeling of stiffness are an indication that the therapist worked on a severe restriction, and more therapeutic improvement can be expected upon correction.

High-Velocity Thrust Manipulation

High-velocity thrust manipulation will stimulate the Type I and II receptors and possibly the high-threshold Type III receptors.[4] Normally, a mobilization of this magnitude would cause massive guarding by the muscles (i.e., increased tone of the antagonist with inhibition of the agonist); however, the high velocity ensures that the procedure is completed before the articular reflexes can prevent the movement.[4]

A high-velocity thrust manipulation can be performed with or without a "pop" (an audible release). The high-velocity thrust with audible release is a quick stretch of the joint capsule that produces a relative vacuum inside the joint, resulting in a release of the dissolved gases from the synovial fluid (see Figure 17.10). All fluids in the human body, including synovial fluid, contain dissolved gases. The strong negative pressure (a relative vacuum) during the manipulation allows the gases to "escape" from the fluid and enter a gaseous state.[4,49] Unsworth and colleagues offer an alternative explanation for this phenomenon in that reduction of pressure causes a reduction in the vapor temperature of fluids.[58] If sufficient pressure reduction takes place, a fluid can boil at ambient temperature, resulting in vapor bubbles in the low-pressure regions.

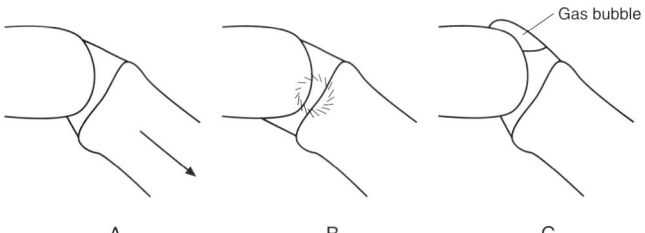

Fig. 17.10 A high-velocity thrust manipulation with audible release.

A. The joint surfaces are separated, creating a relative vacuum inside the joint.

B. The dissolved gases "explode" into a gaseous state.

C. After the manipulation, the joint cavity contains a gas bubble. The gas bubble dissolves back into the synovia in about 20 minutes.

One can expect the intraarticular gas to be predominantly nitrogen, since ambient air contains predominantly nitrogen.[49] However, Unsworth and colleagues analyzed the gas content of the synovial fluid and found the gas volume to be 80 percent carbon dioxide.[58]

Following a manipulation, it takes approximately 20 minutes for the gas to dissolve back into the synovial fluid. Thus, once a manipulation with an audible release has been performed, it cannot be performed on the same joint for about 20 minutes.[4,41,49,58]

The increased range following a manipulation with an audible release is due to two factors:

1. The sudden manipulation, occurring before muscle guarding can prevent it, stretches the joint's periarticular tissue. Following the manipulation, the extensibility of these tissues is greater (the joint is lax), permitting a greater range of motion.[4]

2. The increased extensibility of the periarticular tissues reduces the reflex activity to muscles, allowing the antagonist to relax.[4] In addition, inhibition of pain may also be responsible for reflex inhibition of muscles.[4] The temporary massive firing of the low- and high-threshold articular receptors at the time of the manipulation may inhibit pain.

The clinical importance of a manipulation with an audible release may be that the "pop" is evidence of a successful stretch of the joint's ligamento-capsular tissues; however, a high-velocity thrust without an audible release may be equally effective in increasing joint range.[4] Paris and Loubert suggested that an audible release may be an important psychological event.[49]

Manipulation of a joint alters its articular reflexes and results in muscle tone changes.[63] Cibulka and colleagues found increased strength of the hamstrings after manipulating the sacroiliac joint.[7] In a study of 20 subjects with a hamstrings strain, all were found to have an asymmetric pelvis, with the innominate (os coxae) on the affected side rotated anteriorly. An anterior rotation of the innominate moves the origin of the hamstrings, the ischial tuberosity, farther from its insertion, thus lengthening the hamstrings. The 20 subjects were divided into two groups. Both groups received hot packs and stretching of the hamstrings. The patients in the experimental group also received a manipulation to correct the innominate's anterior rotation.

Before and immediately after the treatment of both groups, a therapist, with no knowledge of the group assignments, tested the hamstrings on an isokinetic station. The experimental group, which had undergone manipulation of the sacroiliac joint, had more strength after the treatment compared with both the control group and the pretreatment test. The manipulation (to correct the innominate's anterior rotation) reduced the length of the hamstrings, but also increased the hamstrings' peak torque. Cibulka and colleagues concluded that the increased strength after the manipulation does not support the concept of the length-tension curve (see Figure 17.11). "According to the length-tension curve concept, shortening the hamstring musculotendinous unit reduces the muscle's ability to develop tension. After reducing the hamstring muscle length, however, we recorded a greater peak torque for the hamstring muscles. The gain in hamstring muscle peak torque, therefore, could not have resulted because of a change in the length-tension curve."[7, p. 1222]

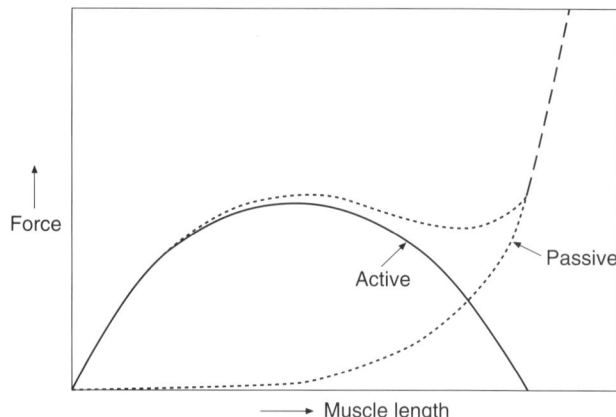

Fig. 17.11 The length-tension curve.

Force is plotted vertically. Muscle length is plotted on the horizontal axis. The muscle's length-tension curve is a summation of its active and passive curves. The active curve is the maximum force a muscle can generate as a result of its contraction. The passive curve is the force the muscle produces when it is completely relaxed (passive tension).

Thabe recorded EMGs of muscles innervated by C1 and muscles innervated by S1 in patients with upper cervical spine dysfunctions and patients with sacroiliac joint dysfunctions.[57] The muscles innervated by C1, from which the EMGs were recorded, were the obliquus capitis superior and the thyrohyoid. The muscles innervated by S1, from which EMGs were recorded, were the S1 portion of the multifidus, the piriformis, and the soleus. The EMG studies showed that nociception from the dysfunctional upper cervical and sacroiliac joints had increased local muscular activity. An anesthetic injected into the joint, a mobilization, or a manipulation abolished or reduced the spontaneous activity recorded from these muscles. The changes in the soleus muscle were not as dramatic as those in the dorsal muscles of the spine.

Murphy and colleagues stimulated the tibial nerve and took a surface EMG recording from the soleus muscle before and after a sacroiliac joint manipulation.[45] Following the manipulation, they found a reduction in motoneuron excitability on the same side as the manipulation.[45]

These last two studies also show an influence of spinal manipulation on the soleus, a segmentally related extremity muscle. The studies seem to support the concept of pseudoradiculopathy (see chapter 13).

Nansel and colleagues studied the effects of C2 and C7 manipulation on lumbopelvic muscle tone.[47] These researchers found decreased muscle tone in the lumbopelvic musculature after the lower cervical spine manipulation. Bringing C7 to tension without the high-velocity thrust also decreased the tone in the lumbar paraspinal muscles, but to a lesser degree. Manipulation of C2 did not create these effects. Nansel and colleagues concluded that spinal manipulation can have significant effects on the tone of the lumbopelvic musculature, presumably via tonic neck reflex pathways. Grice also found a reduction of paraspinal muscle tone following spinal manipulation.[16]

Pain Reduction

Following a mobilization or a manipulation, patients often experience a reduction of pain.[56,65] For example, Terett and colleagues found an increase in pain threshold after spinal manipulation.[56] The following discussion examines what pain reduction mechanisms play a role here.

Manipulation and mobilization result in an adaptation of the articular receptors (including the Type IV receptors), causing decreased afference from these receptors.[1,17,19,24,32,33,42,65] The result is a reduction of articular reflexes.[65] Zusman proposed that the reduction in articular afference and articular reflexes causes an increase in pain-free active and passive movement.[65] This increased pain-free movement may successfully counter the reduced excursion and increased muscle tone caused by articular nociception.[50,51] The improved pain-free range following manual therapy may close the gate for nociception in the spinal cord.

Muscles attach to the ligamento-capsular tissues. The purpose of these attachments may be to pull the capsule away from the articulating surfaces during concentric contractions (see Figure 17.12). Excessive muscle tension produces stress in the ligamento-capsular tissue, and the reduced muscle guarding following manual therapy may reduce pain by decreasing the stress on the joint's ligamento-capsular tissue.[18,19] The reduced tension results in a further decrease in articular afference.[18,19]

Neurotransmitters (e.g., substance P), if released, may produce joint inflammation and lower the threshold of nociceptors.[10,29,30,34,59,64] Joint manipulation may reduce the production of neurotransmitters.[59] Also, the release of endorphins in response to a manipulation may add to the reduction of pain.[59]

Radiculopathy

In patients with C6–7 disc herniations pressing on the C7 nerve root, causing weakness throughout the range of the adductor pollicis muscle and the flexors of

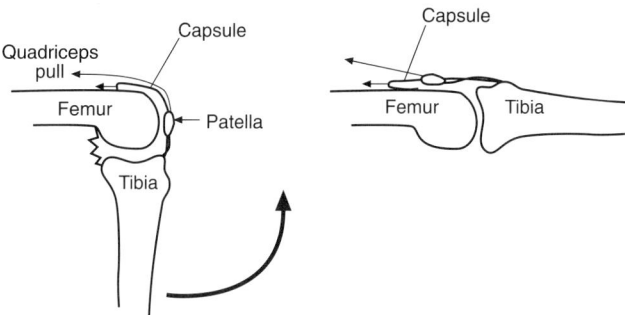

Fig. 17.12 The deepest part of the quadriceps, the articularis genus muscle, attaches to the knee joint capsule proximal to the patella. During active extension, the muscle pulls the capsule proximally, preventing impingement of the capsule between the patella and the femoral condyles.

the wrist, Kabat reported immediate restoration of normal muscle function after cervical traction.[28] Occasionally, cervical traction did not improve the patient's condition because the condition was irreversible or only slowly reversible; however, restoration of normal muscle strength was a routine observation.

Patients with a radiculopathy often have a pseudo-radiculopathy as well. A careful evaluation will reveal the presence of these conditions, as well as the spinal level. The therapist has several treatment options that directly address the physical cause of the dysfunction. The first treatment option is correction of the pseudo-radiculopathy using a mobilization or a manipulation to improve the facet joint's specific mobility. This means that facet flexion and/or extension must be addressed. Any improved (i.e., normalized) mobility of the segment could lead to better nutrition of the disc and other parts of the spinal movement segment.

In the presence of limited mobility, a flexion mobilization of the segment may be the most effective treatment for radiculopathy caused by lateral foraminal stenosis. The opening of the foramen removes pressure from the nerve root. Techniques that bring about lasting improvements in flexion mobility of the segment may allow the nerve to continue to function better. Flexion mobilization or manipulation could also stretch or break scar tissue from a preexisting inflammation within the foramen. Acute posterolateral disc herniations are contraindications for strong flexion mobilizations, as they might cause further tension in the herniated disc or compress the nerve root over the herniation.

If segmental extension causes or aggravates the peripheral muscle weakness, as it often does, the therapist must take care to distinguish between a pseudoradiculopathy and a radiculopathy caused by lateral foraminal stenosis. The treatment for a pseudoradiculopathy could be an extension-improving technique. In the case of lateral foraminal stenosis, extension techniques may be contraindicated, since they could cause more compression of the nerve root. Similarly, if lumbar flexion causes or aggravates the peripheral muscle weakness, the therapist must take care to distinguish between a pseudoradiculopathy and a radiculopathy caused by a herniation. The treatment for the pseudoradiculopathy could be a flexion-improving technique, whereas in the case of a posterolateral disc herniation, flexion-improving techniques are contraindicated. Radiculopathy and pseudoradiculopathy could be separately present at different spinal levels.

A pseudoradiculopathy caused by a facet joint is often corrected by treating the specific segmental level; however, the therapist can also do a generalized treatment of the spine that includes all the cervical or all the lumbar segments. A more generalized treatment circumvents the need for a spinal segmental evaluation. For example, a generalized treatment for the cervical and lumbar spine might consist of parts of the McKenzie method.[39,40]

A second treatment option is positional distraction, or the opening of the foramen through flexion, contralateral side bending, and contralateral rotation of the superior vertebra. Obviously, this technique resembles unilateral facet flexion, but its emphasis is less on the mobility of the segment and more on sustained decompression of the nerve root. The sustained decompression of the nerve root may restore its circulation. Positional distraction must be used carefully in the presence of disc pathology, since this technique stretches the posterolateral part of the disc and may compress the lumbar nerve roots against a herniated nucleus pulposus.

As pressure is removed from the nerve and blood circulation is restored, nerve function slowly returns. Small-diameter nerve fiber function is restored first, and large-diameter nerve fiber function last. With maintained opening of the foramen, muscle function slowly improves. The peripheral muscles often are stronger after only a few minutes of treatment; however, for the treatment to be truly effective, the muscle should remain strong after the positional distraction has been discontinued. Compressed nerves with advanced stages of neural inflammation, possibly with scar tissue interrupting continuity of the axons, may not respond with improved function after opening of the foramen.

A third treatment option is mobilization of the nerve root within its foramen. The mobility of the nerve root should be addressed. For the nerve to function normally, normal nerve mechanics are necessary; therefore, the therapist should check the extensibility of the nerve itself, as well as the mobility of the nerve root in relation to its surroundings (foramen and intervertebral disc). The therapist can address the nerve root mobility in a proximal/cranial and/or a distal/caudal direction. Before mobilizing the nerve root, it is preferable to remove any flexion or extension stiffness at the segment. For more information on nerve mobilization, the reader is referred to other publications.[5,6,8,54] In some cases, nerve mobilizations may be contraindicated by the presence of bony protrusions, such as vertebral osteophytes and cervical ribs, that are pressing on nerves. Generally, nerves do not tolerate being stretched, or repeatedly stretched, over these structures.

A final treatment option is spinal traction. An outcome of spinal traction is that the patient experiences improved muscle function while undergoing traction; however, for this result to be truly beneficial, the muscles should remain strong following treatment. In the cervical spine of a supine patient, frictionless traction forces of 15 kg (150 N) are often most effective in improving muscle function. In the lumbar spine, frictionless traction forces of 25 kg (250 N) are often very effective. In my experience, stronger traction forces are frequently less well tolerated and produce less improvement of muscle function.

Hypertonic Muscles

Receptors in taut ligamento-capsular tissue can inhibit the muscles that stretch this tissue and increase tone in the muscles that oppose the stretch. Therefore, hypertonicity of muscles in orthopedic conditions can result from articular reflex activity, just as can inhibition of muscles. To find the cause of hypertonicity in a muscle, the examiner could test the antagonistic muscles for strength. By finding a weak antagonist and the source of its inhibition in the ligamento-capsular tissue, the therapist can simultaneously correct the stiffness in the joint, the muscle weakness, and the hypertonicity in the muscle that opposed stretch in the tissue.

The following example will clarify the evaluation for the source of hypertonicity in a muscle and its correction. In evaluating a patient complaining of pain and tightness in the right upper trapezius, I found elevation of the right shoulder and increased tone in the right upper trapezius compared with the left trapezius. Muscle strength evaluation revealed a weakness of the right shoulder girdle depressors. Next, a more specific evaluation showed that this weakness was caused by an inhibition of part of the latissimus dorsi deriving from a lumbar segment. A mobilization of the lumbar segment corrected the strength of the latissimus dorsi and the depressors as a group, brought the right shoulder lower than the left, and reduced tone of the right trapezius to less than that of the left trapezius.

Sequencing the Treatment in Orthopedic Conditions

When multiple muscles acting in different directions over the same joint test weak, or when muscles in neighboring joints test weak as well, it may be unclear which dysfunction to address first. Treating dysfunctions in the proper sequence not only saves time, but it may also correct multiple dysfunctions by treating one key element. The following example illustrates the dilemma in choosing the correct dysfunction. When two opposing muscles ("a" and "b") are weak, strengthening muscle "a" might reciprocally inhibit muscle "b" and weaken it even more. Conversely, strengthening muscle "b" may further weaken muscle "a." Therefore, the therapist needs more information to decide which muscle to treat first.

This information is obtained by first retesting all muscles in the maximally loose-packed position of the joint. Many muscles that test weak in a shortened position will test strong in a loose-packed position, and testing in a maximally loose-packed position also eliminates all minor arthrogenic dysfunctions. In orthopedic conditions, weakness in a maximally loose-packed position can only result from a major arthrogenic dysfunction, a radiculopathy, or a pseudoradiculopathy. The muscle(s) testing weak in a maximally loose-packed position should be treated first.

Next, the therapist could test the muscles in a lengthened position. Patients with multiple muscles testing weak in a maximally loose-packed position often have constant pain and can find no position that eases the pain. If multiple muscles test weak in a maximally loose-packed position, testing these muscles in a lengthened position may be indicated. If only one muscle tests weak in a lengthened position, this muscle should be treated first. If the muscle is an extremity muscle, the treatment is often therapy of the spine.

Finally, the therapist could treat a dysfunction in a neighboring joint. If multiple muscles acting over a joint test weak, and only one muscle in a neighboring joint tests weak, the latter could be treated first. Treating inhibitions at neighboring joints is not the fastest way to improve the patient's condition, but it is a safe way to sequence treatment. For example, in a patient who was referred for a tear in his Achilles tendon, evaluation revealed weakness of the ankle dorsiflexors and plantar flexors, the invertors, and the evertors. Evaluation of the knee revealed weakness of the biceps caused by a limited mobility of the proximal tibiofibular joint. After mobilizing this joint, the biceps tested fully strong and three groups of ankle muscles (the evertors, invertors, and dorsiflexors) tested strong as well. These three muscle groups attach to the fibula or the interosseus membrane between the fibula and the tibia. The muscles that required treatment next were the plantar flexors.

Home Exercises

In patients with severe restrictions and muscle tone changes that cannot be corrected in a single visit, home exercises to further improve these conditions can shorten recuperation time. The home exercises described in this book differ from most exercises in that their purpose is to translate the joint surfaces and thus mobilize the joint capsule and ligaments. The exercises can be classified into passive and active home exercises. Active home exercises are modifications of the active mobilizations. In active home exercises, the resistance is provided not by the therapist, but by objects the patient has at home. Examples of passive home exercises are a kneeling knee flexion mobilization and a glenohumeral external rotation mobilization (see chapter 19). The mobilizing forces in these passive exercises are gravity acting on the body and the proximal articular bone and the ground reaction force (the force from the supporting surface) acting on the distal articular bone.

Home exercises are prescribed to address a specific goal. An effective exercise properly executed at home should achieve this goal at some point, and the therapist should continually evaluate the patient's progress and modify the exercise if necessary. Once the goal is achieved, the patient can stop doing the exercise. After the exercise has been discontinued for a few days, the therapist can reevaluate the patient to see if the goal is maintained without the exercise.

Exercises in every direction to improve every function of the joint (as some exercise sheets promote) are not only unnecessary, but may be detrimental to patient progress. Strengthening muscles that have increased tone or stretching tissues with normal extensibility or even hypermobility are definitely contraindicated. All exercises should be tailored to the individual patient. Therapists interested in exercise as an adjunct to manual therapy are referred to *Spinal Manipulation*, by Mark Bookhout, which contains an excellent chapter on exercises as a complement to manual therapy.[3]

Indications and Contraindications

INDICATIONS

The indications for joint mobilization or manipulation include the presence of:

1. Muscle weakness of arthrogenic, radicular, or pseudoradicular origin;
2. Limited mobility;
3. Pain from a joint–muscle complex.

Often these indications are related and are present simultaneously (e.g., a painfully limited movement of a joint, whether actively or passively moved, with weakness, all in the same direction). If only one of these components is present, the value of mobilization or manipulation becomes questionable. For instance, when pain is present but the mobility and strength of the local and related tissues are normal, manual therapy may be of little value. Unless the origin of a dysfunction is found, the dysfunction cannot be treated causally.

Occasionally, a patient presents with gross limitation of motion but with strong agonistic muscles. Limited mobility with strong muscles crossing the joint may occur in patients with long-standing conditions. For instance, limited mobility acquired before birth or in early childhood may present itself with normal muscle strength, often indicating that the patient has fully adapted to the limited range. In other words, the limited mobility (as compared with the contralateral extremity or to average values for an age-related population) is fully integrated in the person's functional physical status. As stated earlier, the value of manual therapy when the muscles are strong is questionable. If a therapist does decide to attempt a joint mobilization, although it may increase range, it might very well disturb the patient's homeostasis.

CONTRAINDICATIONS

Treatment is always contraindicated without a proper evaluation, which should be of an intensity appropriate to the condition. Acute injuries should be evaluated using gentle techniques, whereas an athlete who is ready to return to workouts should be evaluated using techniques that stress the tissues more vigorously. Every muscle test can be performed gently and carefully such that the test itself does not produce or aggravate pain. This requires very careful testing with minimal force to detect the quality of the contraction.

The evaluation should first check for red flags—conditions that should not be treated by a manual therapist, but should be referred to a medical specialist. For instance, red flags for low back pain are signs or symptoms of spinal fracture, tumor, infection, or major neurological compromise, such as a cauda equina syndrome.[2,36] Patients suspected of having any of the above conditions should be referred to a medical specialist immediately.[48]

In addition to red flags that are often contraindications for any manual therapy technique, contraindications for specific techniques exist. For instance, a suspected spinal fracture is a contraindication for any manual therapy technique, and the patient should be referred to a medical specialist. Once the specialist has determined that the fracture is stable and the patient needs pain relief, however, only gentle manual therapy techniques, such as Strain-Counterstrain or the Functional Technique, may be indicated.[27,15,26] Thus, the spinal fracture that initially was a contraindication for any technique later becomes a contraindication for specific techniques (the more forceful ones).

Similar to the intensity of evaluation techniques being dependent on the condition, the intensity of treatment techniques also depends on the condition. Gentle techniques are used for conditions with high reactivity, whereas stronger techniques are indicated for conditions with lower reactivity. For instance, in inflammatory arthritis, the intensity of a mobilization is dependent on the degree of inflammation and the extent of the joint contracture.[12] The intensity of a mobilization should always be lower in patients with rheumatoid arthritis. Children with juvenile chronic arthritis have a decreased threshold for pain, even in areas remote from the inflamed joints or in times of remission.[23] Adults with rheumatoid arthritis may have similar lower thresholds.

Another example of a condition that contraindicates more forceful techniques is pregnancy. Although gentle mobilizations may be well tolerated, high-velocity thrusts are contraindicated in the lower thoracic and upper lumbar areas, as these areas are the origin of the sympathetic innervation of the uterus.

Red flags and contraindications for specific techniques include:

- Weakening of the bone (e.g., due to osteoporosis, fracture, primary or metastatic malignant tumors, or use of steroids).[15]

- Any technique that might harm neurological or vascular structures (e.g., spinal extension in the presence of a radiculopathy or certain upper cervical techniques when the patient has a vertebral artery dysfunction).

- Infection—Infections may also indicate other disorders, and patients with infections may need to be referred to a medical specialist.[14] For example, urogenital infections may be the cause of low back pain.[14]

- Hypermobility—Whereas mobilization and manipulation may be indicated for joints with limited mobility, hypermobility and ligamentous laxity are a contraindication.[22] Certain conditions are often accompanied by hypermobility, including:[22]

    – Athetoid and ataxic forms of cerebral palsy;

    – Down syndrome;

    – Prader-Willi syndrome,

    – Children with generalized developmental delay of unknown etiology may exhibit ligamentous laxity.

Healthy individuals may have generalized hypermobility as well. In these individuals, hypermobility should be individually determined. For instance, a person with generalized hypermobility might have a single joint with limited mobility in one or more directions. This joint could have limited mobility when compared with the patient's other (healthy) joints but might be hypermobile when compared with the joints of other people. Other factors such as muscle strength or pain may assist in detecting the functional status of the tissue.

- Any technique that could damage tissue structure (e.g., in total hip replacements, combinations of internal rotation, adduction, and flexion are contraindicated).

This list of contraindications is not all-inclusive. A proper evaluation is the best method for forming a proper treatment plan. In addition to a standard orthopedic or neurological evaluation, muscle testing as presented in this book is a valuable tool in finding the correct treatment for a patient's complaint.

Summary

Traditional strengthening exercises may not be the most appropriate method for correcting muscle weakness. Therapy for muscle weakness should address the cause of the weakness.

Indications for manual therapy are arthrogenic muscle weakness, weakness caused by radiculopathy or pseudoradiculopathy, limited mobility, and pain from a muscle or a joint. Often, weakness, limited mobility, and musculoskeletal pain are different expressions of the same dysfunction. Contraindications to treatment are the absence of these factors. Other contraindications are listed earlier in this chapter.

The sequence for treating multiple weak muscles often is to treat the greatest dysfunction first. Frequently, the spine causes greater dysfunctions than extremity dysfunctions. In the case of arthrogenous muscle weakness, including pseudoradicular muscle weakness caused by a facet joint, the treatment is a mobilization of the tight ligamento-capsular tissue to increase its extensibility. Following a mobilization, the decreased hyperactivity of articular receptors reduces the dysfunctional muscle tone; however, during the mobilization, before the articular receptors have adapted to the stretch, the dysfunctional muscle tone is more pronounced (e.g., the inhibited muscle is even more inhibited). In some cases, these dysfunctional muscle tone changes may prevent adequate stretching of the joint. A high-velocity thrust manipulation can circumvent these excessive muscle tone changes by completing the mobilization before they can prevent it. Active mobilizations use the inhibited or facilitated muscles to mobilize the joint. In addition, active mobilizations use long lever arms, which results in strong mobilizing forces. Oscillatory mobilizations provide strong stimulation of Type II receptors, possibly increasing the articular reflexes and the gating of nociception.

Home exercises can be used to mobilize a joint if the dysfunction is severe and one office visit cannot resolve it completely. If a weakness is caused by a radiculopathy, the treatment to restore muscle function should be directed at the nerve root or its surrounding tissues. Nerve mobilization may be an effective treatment.

Strong mobilizations over a relatively long period can cause temporary endurance of the articular reflexes evoked by the mobilization such that, following the mobilization, the muscle intended to be strengthened still tests weak. This is termed postmobilization weakness. During this temporary postmobilization weakness, the patient may experience a feeling of stiffness.

References

1. Adelman WJ, Palti Y: The influence of external potassium on the influence of sodium currents in the giant axon of the squid, Loligo Pealei. J Gen Physiol 53:685-703, 1969

2. Bigos S, Bowyer O, Braen G, et al: Acute Low Back Pain in Adults: Clinical Practice Guideline: Quick Reference Guide No. 14. Rockville, MD, Agency for Health Care Policy and Research, AHCPR Pub. No. 95-0643, Dec 1994

3. Bourdillon JF, Day EA, Bookhout MR: Spinal Manipulation, ed 5. London, Butterworth Heinemann, 1992

4. Brodeur R: The audible release associated with joint manipulation. J Manipulative Physiol Ther 18:155-164, 1995

5. Butler DS: Mobilization of the Nervous System. New York, Churchill Livingstone, 1991

6. Butler DS, Shacklock MO, Slater H: Treatment of altered nervous system mechanics. In Boyling JD, Palastanga N (eds): Grieve's Modern Manual Therapy, ed 2. New York, Churchill Livingstone, 1994, pp 693-703

7. Cibulka MT, Rose SJ, Delitto A, et al: Hamstring muscle strain treated by mobilizing the sacroiliac joint. Phys Ther 66:1220-1223, 1986, p 1223

8. Elvey RL: The investigation of arm pain: signs of adverse responses to the physical examination of the brachial plexus and related neural tissues. In Boyling JD, Palastanga N (eds): Grieve's Modern Manual Therapy, ed 2. New York, Churchill Livingstone, 1994, pp 577-585

9. Ferrell WR: The effect of acute joint distension on mechanoreceptor discharge in the knee of the cat. Quarterly Journal of Experimental Physiology 72:493-499, 1987

10. Ferrell WR, Russell NJW: Extravasation in the knee induced by antidromic stimulation of articular C fibre afferents of the anaesthetic cat. J Physiol 379:407-416, 1986

11. Fuglsang-Frederiksen A, Scheel U: Transient decrease in number of motor units after immobilisation in man. Journal of Neurol Neurosurg Psychiatry 41:924-929, 1978

12. Galloway MT, Joki P: The role of exercise in the treatment of inflammatory arthritis. Bulletin on the Rheumatic Diseases 42(1):1-4, 1993

13. Geers A: Kinesiologie: Diagnostiek en Therapie van de Posturale en Fasische Spieren. Eindhoven, Stichting Manuele Geneeskunde, 1984

14. Goodman CC, Snyder TEK: Differential Diagnosis in Physical Therapy, ed 2. London, W.B. Saunders Co, 1995

15. Greenman PE: Principles of Manual Medicine, ed 2. Philadelphia, Williams & Wilkins, 1996

16. Grice AS: Muscle tonus change following manipulation. The Journal of the Canadian Chiropractic Association 74:29-31, 1974

17. Grigg P: Mechanical factors influencing response of joint afferent neurons from cat knee. J Neurophysiol 38:1473-1484, 1975

18. Grigg P: Response of joint afferent neurons in cat medial articular nerve to active and passive movements of the knee. Brain Res 118:482-485, 1976

19. Grigg P, Greenspan BJ: Response of primate joint afferent neurons to mechanical stimulation of knee joint. J Neurophysiol 40:1-8, 1977

20. Halata Z: The ultrastructure of sensory nerve endings in the articular capsule of the knee joint of the domestic cat (Ruffini corpuscles and Pacinian corpuscles). J Anat 124:717-729, 1977

21. Halata Z, Rettig T, Schulze W: The ultrastructure of sensory nerve endings in the human knee joint capsule. Anat Embryol 172:265-275, 1985

22. Harris SR, Lundgren BD: Joint mobilization for children with central nervous system disorders: Indications and precautions. Phys Ther 71:890-896, 1991

23. Hogeweg J: Pain Threshold and Tissue Compliance in Juvenile Chronic Arthritis. Thesis, Utrecht, Netherlands, Universiteit Utrecht, 1995

24. Iggo A, Guilbaud G, Tégner R: Sensory mechanisms in arthritic rat joints. In Kruger L, Liebeskind JC (eds): Advances in Pain Research and Therapy. New York, Raven Press, 1984, vol 6, pp 83-93

25. Johansson H: Role of knee ligaments in proprioception and regulation of muscle stiffness. Journal of Electromyography and Kinesiology 1:158-179, 1991

26. Johnston WL: Functional Technique. In Basmajin JV, Nyberg R (eds): Rational Manual Therapies. Baltimore, Williams & Wilkins, 1993

27. Jones LH: Strain and Counterstrain, ed 12. Indianapolis, The American Academy of Osteopathy, 1993

28. Kabat H: Low Back and Leg Pain from Herniated Cervical Disc. St. Louis, Missouri, Warren H. Green, Inc, 1980

29. Konttinen YT, Grönblad M, Hukkanen M, et al: Pain fibers in osteoarthritis: A review. Semin Arthritis Rheum 18(suppl 23):35-40, 1989

30. Lam FY, Ferrell WR: CGRP modulates nerve-mediated vasoconstriction of rat knee joint blood vessels. Ann N Y Acad Sci 30:519-521, 1992

31. Levick JR: Editorial: Joint pressure-volume studies: Their importance, design and interpretation. J Rheumatol 10:353-357, 1983

32. Loewenstein WR, Mendelson M: Components of receptor adaptation in a pacinian corpuscle. J Physiol 177:377-397, 1965

33. Loewenstein WR, Skalak M: Mechanical transmission in a pacinian corpuscle: An analysis and a theory. J Physiol 182:346-378, 1966

34. Matucci-Cerinic M, Partsch G: The contribution of the peripheral nervous system and the neuropeptide network to the development of synovial inflammation. Clin Exp Rheumatol 10:211-215, 1992

35. Mayer RF, Burke RE, Toop J, et al: The effect of long-term immobilization on the motor unit population of the cat medial gastrocnemius muscle. J Neurosci 6:725-739, 1981

36. Mazanec D: Recognizing malignancy in patients with low back pain. Journal of Musculoskeletal Medicine 13:24-32, 1996

37. McCall WD, Farias MC, Williams WJ, et al: Static and dynamic responses of slowly adapting joint receptors. Brain Res 70:221-243, 1974

38. McClure PW, Flowers KR: Treatment of limited shoulder motion: A case study based on biomechanical considerations. Phy Ther 72:929-936, 1992

39. McKenzie RA: The Lumbar Spine: Mechanical Diagnosis and Therapy. Lower Hutt, New Zealand, Spinal Publications, 1981

40. McKenzie RA: The Cervical and Thoracic Spine: Mechanical Diagnosis and Therapy. Waikanae, New Zealand, Spinal Publications, 1990

41. Méal GM, Scott RA: Analysis of the joint crack by simultaneous recording of sound and tension. Journal of Manipulative and Physiological Therapeutics 9:189-195, 1986

42. Millar J: Flexion-extension sensitivity of elbow joint afferents in cat. Exp Brain Res 24:209-214, 1975

43. Mitchell FL: Elements of Muscle Energy technique. In Basmajian JV, Nyberg R (eds): Rational manual therapies. London, Williams & Wilkins, 1993

44. Mitchell FL: The Muscle Energy manual: Concepts & mechanisms, the musculoskeletal screen, cervical region evaluation and treatment. East Lansing, Michigan, MET Press, 1995, vol 1

45. Murphy BA, Dawson NJ, Slack JR: Sacroiliac joint manipulation decreases the H-reflex. Electromyogr Clin Neurophysiol 35:87-94, 1995

46. Nade S, Newbold PJ: Factors determining the level and changes in intra-articular pressure in the knee joint of the dog. J Physiol 338:21-36, 1983

47. Nansel DD, Waldorf T, Cooperstein R: Effect of cervical spinal adjustments on lumbar paraspinal muscle tone: Evidence for facilitation of intersegmental tonic neck reflexes. J Manipulative Physiol Ther 16:91-95, 1993

48. O'Laoire SA, Crockard HA, Thomas DG: Prognosis for sphincter recovery after operation for cauda equina compression owing to lumbar disc prolapse. BMJ 282:1852-1854, 1981

49. Paris SV, Loubert PV: Foundation of Clinical Orthopaedics. St. Augustine, Institute Press, Division of Patris Inc, 1990

50. Piercy M, Portek I, Shepherd J: The effect of low-back pain on lumbar spinal movements measured by three-dimensional x-ray analysis. Spine 10:150-153, 1985

51. Piercy M, Shepherd J: Is there instability in spondylolisthesis? Spine 10:175-177, 1985

52. Roubal PJ, Dobritt D, Placzek JD: Glenohumeral gliding manipulation following interscalene brachial plexus block in patients with adhesive capsulitis. J Orthop Sports Phys Ther 24:66-77, 1996

53. Schaible HG, Schmidt RF: Activation of groups III and IV sensory units in medial articular nerve by

local stimulation of knee joint. J Neurophysiol 49: 35-44, 1983

54 Slater H, Butler DS, Shacklock MO: The dynamic central nervous system: Examination and assessment using tension tests. In Boyling JD, Palastanga N (eds): Grieve's Modern Manual Therapy, ed 2. New York, Churchill Livingstone, 1994, pp 587-606

55 Stokes M, Young A: The contribution of reflex inhibition to arthrogenous muscle weakness. Clin Sci 67:7-14, 1984

56 Terrett ACJ, Vernon H: Manipulation and pain tolerance: A controlled study of the effect of spinal manipulation on paraspinal cutaneous pain tolerance levels. Am J Phys Med 63:217-225, 1984

57 Thabe H: Electromyography as a tool to document diagnostic findings and therapeutic results associated with dysfunctions in the upper cervical spinal joints and sacroiliac joints. Manual Medicine 2:53-58, 1986

58 Unsworth A, Dowson D, Wright V: 'Cracking joints': A bioengineering study of cavitation in the metacarpophalangeal joint. Ann Rheum Dis 30:348-358, 1971

59 Will THE: The biochemical basis of manipulative therapeutics: Hypothetical considerations. J Manipulative Physiol Ther 1:153-156, 1978

60 Wood L: Pressure-volume relationships in the knee joint of the cat and their effect on the discharge of articular mechanoreceptors (1985), referred to in Ferrell WR: The effect of acute joint distension on mechanoreceptor discharge in the knee of the cat. Quarterly Journal of Experimental Physiology 72: 493-499, 1987

61 Wood L, Ferrell W: Response of slowly adapting articular mechanoreceptors in the cat knee joint to alterations in intra-articular volume. Ann Rheum Dis 43:327-332, 1984

62 Wood L, Ferrell W, Baxendale R: Pressures in normal and acutely distended human knee joints and effects on quadriceps maximal voluntary contractions. Quarterly Journal of Experimental Physiology 73:305-314, 1988

63 Wyke BD: Articular neurology and manipulative therapy. In Glasgow EF, Twomey LT, Scull ER, et al (eds): Aspects of Manipulative Therapy, ed 2. New York, Churchill Livingstone, 1985

64 Zimmermann M: Pain mechanisms and mediators in osteoarthritis. Semin Arthritis Rheum 18(suppl 2):22-29, 1989

65 Zusman M: Spinal manipulative therapy: Review of some proposed mechanisms and a new hypothesis. The Australian Journal of Physiotherapy 32:89-99, 1986

18. Notation

When evaluating a patient, it is important to record the positions the muscle was tested in and the results of the tests. Describing the quadriceps as weak in full extension does not justify an extension mobilization of the knee. The evaluation and the notation should include what made the muscle strong. If the same quadriceps tested strong with a little knee flexion, an extension mobilization of the knee may be indicated, but even then, the antagonists or other muscles may show a larger dysfunction that requires treatment first. For instance, the hamstrings might test weak throughout their range. In testing weak muscles, the notes should reflect the examiner's efforts to find a position where the muscle tests strong. The description of the position in which a muscle tested weak is thus completed by describing the position that improved the muscle's strength or the position(s) that failed to improve the muscle's strength.

To summarize, the notation of muscle tests should include three items:

- The muscles tested,
- The position in which these muscles were tested, and
- The result of each test (weak or strong).

Perhaps the easiest way to record the findings of a muscle strength evaluation is to draw a vertical line in the middle of the page and write "weak" above the left column and "strong" above the right column. Every muscle tested will be categorized in either the left or the right column. Next to the name of the muscle, the therapist should describe the position of the muscle test. The following abbreviations can be used to describe the muscle and the testing position:

- L   Left
- R   Right
- B   Bilateral
- s   tested in a shortened position
- n   tested in a neutral or loose-packed position of the joint
- l   tested in a lengthened position

The notation of weakness in "s, n, l" may indicate extremity muscle weakness resulting from a radiculopathy or a pseudoradiculopathy. The extremity muscle is weak throughout its entire range, and the therapist cannot detect a dysfunctional barrier.

In the case of arthrogenic weakness, the therapist could describe the position of the dysfunctional barrier. For instance, the quadriceps tests weak in a shortened position; the barrier is at 30 degrees of flexion.

Following are three examples of how a therapist could describe the results of a muscle strength evaluation.

Example I

The patient complains of pain in the medial aspect of the right knee. The pain is experienced during the stance phase of walking. The qualitative muscle assessment reveals quadriceps weakness in a shortened position (extended knee) only (arthrogenically inhibited). The adductors are weak throughout the hip range, but become strong with some knee flexion. Obviously, knee extension causes inhibition of the quadriceps and the adductors. The antagonists, the hamstrings, and the abductors test strong in a shortened position.

Weak	Strong
R quadriceps s	R quadriceps 10° knee fl
R adductors s, n, l	R adductors 10° knee fl
	R abductors s
	R hamstrings s

This patient most likely needs an extension improvement of her knee joint (quadriceps). The adductor weakness suggests that the medial knee compartment needs the mobilization.

Example II

The patient has constant right shoulder pain, which worsens when contracting the triceps (cranial glide of the humerus on the glenoid) and when attempting to elevate the arm. The evaluation reveals abductor and external rotator weakness in the shortened and loose-packed positions of the shoulder joint. The muscles test strong in a lengthened position. The other shoulder muscles test strong in a shortened position. This patient has a major (more than 50 percent of the range) abduction and external rotation dysfunction.

Weak	Strong
R abductors s, n	R abductors l
R external rotators s, n	R external rotators l
	R internal rotators s
	R adductors s
	R biceps s
	R triceps s

The most logical treatment is an external rotation and abduction mobilization. These mobilizations can be done separately or combined into one mobilization with the humerus externally rotated and abducted. The mobilization of the humeral head is in an anterior and inferior direction.

Example III

The patient complains of bilateral low back pain, which is reproduced with lumbar extension. Both quadratus lumborum muscles are weak in a shortened position only, revealing a minor bilateral lumbar extension dysfunction. L5 seems responsible for the left lumbar extension dysfunction (weakness of the L5 muscles—hamstrings and hip extensors, while the other leg muscles remain strong in lumbar extension). On the right side, L2 may be responsible for the lumbar extension dysfunction.

Weak	Strong
B quadratus lumborum s, n	B lumbar flexors s
L hamstrings n in L lumbar ext	lumbar fl
L hip extensors n in L lumbar ext	lumbar fl
	in L lumbar ext: L hip fl s,
	L quad s,
	L dorsifl s,
	L plantar fl s
R hip flex n in R lumbar ext	lumbar n
R adductors n in R lumbar ext	lumbar n
	in R lumbar ext: R quad s,
	R dorsifl s,
	R plantar fl s,
	R hamstr s,
	R hip ext s

19. Technique Procedures

Following is the framework that will be used for presentation of the evaluation and treatment techniques. The headings and the descriptions explain the format used to describe the procedures.

Joint(s)

The techniques will be discussed by joint (e.g., hip joint) or group of joints (e.g., cervical spine). A description of joint mechanics may follow.

- *Capsular pattern*—If the joint or joints have an active range, the capsular pattern is listed in sequence of limitation. The most limited movement is listed first, the least limited movement last. If the joint or joints do not have an active range, as in the case of the sacroiliac joints, no restrictions are present, but pain is present when the ligamento-capsular tissue is stretched.

  The cause of a capsular pattern might be trauma (effusion), bacterial, viral, rheumatoid, or osteoarthrosis. In the case of osteoarthrosis, a trauma to the joint (e.g., fracture) may have precipitated the onset of arthrosis.

  Even though capsular patterns are listed for the joints, arthritis and arthrosis are not the only conditions that can cause joint and muscle dysfunctions. Numerous other known and unknown disorders can create joint stiffness with muscle tone changes.

MOVEMENT OF A JOINT OR JOINTS

The sequence in which the movements are listed follows that of a capsular pattern. First, the movement that is most limited when a capsular pattern is present is described, followed by the next most restricted movement, and so on. For example, in the shoulder section, external rotation is described first, followed by abduction, internal rotation, and adduction. This sequence may be followed by combination movements. The advantage of this sequence is that the first test can alert the examiner that a capsular pattern (arthritis or osteoarthrosis) may be present. If the first test yields good range and a strong muscle, no arthritis is suspected in the joint.

Multiple techniques may be available to test certain movements. For instance, several techniques are available to test nutation of S1. If the patient is supine, the examiner could use a supine technique, and if the patient is prone, the examiner could use a prone technique. The second advantage of using multiple techniques is the capability to retest the same movement with different muscles. For instance, in the presence of an L5 radiculopathy, the internal rotators and hamstrings (innervated by L5) are not available to test S1 nutation; however, the iliacus and the adductors are innervated by the upper lumbar spine and are not directly affected by an L5 radiculopathy. Thus, since there are four different muscle tests for S1 nutation, we can test S1 nutation with muscles other than those affected by the L5 radiculopathy.

- *Agonist(s)*—The agonists are the muscles contracting during the test—the muscles being directly tested. This heading will be followed by the name of the muscle group to be tested and the individual muscles within this group, as in the following example:

  Shoulder abductors:
  supraspinatus (suprascapular nerve; C4,5,6)
  deltoid (axillary nerve; C5,6)

The nerves innervating the individual muscles and their segmental origin are listed in parentheses. If the muscle is innervated by cranial nerves, no segmental level applies and none is included. For muscles innervated by the cervical or lumbar plexus, only their segmental levels are listed. Unless stated otherwise, the source for innervation of the muscles is *Gray's Anatomy*.[29]

- ***Muscle test***—Even though a muscle test consists of an excursion into the range (joint movement), a contraction of the agonists, a simultaneous relaxation of the antagonists, and coordination of these changes by the nervous system, evaluation of these complex changes is simply called a "muscle" test. All muscle tests are described for joints on the patient's right side, except the symphysis pubis, which is a midline joint.

For easier reading, the patient and therapist are given different genders. The patient is described as female, the therapist as male. The text "speaks" directly to the reader to provide instructions on how to perform the technique.

Unless the intent is to reproduce pain, the therapist should avoid producing or aggravating any discomfort. Muscle tests should be carefully performed with gentleness. The muscle tests are not geared to grading the patient's maximal strength but to assessing the quality of the contraction. This can always be sensed at the initial stage of a contraction using minimal force.

The text describes testing the muscle(s) in a shortened position. This means that the joint has moved into an end-range position, or toward this position if the end-range position itself is not comfortable. If the contraction quality at the end of the range is good, the therapist can continue on and test a different movement. If the contraction quality is poor, the therapist should retest the muscle in a position where it is likely to test strong. This position is described under "Remarks." Should the muscle be inhibited by the joint it crosses (arthrogenic inhibition), some loose-packed position is likely to allow the muscle to contract without inhibition. If an extremity muscle is not arthrogenically inhibited, and the movement tests weak throughout the range, the inhibition could come from a joint without active range or from a related spinal segment. Evaluation of the spine using extremity muscles is presented in the last part of this chapter.

- ***Passive mobilization***—The treatments described are based on the assumption of arthrogenic inhibition from the local joint(s). Passive mobilization of a joint is the translation of the joint surfaces by the therapist to increase extensibility of the joint's ligamento-capsular tissue. The mobilizations in this book are not performed in maximally loose-packed positions but at or beyond the restrictive barrier. A mobilization can be done at the restrictive barrier, but as the extensibility of the capsule and ligaments increases, the restrictive barrier moves farther into the range, and accordingly, the mobilization must be done farther into the range. Joints with high reactivity (pain is produced before the physiological barrier is engaged) might best be gently mobilized near the restrictive barrier. Joints with low reactivity (only movement well into the physiological barrier produces pain, or no pain is produced at all) might best be mobilized at the end of the available range (i.e., near its physiological barrier).

Frequent evaluation of mobilization by muscle tests can reveal if the quality of contraction has been restored. If the muscle does not strengthen in a reasonable time (often a few minutes), the therapist may not be treating the cause of the inhibition or a second source of inhibition may be present. A reevaluation could provide more information.

- ***Active mobilization***—The active mobilization is done by the patient's agonist or antagonist to the restricted movement and is often identical to the muscle test, except that the contraction is gentler and of longer duration. Often while performing the active mobilization with an agonist, the therapist can feel the muscle strength increasing without having to do a formal muscle test.

- ***Postisometric stretch technique***—The postisometric stretch techniques described in this book are modifications of similarly named Osteopathic Muscle Energy techniques. The techniques involve bringing the patient's joint three-dimensionally to the feather edge of its restricted barrier. In this position, the patient is invited to gently contract the antagonist to the restricted movement. This invitation could be a request to "hold the body part in place" while the therapist gently attempts to push the body part farther into the range, or a request to "gently move the body part in a specific direction" while the therapist resists. After the therapist provides unyielding resistance to the patient's contraction for about 6 seconds, the patient relaxes completely, and the therapist moves the joint farther into the range to find the new restrictive barrier, and so on. It may be that the isometric contractions allow more movement by resetting the gamma bias.[14,15]

After any treatment technique, the therapist should reexamine the patient. The reexamination will reveal if the treatment was successful or if more of the same or a different technique are required.

Cervical Spine

The cervical spine is the most densely innervated part of the spine.[30] It contains two regions with distinctly different mechanics. The first region, or subcranial area, runs from the cranium through C2, and the second region runs from C2 though C7. Biederman reported cervical conditions in newborns, such as torticollis, to result from limited mobility in the subcranial joints.[2]

SUBCRANIAL REGION

Cranium–C1—The facet joints of occiput–C1 are located laterally to the foramen magnum (see Fig. 19.1). The orientation of the facet joints is diagonal. The anterior parts of the facet joints are closer to the midline, whereas the posterior parts are farther apart. Thus, the facet joints run from anteromedial to posterolateral.

The occipital condyles are convex on concave C1 facet joints. The right occipital condyle translates posteriorly on C1 during flexion and/or right rotation and left side bending. The right occipital condyle translates anteriorly on C1 during extension and/or left rotation and right side bending. The orientation of the facet joints and ligamentous attachments couple rotation of the cranium with side bending to the opposite side. The coupling of rotation with side bending to the opposite side is independent of the cranium flexing or extending.

C1–C2—The joints between the atlas and axis allow predominantly rotation. The center for rotation is the dens. Ligaments that run from the dens to the occiput tighten during atlas-on-axis rotation, since the cranium rotates with the atlas. Tightening of the ligaments during rotation causes the cranium to side bend to the opposite side. For instance, atlas (with cranium)-on-axis left rotation is coupled with cranium-on-atlas right side bending.

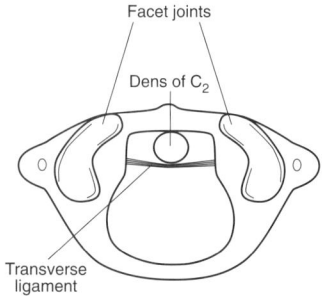

Fig. 19.1 The superior facet joints of C1.

C2–C7—The angle of the facet joints in the sagittal plane is about 45 degrees (Fig. 19.2). The movements that occur at the C2 through C7 facet joints are flexion and extension. When both the right and left facet joints of the superior vertebra translate cranially and anteriorly, the superior vertebra flexes. When both the right and left facet joints of the superior vertebra translate posteriorly and caudally, the vertebra extends. The movement whereby the right facet joint translates cranially and anteriorly (flexion of the right facet joint) while the left facet joint translates posteriorly and inferiorly (extension of the left facet joint) is called left rotation or left side bending. No difference exists between side bending and rotation at the levels of C2 through C7, because these movements are coupled to the same side. The difference between left side bending and left rotation of the entire neck is the movement of the subcranial spine.

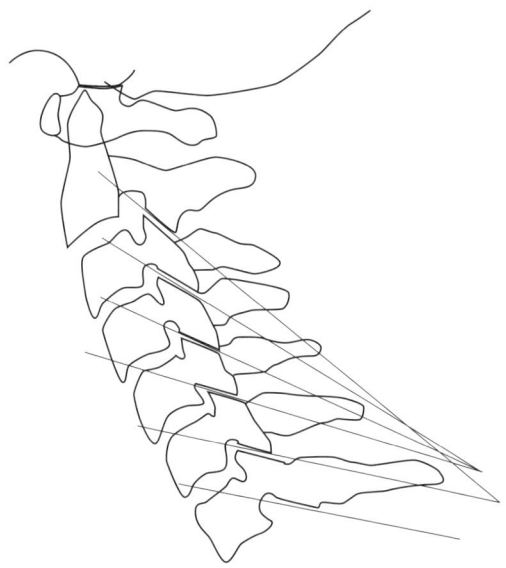

Fig. 19.2 The angle of the facet joints of C2–C7.

If the right facet joints of C2 and lower are flexed, the left facet joints are extended, and the subcranial spine is rotated to the left, the neck is rotated to the left. The subcranial left rotation is coupled with subcranial right side bending. The subcranial right side bending counters the left side bending of C2 and lower. Thus, the head is in neutral side bending but rotated to the left.

If the right facet joints of C2 and lower are flexed and the left facet joints are extended, the subcranial spine could rotate to the right to create left side bending of the neck. The subcranial right rotation is coupled with subcranial left side bending. The right rotation counters the left rotation of C2 and lower. Thus, the head is in neutral rotation but side bent to the left.

Movements of the head and neck do not stop at C7. Often, head and cervical movements extend into the upper thoracic spine. For example, left rotation of the head might involve left rotation of all the cervical segments and T1 through T3.

Integrity of subcranial ligaments — The subcranial ligaments might be lax or might be damaged by trauma or disease. The danger of lax or torn subcranial ligaments is that the dens is not kept in place and compresses the spinal cord. Also, instability of the subcranial spine might compromise vertebral artery function. Symptoms of lost integrity of the subcranial ligaments include[23]:

- balance disturbance with head movements,
- bilateral or quadrilateral limb paresthesia associated with head movements, or
- nystagmus associated with head movements.

However, if a patient complains of all or any of these symptoms, the cause is not necessarily always a subcranial ligament dysfunction. A vertebral artery dysfunction can also cause these symptoms. Referral to an appropriate specialist may be crucial. More information on this topic can be found in "Stress tests of the craniovertebral joints" by Pettman.[23]

Vertebral artery insufficiency — The two vertebral arteries supply blood to part of the brain and brain stem. Certain neck positions could compromise circulation in a vertebral artery. Vertebral artery insufficiency may become evident in cervical extension with side bending and rotation to the same side; however, other positions could also cause symptoms. These symptoms often include dizziness and/or acute anxiety, but could also include blurred vision, sleepiness, or nystagmus. In response to a cervical position, any sensation by the patient other than a normal stretch of the neck could be a symptom of insufficiency. Vertebral artery insufficiency symptoms could occur as soon as the neck assumes the pathogenic position, but could also occur after the position has been sustained for a short time.

Referral of these patients to an appropriate specialist might be in order. More information on this topic can be found in publications by Grant, Greenman, and Oostendorp et al.[8,9,17,18,19]

- ***Capsular pattern*** — Cervical extension, ipsilateral rotation, and ipsilateral side bending (facet extension) are limited. Cervical flexion is full but often painful.

OCCIPUT RIGHT FACET EXTENSION

Anterior translation of the right occipital condyle on C1 occurs with cranium extension, left rotation, and right side bending.

- **Agonists**

    Subcranial extensors:
    rectus capitis lateralis (C1,2)
    obliquus capitis superior (C1)
    rectus capitis posterior minor (C1)
    sternocleidomastoid, right (accessory nerve and C2,3)

- **Muscle test** (Figs. 19.3 and 19.4)

    1. With the patient seated, stand at her left side.
    2. Place your right hand on the left side of her head and hold her chin in your left hand.
    3. Position her head in extension, right side bending, and left rotation. Relaxing the jaw (mouth relaxed open) might allow more extension of the cranium on the atlas (Fig. 19.3).
    4. Reposition yourself to stand behind the patient.
    5. Place the fingers of your stabilizing left hand on the left side of her cervical spine, with your left index finger contacting the lateral part of the transverse process of C1.
    6. Place your right hand on the right side of her head, above and slightly behind her right ear (Fig. 19.4).
    7. Ask her to keep her head in the same position as you attempt to push her head into flexion and left side bending.

    Remarks—If the test yields weakness, compare the result with a test in which the patient's right occipital condyle is less extended.

- **Active mobilization**—The active cranium extension is a modification of a bilateral extension mobilization taught by Paris.[20] The mobilization described here is a unilateral technique. It is identical to the test described above, except the treatment contraction is gentler and sustained. Retest after the mobilization.

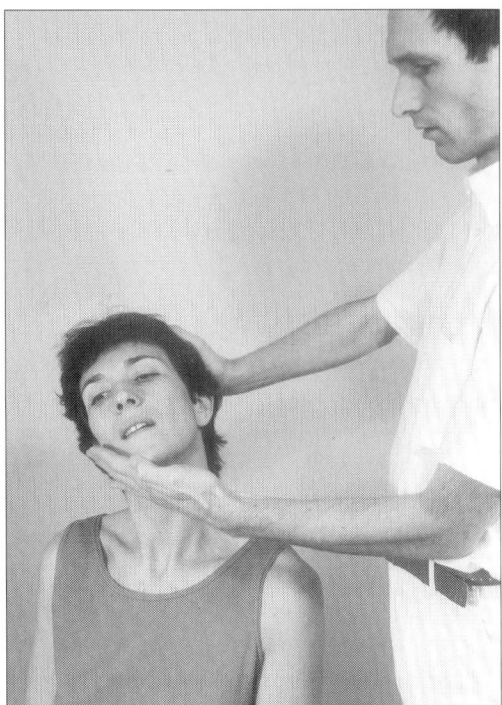

Fig. 19.3

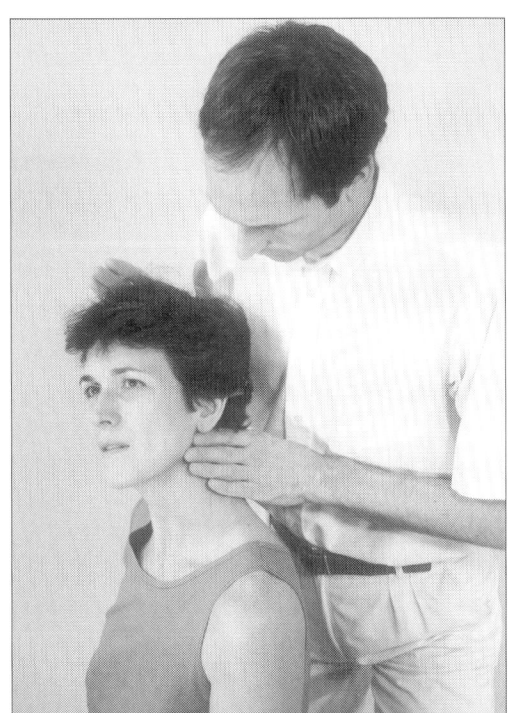

Fig. 19.4

OCCIPUT RIGHT FACET FLEXION

Posterior translation of the right occipital condyle on C1 occurs with cranium flexion, right rotation, and left side bending.

- **Agonists**

    Subcranial flexors:
    platysma (facial nerve)
    supra- and infrahyoid muscles (alveolar, facial, and hypoglossal nerve and C1,2,3)
    sternocleidomastoid, left (accessory nerves and C2,3)
    rectus capitis anterior (C1,2)
    rectus capitis lateralis, left (C1,2)
    longus capitis (C1,2,3)

- **Muscle test** (Figs. 19.5 and 19.6)

    1. With the patient seated, stand at her right side.

    2. Place your left hand on the back of her head and hold her chin with your right hand.

    3. Position her head in flexion (ask her to tuck her chin), left side bending, and right rotation (Fig. 19.5).

    4. Reposition yourself to stand behind her left shoulder.

    5. Place your stabilizing right hand on the right side of her cervical spine, with your right index finger over the posterior aspect of the right transverse process of C1.

    6. Place your left hand on the left side of her forehead (Fig. 19.6).

    7. Ask her to keep her head in the same position as you attempt to push her head into extension with your left hand.

    Remarks—If the test yields weakness, compare the result of this test with one in which the patient's right occipital condyle is less flexed.

- **Active mobilization**—The active mobilization is identical to the test described above, except that the treatment contraction is gentler and sustained. After the mobilization has been performed for a moment and more range is obtained, perform the technique again in the newly acquired range. Repeat this procedure until normal flexion is available.

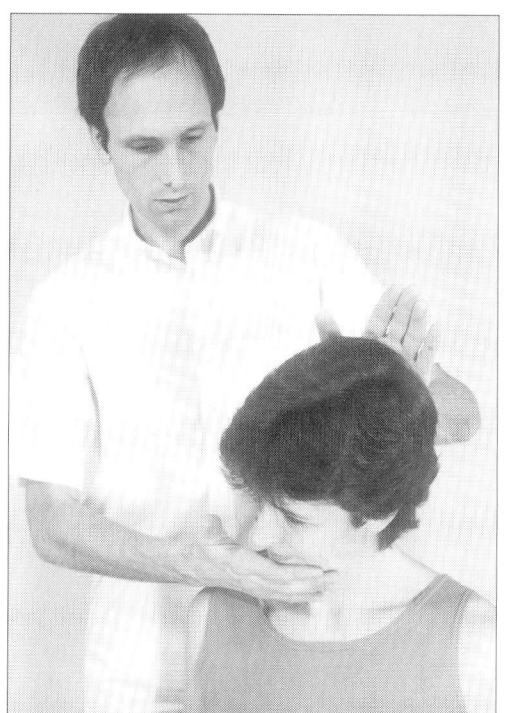

Fig. 19.5

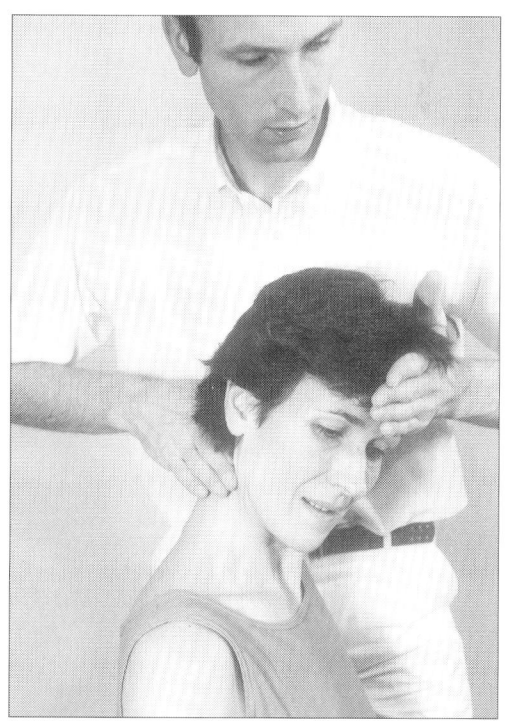

Fig. 19.6

C1 RIGHT ROTATION

The evaluation of atlas rotation on the axis makes no distinction between the left and right facet joints. Atlas rotation is coupled with cranium contralateral side bending. Limited C1 rotation could thus be caused by restricted movement of the occipital condyles. Since it is possible to evaluate and treat the left and the right occipital facets separately, it is better to treat occipital facet restrictions before treating restricted atlas rotation. After correcting the occipital facet movements, evaluate atlas rotation again.

- *Agonists*

    Subcranial rotators:
    rectus capitis anterior, right (C1,2)
    rectus capitis posterior major, right (C1)
    obliquus capitis inferior, right (C1)
    sternocleidomastoid, left (accessory nerve and C2,3)

- *Muscle test* (Figs. 19.7 and 19.8)

    1. With the patient seated, stand at her right side.

    2. Place your left hand on the back of her head and hold her chin with your right hand.

    3. Position her head in left side bending and right rotation while maintaining neutral flexion/extension at her subcranial spine (Fig. 19.7).

    4. Keeping your left hand on the back of her head, place your right hand on the right side of her forehead (Fig. 19.8).

    5. Ask her to keep her head in the same position as you gently attempt to derotate her head (i.e., rotate her head to the left).

    Remarks—If the test yields weakness, compare the result of the test with one in which the patient's atlas is rotated less to the right.

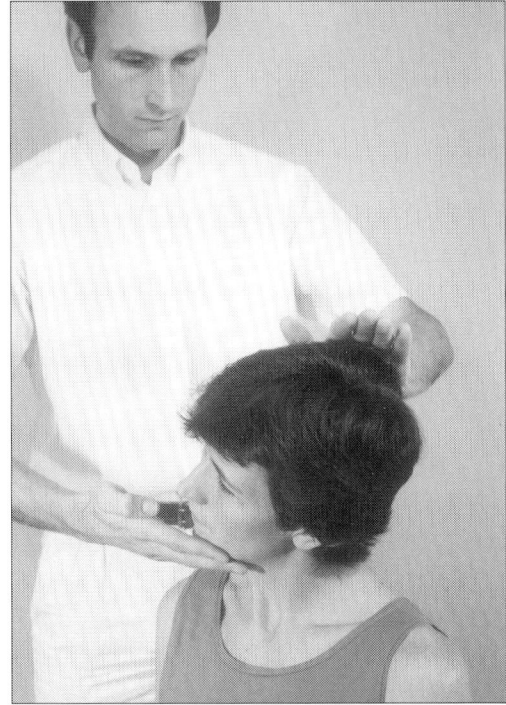

Fig. 19.7

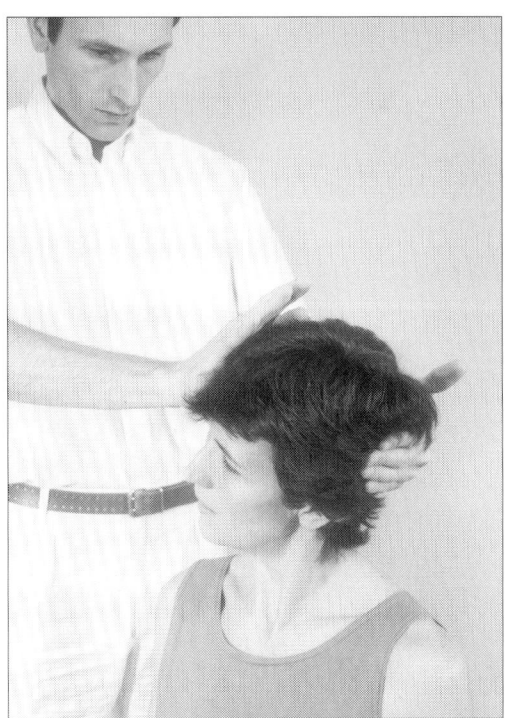

Fig. 19.8

- **Passive mobilization** (Fig. 19.9)

    1. With the patient seated, stand at her right side.

    2. Place your stabilizing left thumb posterior on the right transverse process of C2.

    3. Hold the patient's head between your right hand and your chest (or shoulder).

    4. While maintaining stabilization of C2, gently translate the head posteriorly (to maintain forward position of the dens in relation to the axis) and rotate it to the right (with left side bending), so the arch of C2 presses into your left thumb.

    5. Retest.

        Remarks—This technique can be helpful after the occipital facet movement has been restored. The technique improves right rotation regardless of whether the restriction is at the right or left facet joint.

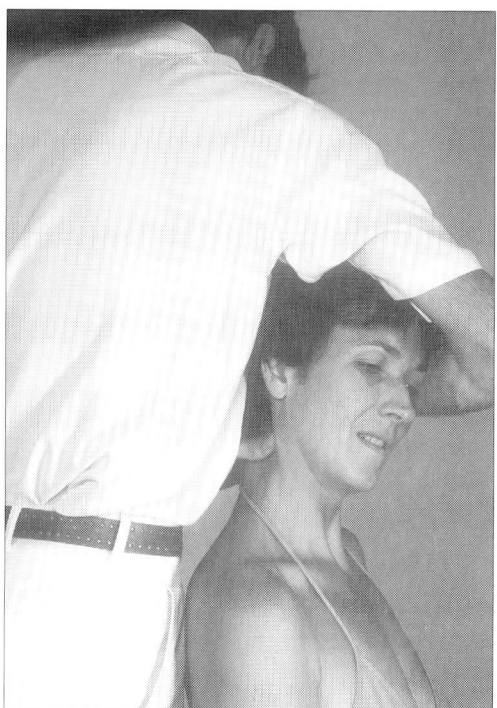

Fig. 19.9

C2 THROUGH T4 RIGHT FACET EXTENSION

Unilateral extension, whether performed as a muscle test or as a treatment, consists of movements of the cervical spine that are similar to the vertebral artery test (cervical extension with side bending and rotation to the same side). Note: Watch for symptoms of vertebral artery dysfunction.

- *Agonists*

    Cervical extensors:
    posterior scalenus, right (C6,7,8)
    semispinalis cervicis, left (C6,7,8)
    multifidi, left (C3–8)
    rotatores cervicis, left (C3–8)
    interspinalis cervicis (C3–8)
    trapezius, superior part, right (accessory nerve, C3,4)
    splenius capitis, right (C2,3,4)
    splenius cervicis, right (C2–5)
    levator scapulae (C3,4, and the dorsal scapular nerve; C5)
    erector spinae (lower cervical nerves)

- *Muscle test* (Figs. 19.10 and 19.11)

    1. With the patient seated, stand behind her.

    2. Bring her cervical spine into right facet extension by asking, "Could you bring the back of your head toward your right shoulder blade?" or by passively positioning her (Fig. 19.10). This extends C2 and lower on the right while maintaining the subcranial spine in a neutral position.

    3. Place your right hand on the back of her head.

    4. Place your left hand on her left upper ribs for stabilization (Fig. 19.11).

    5. Ask her to keep her head in this position while you attempt to push her head forward and to the left.

    Remarks—If the test yields weakness, compare the result with a test in which the right cervical or upper thoracic spine is less extended.

    The muscle test can be made specific for one segment by placing your right index finger on the lamina of a specific vertebra (Fig. 19.12). Your index finger fixates this segment. Extend the

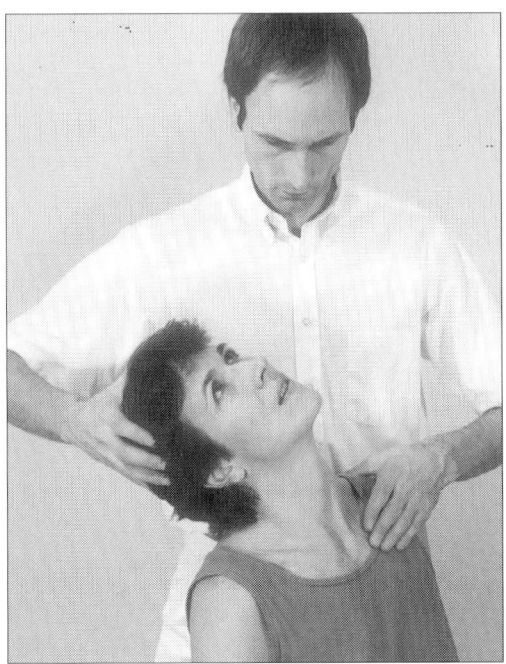

Fig. 19.10

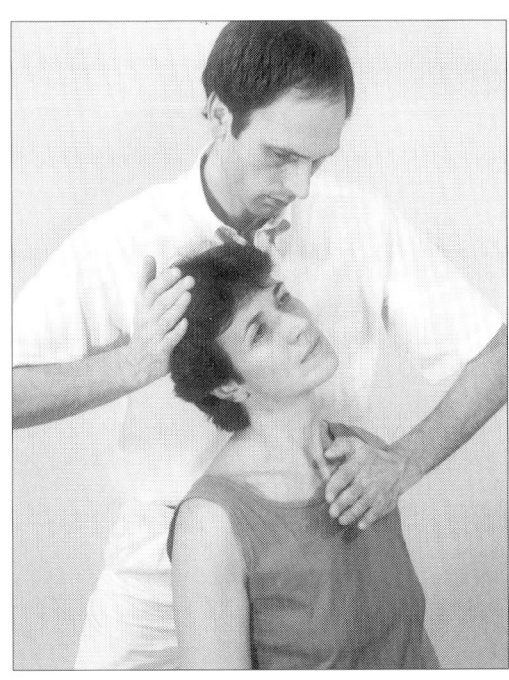

Fig. 19.11

right cervical spine down to this level. Wrap your left arm around the patient's head and test the right cervical extensors. Test the more cranial segments first, working your way caudally. If, for instance, C3's right facet joint has a limited extension on C4, fixating the right C3 lamina will keep C3 in a flexed position and the extensors will test strong. Fixating the right C4 lamina allows C3 to extend into the restriction and will yield weakness of the extensors.

With a modification, this test can be used down to T4. Placing your left forearm against the left side of the patient's head, translate her head posteriorly, and extend her right thoracic spine (Fig. 19.13). Placing your right thumb on the posterior aspect of a right transverse process blocks extension up to this level. The right thoracic extensors are tested with your left fingertips on the back of the patient's head.

- ***Active Mobilization*** — The active mobilization is identical to the test described above, except that the treatment contraction is gentler and sustained. When more range is obtained, repeat the active mobilization farther into the range until full correction of the dysfunction is achieved.

- ***Postisometric stretch technique*** (Fig. 19.14)

    1. With the patient seated, stand at her left side with her head between your left hand and your upper arm.

    2. Your right index finger could palpate motion at a specific segment or rest on her right shoulder.

    3. Passively and very slowly move her neck into right extension, stopping at the first perception of a restrictive barrier.

    4. Ask her to attempt to move her head gently forward and to the left while you provide unyielding resistance for approximately 6 seconds.

    5. After the patient fully relaxes, passively move her neck farther into right facet extension to find the next restrictive barrier.

    6. Repeat steps 3 through 5 until full extension motion and extensor strength are obtained.

    7. Retest.

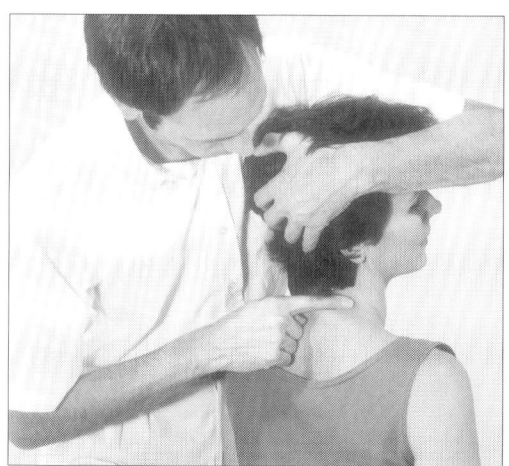

Fig. 19.12

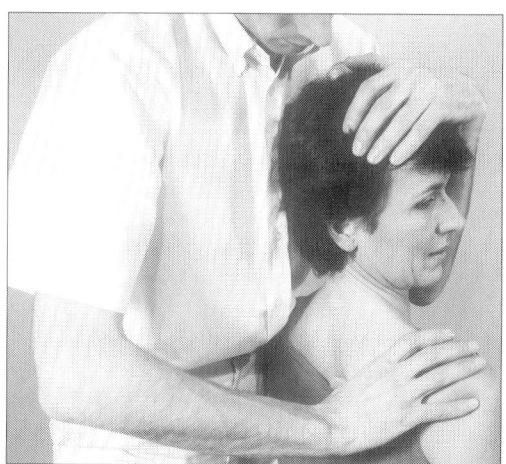

Fig. 19.13

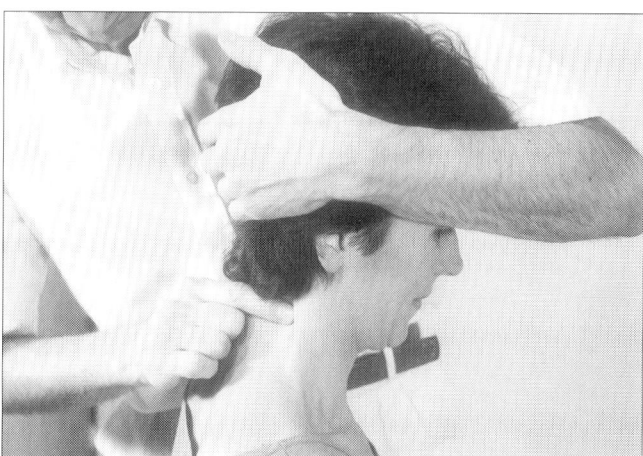

Fig. 19.14

C7 AND T1 RIGHT FACET EXTENSION

- *Agonist*

    Rhomboid:
    rhomboideus minor, left (dorsal scapular nerve; C4,5)

- *Muscle test* (Fig. 19.15)

    1. Have the patient be seated with her left elbow flexed and her left shoulder externally rotated and extended (her elbow close to her spine).
    2. The patient's head should be retracted, rotated, and side bent to her right (right ear brought in the direction of her right scapula).
    3. Sit to her left with the base of your left hand anteriorly stabilizing her left upper ribs.
    4. Hold her left elbow in your right hand.
    5. Ask her to keep her arm in this position while you attempt to move her elbow away from her spine (into left shoulder flexion and abduction).

    Remarks—This test assesses the left rhomboid's function to extend the right C7 and T1 facet joints. The therapist compares the results of this test with one in which the patient's right cervical facets are less extended (more flexed). If the muscle does not strengthen with right cervical facet flexion, the inhibition is likely to derive from a limited retraction of the left side of the shoulder girdle.

- *Active mobilization* (Fig. 19.16)

    1. Have the patient be seated with her right arm maximally elevated.
    2. Her left arm should should be extended, adducted, and externally rotated with the elbow extended.
    3. Have her look, or attempt to look, behind her right shoulder.
    4. Seated behind her, provide resistance to both her wrists. Ask her to bring both arms farther backward (right shoulder farther into flexion and abduction, left shoulder farther into extension and adduction).
    5. Ask her to look farther behind her shoulder.
    6. Retest.

    Remarks—The entire mobilization often takes only about 10 seconds to complete. The active mobilization consists of two parts. First, the segment with limited extension mobility is brought to the end of its available range. This is done by having the patient flex her right shoulder and look in the direction of her right shoulder blade. Second, the left rhomboid muscle is contracted to actively mobilize the segment.

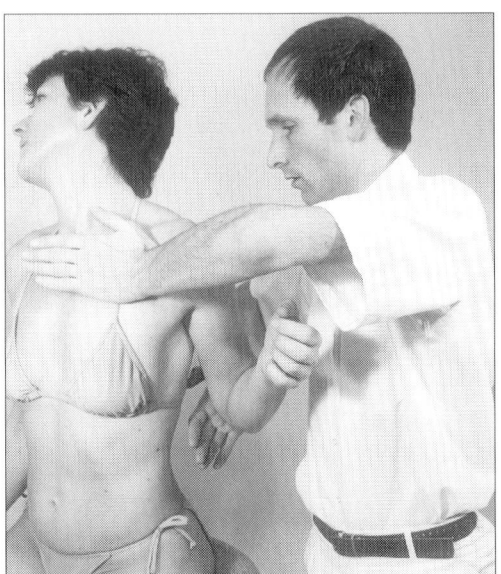

Fig. 19.15

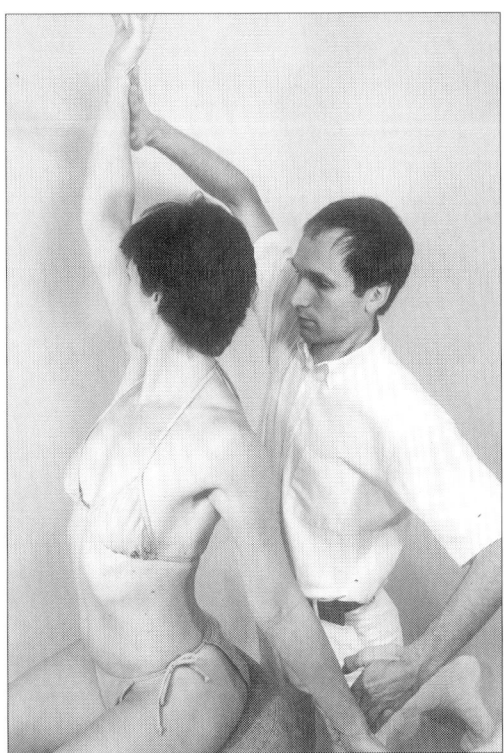

Fig. 19.16

C2 THROUGH T4 RIGHT FACET FLEXION

- **Agonists**

    Cervical flexors:
    supra- and infrahyoid muscles (alveolar, facial, hypoglossal, and C1,2,3 nerves)
    sternocleidomastoid muscle, right (accessory nerve, C2,3)
    longus capitis (C1,2,3)
    longus colli (C2–6)
    scalenus anterior, right (C4,5,6)

- **Muscle test** (Fig. 19.17)

    1. Have the patient be seated with her right cervical facets in flexion (ask her to look at her left hip).
    2. Stand behind her left shoulder with your right hand stabilizing her right upper thoracic spine and ribs.
    3. Place your left hand on her forehead.
    4. Ask her to keep her head in this position while you push with your left hand to attempt to extend her right cervical spine.

    Remarks—If the test yields weakness, compare the result with a test in which the patient's right cervical spine is less flexed.

    The muscle test can be made specific for one segment by placing your right index finger (for the cervical spine) or thumb pad (for the thoracic spine) on the right lamina of a vertebra (Figs. 19.18 and 19.19). The pressure of your right index finger or thumb on the vertebra will translate this vertebra into more flexion. When a limited flexion mobility exists, the muscle will test weak. In contrast, the pressure of your index finger or thumb on the vertebra caudal to the restriction will reduce flexion at the segment, and the muscle will test strong or stronger.

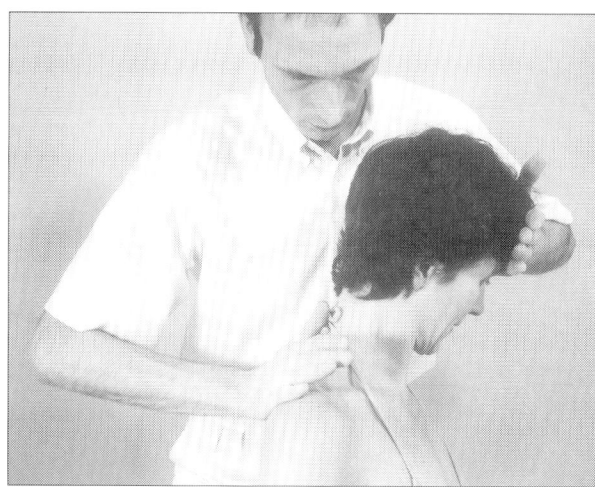

Fig. 19.17

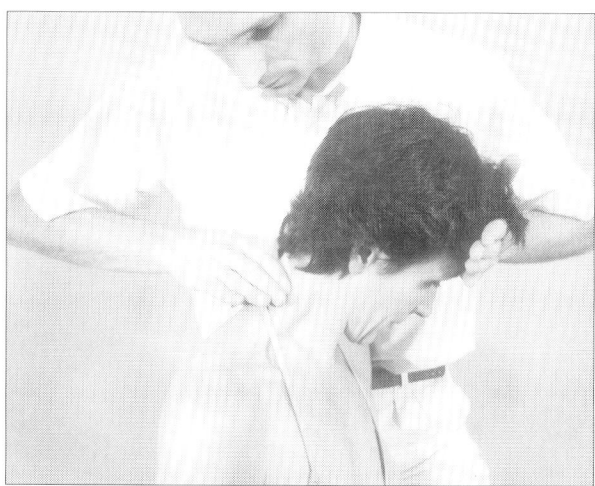

Fig. 19.18

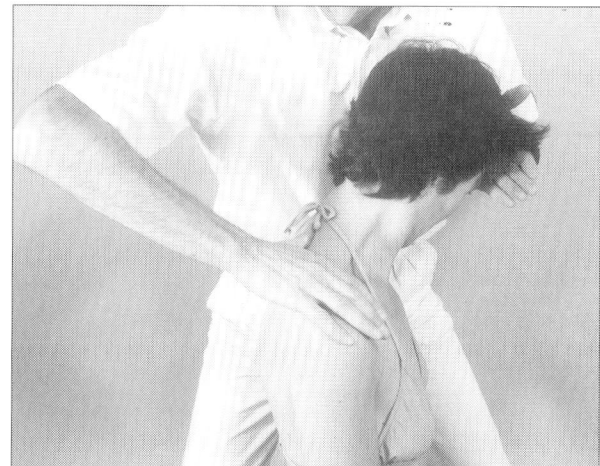

Fig. 19.19

- ***Postisometric stretch technique*** (Fig. 19.20)

  1. With the patient seated, stand behind her left shoulder with your right hand resting on her right shoulder.

  2. Hold her head between your left forearm/hand and your left upper arm.

  3. Passively and very slowly bring her right cervical facets into flexion. Stop at the first perception of a restrictive barrier and invite her to gently contract her right cervical extensors. Provide unyielding resistance to this gentle isometric contraction for about 6 seconds.

  4. Following the contraction, passively move her head farther into flexion to find the next barrier, repeating steps 3 and 4.

  5. Retest.

Remarks—The above technique will simultaneously treat all flexion restrictions between C2 and T4; however, it is also possible to treat one level at a time. Place your right index finger caudal to the lamina of the vertebra with the restricted flexion mobility Fig. 19.21). During movement of the cervical spine, you can feel the restrictive barriers with your left hand holding the patient's head and your right index finger palpating intersegment motion. Move the segment to the feather edge of the restrictive barrier and ask the patient to gently contract her right cervical extensors. Repeat this mobilization until satisfactory movement is obtained.

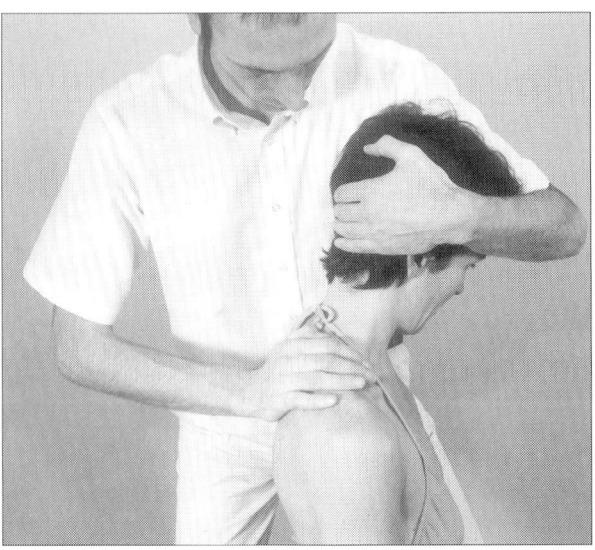

Fig. 19.20

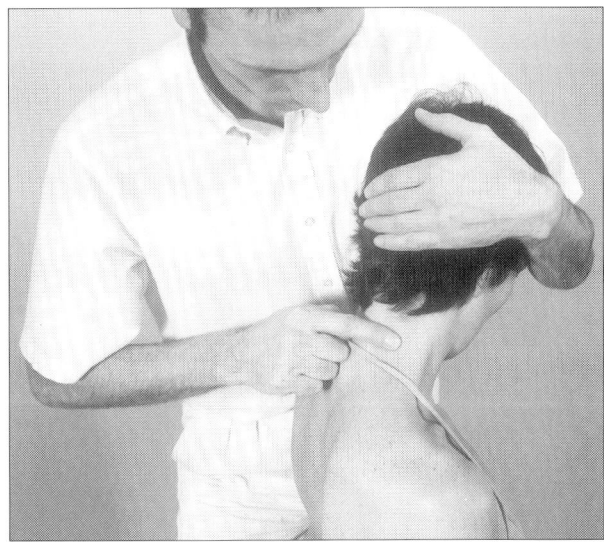

Fig. 19.21

Thoracic Spine

Besides some gapping (a separation of the facet joints), the thoracic facet joints can flex and extend. Maximal right facet joint flexion is accomplished by spinal flexion, left side bending, and left rotation. Maximal right facet joint extension is accomplished by spinal extension, right side bending, and right rotation.

- *Capsular pattern*—Bilateral rotation is limited.

C7 THROUGH T12 RIGHT FACET EXTENSION

- *Agonists*

    Trapezius:
    middle and lower trapezius (accessory nerve and C3,4)

- *Muscle test* (Fig. 19.22)

    1. Have the patient be seated with her neck and trunk in thoracic extension and right rotation.

    2. Her right shoulder should be abducted to 90 degrees.

    3. Stand in front of her right arm, holding her right distal forearm with your left hand.

    4. Place your stabilizing right hand on her right upper ribs.

    5. Ask her to keep her arm in this position while you attempt to adduct her arm horizontally.

    Remarks—This test is repeated several times, each time with more abduction of the right shoulder. The test is positive for a thoracic dysfunction if a weak trapezius becomes strong with less thoracic right rotation or extension. The test is also positive for a thoracic dysfunction if a weak trapezius becomes strong with slightly more or slightly less abduction of the right shoulder.

    As the patient abducts her shoulder more, the lower vertebrae are tested. The specific thoracic segment tested is the one that lies in line with her right arm. If the trapezius tests weak at all angles of abduction, evaluate the retraction function of the right shoulder girdle.

    Instead of performing multiple muscle tests to evaluate each thoracic segment, a single passive technique could provide the same information.

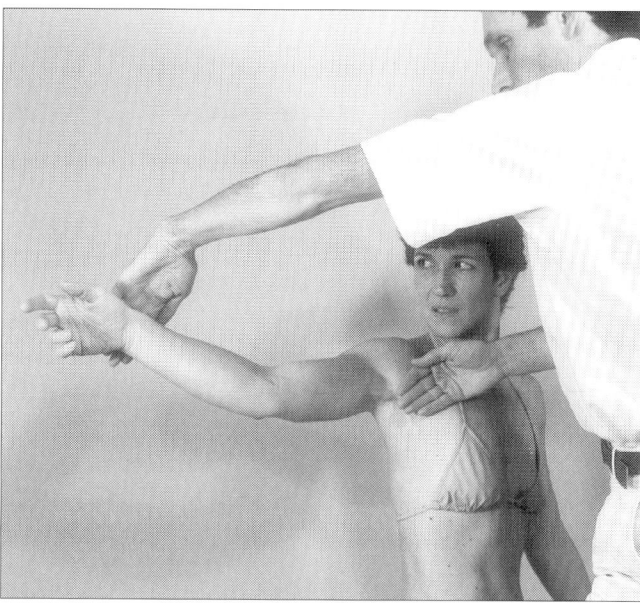

Fig. 19.22

- *Passive test* (Fig. 19.23)

    1. Have the patient be seated with her thoracic spine extended and rotated to the right.

    2. Her right shoulder should be maximally flexed and abducted.

    3. Stand in front of her, holding her right forearm with your left hand.

    4. Place your right hand on her thoracic spine to assist in thoracic extension and right rotation.

    5. Maintaining some degree of right shoulder retraction, passively adduct her right shoulder, feeling for a point in the range where her arm moves anteriorly. An anterior movement of her arm is the result of inhibition of specific trapezius fibers with simultaneous facilitation of antagonistic pectoralis major fibers.

    Remarks—If T4 has limited right rotation on T5 (stiffness of the right T4 facet for extension), the trapezius fibers that pull the T4–5 segment into the restriction will be inhibited. The right trapezius muscle fibers that attach to the spinous process of T5 and the left trapezius fibers that attach to the T4 spinous process will be inhibited (Fig. 19.24). A strong contraction of these fibers while the patient attempts to further extend the right T4 facet joint (thoracic right rotation and extension) is an active mobilization for this thoracic dysfunction.

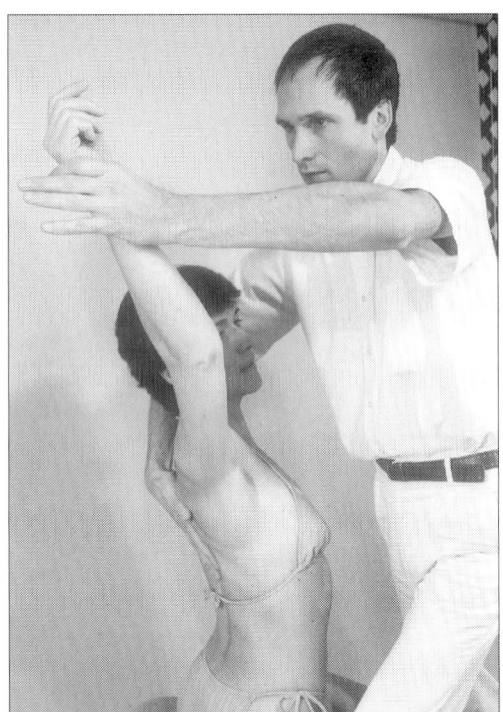

Fig. 19.23

- **Active mobilization** (Fig. 19.25)

  1. Find the inhibited trapezius fibers on the right and left sides. If the patient's thoracic spine is rotated right, her left shoulder is always a little less abducted than the right.

  2. Holding both of her wrists, provide resistance to the trapezius contraction on both sides, maintaining the previous abduction angle at both shoulders.

  3. As the patient attempts to rotate (and extend) her spine farther while bringing her arms more posterior, make sure she maintains the exact initial abduction angle at her shoulders.

  4. A strong contraction confirms that the inhibition from the thoracic segment has disappeared.

  5. Retest.

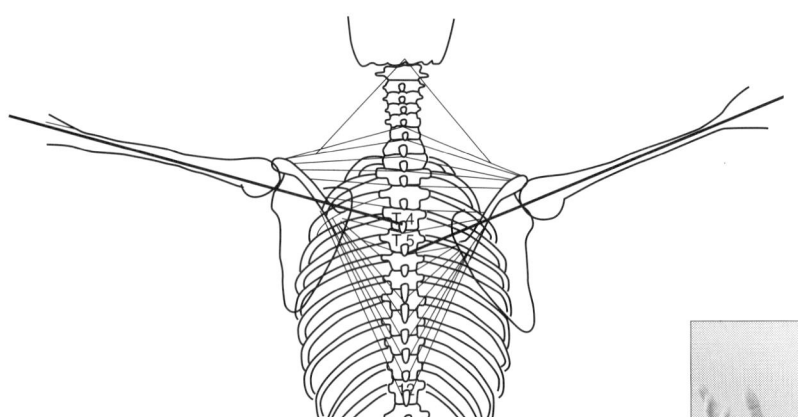

Fig. 19.24 T4 limited right rotation on T5.

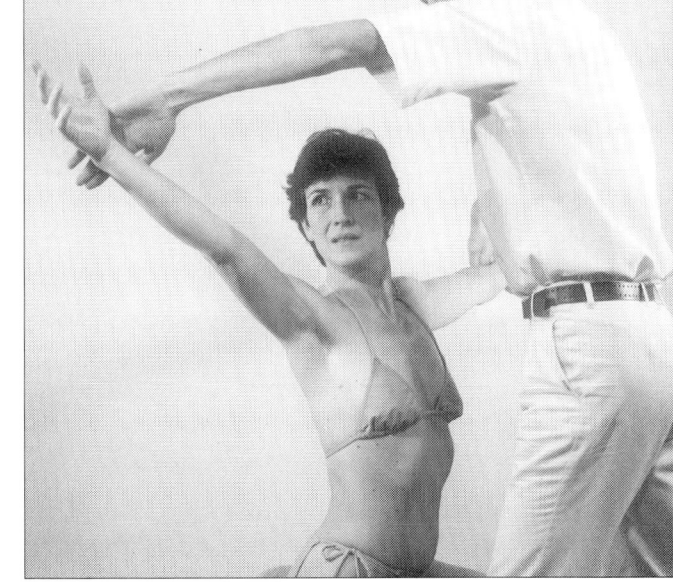

Fig. 19.25

C7 THROUGH T7 FACET EXTENSION

- **Agonists**

    Thoracic extensors:
    intertransversarii (dorsal rami of spinal nerves)
    interspinales (dorsal rami of spinal nerves)
    rotatores thoracis (dorsal rami of spinal nerves)
    multifidi, left (dorsal rami of spinal nerves)
    semispinalis thoracis, left (dorsal rami of lower cervical and thoracic spinal nerves)
    spinalis thoracis (dorsal rami of lower cervical and thoracic spinal nerves)
    longissimus thoracis (dorsal rami of cervical, thoracic, and lumbar spinal nerves)

- **Muscle test** (Fig. 19.26)

    1. Have the patient be seated with her left hand holding her neck and her head slightly rotated to the right.
    2. Sit behind her left shoulder.
    3. With your right thumb, fixate the lamina of T1.
    4. Support her left upper arm with your left arm and place your left hand on top of her head. Extend C7 over T1 with your left arm by lifting her left arm and extending her neck.
    5. Ask her to stay in this position and, with the fingers of your left hand, attempt to push her head into flexion and left side bending.

Remarks—The above muscle test can be used to evaluate C7 through T6. If this test yields weakness, compare the result with with a test in which the patient's spine is less extended to the right or to one in which your thumb fixates the lamina of the vertebra above.

- **Active mobilization**—The treatment contraction is identical to the testing contraction, except the treatment contraction is gentler and sustained. Retest after the mobilization.

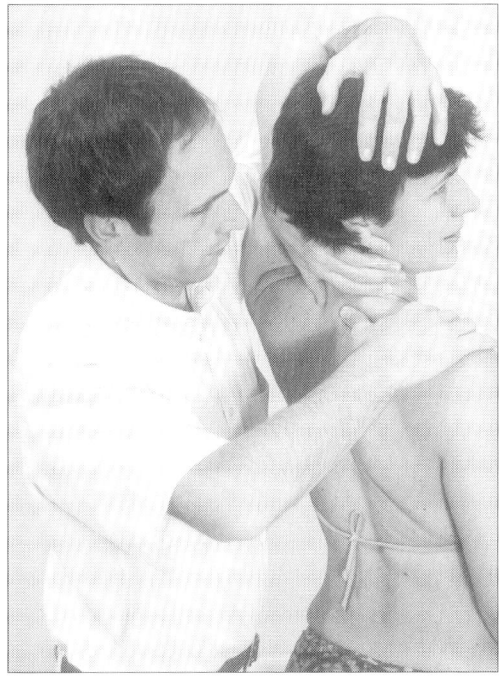

Fig. 19.26

T7 THROUGH L5 RIGHT FACET EXTENSION

- **Agonists**

    Thoracic and lumbar extensors:
    - intertransversarii (dorsal rami of spinal nerves)
    - interspinales (dorsal rami of spinal nerves)
    - rotatores thoracis (dorsal rami of spinal nerves)
    - multifidi, left (dorsal rami of spinal nerves)
    - semispinalis thoracis, left (dorsal rami of lower cervical and thoracic spinal nerves)
    - spinalis thoracis (dorsal rami of lower cervical and thoracic spinal nerves)
    - longissimus thoracis (dorsal rami of cervical, thoracic, and lumbar spinal nerves)
    - trapezius, middle and lower part (accessory nerve and C3,4)

- **Muscle test** (Fig. 19.27)

    1. Have the patient be seated with her left hand holding her right shoulder.
    2. Sit behind her right shoulder.
    3. Place your right arm over her right shoulder and hold her left upper arm with your right hand.
    4. With your left thumb, fixate the right lamina of T7.
    5. With your right arm, move her upper thoracic spine into extension, right side bending, and right rotation down to the level of T7.
    6. Ask her to stay in this position.
    7. While maintaining the pressure on the lamina of T7 with your left thumb, attempt to flex, left side bend, and left rotate her thoracic spine with your right hand.

    Remarks—The above muscle test can be repeated with your left thumb on T8 to test the extension mobility of T7. All vertebra between T7 and S1 can be tested this way, always testing the vertebra above your thumb. If the test yields weakness, compare the result with a test in which the patient's facet joint is less extended on the right or to one in which your thumb fixates the lamina of the vertebra above.

- **Active mobilization**—The treatment contraction is identical to the testing contraction, except the treatment contraction is gentler and sustained. Retest after the mobilization.

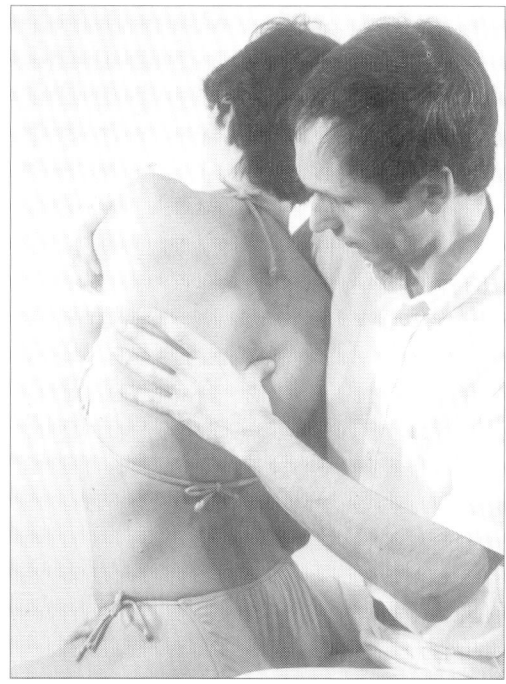

Fig. 19.27

T7 THROUGH L5 RIGHT FACET EXTENSION

- *Agonist*

    Latissimus dorsi:
    latissimus dorsi (thoracodorsal nerve; C6,7,8)

- *Muscle test* (Fig. 19.28)

    1. Have the patient be seated with her thoracic and lumbar spine in extension, right rotation, and right side bending, with her right arm alongside her body and internally rotated.

    2. Stand at her right side.

    3. Hold her right wrist with your left arm.

    4. Stabilize her iliac crest with your right hand.

    5. Ask her to resist your movement while you attempt to abduct and flex her right shoulder.

Remarks—This technique tests for the presence of right facet extension dysfunctions in the lower thoracic spine. The latissimus dorsi fibers that run perpendicular to the humerus are most effective in moving the arm. This means that, with the shoulder adducted, the fibers running to the thoracic spine are emphasized. When the shoulder is abducted 45 degrees, the muscle fibers attaching to the lumbar spine are emphasized. Repeat the above test with the right shoulder abducted 45 degrees to gain information about the lumbar spine.

The test is positive for a right facet extension dysfunction if a weak latissimus becomes strong after the spine moves out of its position with right facet extension. For instance, slouching or left rotation might remove inhibition from the latissimus dorsi fibers.

This technique is not useful for finding the precise level of a spinal extension dysfunction, but it is useful as an overview of lower spine functioning and for evaluating the latissimus dorsi itself.

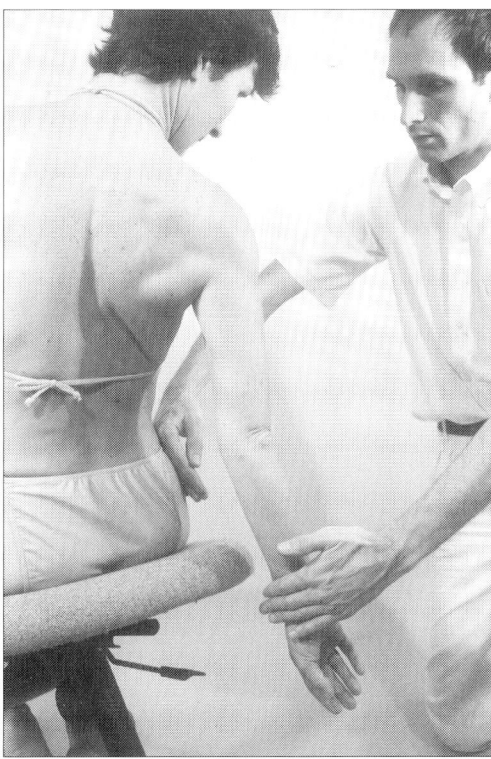

Fig. 19.28

T7 THROUGH L5 RIGHT FACET FLEXION

- *Agonist*

    Latissimus dorsi:
    latissimus dorsi (thoracodorsal nerve; C6,7,8)

- *Muscle test* (Fig. 19.29)

    1. Have the patient be seated with her thoracic and lumbar spine flexed, rotated left, and side bent to the left, with her right arm alongside her body and internally rotated.

    2. Stand at her right side.

    3. Hold her right wrist with your left arm.

    4. Stabilize her iliac crest with your right hand.

    5. Ask her to resist your movement while you attempt to abduct and flex her right shoulder.

Remarks—This technique tests for the presence of right facet flexion dysfunctions in the lower thoracic spine. The latissimus dorsi fibers that run perpendicular to the humerus are most effective in moving the arm. This means that, with the shoulder adducted, the fibers running to the thoracic spine are emphasized. When the shoulder is abducted 45 degrees, muscle fibers attaching to the lumbar spine are emphasized. Repeat the above test with the shoulder abducted 45 degrees to gain information about the lumbar spine.

The test is positive for a right facet flexion dysfunction if a weak latissimus dorsi becomes strong after the spine moves out of its position with right facet flexion. For instance, extending the spine or right rotation might remove inhibition from the latissimus dorsi fibers. The test is also positive for a spinal dysfunction if the latissimus dorsi tests strong in more or less abduction.

This test can also be used to find left facet joint dysfunctions. Repeat the test using the right latissimus dorsi with the patient seated in spinal flexion, right rotation, and right side bending. Similarly, with the patient seated in spinal extension, left rotation, and left side bending, you can test for left facet joint extension dysfunctions. Of course, the left latissimus dorsi can also be used to evaluate facet joint dysfunctions on either side.

This technique is not useful for finding the precise level of a spinal flexion dysfunction, but it is useful as an overview of lower spine functioning and for evaluating the latissimus dorsi itself.

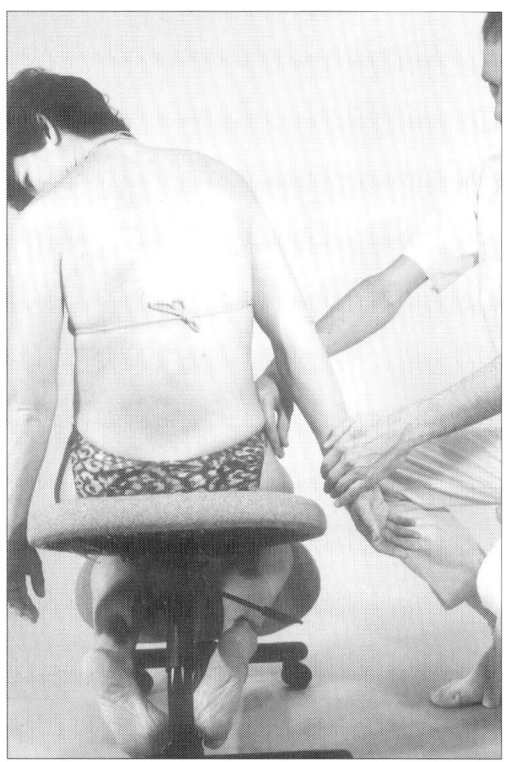

Fig. 19.29

T6 THROUGH L5 FLEXION

- **Agonists**

    Thoracic and lumbar flexors:
    external oblique abdominal, right (ventral rami of spinal nerves; T7–T12)
    internal oblique abdominal, left (ventral rami of spinal nerves; T7–L2)
    rectus abdominis (ventral rami of spinal nerves; T6–T12)
    pectoralis major, right (lateral and medial pectoral nerves; C5–T1)
    pectoralis minor, right (lateral and medial pectoral nerves; C6,7,8)
    serratus anterior, right (long thoracic nerve; C5,6,7)

- **Muscle test** (Fig. 19.30)

    1. Have the patient be seated with her spine flexed, left rotated, and side bent to the left and her right hand holding her left shoulder.
    2. Sit behind her left shoulder.
    3. Place your left arm over her left shoulder and hold her right upper arm with your left hand.
    4. Place your right index finger or thumb on the right lamina of T6, pushing anteriorly.
    5. Ask her to stay in this position and, while sustaining pressure on the right lamina of T6, push her shoulder posteriorly and to her right with your left hand.

    Remarks—If the test yields weakness, compare the result with a test in which the patient's right thoracic spine is less flexed or in which your right hand fixates the lamina of the vertebra below.

    Repeat the test, fixating the right lamina of T7 through L5.

- **Passive mobilization** (Fig. 19.31)

    1. Have the patient be seated with her spine flexed and her right hand holding her left shoulder.
    2. Sit behind her left shoulder.
    3. Place your left arm over her left shoulder and hold her right upper arm with your left hand.
    4. With your left arm, move her spine into flexion, left side bending, and left rotation.

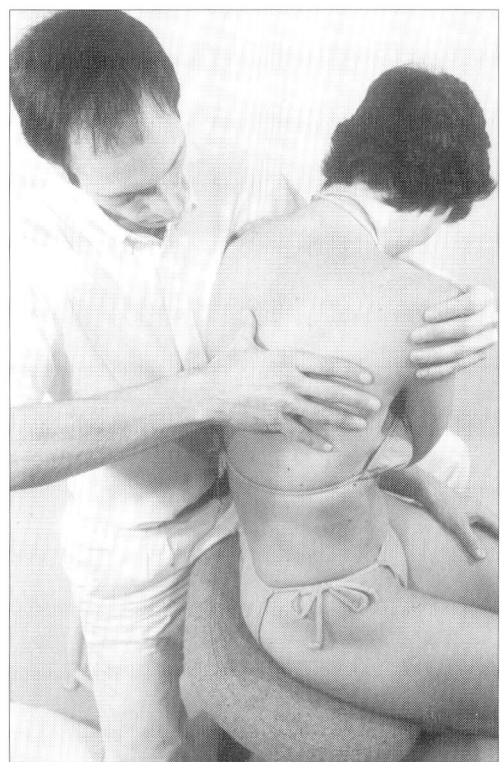

Fig. 19.30

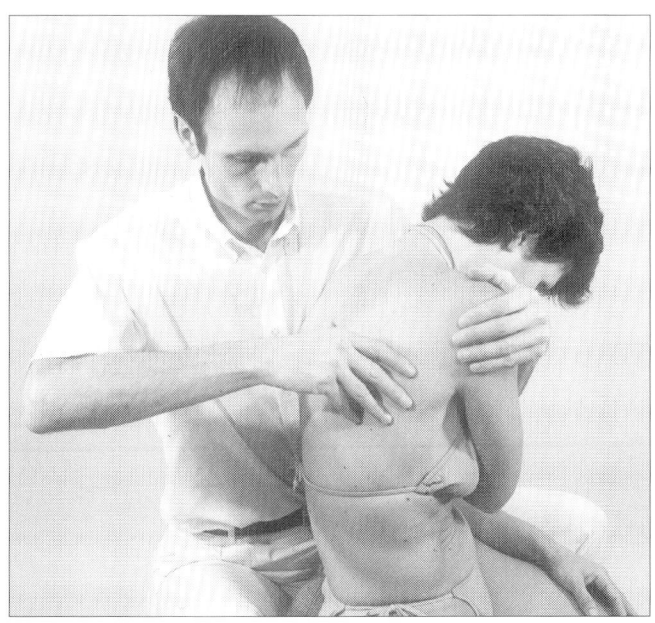

Fig. 19.31

5. Use your right thumb or index finger to mobilize T6 on the right into flexion (push the right lamina or transverse process anteriorly and cranially).
6. Retest.

Rib Cage

The anterior and lateral parts of the ribs move cranially during inhalation and caudally during exhalation. The lateral part of the ribs moving cranially and caudally is likened to the movement of a bucket handle. The caudal ribs move predominantly with this bucket handle movement. The anterior part of the ribs moving cranially and caudally is likened to the movement of a pump handle, and the cranial ribs move predominantly with this pump handle movement.

Ribs move in the costotransverse joints and the costovertebral joints. Both constitute movement of the ribs in comparison to the spine. Besides moving in comparison to the spine, the ribs move in comparison to each other. During inhalation, the lateral parts of the ribs separate from each other (the intercostal spaces increase in size), and during exhalation, the lateral parts of the ribs approach each other.

In the sagittal plane, without spinal movement, the superior rib translates anteriorly in relation to its caudal neighbor during inhalation and posteriorly over the inferior rib during exhalation. A similar movement occurs during rotation of the trunk. During left rotation of the trunk, the superior ribs on the right side move anteriorly over the inferior ribs. The right ribs could be said to move into inhalation during left rotation. In addition, during left rotation, the lateral parts of the right ribs separate from each other. The opposite occurs during right rotation; the right ribs move into an exhalation position.

While the right ribs move into an inhalation position during left rotation, the left ribs move into an exhalation position. The reverse occurs during right rotation.

The techniques described in this section on the rib cage do not follow that of a capsular pattern. The capsular pattern of the rib joints is not known. Inhalation techniques are discussed first, followed by exhalation techniques and techniques that test the posterior mobility of the ribs.

RIBS 1 AND 2 INHALATION

- **Agonists**

    Scaleni:
    Scaleni (ventral rami of spinal nerves; C3–8)

- **Muscle test** (Fig. 19.32)

    1. Have the patient be seated with her neck slightly rotated to the left.

    2. Stand behind her, stabilizing her left shoulder with your left hand.

    4. Place your right hand on the right side of her head.

    5. Ask the patient to inhale deeply, then hold her breath and maintain her head position while you attempt to side bend her head to the left.

    Remarks—If the scaleni test weak, compare the results of this test with one in which the patient has exhaled. If the scaleni test weak with inhalation but strong with exhalation, an inhalation restriction from the first or second rib is likely.

    Right extension restrictions in the cervical spine may also cause the right scaleni to test weak. If cervical restrictions inhibit the scaleni, repeating the test after the patient has exhaled will yield the same weak result. In this case, test the scaleni in a position with more flexion of the right cervical facet joints.

    It is possible for the right scaleni to test weaker with exhalation than with inhalation. This could

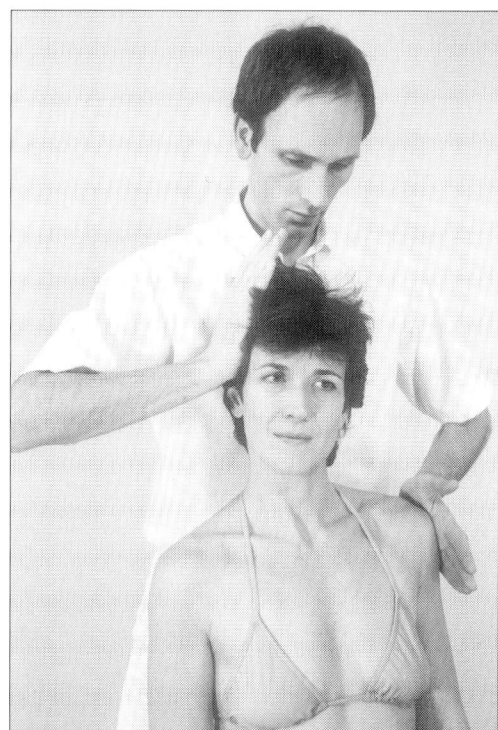

Fig. 19.32

be caused by limited depression mobility at the right sternoclavicular joint (see shoulder girdle depression; sternocleidomastoid muscle).

- **Active mobilization**—The active mobilization is identical to the test described above, except the treatment contraction is gentler and sustained. Ask the patient to hold her breath as long as is comfortable. Repeat the mobilization if needed. Retest after the mobilization.

RIBS 7 THROUGH 9 INHALATION

- **Agonists**

    Serratus anterior:
    serratus anterior, inferior part (long thoracic nerve; C5,6,7)

- **Muscle test** (Fig. 19.33)

    1. Have the patient be seated with her right shoulder flexed to 90 degrees.
    2. Stand at her right side.
    3. Place your left hand on her right shoulder.
    4. Hold her distal forearm with your right hand.
    5. Ask her to inhale and resist your movement while you attempt to push her right arm caudally to bring her shoulder out of the flexed position.

Remarks—This muscle test can be sensitized by having the patient's thoracic spine side bent and/or rotated to the left (Fig. 19.34). Compare the result with a test test in which the patient has exhaled.

Although the serratus anterior muscle attaches to ribs 1 through 9, this test places emphasis on the inferior fibers that attach to ribs 7 through 9. The more superior serratus anterior fibers are better evaluated by testing resisted shoulder girdle protraction.

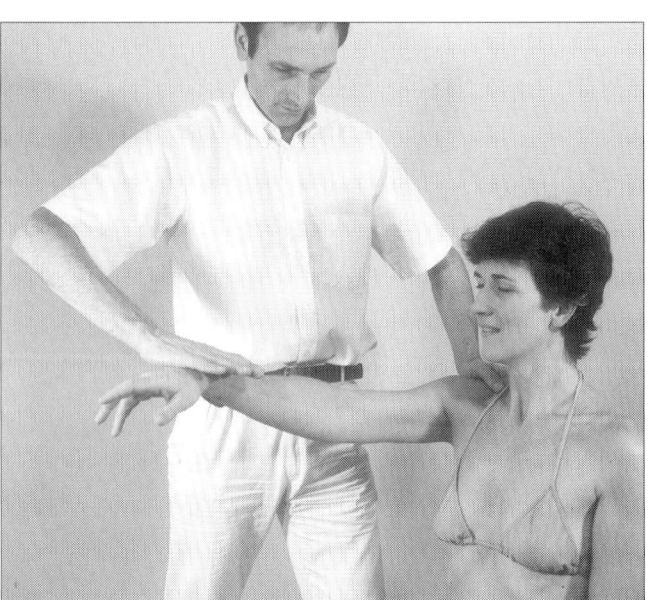

Fig. 19.33

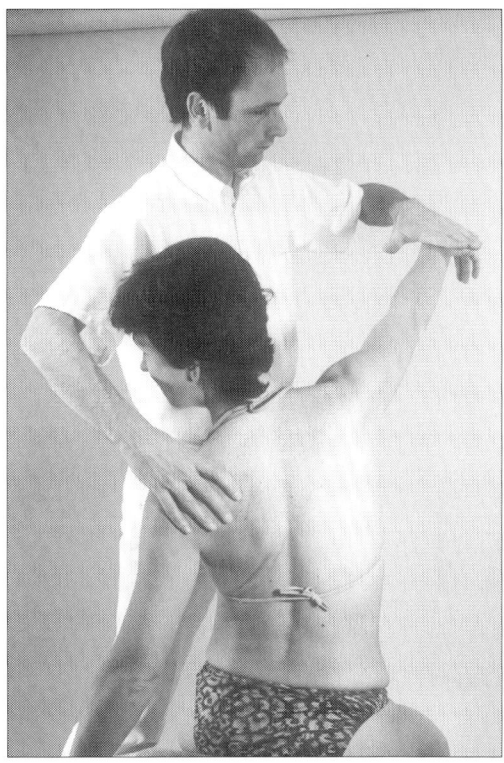

Fig. 19.34

TECHNIQUE PROCEDURES

- **Passive mobilization (seated)** (Fig. 19.35)

  1. Have the patient be seated with her right hand holding her left shoulder.

  2. Sit behind her left shoulder.

  3. Place your left arm over her right arm, with your left hand holding her right shoulder, placing her thoracic spine in left rotation and some left side bending.

  4. With your right index finger, palpate posterolaterally on ribs 7 through 9 for their mobility in an anterior direction.

  5. The treatment consists of a sustained anterior mobilization of the rib with restricted mobility.

  6. Retest.

- **Passive mobilization (side lying)** (Fig. 19.36)

  1. Have the patient lie on her left side with her right arm in front of her.

  2. Stand in front of her, facing her feet.

  3. With your thumbs, palpate the exhalation mobility of her right lower ribs.

  4. The rib with restricted exhalation mobility should be mobilized by a gentle sustained pressure in a caudal direction.

  5. Ask the patient to inhale deeply and hold her breath as long as is comfortable.

  6. Retest.

  Remarks—Two adjacent ribs with restricted movement away from each other can be thought of as a restricted inhalation mobility of the superior rib or as a restricted exhalation mobility of the inferior rib. This treatment technique increases the mobility of the intercostal tissue.

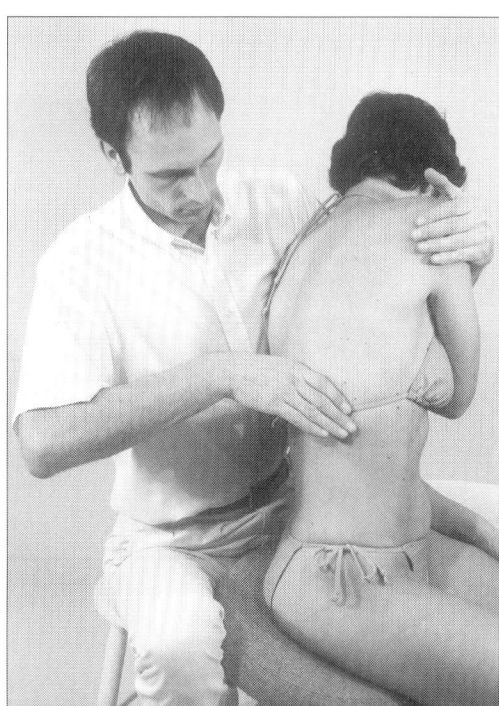

Fig. 19.35

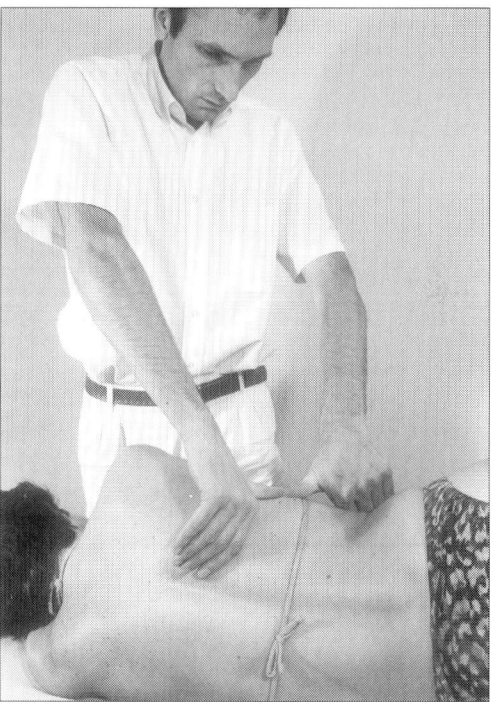

Fig. 19.36

RIBS 4 THROUGH 12 INHALATION

- **Agonists**

    Rib elevators:
    external intercostals (intercostal nerves of related segments; T1–T11)
    levatores costarum (lateral branches of the dorsal rami of the corresponding nerves; C7–T11)

- **Muscle test** (Fig. 19.37)

    1. Have the patient lie prone with her arms alongside her body.
    2. Stand on either side of the table, placing your index fingers laterally in the intercostal space between her eleventh and twelfth ribs on both sides.
    3. Ask her to breathe in maximally and maintain the effort of breathing in.
    4. To test the respiratory muscles for inhalation, feel the "give" of the twelfth rib in a caudal (expiratory) direction.
    5. Repeat this test on the eleventh rib, gradually working your way up to the fourth rib.

    Remarks—This test can be performed bilaterally by comparing the amount of give at each level with the contralateral side and with the superior and inferior ribs.

    Quickly test as many ribs as you can while the patient holds her breath. Repeat the test for the remaining ribs, again asking the patient to inhale maximally.

    Even though the ribs are pushed into exhalation, the test assesses the ability of the respiratory muscles to maintain the ribs in an inhalation position.

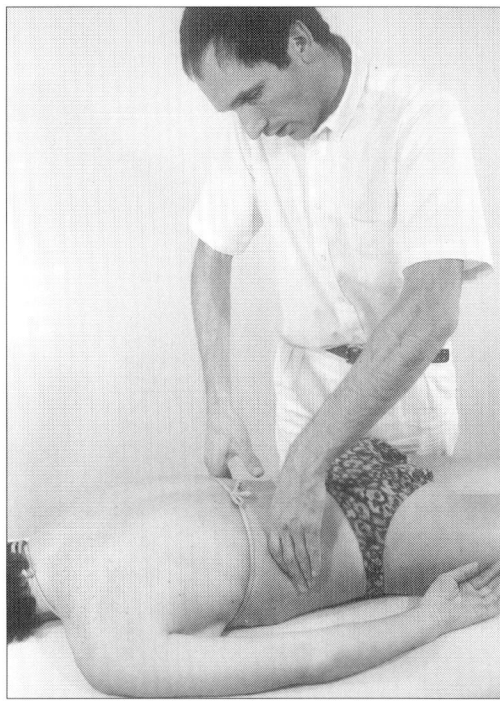

Fig. 19.37

- **Passive mobilization** (seated) (Fig. 19.35)

    1. Have the patient be seated with her right hand holding her left shoulder.
    2. Sit behind her left shoulder.
    3. Place your left arm over her right arm and hold her right shoulder with your left hand. The position of your left arm places her thoracic spine in left rotation and some left side bending.
    4. Place the radial side of your right index finger posterolaterally in the intercostal space caudal to the rib that had excessive give.
    5. Use your right index finger to mobilize the rib in an anterior and slightly cranial direction.
    6. Retest.

RIBS 6 THROUGH 12 INHALATION

- **Agonists**

    Diaphragm:
    diaphragm (phrenic nerve; C3,4,5)

- Muscle test (Fig. 19.38)

    1. Have the patient lie supine.

    2. Place both of your hands, fingers spread, on her abdomen.

    3. Ask her to inhale, hold her breath, and "make a belly."

    4. As she maintains this position, attempt to push her belly dorsally.

    Remarks—If the muscle tests weak, compare this test with one in which the patient has "made a belly" after exhalation.

    Pregnancy might be a contraindication. To prevent a Valsalva maneuver when testing the patient after inhalation, ask her to make a humming sound while you test the muscle.

    This test does not differentiate among the individual six lower ribs, nor does it differentiate between the left and right sides. Other muscles can be tested to get information regarding individual ribs. This test is useful as an overview of inhalation mobility of the bilateral lower six ribs.

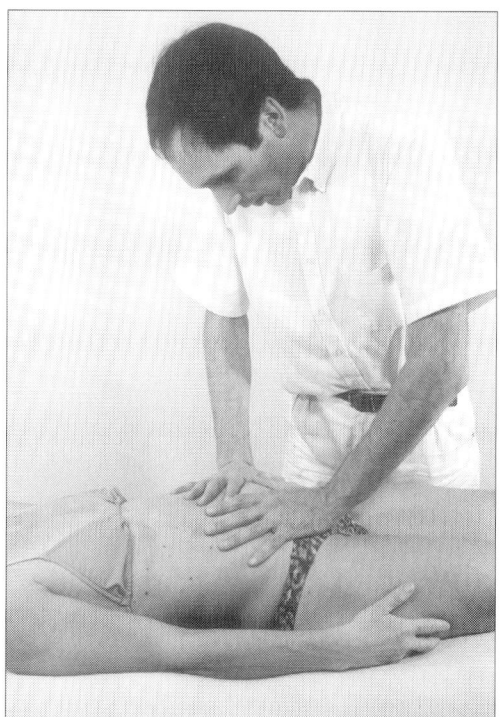

Fig. 19.38

RIBS 7 THROUGH 12 EXHALATION

- **Agonists**

    Abdominals

- **Muscle test**—See Lumbar Spine; Flexion L1 through L5; Lumbar flexors.

RIBS 9 THROUGH 12 EXHALATION

- *Agonists*

    Rib depressors:
    internal intercostals (intercostal nerves of related segments; T1–T12)
    innermost intercostals (intercostal nerves of related segments; T1–T12)
    subcostals (intercostal nerves of related segments; T1–T12)
    inferior posterior serratus (ventral rami; T9–12)

- *Muscle test* (Fig. 19.39)

    1. Have the patient lie prone.
    2. Stand at the top of the table with your index fingers placed below the lateral aspects of her twelfth ribs.
    3. Ask her to breathe out maximally and maintain the effort of breathing out.
    4. To test the strength of the rib depressors, feel the amount of give of the twelfth rib in a cranial (inspiratory) direction.
    5. Repeat this test with your fingers below the lateral aspect of the eleventh rib, gradually working your way up to the ninth rib.

    Remarks—This test can be performed bilaterally by comparing the amount of give at each level with the contralateral side and with the other ribs.

    Quickly test as many ribs as you can while the patient holds her breath. Because the ribs are pushed into inhalation, the test assesses the ability of the respiratory muscles to maintain the ribs in an expiratory position.

- *Passive mobilization* (Fig. 19.40)

    1. Have the patient lie prone on the treatment table.
    2. Stand at the side of the patient with the rib that has restricted exhalation mobility.
    3. Face toward the patient's feet and place your thumbs on the rib that had too much give.
    4. Using your thumbs, mobilize the rib in a caudal and medial direction.
    5. Retest.

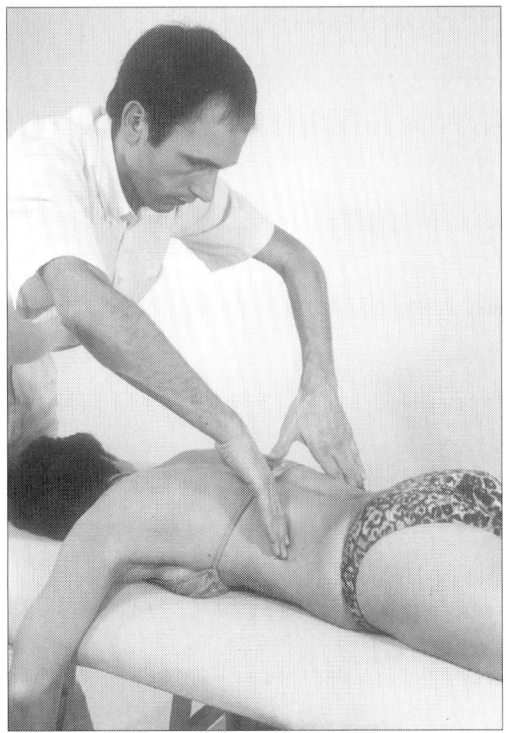

Fig. 19.39

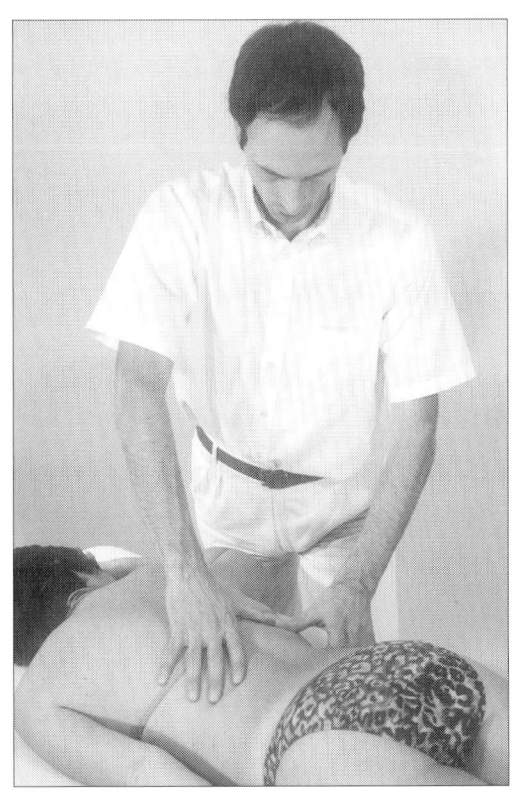

Fig. 19.40

TWELFTH RIB EXHALATION

- *Agonist*

    Quadratus lumborum:
    quadratus lumborum (T12–L4)

- *Muscle test* (Fig. 19.41)

    1. Have the patient lie supine with her hands holding the sides of the treatment table.
    2. Stand at her left side.
    3. Stabilize her left iliac crest with your right hand.
    4. Support her ankles with your left forearm, holding her right ankle with your left hand.
    5. Ask her to exhale deeply, maintain the exhalation phase, and hold her legs in place while you attempt to move her ankles toward you (to her left).

    Remarks—Compare the results of this test with one in which the patient has inhaled. If inhalation does not improve the muscle's strength, the inhibition might derive from a right lumbar extension dysfunction or from the right sacroiliac joint.

- *Passive mobilization* (Fig. 19.42)

    1. Have the patient lie prone.
    2. Stand at her right side, facing her feet.
    3. Place your thumbs in the intercostal space between the eleventh and twelfth ribs.
    4. Ask her to exhale deeply and to maintain the exhalation phase as long as is comfortable while you mobilize her twelfth rib caudally.
    5. Retest.

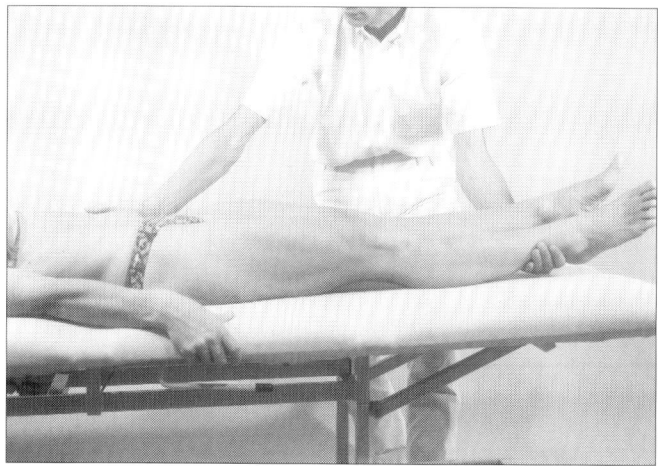

Fig. 19.41

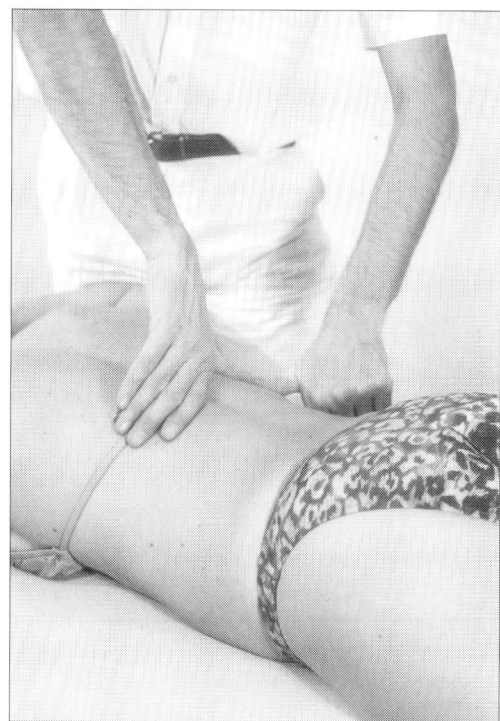

Fig. 19.42

RIBS 1 THROUGH 5 MOBILITY IN A POSTERIOR DIRECTION

- *Agonists*

    Shoulder protractors:
    pectoralis minor (lateral and medial pectoral nerves; C6,7,8)
    serratus anterior, upper part (long thoracic nerve; C5,6,7)

- *Muscle test* (90 degrees shoulder flexion) (Fig. 19.43)

    1. Have the patient lie supine with her right shoulder flexed to 90 degrees and slightly protracted.
    2. Have her extend her right elbow and make a fist with her hand.
    3. Stabilize her wrist with your right hand and place your left hand on her fist.
    4. Ask her to keep her arm in place while you attempt to push it in a dorsal direction (into retraction of the shoulder girdle).

- *Muscle test* (0 degrees shoulder flexion) (Fig. 19.44)

    1. Have the patient lie supine with her right arm alongside her body but off the table.
    2. Have her protract her right shoulder submaximally.
    3. Support her arm by holding her right wrist with your right hand.
    4. Place your left hand on the anterior surface of her humeral head and coracoid process.
    5. Ask her to maintain this position while you attempt to push her shoulder girdle posteriorly.

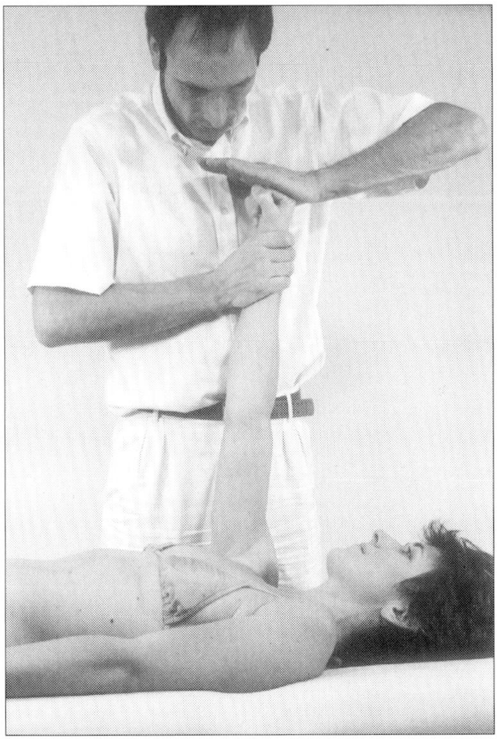

Fig. 19.43

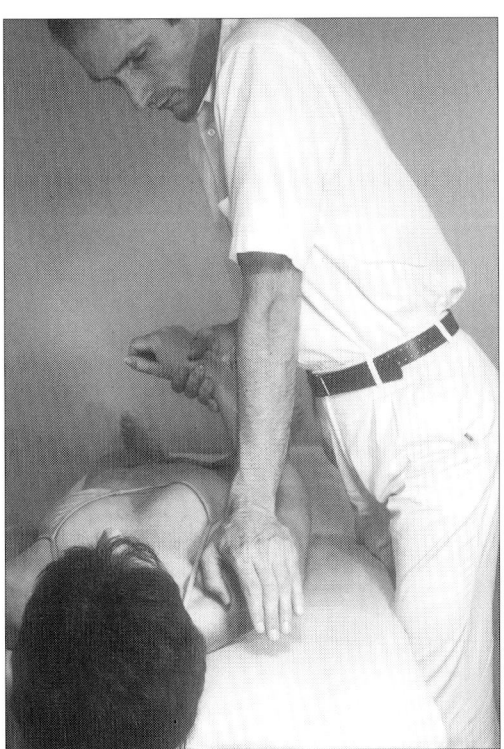

Fig. 19.44

Remarks—Both tests can be sensitized for rib dysfunctions by repeating them after the patient has inhaled or exhaled deeply. Similarly, the tests can be sensitized for rib dysfunctions by repeating them after the patient has rotated her trunk to the right (knees bent and rotated to the left, head rotated to the right [Fig. 19.45]). Testing the protractors in spinal left rotation also sensitizes the test (knees rotated right, head rotated left [Fig. 19.46]). The sensitization maneuvers might place additional stress on the rib structures and further inhibit the protractors.

The test is positive for a dysfunction in the area of the first five ribs if protractor strength alters with the respiratory phase or with rotation of the thorax. If rotation of the thoracic spine or a change in the respiratory phase do not improve protractor strength, the cause of weakness may be limited shoulder girdle protraction.

- **Passive mobilization** (Fig. 19.47)

    1. Place the patient in the position where the muscle tested maximally weak.

    2. Gently press on the anterior aspect of ribs 1 through 5 and compare their end-feels.

    3. Gently mobilize the rib with stiff posterior translation in a posterior direction.

    4. Retest.

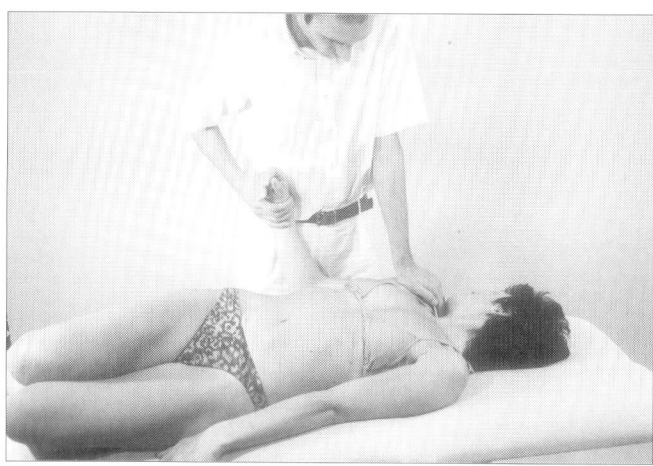

Fig. 19.45

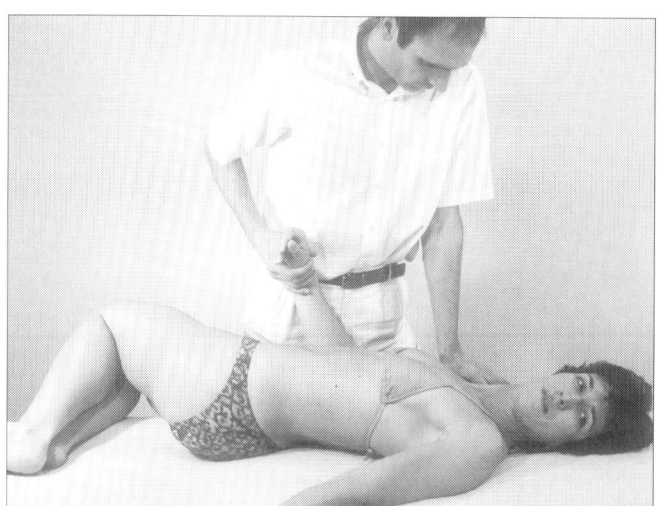

Fig. 19.46

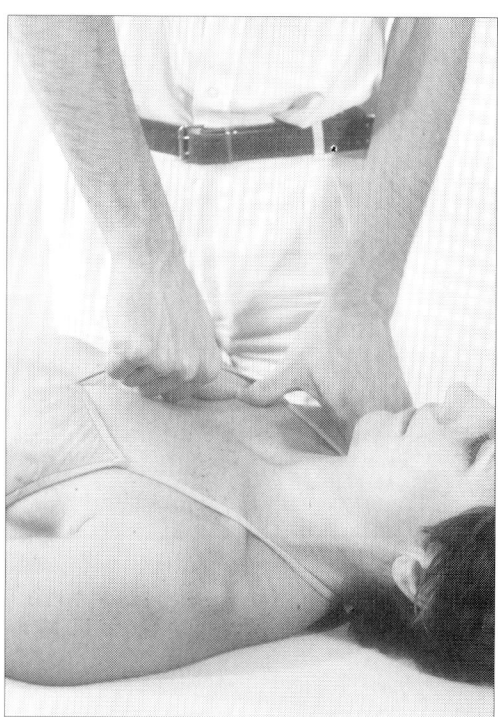

Fig. 19.47

CLAVICLE THROUGH RIB 6 MOBILITY IN A POSTERIOR DIRECTION

- *Agonists*

    Pectoralis major:
    pectoralis major, clavicular and sternocostal part (lateral and medial pectoral nerves; C5–T1)

- *Muscle test* (Fig. 19.48)

    1. Have the patient lie supine with her right shoulder flexed to 90 degrees and internally rotated.

    2. Stand at her right side.

    3. Hold her distal forearm with your right hand.

    4. Ask her to keep her arm in this position while you attempt to move the arm laterally, away from the medial part of her clavicle (attempt to horizontally abduct her shoulder).

    5. Repeat this test, attempting to move her arm away from the sternocostal junction of the individual ribs, one rib at a time.

Remarks—Instead of performing seven separate tests (clavicle through rib 6), the advanced examiner can do a single test for all seven structures. During this test, the examiner changes the angle of pull from lateral to cranial or vice versa.

The test can be sensitized for rib dysfunctions by rotation of the thoracic spine or by changing the respiratory phase. The patient could bend her knees, rotate them to the right, and look left (Fig. 19.49). In this position, the muscle test can be performed after inhalation and after exhalation. Repeat these tests with thoracic rotation to the opposite side.

Rotation of the thoracic spine and the respiratory phase can add stress to the dysfunction. Minimal dysfunctions may show up only after these stresses have been added.

If the test yields weakness, it can be compared with one in which the examiner pulls in a slightly different direction (testing muscle fibers inserting to other structures) or to one in which the patient is in a different respiratory phase or in opposite thoracic rotation.

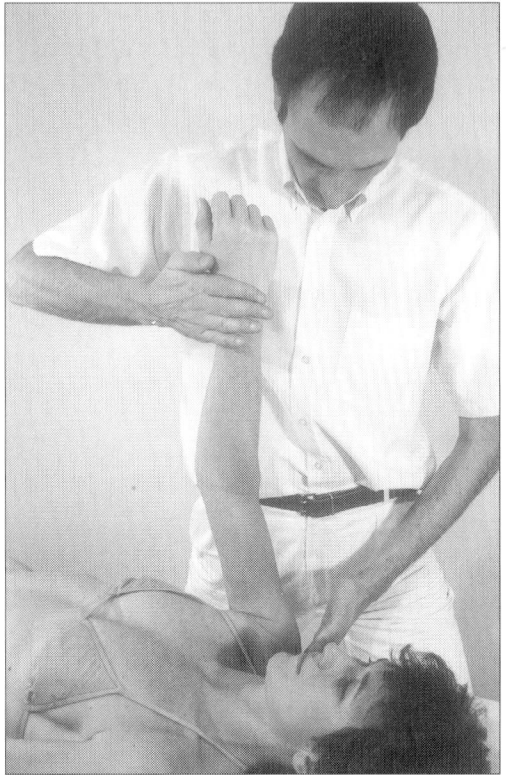

Fig. 19.48

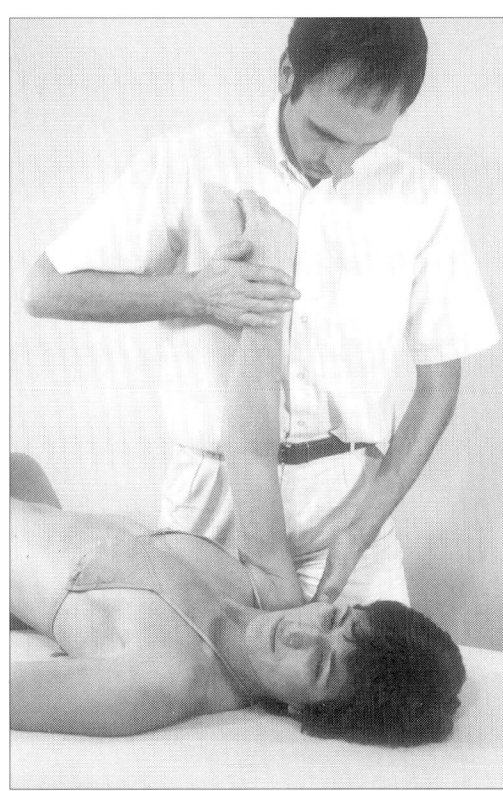

Fig. 19.49

- **_Passive mobilization_** (Fig. 19.50)

    1. Have the patient lie supine with her thorax rotated into the position that yielded maximum inhibition of the pectoralis major.

    2. Stand at her right side, facing toward her head.

    3. Place your right thumb on the structure responsible for inhibition of her pectoralis major (the medial clavicle or the costal cartilage of one of the upper six ribs).

    4. Ask her to inhale or exhale deeply (whichever increases inhibition of her pectoralis major) and hold her breath for as long as is comfortable. The pressure of your thumbs should be in a posterior direction.

    5. Retest.

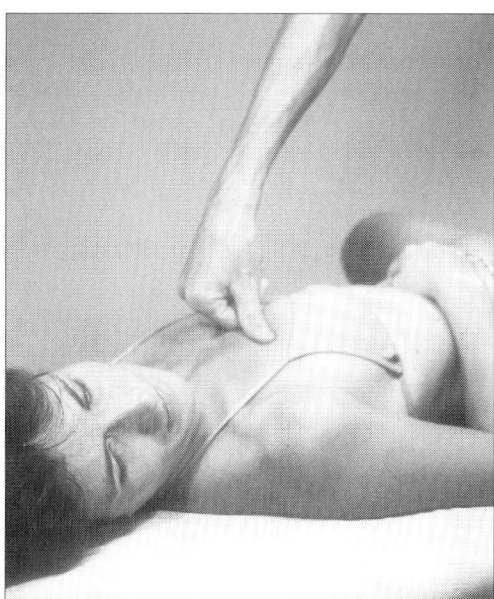

Fig. 19.50

Lumbar Spine

In addition to some natural gapping (separation) of the lumbar facet joints, they can flex and extend. Maximal right facet joint flexion is accomplished by spinal flexion, left side bending, and left rotation. Maximal right facet joint extension is accomplished by spinal extension, right side bending, and right rotation. Many patients with low back pain present with extension dysfunctions.[11] Patients with an extension dysfunction present with extension range limitation as well as extension weakness. McNeill and colleagues tested the strength of flexion, extension, and left and right side bending in patients with low back pain. Their study revealed that extension was the weakest of these four movements.[12]

- **Capsular pattern**—Extension and side bending are limited. Flexion is full but often painful.

T6 THROUGH L5 RIGHT FACET EXTENSION

- **Agonists**

    Thoracic and lumbar extensors and latissimus dorsi

See "Thoracic Spine."

L1 THROUGH L5 RIGHT FACET EXTENSION

- **Agonist**

    Psoas major:
    psoas major (L1,2,3)

- **Muscle test** (Fig. 19.51)

    1. Have the patient lie supine with her right hip and knee bent.
    2. Her right knee should be moved in the direction of her left shoulder.
    3. Her left leg should be adducted such that her lumbar spine is in a right side bent position.
    4. Stand at her right side and fixate her left anterior superior iliac spine (ASIS) with your right hand.
    5. Hold her knee with your left hand.
    6. Ask her to keep her knee in place while you attempt to push it (about 30 degrees) in a caudal and lateral direction, away from her lumbar vertebrae.

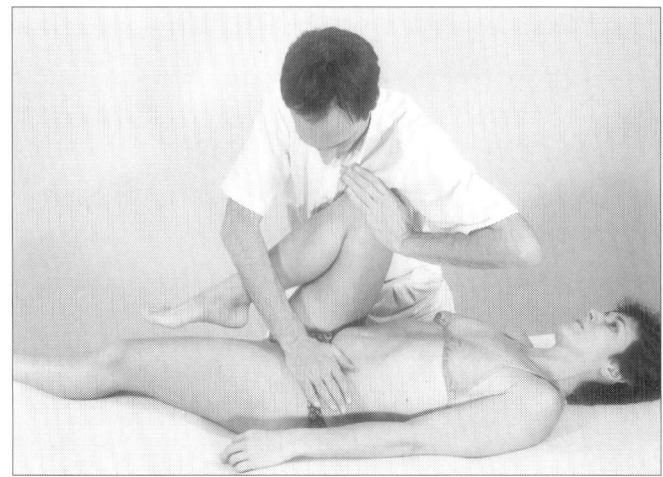

Fig. 19.51

Remarks—If you have trouble reaching the left ASIS with your right hand, switch hands (Fig. 19.52).

Compare the result of this test with one in which the patient decreases the amount of right side bending of her lumbar spine by abducting her left leg or bringing her left knee to her chest.

The right psoas major attaches to the right side of the lumbar spine. A contraction of the right psoas major creates a force that increases right lumbar side bending and thus increases right lumbar facet extension.

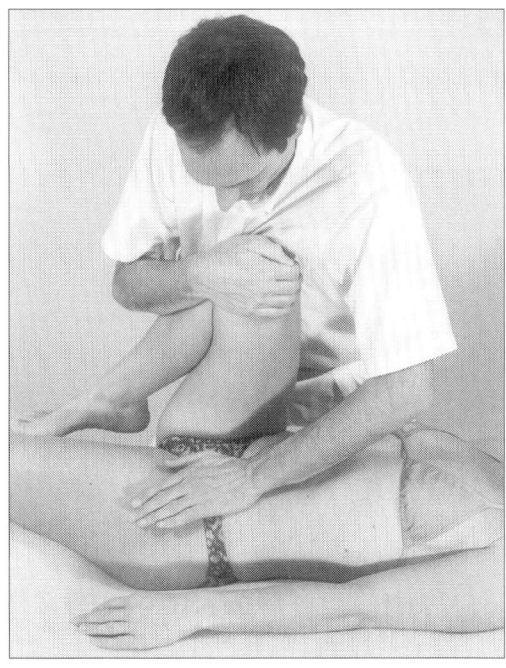

Fig. 19.52

- **_Passive mobilization_** (prone) (Fig. 19.53)

    1. With the patient lying prone, raise the head end of the table to bring her lumbar spine into extension. You can also create right lumbar extension on a flat table by having the patient push up onto her elbows or side bend her lumbar spine to the right.

    2. Stand at her right side.

    3. Palpate her right lumbar transverse processes with an anteromedially directed pressure. Compare the end-feel of the different segments.

    4. Mobilize the stiff segment(s) with an anteromedially directed pressure. This mobilization can be done statically or with oscillations of different frequencies. High-frequency oscillations cause local movement, whereas low-frequency oscillations can result in relaxed movements of the entire body.

    5. Retest.

- **_Postisometric stretch technique_** (Fig. 19.54)

    1. Have the patient lie on her left side with her hips and knees bent and her lumbar spine flexed.

    2. Stand in front of her, supporting her knees, which are off the table, with your proximal left thigh and her ankles with your left hand.

    3. Monitor her lumbar spine with your right hand.

    4. Move her hips and lumbar spine slowly into extension, stopping at the first perception of a restrictive barrier.

    5. Lift her ankles (causing her lumbar spine to side bend to the right), again stopping at the first perception of a barrier.

    6. Ask the patient to gently pull her feet toward the floor for 6 seconds while you provide unyielding resistance.

    7. After the patient completely relaxes, passively move her lumbar spine farther into right side bending and extension, each time stopping at the first perception of a restrictive barrier. Rotate her right shoulder backward (right rotation of her spine), stopping at the first perception of a barrier.

    8. Repeat the contraction.

    9. Repeat steps 7 and 8 until all restrictive barriers have disappeared.

    10. Retest.

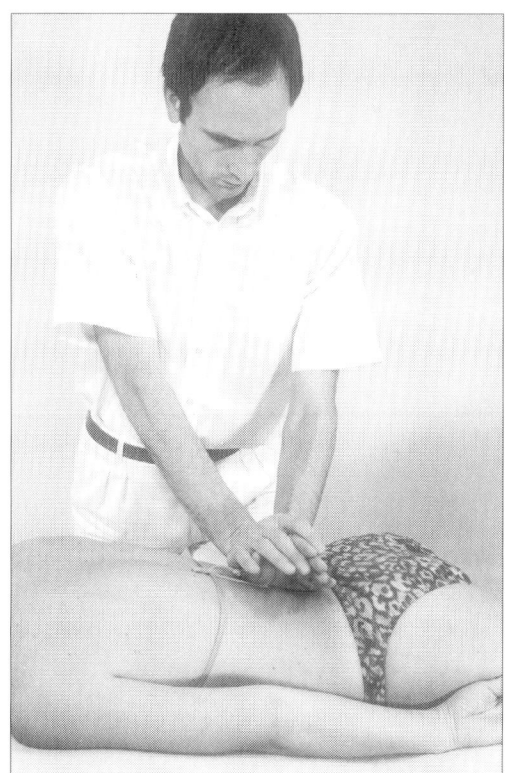

Fig. 19.53

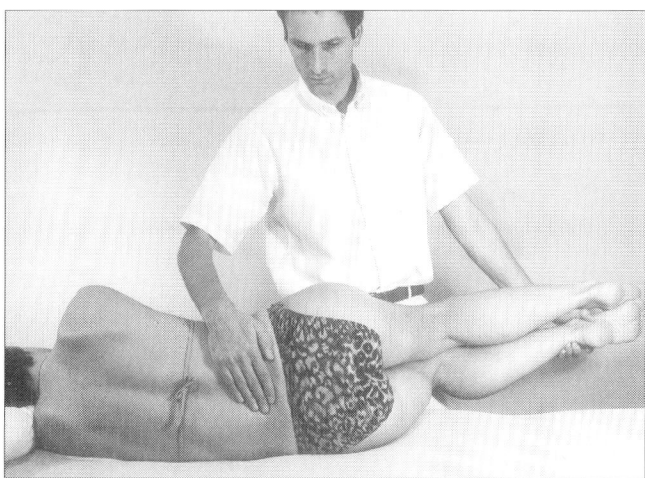

Fig. 19.54

Remarks—This technique can be used to treat several extension dysfunctions simultaneously on one side of the lumbar spine. To treat an extension dysfunction at a specific level, your right hand can monitor the barriers at this level by palpating movement at this segment. Contraction of the muscle is again at the feather edge of the restricted barrier.

- ***Passive mobilization*** (side lying) (Fig. 19.55)

  1. Have the patient lie on her left side. If possible, raise the head of the table such that her lumbar spine side bends to the right.

  2. The patient's left leg should be extended, her right hip and knee flexed, with her foot resting behind her left knee.

  3. Her right elbow should be bent, pointing dorsally, with her right hand resting on her waist.

  4. Stand in front of her.

  5. Monitor the movement at a specific lumbar segment with your right hand.

  6. The patient's right knee should rest on your left upper leg (in your left groin). Move her right knee to your left or right until you have achieved submaximal extension at the level you want to mobilize.

  7. Maintaining the same degree of lumbar extension, rest her right leg on the table.

  8. With your right elbow, rotate her right shoulder posteriorly (patient rotates to her right).

  9. The mobilization is performed by bringing her right shoulder more posterior with your right forearm on her rib cage and rotating her right pelvis forward with your left forearm.

  10. Retest.

Remarks—The effect of the mobilization can be monitored by stabilizing the patient's right leg, asking her to keep her right shoulder back, and attempting to bring her right shoulder forward with your right hand (derotate her spine). Limited right lumbar facet extension causes inhibition of the right rotators in this position. This technique is described in more detail on the following page.

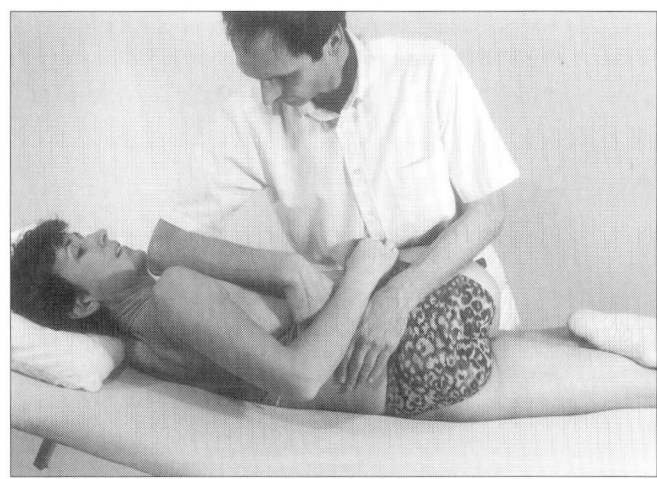

Fig. 19.55

L1 THROUGH L5 FACET EXTENSION

- **Agonists**

    Lumbar rotators:
    multifidi (dorsal rami of related spinal nerves)
    rotatores lumborum (dorsal rami of related spinal nerves)

- **Muscle test** (Fig. 19.56)

    1. Have the patient lie on her left side. If possible, raise the head end of the table such that her lumbar spine side bends to the right.
    2. Her left leg should be extended.
    3. Monitor the movements at her lumbar spine with your right hand.
    4. Have the patient flex her right hip and knee and rest the knee on your left upper leg (in your left groin). Move her right knee to your left and right until you have achieved submaximal lumbar extension.
    5. Have the patient bend her right elbow, pointing it posteriorly, and rest her right hand on her waist.
    6. Rotate her right shoulder posteriorly (the patient rotates to her right).
    7. Ask her to keep her right shoulder back and, while stabilizing her right knee with your groin, place your right hand on her scapula and attempt to rotate her shoulder forward (derotate her spine).

    Remarks—The above technique assesses extension of the entire right lumbar spine. The test can be made more specific by extending and rotating to a specific lumbar level. In this case, your right hand should monitor movement at this level. Another method of making the technique specific to one segment is to passively emphasize extension of one segment with your left hand while testing. This could be done by pressing the spinous process of the superior vertebra to the patient's left while pulling the spinous process of the inferior vertebra to her right. Extension at a specific level can also be emphasized by exerting an anteriorly directed pressure on the right transverse process of the vertebra inferior to the restriction. If a restriction is present at this segment, this specific test will yield weakness, whereas removing stress from the segment yields more strength. Reducing extension stress at a specific segment could be done by pulling the spinous process of the superior vertebra to the patient's right while pushing the spinous process of the inferior vertebra to her left.

- **Active mobilization**—The active mobilization is identical to the testing procedure, except the treatment contraction is gentler and sustained. Retest after the mobilization.

- **Passive mobilization** (side lying) (Fig. 19.55)—See L1 through L5 right facet extension; Agonist; Psoas major.

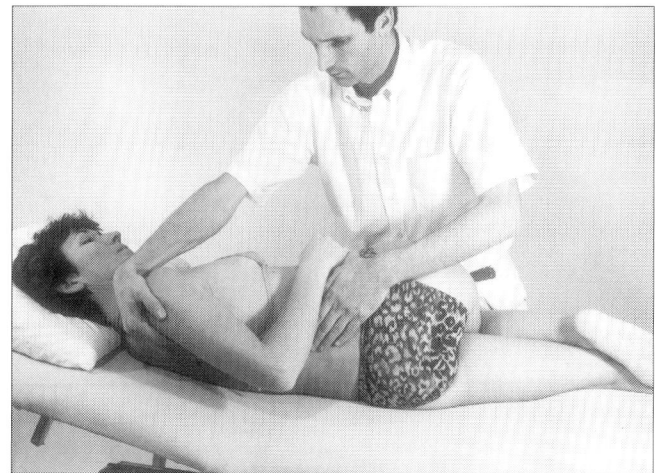

Fig. 19.56

L1 THROUGH L5 FACET EXTENSION

- **Agonists**

    Lumbar extensors:
    quadratus lumborum (ventral rami; T12–L4)
    erector spinae (dorsal rami lumbar spinal
        nerves)
    intertransversarii lateralis lumborum
        (ventral rami spinal nerves)
    intertransversarii medialis lumborum
        (dorsal rami spinal nerves)
    external oblique abdominis, right (T7–12)

- **Muscle test** (supine) (Fig. 19.57)

    1. Have the patient lie supine with her hands holding the sides of the treatment table.
    2. Stand at her left side.
    3. Fixate her left iliac crest with your right hand.
    4. Support her ankles with your left forearm, holding her right ankle with your left hand.
    5. Position her legs to the right (right lumbar side bending).
    6. Ask her to keep her legs to the right while you attempt to move her ankles toward you (to her left).

Remarks—This test can be sensitized by placing a small, firm support (e.g., a towel roll) under the patient's lumbar spine before testing. This will increase her lumbar extension.

If the patient's right lumbar extensors test weak, the result of this test should be compared with one in which her right lumbar facet joints are less extended (less lumbar side bending to the right).

- **Muscle test** (prone, two legs) (Fig. 19.58)

    1. Have the patient lie prone with her hands holding the sides of the treatment table.
    2. Stand at her left side.
    3. Stabilize her left iliac crest with your left hand.
    4. Support her left lower leg with your right forearm and, with your right hand, hold her right lower leg just distal to the knee.
    5. Position her legs to her right (right lumbar side bending).
    6. Ask her to keep her legs in place while you attempt to pull them to her left.

Remarks—See remarks below.

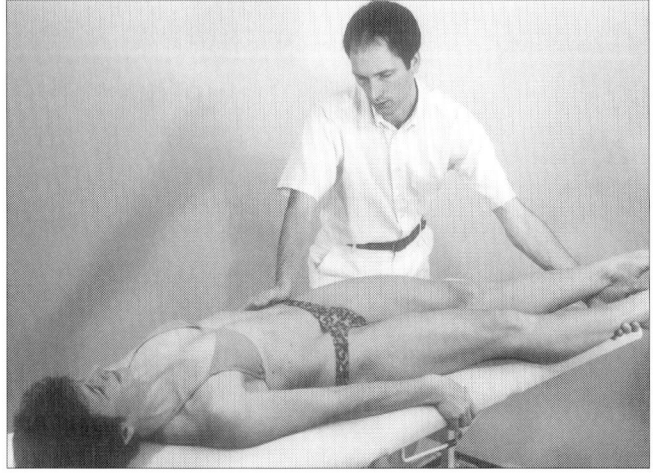

Fig. 19.57

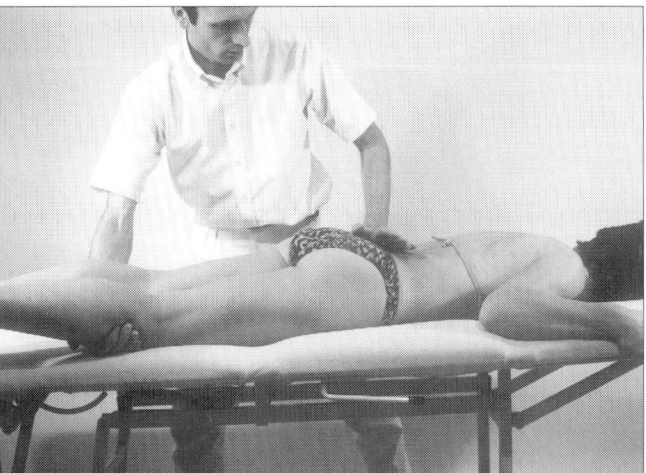

Fig. 19.58

- **Muscle test** (prone, one leg) (Fig. 19.59)

  1. Have the patient lie prone with her hands holding the sides of the treatment table.
  2. Stand at her right side, facing toward her feet.
  3. Stabilize her left iliac crest with your right hand.
  4. Abduct and extend her right leg with your left hand.
  5. Ask her to keep her leg in place.
  6. Reposition your left hand to the dorsum of her knee and attempt to adduct and flex her right leg (push her right leg toward her left leg).

  Remarks—These two prone tests can be sensitized for a right lumbar facet extension dysfunction by side bending the patient (farther) to the right and/or having the patient push herself up on her elbows and hold the table in front of her (Fig. 19.60).

  If the patient's lumbar extensors test weak, the result of this test should be compared with one in which her right lumbar facet joints are less extended (less lumbar side bending to the right or no extension).

FLEXION T6 THROUGH L5

- *Agonists*

  Latissimus dorsi and thoracic and lumbar flexors

  See Thoracic Spine.

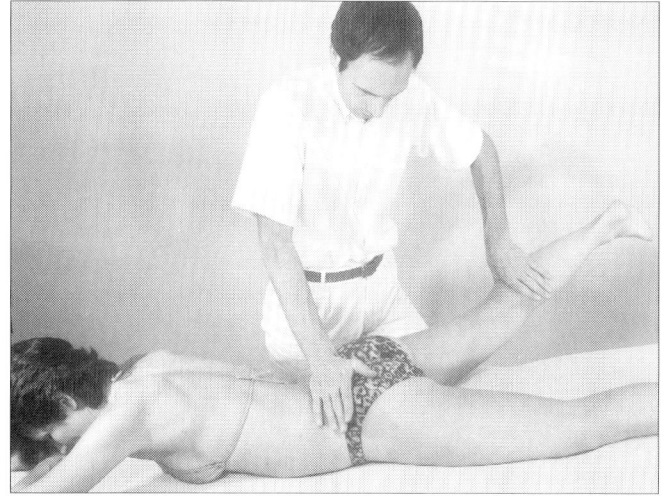

Fig. 19.59

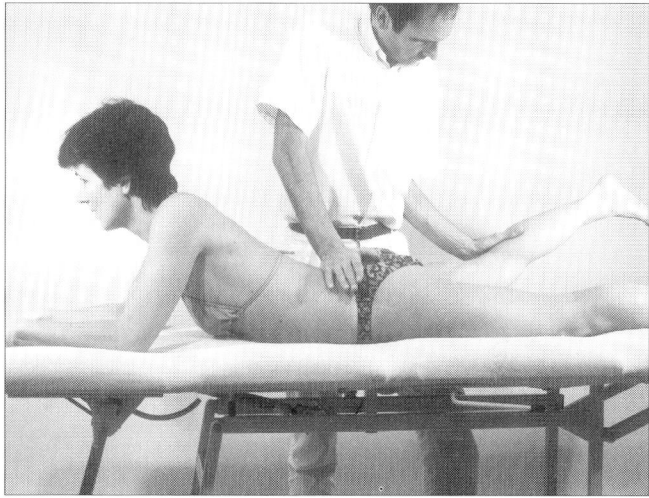

Fig. 19.60

FLEXION T10 THROUGH L5

- *Agonists*

    Lumbar flexors:
    external oblique abdominis, right (T7–12)
    internal oblique abdominis, left (T7–L1)
    rectus abdominis (T6–12)

- *Muscle test* (Fig. 19.61 and 19.62)

    1. Have the patient lie supine with her hips maximally bent and her knees bent as well.

    2. Stand at her left side.

    3. Stabilize her shoulders with your right forearm.

    4. Support her knees with your left hand and forearm.

    5. Keeping her knees close to her chest, bring them to her right (Fig. 19.61).

    6. Straighten your right elbow and stabilize her right shoulder with your hand (Fig. 19.62).

    7. Ask her to keep her knees in place while you attempt to bring them to her left and caudally (lumbar derotation and extension).

Remarks—If this test yields weakness, compare the result with a test in which the patient's knees are less rotated to the right (right lumbar spine is less flexed). Weakness in right lumbar flexion only may be caused by a right lumbar facet flexion restriction.

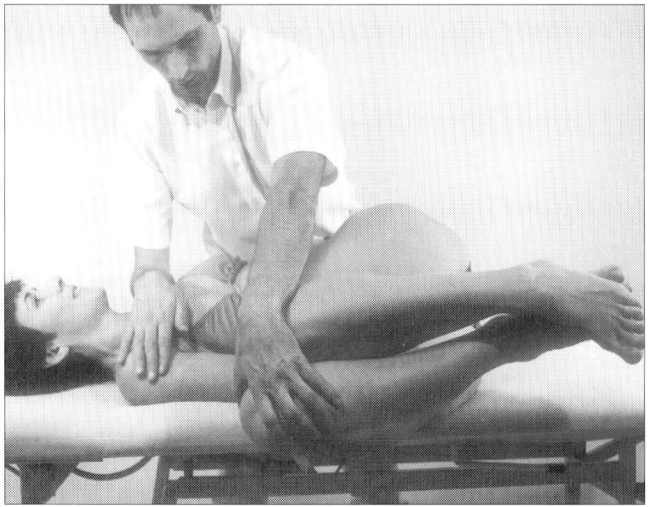

Fig. 19.61

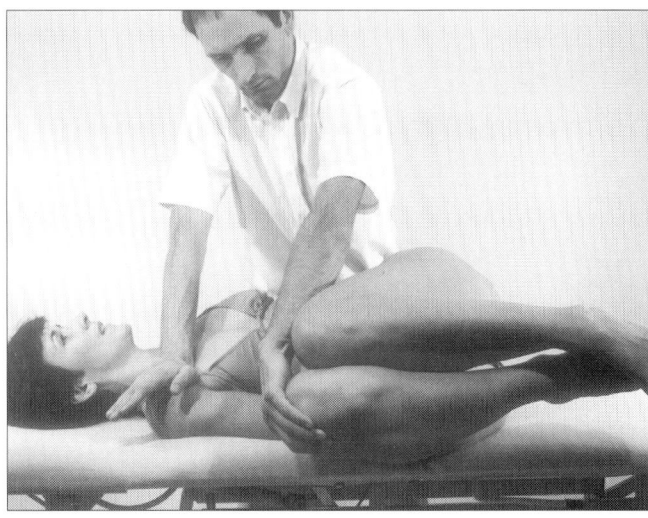

Fig. 19.62

Weakness throughout the range may be caused by a pubis with limited cranial translation mobility on the right side. The restricted cranial translation mobility of the right pubis inhibits the abdominal muscles on the right side.

Right lower rib dysfunction can also cause weakness of the abdominals throughout their lumbar range. Ribs with exhalation restrictions will inhibit the abdominals after exhalation, whereas inhalation restores strength. A posteriorly dislocated rib inhibits the abdominal muscles regardless of the respiratory cycle. The posteriorly dislocated rib can be felt dorsally on the rib cage as being more posterior. The correction is a mobilization of the rib in an anterior and medial direction. Unless other sources of inhibition are present, the correction of a posteriorly dislocated rib will normalize the strength of the lumbar flexors.

- *Postisometric stretch technique* (Fig. 19.63)

    1. Have the patient lie on her right side with her hips and knees bent.

    2. Stand in front of her.

    3. Her knees should be off the edge of table, supported by your proximal right thigh. Support her ankles with your right hand.

    4. Monitor her lumbar spine with your left hand.

    5. From a position of lumbar extension, move her hips and lumbar spine into flexion, stopping at the first perception of a restrictive barrier.

6. Slowly lift her ankles (bringing her lumbar spine into left side bending), again stopping at the first perception of a barrier.

7. Ask her to gently push her feet toward the floor for 6 seconds as you provide unyielding resistance.

8. After the patient relaxes, passively move her lumbar spine farther into left side bending and flexion, stopping each time at the first perception of a restrictive barrier. Rotate her left shoulder backward, again stopping at the first perception of a barrier.

9. Repeat the contraction.

10. Repeat steps 8 and 9 until all restrictive barriers have disappeared.

11. Retest.

Remarks—This technique can be used to treat flexion dysfunctions simultaneously at multiple levels in the lumbar spine. To treat a flexion dysfunction at a specific segment, monitor the flexion excursion and barriers at this level with your left hand while performing the same treatment.

- **Passive mobilization** (Fig. 19.64)

  1. Have the patient lie on her right side. If possible, raise the head end of the table such that her lumbar spine side bends to her left.

  2. Have the patient bend her left elbow, pointing it posteriorly, and rest her left hand on her waist.

  3. Her right knee should be extended.

  4. Her left hip and knee should be flexed.

  5. Stand in front of her, placing her left knee on your proximal right thigh (in your groin).

  6. Monitor the movements of her lumbar spine with your right hand. Move her left knee to your left or right until you have achieved submaximal flexion at the level you want to mobilize.

  7. Maintaining the same degree of lumbar flexion, place her knee on the treatment table.

  8. With your left forearm contacting her left ribs, rotate her left shoulder posteriorly (patient rotates to the left).

  9. The mobilization is performed by using your left forearm to bring her left shoulder more posterior and rotating her left pelvis forward with your right forearm.

  10. Retest.

Remarks—The effect of the mobilization can be monitored by stabilizing the patient's left knee with your right groin, asking her to keep her left shoulder back, and attempting to bring her shoulder forward with your left hand (derotate her spine).

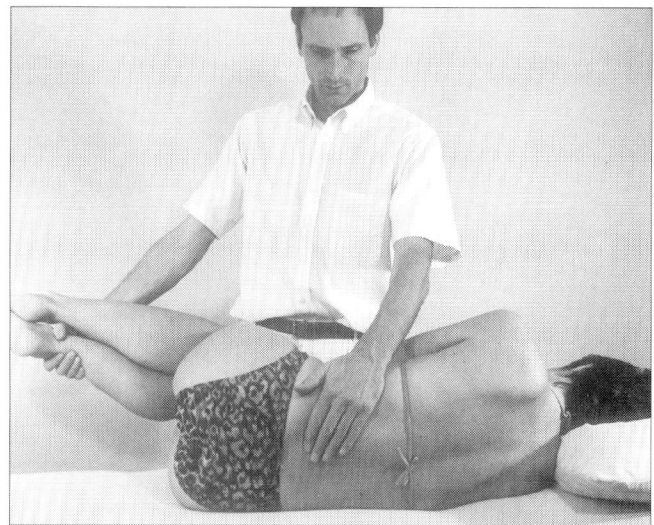

Fig. 19.63

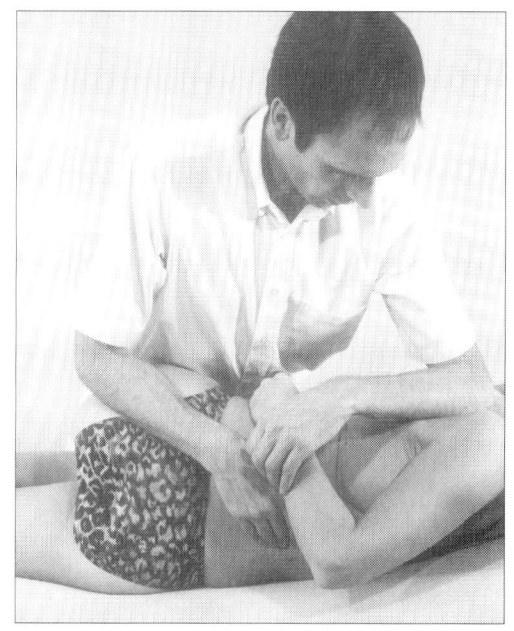

Fig. 19.64

Sacroiliac Joints

The pelvis consists of a sacrum and two innominates (also called the os coxae). Each innominate comprises an ilium, ischium, and pubis. Sacroiliac joint dysfunctions can express themselves in the muscles attaching to the innominate or the sacrum.[5,6,7] These include hip muscles (e.g., hip abductors) and spinal muscles (e.g., quadratus lumborum).

Dorman and colleagues found differences in hip abductor strength in response to sacroiliac joint dysfunctions.[7] The hip abductor strength, measured with a commercial measurement device, could be altered by sacroiliac joint mobilizations. Dorman and colleagues wrote that "the abductors of the hip, the gluteus medius muscle mainly, are facilitated or inhibited by the position of the sacroiliac articulation. We presume this action is facilitated through a reflex arc modulated by joint position sensors in proximity to the sacroiliac articulations."[7, p. 89]

All muscles affecting the sacroiliac joint cross neighboring joints such as the hip joint or the lumbar spine. Therefore, changes in tone of any of the muscles affecting the sacroiliac joint will alter the function of these neighboring joints. Thus, we can use lumbar, hip, or other muscles to evaluate sacroiliac joint function.

The innominate can move in any direction on the sacrum. All directions of innominate movement can be grouped into six categories. The innominate can translate cranially and caudally. When the innominate translates cranially, the anterior superior iliac spine (ASIS) and the posterior superior iliac spine (PSIS) translate cranially. During a caudal movement of the innominate, the ASIS and PSIS translate caudally. The innominate can also rotate anteriorly and posteriorly on the other parts of the pelvis. During anterior rotation of the innominate, the ASIS moves caudally while the PSIS moves cranially. Posterior rotation is the reverse. Internal and external rotation of the innominate are the lateral and medial movements of the innominate on the sacrum. Internal rotation of the innominate is the lateral translation of the PSIS and/or the posterior inferior iliac spine (PIIS) on the sacrum. External rotation of the innominate is the medial movement of the PSIS and PIIS over the sacrum. All movements of the innominate can always be reduced to these six movements or combinations thereof.

Decreased mobility of the innominate on the sacrum might be associated with an innominate displacement. For instance, an innominate with limited mobility in a cranial direction might be positioned caudally (i.e., the PSIS, ASIS, and iliac crest might be lower), and the leg on that side might seem longer in the supine patient. This condition is called a "downslip." If sufficient mobility existed in a cranial direction, the downslip would correct itself. Thus, a downslip is a caudally displaced innominate that has stiffness in a cranial direction.

The reverse displacement signifies an "upslip." The innominate has translated cranially on the sacrum and the other innominate. In the supine patient, the leg on the side of the upslip might seem short. The upslip presents as limited mobility of the innominate in a caudal direction.

An "outflare" is an externally rotated innominate with limited internal rotation mobility, and an "inflare" is an internally rotated innominate with limited external rotation mobility.

Following is a listing of innominate conditions, the accompanying stiffness, and the associated inhibited muscles (Table 19.1). The table does not list all muscles acting on the sacroiliac joint, but only those that are useful for testing of arthrogenic weakness. For instance, the sartorius has an anterior rotation function on the innominate, but it is not listed because the sartorius also functions as a hip flexor, external rotator, and abductor, a knee flexor, and an internal rotator of the tibia. If we were to test this muscle and find weakness, we would know little about the cause of the weakness. Thus, only muscles practical to test are listed.

Table 19.1 Innominate dysfunctions and corresponding muscle weakness.

Condition	Innominate stiffness	Muscle weakness
Downslip	Cranial direction	Quadratus lumborum Latissimus dorsi
Upslip	Caudal direction	Anterior adductors Hamstrings
Posteriorly rotated innominate	Anterior rotation	Iliacus Quadratus lumborum Latissimus dorsi
Anteriorly rotated innominate	Posterior rotation	Hamstrings Hip extensors
Inflare	External rotation	Hip internal rotators (anterior abductors)
Outflare	Internal rotation	Hip external rotators

All the above dysfunctions cause corresponding changes in the sacrum. For instance, a right innominate downslip causes the sacrum to be high (cranial) on the side of the dysfunction. Another frequently encountered dysfunction involving combined innominate and sacrum movement is limited sacral nutation. Sacral nutation is the movement whereby S1 moves anteriorly and caudally and S3 moves posteriorly and cranially, possibly around an axis through S2 (Fig. 19.65). The nutation movement may increase lumbar lordosis.

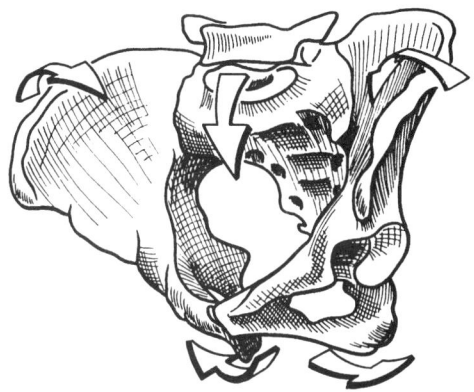

Fig. 19.65 Sacral nutation with the accompanying movements of the innominates.

Reprinted with permission from Kapandji IA: The Physiology of the Joints. New York, Churchill Livingstone, 1974, Vol 3

The innominate movements accompanying sacral nutation are a medial movement of the innominate over S1 and a lateral movement of the innominate away from S3. The sacroiliac joint consists of the sacrum anteriorly and the innominate posteriorly. When S1 moves forward in relation to the innominate, a gap would theoretically be created between S1 and the innominate covering S1 posteriorly (PSIS). This gap is filled by the PSIS moving medially, thus making contact again with S1. This medial movement of the innominate over S1 could be termed external rotation of the innominate.

If the axis of sacral nutation runs through S2, nutation is accompanied by S3 moving posteriorly (and cranially). The part of the innominate making contact with S3 (PIIS) moves laterally, out of the way. The lateral movement of this part of the innominate could be called an internal rotation.

During nutation, the part of the innominate making contact with S1 moves medially and the part making contact with S3 moves laterally. These two movements could be thought of as an abduction-like movement of the innominate.

Counternutation is the opposite of nutation: S1 moves posteriorly (and cranially) while the innominate covering S1 moves laterally, out of the way. S3 moves anteriorly (and caudally) while the part of the innominate making contact with S3 moves medially. The innominates make an adduction-like movement. Table 19.2 depicts the muscle weakness associated with stiff sacral nutation and counternutation. The muscles inhibited by stiff S1 nutation would create S1 nutation if they were not inhibited. The same applies for the other movements.

Table 19.2 Muscle weakness in relation to nutation and counternutation dysfunctions.

Stiffness	Muscle weakness
S1 nutation	Iliacus, S1 part Adductors, at 45-degree hip flexion Internal rotators in hip abduction Hamstrings in hip abduction Quadratus lumborum
S3 nutation	External rotators in hip abduction
S1 counternutation	External rotators in hip adduction
S3 counternutation	Internal rotators in hip adduction

It is important to remember that the only dysfunction at the sacroiliac joint that causes muscle inhibition is altered joint receptor functioning, which is related to stiffness for specific movements and to sacroiliac malposition. For instance, a downslipped innominate has limited cranial translation. If sufficient cranial translation existed, the downslip would correct itself. Therefore, it is the limited cranial translation that inhibits the muscles which translate the innominate cranially. Similarly, in case of an upslip, it is the innominate's stiffness for translating caudally that inhibits specific muscles. The same applies for all other sacroiliac dysfunctions.

- *Capsular pattern*—Pain when the sacroiliac ligamento-capsular tissue is stressed.

INNOMINATE CRANIAL TRANSLATION ON THE SACRUM

The cranial translation of the innominate on the sacrum is disturbed in the case of both a downslip and a posteriorly rotated innominate. Common to both conditions is that the posterior part of the innominate is stiff for moving cranially. The PSIS and iliac crest may be low on the affected side. The difference between a downslip and a posteriorly rotated innominate is the function of the anterior part of the innominate. In a downslip, the anterior part of the innominate (acts as if it) has moved caudally, whereas in a posteriorly rotated innominate, the anterior part (acts as if it) has moved cranially. Both the downslip and the posteriorly rotated innominate have in common a limited cranial glide of the innominate at the sacroiliac joint. The quadratus lumborum and the iliac fibers of the latissimus dorsi are inhibited in both conditions. In terms of muscles, the difference between a downslip and a posteriorly rotated innominate is the function of the iliacus, an anterior rotator of the innominate. The iliacus is strong in the case of a downslip but weak in the case of a posteriorly rotated innominate.

- *Agonists*

    Posterior innominate hikers:
    quadratus lumborum (ventral rami of spinal nerves; T12–L4)
    latissimus dorsi (thoracodorsal nerve; C6,7,8)

- *Muscle test* (Fig. 19.66)

    1. Have the patient lie supine with her hands holding the sides of the treatment table.

    2. Stand at her left side, supporting her left ankle with your left forearm and holding her right ankle with your left hand.

    3. Fixate her left iliac crest with your right hand.

    4. Ask her to maintain the position of her legs while you attempt to move her ankles toward you (to her left).

    Remarks—If this test yields weakness, compare the result with a test in which the patient's lumbar spine is side bent left and she has inhaled. The left side bending and the inhalation eliminate arthrogenic inhibition from the lumbar spine and the twelfth rib. In the case of a limited cranial translation of the innominate on the sacrum, the quadratus lumborum will test weak regardless of the position of the lumbar spine or the breathing phase. A downslip will also inhibit the latissimus dorsi, which has a similar function over the sacroiliac joint.

The quadratus lumborum could also be inhibited by limited S1 nutation mobility. The test assesses the contraction quality of the diagonal fibers of the quadratus lumborum, which run from the iliac crest cranially and medially to the lumbar spine. A contraction of these fibers creates an external rotation force of the innominate over S1 (S1 nutation). These diagonal quadratus fibers are inhibited in the case of limited nutation mobility. The quadratus lumborum will test weak throughout its lumbar range (i.e., in a lumbar right, left, or neutral side bent position). The iliacus and latissimus dorsi will test strong.

Weakness throughout the range could also result from a combination of two lumbar spine dysfunctions. One dysfunction causes weakness in a right side bent position; the other causes weakness in a left side bent position. The first dysfunction is an extension dysfunction of any lumbar facet joint on the right. The facet joint extension dysfunction will produce an arthrogenic inhibition of the right

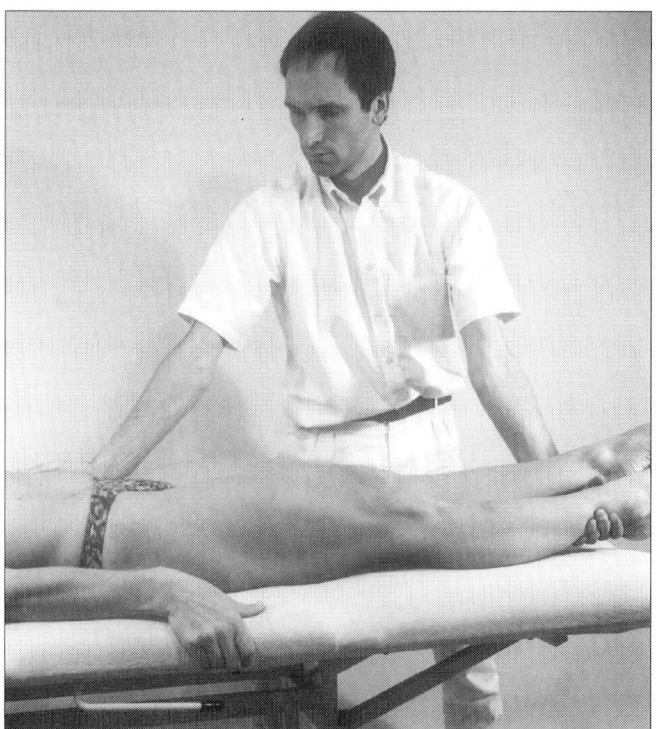

Fig. 19.66

quadratus lumborum when the lumbar spine is in right side bending. The second dysfunction is a pseudoradiculopathy of the right upper lumbar facet joints with stiffness for flexion. This will produce a pseudoradicular inhibition of the quadratus lumborum when the facet joint is flexed. Together, the flexion and extension dysfunctions of the lumbar spine will cause weakness of the quadratus lumborum throughout its working range. Other muscle tests can confirm this condition.

- **High-velocity thrust manipulation to cranially translate the innominate** (Fig. 19.67)

  1. Have the patient lie on her left side with her hips and knees bent 90 degrees. Her spine should be straight.

  2. Stand in front of her and hold her right trochanter with your right hand, pulling it cranially.

  3. With the base of your left hand, thrust her right ischial tuberosity in a cranial direction.

  4. Repeat the manipulation if needed.

- **High-velocity thrust manipulation to anteriorly rotate the innominate**—See Innominate anterior rotation; Iliacus.

- **Postisometric stretch technique to anteriorly rotate the innominate**—See Innominate anterior rotation; Iliacus.

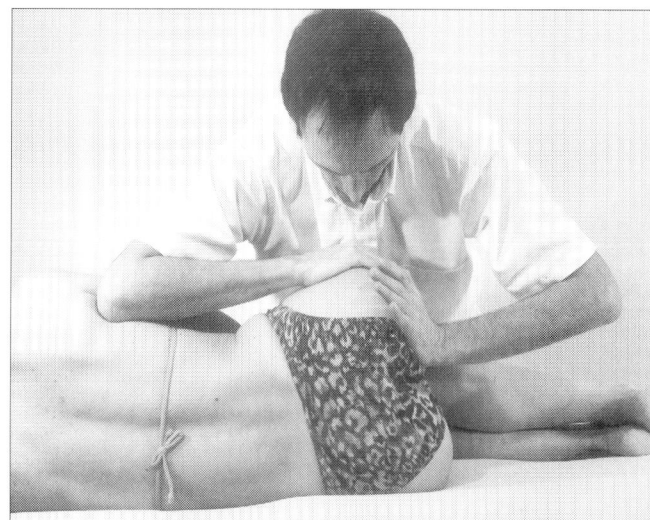

Fig. 19.67

INNOMINATE CRANIAL TRANSLATION ON THE SACRUM

- *Agonists*

    Latissimus dorsi:
    latissimus dorsi, lateral fibers running to the iliac crest (thoracodorsal nerve; C6,7,8)

- *Muscle test* (Fig. 19.68)

    1. Have the patient lie supine with her right shoulder abducted to 90 degrees and internally rotated.

    2. Stand cranial to her right arm, holding her distal forearm with your right hand.

    3. Ask her to maintain her arm position while you attempt to further abduct her shoulder.

    Remarks—If the muscle tests weak, compare the result of this test with one in which the patient's shoulder is adducted.

    The latissimus dorsi attaches to the spinous processes of T7 through the sacrum and to the iliac crest. The muscle fibers running perpendicular to the humerus are most effective in moving the arm. With the shoulder abducted to 90 degrees, the muscle fibers most effective in adducting the shoulder are those attaching to the iliac crest.

- ***High-velocity thrust manipulation to cranially translate the innominate***—See Innominate cranial translation on the sacrum; Posterior innominate hikers.

- ***High-velocity thrust manipulation to anteriorly rotate the innominate***—See Innominate anterior rotation; Iliacus.

- ***Postisometric stretch technique to anteriorly rotate the innominate***—See Innominate anterior rotation; Iliacus.

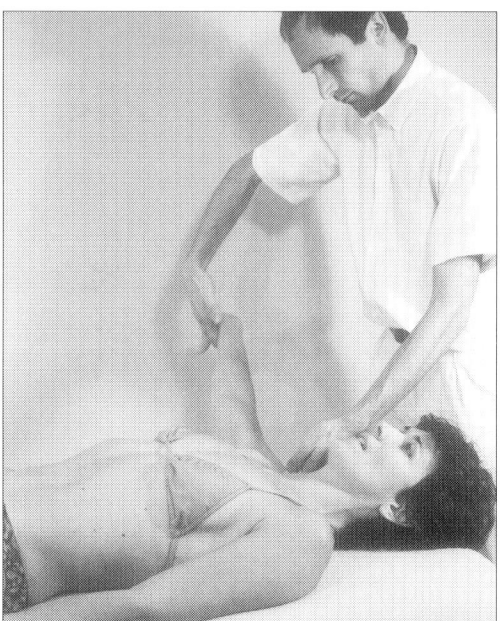

Fig. 19.68

INNOMINATE ANTERIOR ROTATION

- **Agonists**

    Iliacus:
    iliacus, ilial part (femoral nerve; L2,3)

- **Muscle test** (Fig. 19.69)

    1. Have the patient lie supine with her right hip maximally flexed (and slightly abducted) and her knee bent.

    2. Stand at her right side and fixate her left ASIS with your right hand.

    3. Place your left hand on her right knee.

    4. Ask her to maintain this knee position while you attempt to push her knee in a caudal and slightly medial direction.

    Remarks—If the test yields weakness, compare the result with a test in which the patient's hip is less flexed. If less hip flexion yields a strong iliacus, inhibition from limited hip flexion may be the cause. Testing the psoas major might confirm a hip flexion dysfunction. A limited hip flexion will inhibit the iliacus as well as the psoas when tested in a shortened position. Less hip flexion will yield strong muscles.

    If the iliacus inhibition is due to limited anterior rotation of the innominate the quadratus lumborum and the latissimus dorsi in 90 degrees of shoulder abduction will also test weak. The psoas usually tests strong.

- **High-velocity thrust manipulation to anteriorly rotate the innominate** (Fig. 19.70)

    1. Have the patient lie on her left side holding her bent left knee with her right hand. Her right hip should be extended.

    2. Stand behind her, facing cranially. Position your right knee on the table, with the patient's right leg resting on your lower leg.

    3. Place your right hand on her proximal anterior thigh.

    4. Cover her right PSIS with your left hand.

    5. The manipulation is a quick thrust with both hands to produce an anterior rotation of the innominate.

    6. Repeat the manipulation if needed.

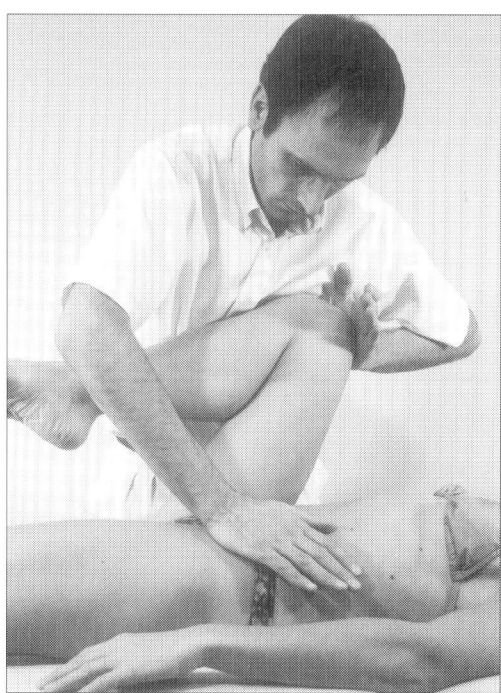

Fig. 19.69

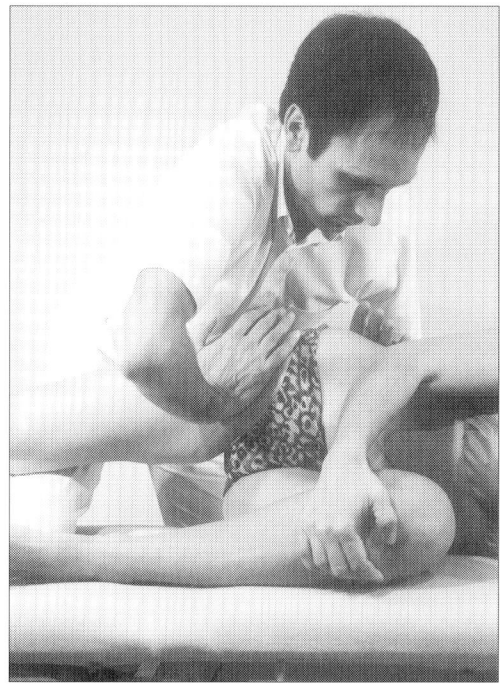

Fig. 19.70

- **Postisometric stretch technique to anteriorly rotate the innominate** (Fig. 19.71)

  1. Have the patient lie supine, holding her left knee close to her chest. Her right leg should be off the table, with the hip passively extended.

  2. Stand at her right side, stabilizing her left lower leg with your left forearm.

  3. Place your right hand on her distal femur.

  4. Ask her to push her right knee into your hand (in an attempt to flex her hip) while you provide unyielding resistance. After about 6 seconds of contracting her hip flexors, ask her to relax.

  5. Repeat step 4 two times.

  6. Retest.

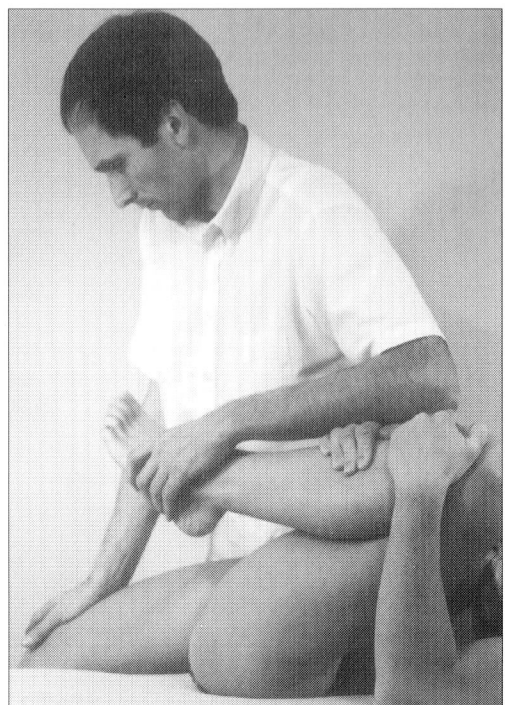

Fig. 19.71

INNOMINATE POSTERIOR ROTATION

- **Agonists**

    Hamstrings:
    hamstrings (sciatic nerve; L5–S2)

- **Muscle test** (Fig. 19.72)

    1. Have the patient lie supine with her right hip and knee bent. Ask her to place her hands beneath her lumbar spine to maintain a lumbar lordosis.
    2. Stand at her right side, with your left hand holding her knee (femoral condyles), and flex her hip into its physiological barrier so that her knee is passively flexed.
    3. Hold her ankle with your right hand.
    4. Ask her to maintain this ankle position while you attempt to extend her knee.

    Remarks—In step 1, the patient's hands must maintain her lumbar lordosis. Otherwise her lumbar spine will move into flexion during the test and may cause hamstring weakness if an L5 flexion restriction is present (L5 pseudoradiculopathy).

    The maximal hip flexion and the hamstrings contraction each cause a posterior rotation force on the innominate. If the test yields weak hamstrings, compare this result with a test in which the patient's hip is less flexed.

    The hamstrings may also be weak in the case of an upslip.

- **Active mobilization** (Fig. 19.72)—The treatment is identical to the evaluation, except the treatment contraction is gentler and sustained. Retest after the mobilization.

- **High-velocity thrust manipulation** (Fig. 19.73)

    1. Have the patient lie on her left side with her left leg extended and her right hip and knee flexed.
    2. Stand in front of her with the sole of her right foot against your right hip.
    3. Position your left knee on the treatment table to stabilize her anterior left thigh.
    4. Place your right hand on her right ASIS and iliac crest.
    5. Place your left hand against the posterior aspect of her right ischial tuberosity.
    6. Thrust with both hands to rotate her right innominate posteriorly.
    7. Repeat the manipulation if needed.

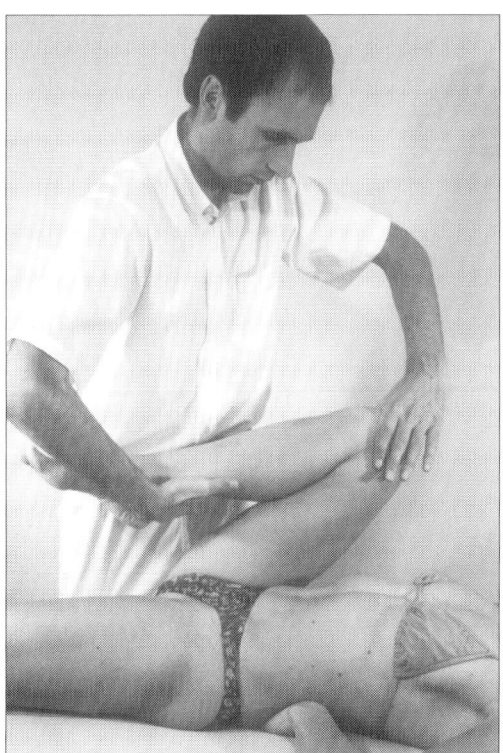

Fig. 19.72

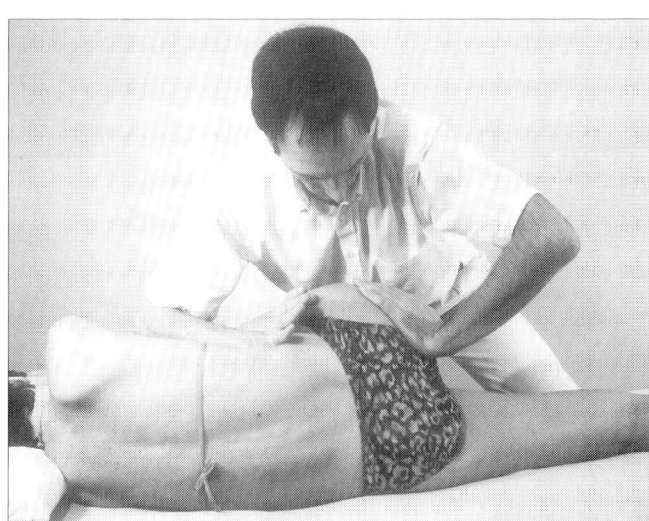

Fig. 19.73

INNOMINATE POSTERIOR ROTATION

- **Agonists**

    Gluteus maximus:
    gluteus maximus (inferior gluteal nerve; L5–S2)

- **Muscle test** (Fig. 19.74)

    1. Have the patient lie supine. Stand at her right side with your right hand beneath her right knee (your wrist medial to her knee).

    2. Use the base of your left hand to stabilize her right ASIS.

    3. Ask her to maintain this knee position while you attempt to lift her knee.

- **Active mobilization** (Fig. 19.72)—See Innominate posterior rotation; Hamstrings.

- **High-velocity thrust manipulation** (Fig. 19.73)—See Innominate posterior rotation; Hamstrings.

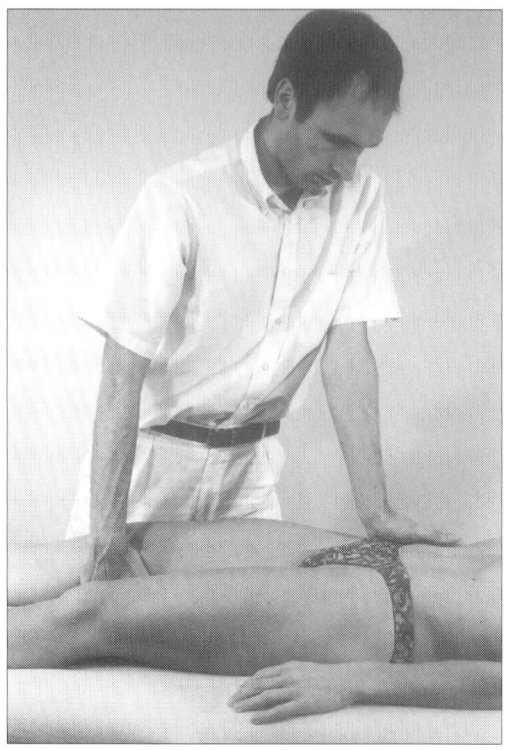

Fig. 19.74

INNOMINATE CAUDAL TRANSLATION

- *Agonists*

    Anterior and posterior innominate rotators:
      anterior adductors
      hamstrings

- **Muscle test**—See Symphysis pubis; Caudal translation of the right pubis; Hip adductors, and see Sacroiliac joint; Innominate posterior rotation; Hamstrings.

    Remarks—An upslip, or an innominate with limited mobility in a caudal direction, cannot be tested directly with muscles. The human body has no muscle that can translate the innominate caudally. However, anteriorly, the upslip behaves as a posteriorly rotated innominate with limited caudal translation of the pubis. Posteriorly, the upslip behaves as an anteriorly rotated innominate with a limited caudal translation of the innominate on the sacrum. In other words, the innominate has a limited caudal translation, both anteriorly and posteriorly. Therefore, an upslip can be recognized by weakness of the right anterior adductors together with weakness of the right hamstrings.

- *Passive mobilization* (Fig. 19.75)

    1. Have the patient lie on her left side with her hips and knees bent 90 degrees.

    2. Stand in front of her, facing toward her feet. Place both your hands on her right iliac crest.

    3. Push her right innominate caudally.

    4. Retest.

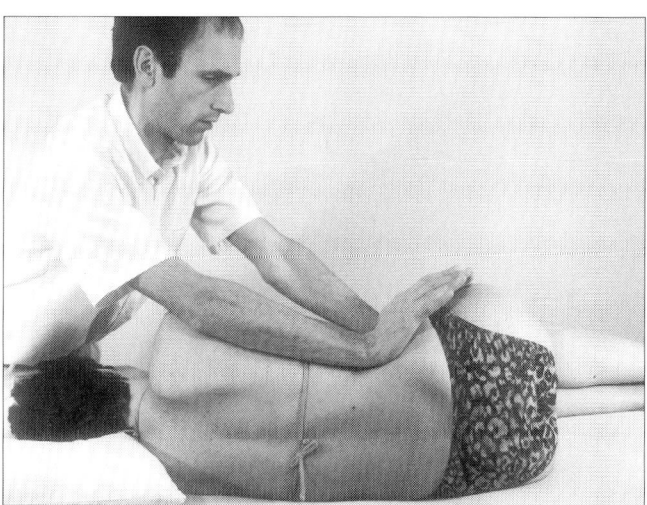

Fig. 19.75

S1 NUTATION

- **Agonists**

    Iliacus:
    iliacus, sacral part (femoral nerve; L2,3)

- **Muscle test** (Fig. 19.76)

    1. Have the patient lie supine, holding her left knee close to her chest with both hands. Her right hip should be flexed to 90 degrees, and her right knee should be relaxed and passively bent.
    2. Stand at her right side with your right hand stabilizing her left lower leg.
    3. Place your left hand on her knee.
    4. Ask her to maintain this knee position while you attempt to push her knee caudally and laterally (45 degrees out).

    Remarks—The medial part of the iliacus attaches to the lateral tip of S1. A contraction of these iliacus fibers pulls S1 anteriorly (S1 nutation). If this test yields weakness, compare the result with one with less hip flexion and adduction (to eliminate possible inhibition from the hip).

    The result of the test can also be compared with one with the patient's left hip extended. Extension of the left hip brings the caudal part of the sacrum down onto the treatment table, counternutating the sacrum. The counternutation of the sacrum reduces the possibility of the iliacus pulling S1 into nutation or into restricted nutation. Therefore, if the iliacus becomes strong with the left hip extended, limited S1 nutation mobility is confirmed.

    The result of the test can also be compared with the psoas test (see L1 through L5 right facet extension; Psoas major). The iliacus test described above resembles the psoas test for lumbar extension; however, if the above test yields weakness that originates in the sacroiliac joint, no lumbar position can alter the strength.

- **Active mobilization** (Fig. 19.76)—The treatment is identical to the evaluation, except the treatment contraction is gentler and sustained. Retest after the mobilization.

    Remarks—Other treatment techniques to improve S1 nutation are described below under the three headings "Innominate external rotation on S1 (S1 nutation)."

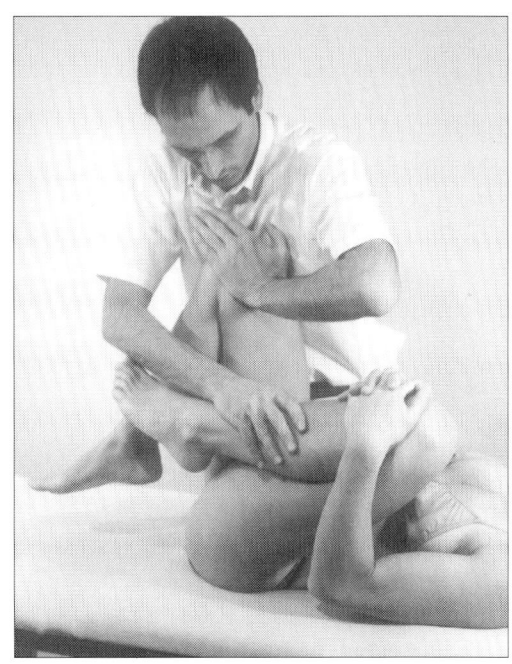

Fig. 19.76

INNOMINATE EXTERNAL ROTATION ON S1 (S1 NUTATION)

- **Agonists**

    Adductors in 40 degrees of flexion:
    pectineus (femoral nerve; L2,3, and accessory obturator; L3)
    adductor longus (obturator nerve; L2,3,4)
    gracilis (obturator; L2,3)
    adductor brevis (obturator; L2,3,4)
    adductor magnus (obturator and sciatic nerve; L2,3,4)

- **Muscle test** (Fig. 19.77)

    1. Have the patient lie supine, holding her left knee close to her chest with both hands. Her right hip should be flexed to 40 degrees, and her right knee should be bent.
    2. Stand at her right side with your left hand stabilizing her left lower leg.
    3. Support her right lower leg with your right forearm, placing your right hand on the medial aspect of her knee.
    4. Ask her to maintain this knee position while you attempt to move her knee laterally.

- **Active mobilization** (Fig. 19.77)—The treatment is identical to the evaluation, except the treatment contraction is gentler and sustained. To enhance the effect of the mobilization, the patient could be asked to exhale as deeply as is comfortable for her. Retest after the mobilization.

- **Home exercise active mobilization** (Fig. 19.78)

    1. Have the patient lie comfortably on either side with her hips bent 40 degrees and a pillow between her knees.
    2. Ask her to adduct her knees and maintain the contraction for several minutes.

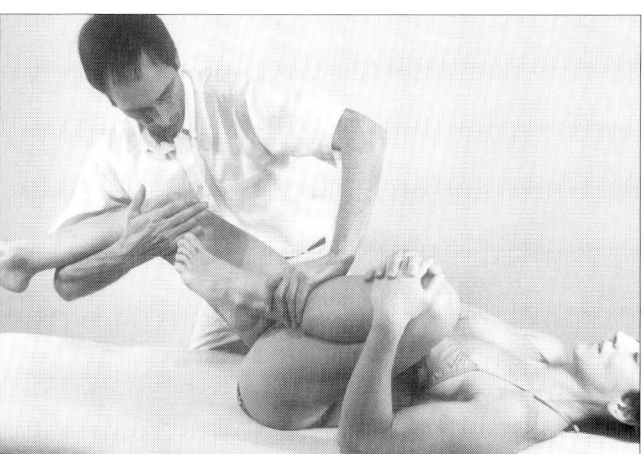

Fig. 19.77

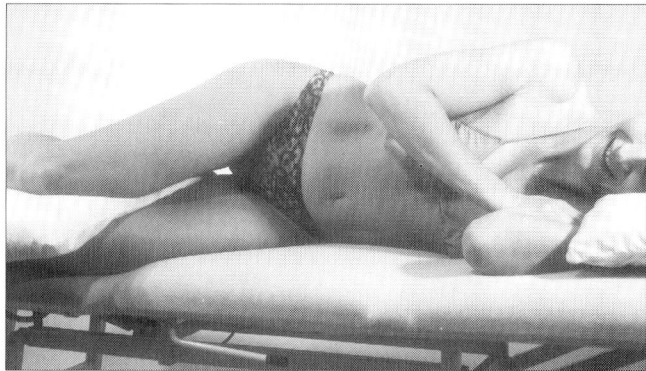

Fig. 19.78

Remarks—When the hips are flexed to 40 degrees, the cranial part of the sacroiliac joint (S1 and its counterpart on the ilium), the acetabulum, the pubis, and the adductors are in line (Fig. 19.79). A contraction of the adductors pulls the pubis caudally and anteriorly (Fig. 19.80). The acetabulum acts as the fulcrum, resulting in a mobilization of the innominate in a medial direction (external rotation) over S1.

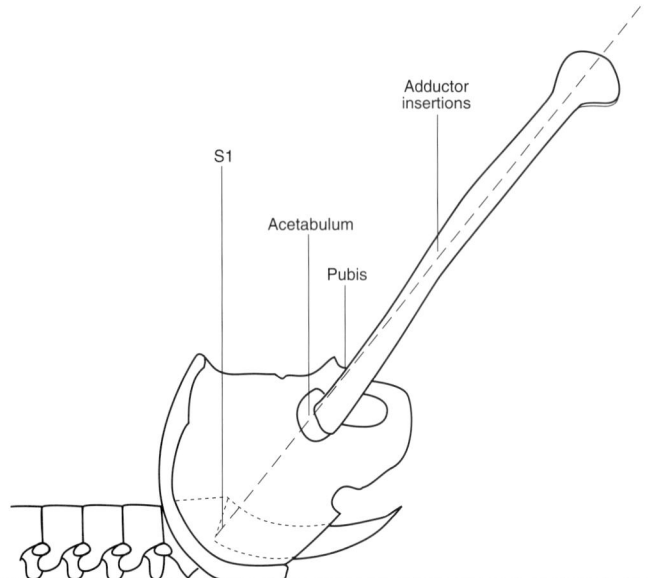

Fig. 19.79 In 40 degrees of hip flexion, the cranial part of the sacroiliac joint, the acetabulum, the pubis, and the adductors are in one line.

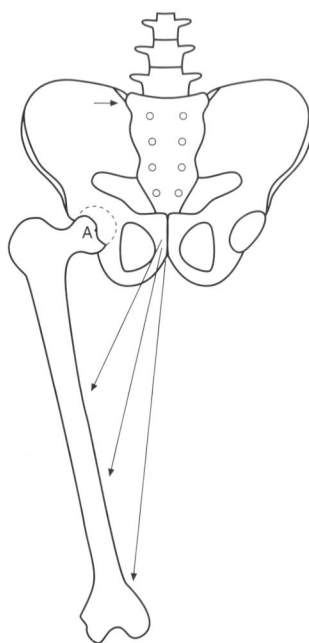

Fig. 19.80 In 40 degrees of hip flexion, a contraction of the adductors pulls the pubis caudally and anteriorly. The acetabulum acts as the fulcrum (A), resulting in a mobilization of the innominate in a medial direction over S1.

INNOMINATE EXTERNAL ROTATION ON S1 (S1 NUTATION)

- *Agonists*

    Hamstrings:
      hamstrings (sciatic nerve; L5–S2)

- *Muscle test* (Fig. 19.81)

    1. Have the patient lie supine, holding her left knee close to her chest with both hands. Her right hip should be flexed to 90 degrees and maximally abducted. Her leg should be passively internally rotated.

    2. Stand at her right side, holding her right femoral condyles with your left hand.

    3. While maintaining the abduction in her hip, flex her hip to its physiological barrier.

    4. Hold her ankle with your right hand.

    5. Ask her to maintain this ankle position while you attempt to extend her right knee.

    Remarks—If this test yields weakness, compare the result with a test with the patient's hip in less flexion and less abduction.

    Since this muscle test involves lumbar flexion, a right L5 facet limitation for flexion may also weaken the hamstrings (pseudoradiculopathy).

    Treatment—Any of the above-mentioned techniques to improve S1 nutation can be used.

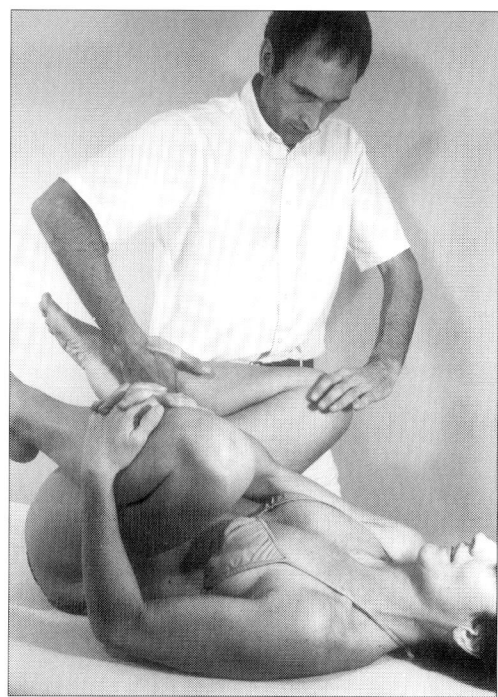

Fig. 19.81

INNOMINATE EXTERNAL ROTATION ON S1 (S1 NUTATION)

- *Agonists*

    hip internal rotators:
    anterior part of the gluteus medius (superior gluteal nerve; L5,S1)
    anterior part of the gluteus minimus (superior gluteal nerve; L5,S1)
    tensor fascia latae (superior gluteal nerve; L4,5)

- *Muscle test* (Fig. 19.82)

    1. Have the patient lie prone with both legs abducted. Her right knee should be flexed to 90 degrees, and her right hip should be internally rotated.
    2. Position yourself distal to her right leg, stabilizing her medial knee with your left hand.
    3. Hold her ankle with your right hand.
    4. Ask her to maintain this leg position while you attempt to externally rotate her leg (push her ankle medially).

    Remarks—Of all the tests that evaluate S1 nutation, the internal rotator test described above may be the most sensitive. If the test yields weakness, compare the result with a test in which the patient's hip is less internally rotated. Limited S1 nutation causes internal rotator weakness that gradually improves as the hip is positioned in less internal rotation. In contrast, weakness deriving from limited hip internal rotation causes the weak internal rotators to become acutely strong when the inhibiting barrier has been passed. A spinal cause for inhibition of the internal rotators (radiculopathy or pseudoradiculopathy) causes weakness throughout the complete internal–external rotation range.

- *Passive mobilization* (prone) (Fig. 19.83)

    1. Have the patient lie prone with her right knee bent 90 degrees and her right hip abducted and externally rotated.
    2. Stand at her left side. With your left pisiform, push the right side of her S1 anteriorly.
    3. With your right hand, hold her iliac crest just cranial to the ASIS. Use this hand to pull her iliac crest dorsally (to externally rotate her right innominate over S1).
    4. Retest.

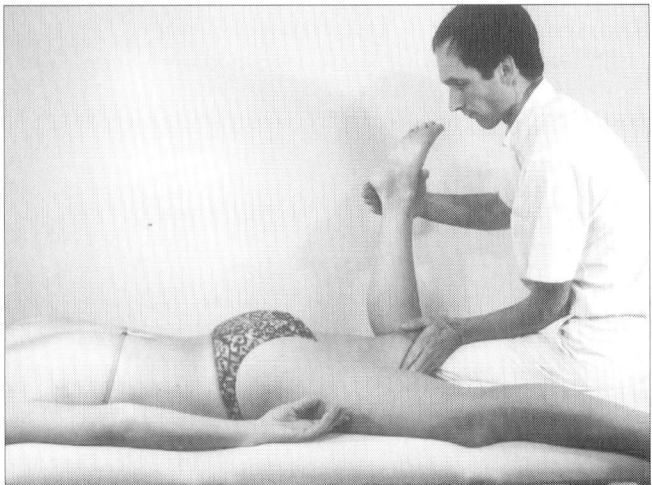

Fig. 19.82

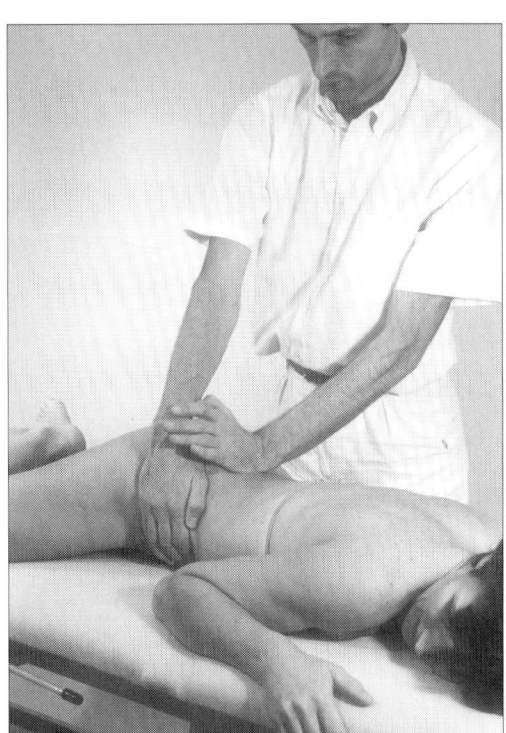

Fig. 19.83

- **Combined active and passive mobilization** (side lying) (Fig. 19.84)

  1. Have the patient lie on her left side with her hips flexed to 45 degrees and a pillow between her knees. The right part of her pelvis should be rotated 30 degrees forward, causing her to bear the weight of her pelvis equally on her left trochanter and left iliac crest. Place a folded towel on her right iliac crest.

  2. Stand behind her and place your chest on the towel on the posterior part of her iliac crest.

  3. Position your hands in front of her, holding the edge of the table.

  4. Press with your chest on the posterior part of her right iliac crest (PSIS) to move it medially (to create an external rotation of her innominate) over S1.

  5. The patient should squeeze the pillow comfortably with her knees.

  6. Retest.

  Remarks—This technique is a modification of one described by Bakker, which he used in an oscillatory manner for all sacroiliac problems.[1] The high incidence of limited S1 nutation mobility in our slouching society may explain the high success rate of this technique. The above-described technique can be maintained for several minutes.

- **Combined active and passive mobilization** (prone) (Fig. 19.85)

  1. Have the patient lie prone with her right knee bent 90 degrees and her right hip abducted, externally rotated, and slightly flexed such that her femur makes a 45-degree angle with her trunk. Due to the position of her right leg, her right innominate will come off the table slightly.

  2. Stand at her right side, placing the base of your right hand lateral to her right PSIS. Use this hand to push her PSIS medially. Hold the medial side of her right knee with your left hand.

  3. Ask her to push her right knee toward the floor (into your hand) while you provide unyielding resistance.

  4. Repeat if necessary.

  5. Retest.

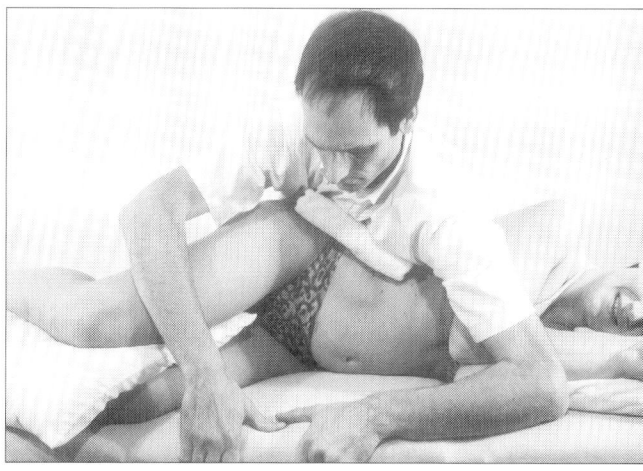

Fig. 19.84

Remarks—During the mobilization, the base of your right hand can feel for external rotation movement of the right innominate (S1 nutation movement).

This mobilization may be the most effective technique for S1 nutation described in this book. Its effectiveness may be partly due to the sacral portion of the iliacus muscle pulling the sacrum into nutation on the ipsilateral side.

This technique can also be used to evaluate for limited S1 nutation.

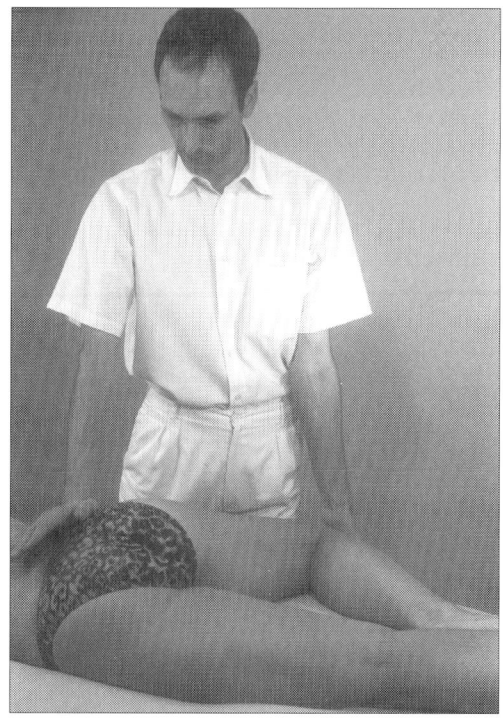

Fig. 19.85

INNOMINATE EXTERNAL ROTATION ON S3 (S3 COUNTERNUTATION)

- *Agonists*

    hip internal rotators:
    gluteus medius, anterior part (superior gluteal nerve L5,S1)
    gluteus minimus, anterior part (superior gluteal nerve; L5,S1)
    tensor fascia latae (superior gluteal nerve; L4,5)

- *Muscle test* (Fig. 19.86)

    1. Have the patient lie prone with her right leg adducted and internally rotated. Her right knee should be flexed to 90 degrees.

    2. Position yourself distal to her right leg and stabilize her medial knee with your left hand.

    3. Hold her ankle with your right hand.

    4. Ask her to maintain this leg position while you attempt to externally rotate her leg (push her ankle medially).

    Remarks—If the test yields weakness, compare the result with a test in which the patient's hip is less internally rotated.

- *Passive mobilization* (Fig. 19.87)

    1. Have the patient lie prone with her right hip adducted and externally rotated.

    2. Stand at her left side, placing your stabilizing left hand on the right side of S3 and S4.

    3. Hold her right proximal tibia with your right hand, with your wrist on the lateral side.

    4. Pull her knee toward you (externally rotate, extend, and adduct her leg).

    5. Retest.

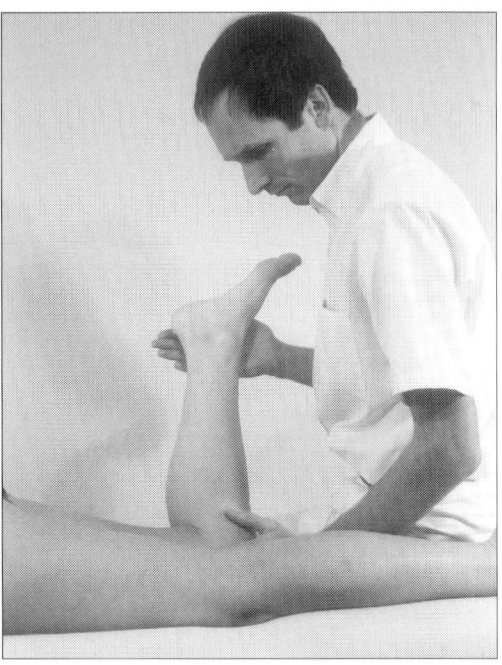

Fig. 19.86

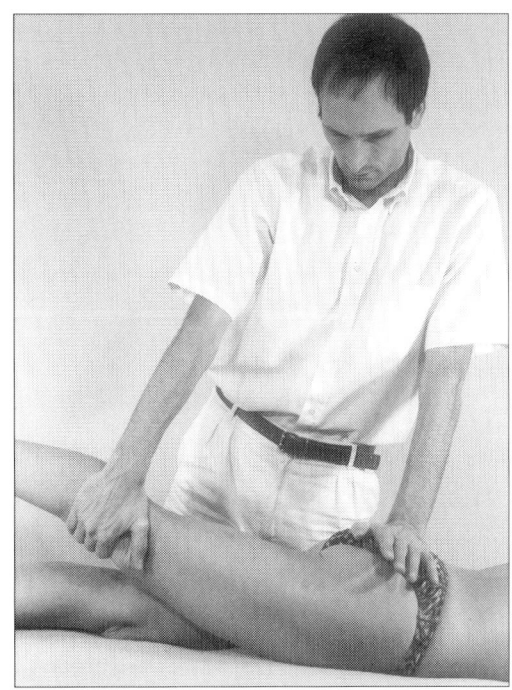

Fig. 19.87

INNOMINATE INTERNAL ROTATION ON S1 (S1 COUNTERNUTATION)

- *Agonists*

    Hip external rotators:
    gluteus maximus (inferior gluteal nerve; L5–S2)
    gluteus medius, posterior part (superior gluteal nerve; L5,S1)
    gluteus minimus, posterior part (superior gluteal nerve; L5,S1)
    piriformis (L5–S2)
    obturator internus (nerve to the obturator internus; L5,S1)
    obturator externus (obturator nerve; L3,4)
    gemelli superior and inferior (nerve to the quadratus femoris; L5,S1)
    quadratus femoris (nerve to the quadratus femoris; L5,S1)

- *Muscle test* (Fig. 19.88)

    1. Have the patient lie prone with her right hip adducted and externally rotated. Her right knee should be flexed to 90 degrees.
    2. Position yourself distal to her right knee, stabilizing the lateral side of her right knee with your right hand.
    3. Hold her right ankle with your left hand.
    4. Ask her to maintain this leg position while you attempt to internally rotate her leg (push her ankle laterally).

    Remarks—If the test yields weakness, compare the result with a test in which the patient's leg is less externally rotated.

- *Passive mobilization* (Fig. 19.89)

    1. Have the patient lie prone with her right knee flexed to 90 degrees and her right hip adducted and internally rotated.
    2. Stand at her right side, using your left hand to hold her right ankle and maintain the position of this leg.
    3. Place your right thumb against the spinous process of S1 and push S1 to her left.
    4. Retest.

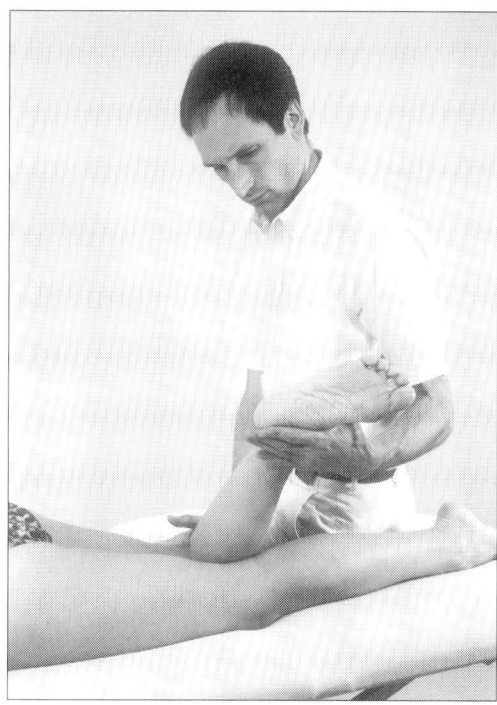

Fig. 19.88

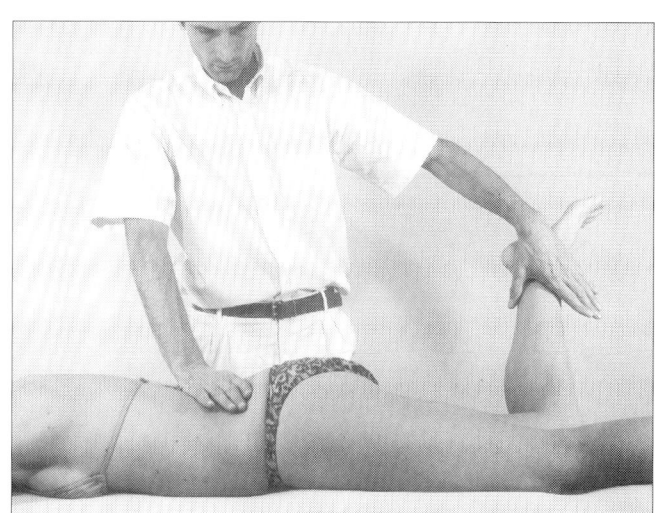

Fig. 19.89

INNOMINATE INTERNAL ROTATION ON S3 (S3 NUTATION)

- *Agonists*

    Hip external rotators:
    gluteus maximus (inferior gluteal nerve; L5–S2)
    gluteus medius, posterior part (superior gluteal nerve; L5,S1)
    gluteus minimus, posterior part (superior gluteal nerve; L5,S1)
    piriformis (L5–S2)
    obturator internus (nerve to the obturator internus; L5,S1)
    obturator externus (obturator nerve; L3,4)
    gemelli superior and inferior (nerve to the quadratus femoris; L5,S1)
    quadratus femoris (nerve to the quadratus femoris; L5,S1)

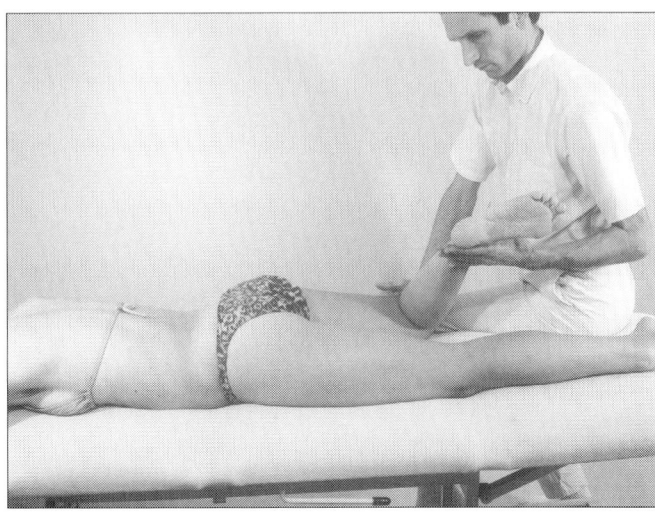

Fig. 19.90

- *Muscle test* (Fig. 19.90)

    1. Have the patient lie prone with her right hip abducted and externally rotated and her right knee flexed to 90 degrees.
    2. Position yourself distal to her right knee, stabilizing the lateral aspect of the knee with your right hand.
    3. Hold her ankle with your left hand.
    4. Ask her to maintain this leg position while you attempt to internally rotate her leg (push her ankle laterally).

- *Passive mobilization* (Fig. 19.91)

    1. Have the patient lie prone with her right knee flexed to 90 degrees and her right hip abducted and internally rotated.
    2. Stand at her right side, using your left hand to hold her right ankle and maintain the position of this leg.
    3. Place your right hand on the right side of S4 and push S4 to her left.
    4. Retest.

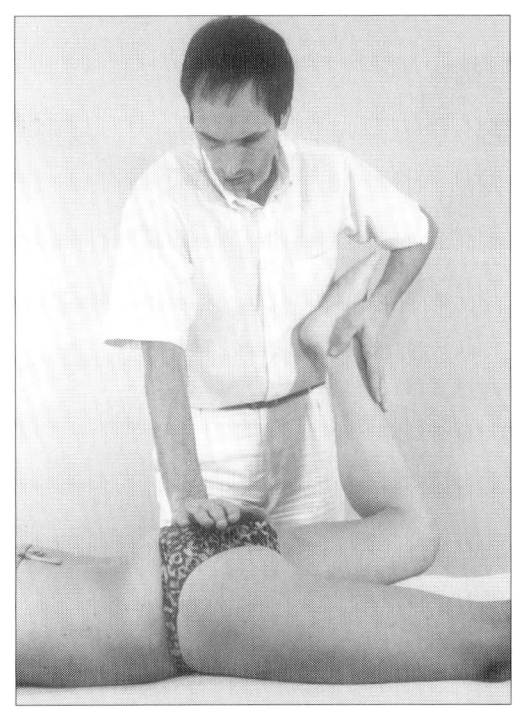

Fig. 19.91

Symphysis Pubis

A small amount of movement in different directions is available at the symphysis pubis. The movements at the symphysis pubis can be categorized into anterior–posterior movements, cranial–caudal movements, and combination movements. An anterior movement of the right pubis on the left pubis could also be called a posterior movement of the left pubis on the right pubis. Therefore, testing of a particular movement on the right side automatically tests the opposite movement on the left side.

Testing for inhibition from the symphysis pubis could be done by testing the hip adductors; however, when the adductors test weak, the examiner must consider the possibility of inhibition from the hip joint, the lumbar spine, and the sacroiliac joints as well. Each cause of inhibition has its own specific pattern of weakness. The treatments listed here assume inhibition from the pubis.

Symphysis pubis movements always affect the sacroiliac joints. An example is the adductor muscle test at 40 degrees of hip flexion to evaluate and treat S1 nutation mobility.

- *Capsular pattern*—Pain when the joint is stressed.

CAUDAL TRANSLATION OF THE RIGHT PUBIS/ CRANIAL TRANSLATION OF THE LEFT PUBIS

- *Agonists*

    Hip adductors:
    pectineus (femoral nerve; L2,3, and accessory obturator; L3)
    adductor longus (obturator nerve; L2,3,4)
    gracilis (obturator nerve; L2,3)
    adductor brevis (obturator nerve; L2,3,4)
    adductor magnus (obturator nerve; L2,3,4, and sciatic nerve; L2,3,4)

- *Muscle test* (Fig. 19.92)

    1. Have the patient lie supine with her left knee drawn up to her chest, holding the knee with both hands.
    2. Stand at her feet, with your right hand on the sole of her left foot, pushing it cranially for stabilization.
    3. Hold her right ankle with your left hand.
    4. Externally rotate her right leg and lift it about 30 cm off the table.
    5. Ask her to maintain the position of her right leg. Then attempt to abduct and extend it.

    Remarks—The passive flexion of the left hip rotates the left innominate posteriorly, translating the left pubis cranially. The right pubis is pulled caudally by the contraction of the right adductors. This procedure tests the cranial mobility of the left pubis and the caudal mobility of the right pubis.

    If this test yields weakness, compare the result with a test in which the patient's left leg is laying extended on the table (less stress on the symphysis pubis) or to one in which her right leg is more abducted (less tension on the right hip joint capsule).

    A right inferior pubis (limited cranial translation mobility) will not only inhibit the left adductors, but may also inhibit the right abdominal muscles throughout the lumbar range (see Lumbar spine; Flexion T10 through L5; Lumbar flexors).

- *Active mobilization* (Fig. 19.92)—The treatment is identical to the test described above, except the treatment contraction is gentler and sustained. Retest after the mobilization.

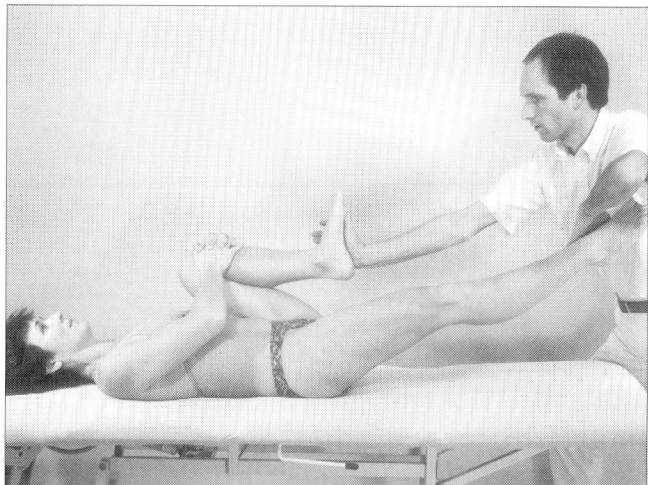

Fig. 19.92

ANTERIOR TRANSLATION OF THE RIGHT PUBIS/ POSTERIOR TRANSLATION OF THE LEFT PUBIS

- *Agonists*

  Hip adductors:
  - pectineus (femoral nerve; L2,3, and accessory obturator; L3)
  - adductor longus (obturator nerve; L2,3,4)
  - gracilis (obturator; L2,3)
  - adductor brevis (obturator; L2,3,4)
  - adductor magnus (obturator nerve; L2,3,4, and sciatic nerve; L2,3,4)

- *Muscle test* (Fig. 19.93)

  1. Have the patient lie supine with her right hip flexed to 90 degrees and her right knee passively bent.
  2. Stand on her right side with your stabilizing left hand on her left ASIS.
  3. Place your right hand medially on her knee.
  4. Ask her to maintain the position of her knee. Then attempt to push it laterally. Allow the left side of her pelvis to come slightly off the table.
  5. After the left side of the pelvis has come off the table, prevent it from moving farther by pressing the ASIS posteriorly and medially with your left hand.

  Remarks—This procedure tests the anterior mobility of the right pubis and the posterior mobility of the left pubis. If this test yields weakness, compare the result with a test with the hip in less horizontal adduction (reduces tension on the right hip joint). In the case of a right posterior pubis, the test will continue to yield weakness.

- *Active mobilization* (Fig. 19.93)—The treatment is identical to the evaluation, except the treatment contraction is gentler and sustained. Retest after the mobilization.

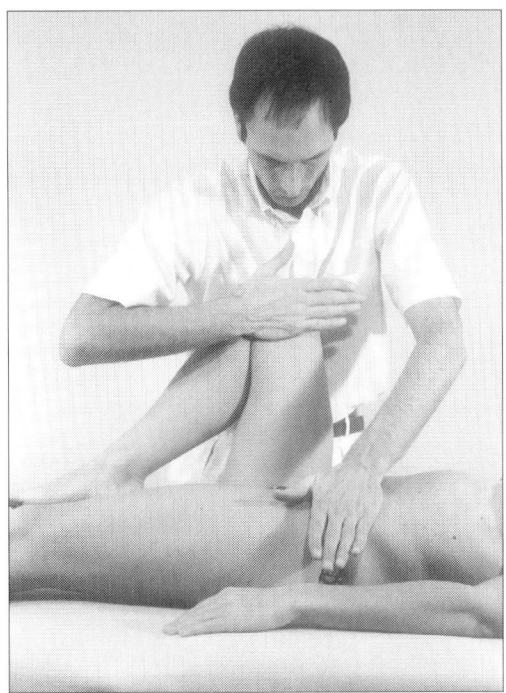

Fig. 19.93

Shoulder Girdle

The muscle tests for the shoulder girdle often involve testing glenohumeral muscles. It is often easier to evaluate (and treat) the patient's shoulder girdle after any glenohumeral joint dysfunctions have been corrected.

In the frontal plane, the medial clavicle is convex. During elevation, the medial clavicle translates caudally on the sternum, and during depression it translates cranially.

In the transverse plane the medial clavicle is generally considered concave, making the sternoclavicular joint a saddle joint.[29] However, perhaps due to the presence of the discus, the clavicle acts as a biconvex joint on the sternum.[3,9] This can easily be palpated on oneself. Place the tip of your left index finger on the anterior part of your right sternoclavicular joint and palpate both the sternum and the medial part of the clavicle, then actively retract your right shoulder. The clavicle can be felt to translate anteriorly during retraction and posteriorly during protraction.

The lateral clavicle is convex in the transverse plane. In this plane, the clavicle and the scapula approach each other (become more parallel) during protraction. During shoulder girdle elevation, the clavicle and scapula approach each other as well. Both movements can easily be palpated on oneself. Because the clavicle and scapula approach each other, the convex clavicle translates anteriorly on the acromion during protraction and elevation. The reverse occurs during retraction and depression of the shoulder girdle; the angle between the clavicle and the scapula increases, and the clavicle translates posteriorly on the acromion.

- *Capsular pattern* — The sternoclavicular and acromioclavicular joints have no restricted movement, but pain is felt at the extremes of the range.

SHOULDER GIRDLE RETRACTION

- *Agonist*

    Rhomboids:
       rhomboideus major and minor (dorsal scapular nerve; C4,5)

- *Muscle test* (Fig. 19.94)

    1. Have the patient be seated with her right shoulder retracted and her elbow bent. Her glenohumeral joint should be maximally extended, adducted, and externally rotated (her elbow close to her spine).
    2. Sit to her right side and place the base of your stabilizing right hand on her ribs, just inferior to her clavicle.
    3. Hold her elbow with your left hand.
    4. Ask her to keep her arm in this position. Then attempt to push her elbow anteriorly and laterally (away from her spine).

Remarks — If a limited retraction is inhibiting the rhomboids, the limitation can inhibit the middle and lower trapezius as well. If the patient has limited retraction mobility, the rhomboids and the trapezius will test weak only when the shoulder is retracted. They will test strong with less shoulder girdle retraction.

A limited extension of the cervicothoracic junction may also yield weakness during this test (see Cervical spine; C7 and T1 right facet extension; Rhomboid).

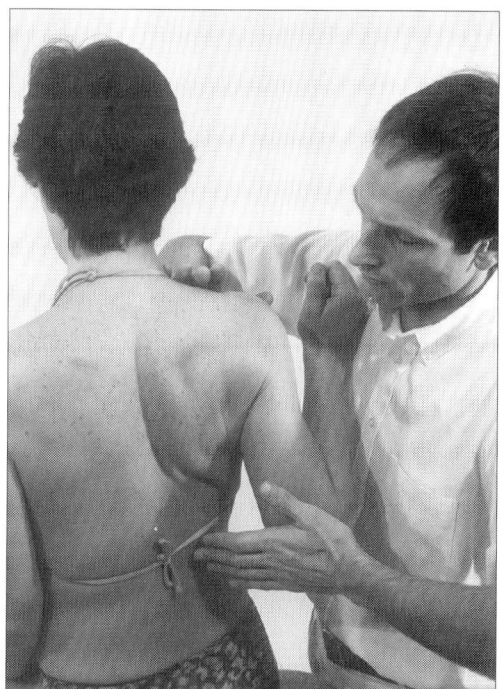

Fig. 19.94

- ***Passive posterior mobilization of clavicle on acromion*** (Fig. 19.95)

    1. Have the patient be seated with both shoulders maximally retracted.

    2. Sit to her right, placing your left hand on the spine of her scapula to check how she maintains the contraction of her retractors.

    3. Place the thenar eminence of your right hand against the anterior surface of the lateral clavicle and mobilize the clavicle posteriorly on her scapula.

    4. Retest.

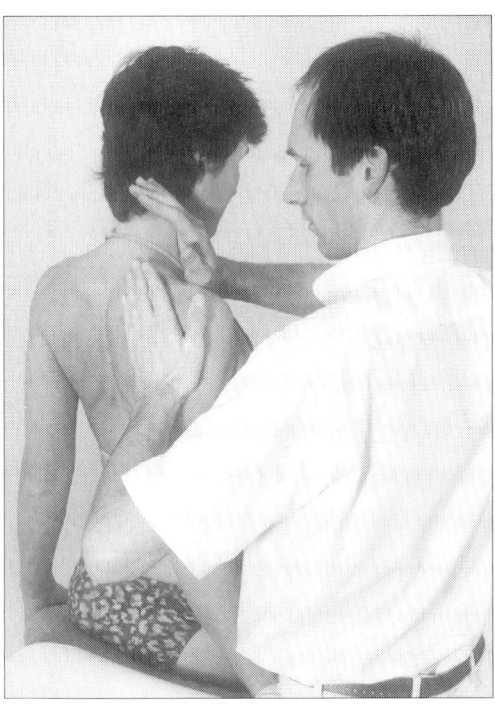

Fig. 19.95

- ***Passive anterior mobilization of clavicle on sternum*** (Fig. 19.96)

    1. Have the patient be seated with both shoulders retracted and her neck slightly flexed and side bent to the right to relax her right sternocleidomastoid muscle.

    2. Sit to her right, placing your left hand on the spine of her scapula to check the strength of the retractors' contraction.

    3. Place your right index finger flat on the posterior surface of the medial clavicle to push the clavicle anteriorly on the sternum.

    4. Retest.

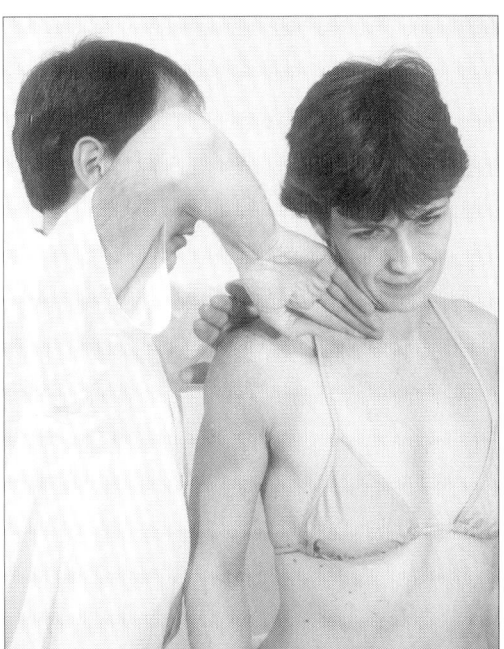

Fig. 19.96

- **Active anterior mobilization of clavicle on sternum** (Fig. 19.97)

    1. Have the patient be seated with both shoulders retracted.

    2. Sit at her right side. Ask her to maintain the retraction. Then internally rotate and flex her right shoulder to 90 degrees.

    3. Ask her to maintain the position of her arm. Then gently attempt to abduct her shoulder horizontally (move her arm to her right).

    4. Retest.

    Remarks—This active mobilization uses the clavicular fibers of the pectoralis major and the anterior deltoid to pull the clavicle anteriorly.

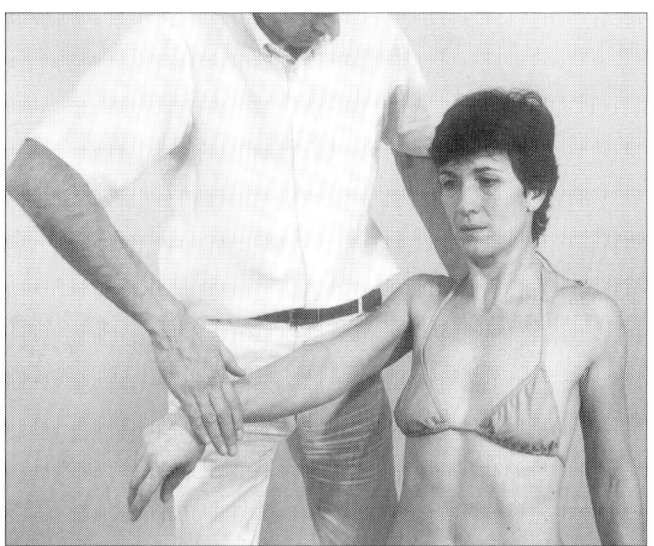

Fig. 19.97

SHOULDER GIRDLE RETRACTION

- *Agonist*

    Trapezius:
    middle and lower trapezius (accessory nerve and C3,4)

- *Muscle test* (Fig. 19.98)

    1. Have the patient be seated with her right shoulder abducted 90 degrees horizontally (moved posteriorly).

    2. Stand in front of her right arm with your left hand holding her right distal forearm and your stabilizing right hand on her right upper ribs.

    3. Ask her to maintain this position. Then attempt to adduct her shoulder horizontally (move her arm anteriorly).

    Remarks—The above test assesses the strength of the middle trapezius fibers. The strength of the lower trapezius fibers is assessed with the shoulder maximally flexed. If any of these tests yields weakness, compare the results with a test in which the patient's shoulder girdle is less retracted. If a retraction restriction is responsible for the inhibition, both trapezius tests and the rhomboid test will yield weakness in retraction.

Inhibition from the thoracic spine may also inhibit the trapezius muscle. In this case, protraction of the shoulder will not improve trapezius strength (see Thoracic spine; C7–T12 right facet extension; Trapezius).

Treatment—If the trapezius muscle is inhibited by limited retraction at the acromioclavicular or sternoclavicular joints, use the treatment described under Shoulder girdle retraction; Rhomboid.

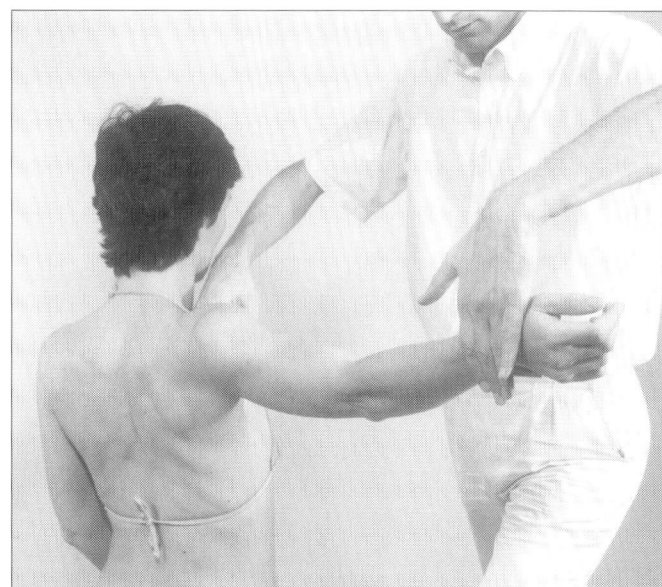

Fig. 19.98

SHOULDER GIRDLE PROTRACTION

- *Agonists*

    Shoulder girdle protractors:
    pectoralis major (lateral and medial pectoral nerves; C5–T1)
    pectoralis minor (lateral and medial pectoral nerves; C6,7,8)
    serratus anterior, upper fibers (long thoracic nerve; C5,6,7)

- *Muscle test (90 degrees flexion)* (Fig. 19.99)

    1. Have the patient lie supine with her right shoulder flexed to 90 degrees and protracted. Her right elbow should be extended, and she should make a fist with her hand.

    2. Stand close to her right elbow, stabilizing her wrist with your right hand, and place your left hand on top of her fist.

    3. Ask her to maintain this position. Then push her arm in a dorsal direction (toward retraction).

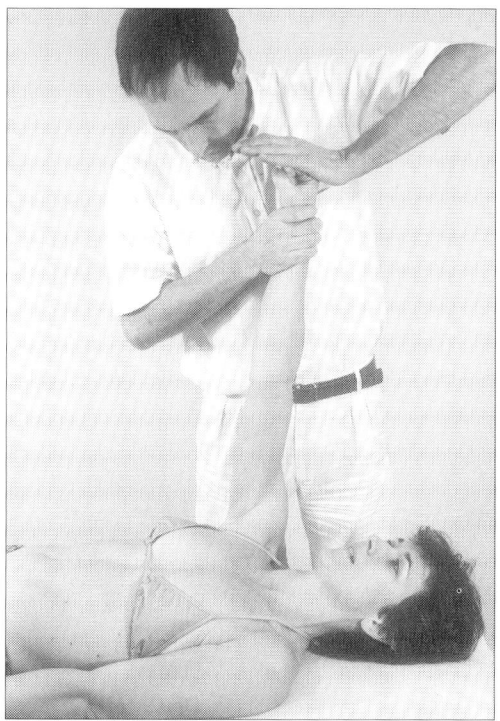

Fig. 19.99

- *Muscle test (0 degrees flexion)* (Fig. 19.100)

    1. Have the patient lie supine with her right shoulder protracted and right arm alongside her body but off the table.

    2. Support her right elbow with your right hand.

    3. Place your left hand on the anterior surface of the humeral head and coracoid process.

    4. Ask her to keep her shoulder in place. Then attempt to push her shoulder girdle posteriorly.

Remarks—If either of the above two tests yields weakness, compare the result with a test in which the patient's shoulder girdle is less protracted. If the protractors test strong in less protraction, the cause of the inhibition is likely a protraction dysfunction at the sternoclavicular or acromioclavicular joint. If the muscle exhibits weakness throughout its protraction range, the cause may be located in the C5–T1 spine or in the mobility of the first five ribs.

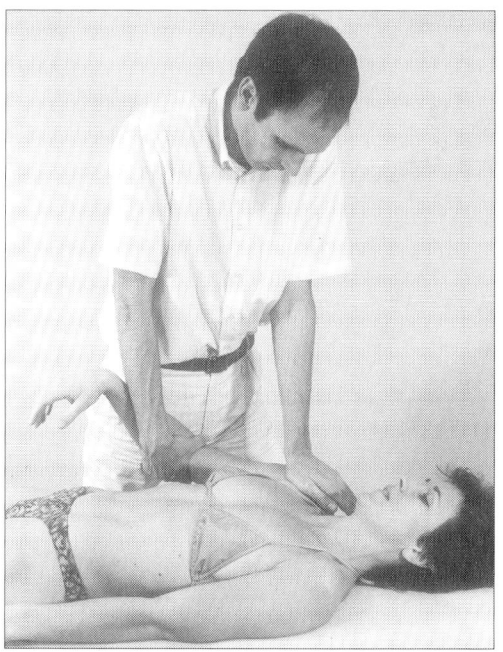

Fig. 19.100

- ***Passive posterior mobilization of clavicle on sternum*** (Fig. 19.101)

    1. Have the patient lie supine with a towel roll under her right scapula such that her scapula is protracted.

    2. Stand at her right side, placing your right thumb anteriorly on the medial part of her right clavicle. Your fingers should be bent and resting on her sternum or ribs.

    3. Mobilize her clavicle posteriorly with your thumb.

    4. Retest.

- ***Active mobilization of the acromioclavicular joint***—See Shoulder girdle protraction and elevation; Anterior deltoid.

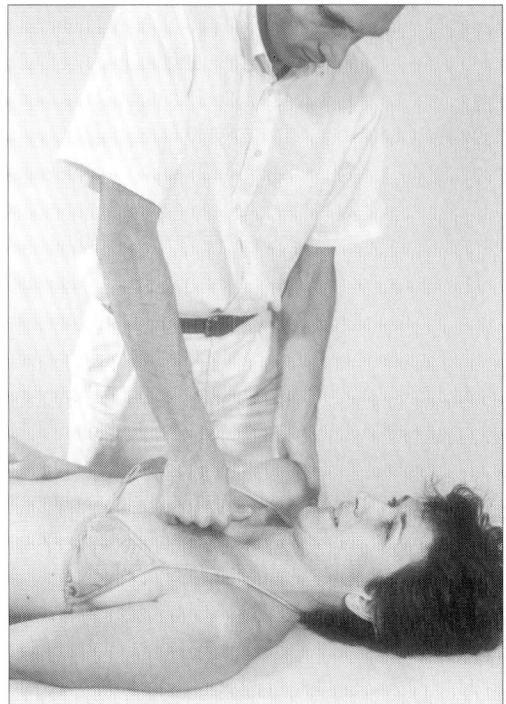

Fig. 19.101

SHOULDER GIRDLE PROTRACTION AND ELEVATION

- **Agonist**

    Anterior deltoid:
    anterior deltoid (axillary nerve; C5,6)

- **Muscle test** (Fig. 19.102)

    1. Have the patient be seated with her right shoulder girdle protracted and elevated. Her right shoulder should be flexed (humerus makes a 100-degree angle with her trunk) and internally rotated.

    2. Stand at her right side, stabilizing the dorsum of her scapula with your left hand.

    3. Hold her elbow with your right hand.

    4. Ask her to maintain this arm position. Then attempt to abduct and extend her elbow horizontally (bring her elbow dorsally and caudally).

    Remarks—If the test yields weakness, compare the result with a test in which the patient's shoulder girdle is less protracted and elevated (less movement of the acromioclavicular joint) or with one in which her shoulder is less internally rotated and flexed (less glenohumeral stress).

- **Active mobilization** (Fig. 19.102)—The treatment is identical to the testing procedure, except the treatment contraction is gentler and sustained. Retest after the mobilization.

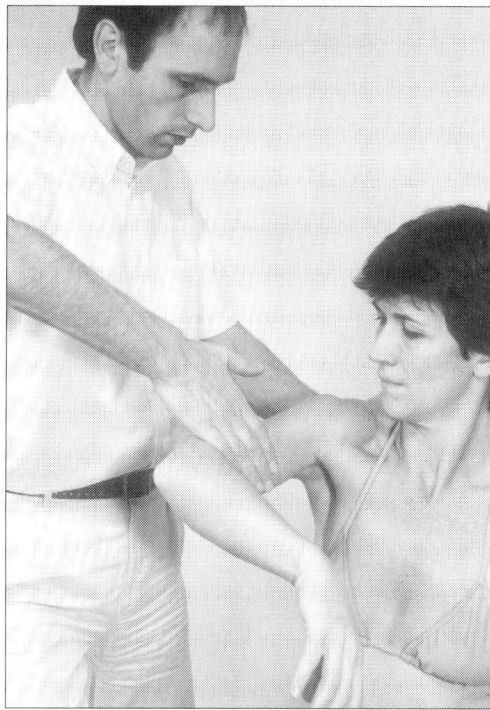

Fig. 19.102

SHOULDER GIRDLE ELEVATION

During shoulder girdle elevation, the lateral clavicle moves cranially while its medial end translates caudally at the sternoclavicular joint. Insufficient shoulder girdle elevation inhibits the elevators. In addition, the clavicular part of the pectoralis major can be inhibited because of its ability to translate the medial clavicle caudally. Thus, weakness of both the shoulder girdle elevators and the pectoralis major suggests limited shoulder girdle elevation mobility at the sternoclavicular joint.

- *Agonists*

    Shoulder girdle elevators:
    upper trapezius (accessory nerve and C3,4)
    levator scapulae (dorsal scapular nerve and spinal nerves; C3,4,5)
    rhomboids (dorsal scapular nerve; C4,5)

- *Muscle test (elevators)* (Fig. 19.103)

    1. Have the patient lie supine with her right elbow flexed to 90 degrees.
    2. Sit at her right side with your right hand holding her right elbow and your left holding her shoulder.
    3. Passively elevate her right shoulder girdle by pushing her right elbow cranially.
    4. Ask her to maintain this shoulder position. Then attempt to bring her shoulder girdle caudally.

- *Agonist*

    Pectoralis major:
    pectoralis major, clavicular part (medial and lateral pectoral nerves; C5,6)

- *Muscle test (clavicular fibers of the pectoralis major)* (Fig. 19.104)

    1. Have the patient lie supine with her right shoulder girdle elevated. Her right shoulder should be adducted and internally rotated, and her arm should be in front of her.
    2. Stand at her right side and hold her proximal humerus with your left hand, maintaining her shoulder girdle elevation.
    3. Hold her wrist with your right hand.
    4. Ask her to maintain her arm position. Then attempt to abduct her shoulder.

Remarks—If either of these tests yields weakness, compare the results with a test in which the shoulder girdle is less elevated. If both tests yield full strength with less shoulder girdle elevation, inhibition from the sternoclavicular joint is likely.

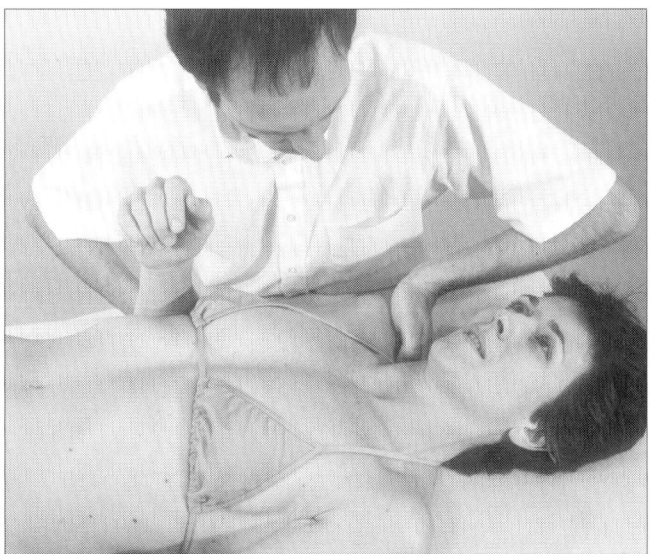

Fig. 19.103

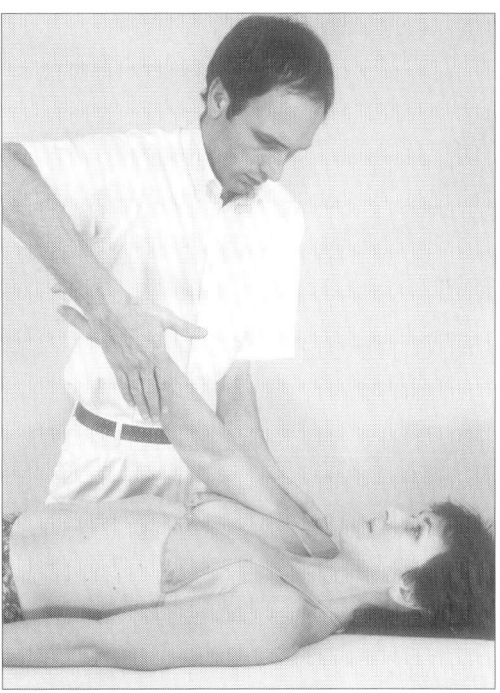

Fig. 19.104

- *Passive caudal mobilization of clavicle on sternum* (Fig. 19.105)

  1. Have the patient lie supine with her right elbow flexed to 90 degrees and her right shoulder girdle elevated.

  2. Sit cranial to her head and hold her right elbow with your right hand, maintaining her shoulder elevation.

  3. Place your left thumb on the medial aspect of her right clavicle, mobilizing it caudally, with your fingers bent and resting on her sternum.

  4. Retest.

- *Active caudal mobilization of clavicle on sternum* (Fig. 19.104)—The active mobilization is identical to the pectoralis major muscle test, except the treatment contraction is gentler and sustained. Retest after the mobilization.

  Remarks—During the active and passive mobilization, make sure the patient's right sternocleidomastoid muscle is relaxed. Relaxation of this muscle is sometimes accomplished by rotating the head slightly to the right.

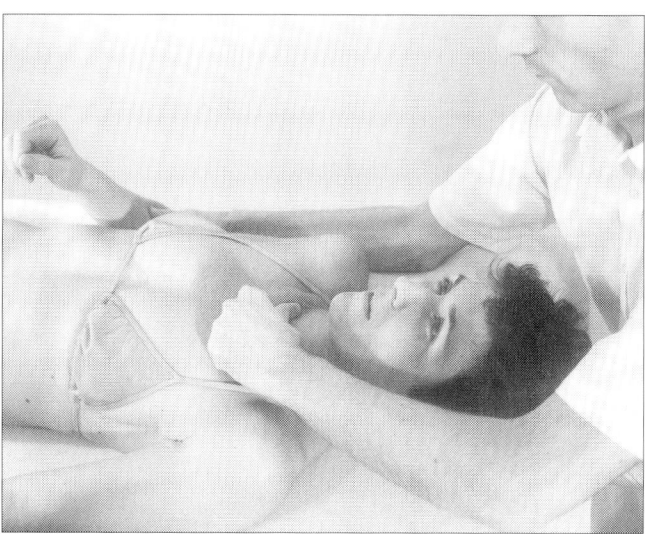

Fig. 19.105

SHOULDER GIRDLE DEPRESSION

During shoulder girdle depression, the lateral clavicle moves caudally while its medial end translates cranially on the sternum. A contraction of the sternocleidomastoid muscle (clavicular part) actively translates the medial clavicle cranially on the sternum. Limited superior translation mobility of the clavicle at the sternoclavicular joint will inhibit both the depressors and the clavicular part of the sternocleidomastoid.

- *Agonists*

    Depressors:
    pectoralis major, sternocostal part (medial and lateral pectoral nerves; C7–T1)
    pectoralis minor (medial and lateral pectoral nerves; C6,7,8)
    latissimus dorsi (thoracodorsal nerve; C6,7,8)

- *Muscle test* (Fig. 19.106)

    1. Have the patient lie supine with her right arm alongside her body and her elbow flexed to 90 degrees.
    2. Sit at her right side with your right hand holding her right elbow and your left hand holding her shoulder.
    3. Depress her right shoulder girdle with your left hand.
    4. Ask her to maintain this arm position, then attempt to elevate her shoulder girdle by pressing cranially with your right hand.

- *Agonist*

    Sternocleidomastoid:
    sternocleidomastoid, clavicular part (C2,3,4, and accessory nerve)

- *Muscle test* (Fig. 19.107)

    1. Have the patient lie supine, actively keeping her right shoulder girdle in depression. Her head should be rotated 30 degrees to the left and lifted off the table.
    2. Ask her to maintain this head position. Then carefully attempt to push her head to her left and posteriorly (into cervical left side bending and extension).

Remarks—If either test yields weakness, compare this result with a test in which the patient's shoulder girdle is less depressed. Weakness throughout the shoulder girdle range does not derive from the acromioclavicular joint, but could derive from the acromioclavicular joint or cervical spine.

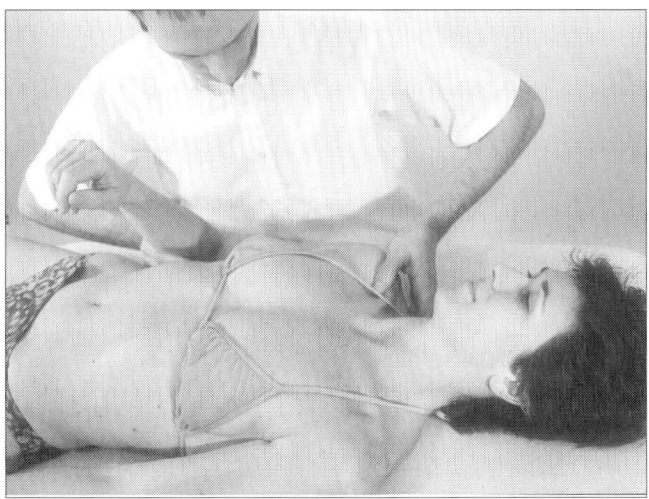

Fig. 19.106

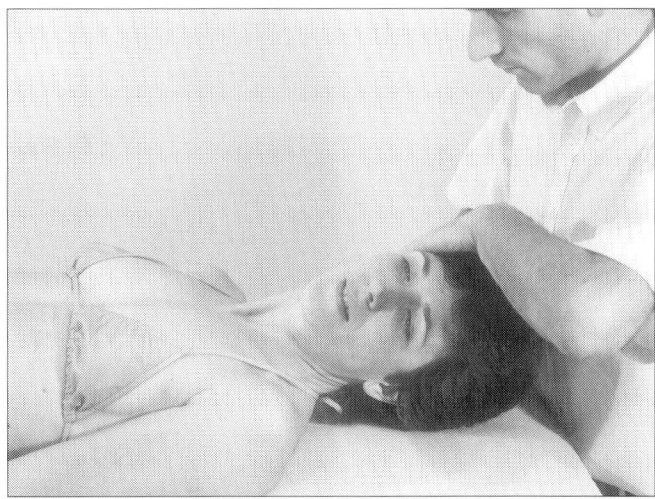

Fig. 19.107

- *Passive cranial mobilization of clavicle on sternum* (Fig. 19.108)

  1. Have the patient lie supine with her right shoulder girdle depressed.
  2. Sit at her right side, placing your left hand on the cranial aspect of her shoulder girdle to maintain the depression.
  3. Place your right thumb on the inferior aspect of her medial clavicle, with your fingers bent and resting on her sternum.
  4. Gently mobilize her medial clavicle cranially.
  5. Retest.

- *Active cranial mobilization of clavicle on sternum* (Fig. 19.109)

  1. Have the patient lie supine with her head turned 30 degrees to her left.
  2. Sit cranial to her right shoulder and depress her right shoulder girdle with your right hand.
  3. Place your left hand on the right side of her head.
  4. Ask her to maintain this head position. Then gently push her head to the left.
  5. Retest.

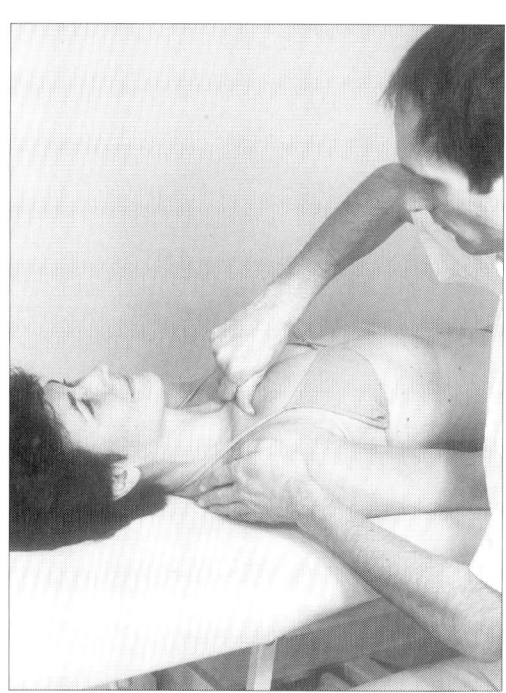

Fig. 19.108

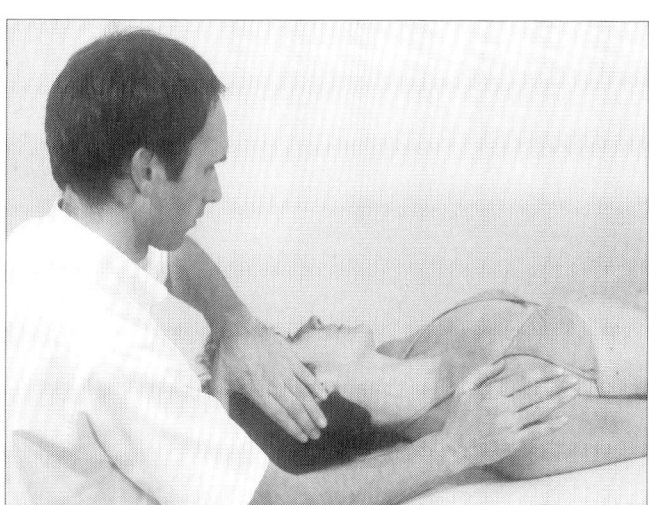

Fig. 19.109

Glenohumeral Joint

The convex head of the humerus articulates with the concave glenoid. During movement, humeral head translation is opposite the humerus' angular movement. For instance, on abduction, the humerus moves cranially and the humeral head translates caudally. The easiest way to visualize the translation accompanying internal and external rotation is to picture the greater tuberculum moving anteriorly or posteriorly. The translation of the humeral head is opposite the tuberculum movement.

- *Capsular pattern*—External rotation is most limited. Abduction is less limited. Internal rotation is less limited than abduction. Adduction is free.

SHOULDER EXTERNAL ROTATION

- *Agonists*

    Shoulder external rotators:
    Infraspinatus (suprascapular nerve; C4,5,6)
    teres minor (axillary nerve; C4,5,6)

- *Muscle test* (Fig. 19.110)

    1. Have the patient be seated with her right elbow flexed to 90 degrees and her shoulder externally rotated.
    2. Stand at her right and stabilize the medial part of her elbow with your left hand.
    3. Hold her distal forearm with your right hand.
    4. Ask her to keep her arm in place. Then attempt to internally rotate her shoulder.

    Remarks—If this test yields weakness, compare the result with a test in which the patient's shoulder is less externally rotated.

    A limited retraction mobility (posterior translation of the clavicle) at the acromioclavicular joint may also inhibit the external rotators.

- *Passive mobilization (seated)* (Fig. 19.111)

    1. Have the patient be seated with her elbow flexed and her shoulder externally rotated.
    2. Sit at her right side with your left hand on her right shoulder, placing your middle finger anterior to her coracoid process. Stabilize the scapula with this finger during the mobilization.
    3. Place your mobilizing right hand in her axilla, holding her humerus, with your fingers pointing posteriorly. The patient's forearm should rest on your right forearm, with her elbow against your trunk.
    4. With your right hand, translate her humerus anteriorly, controlling the amount of external rotation with your right arm.
    5. Retest.

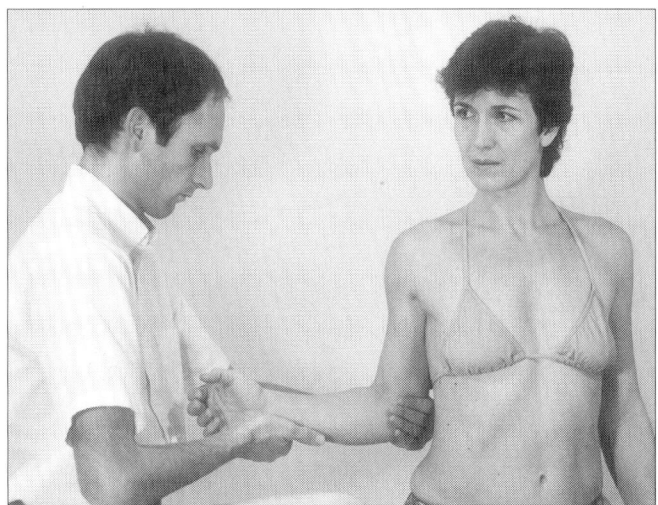

Fig. 19.110

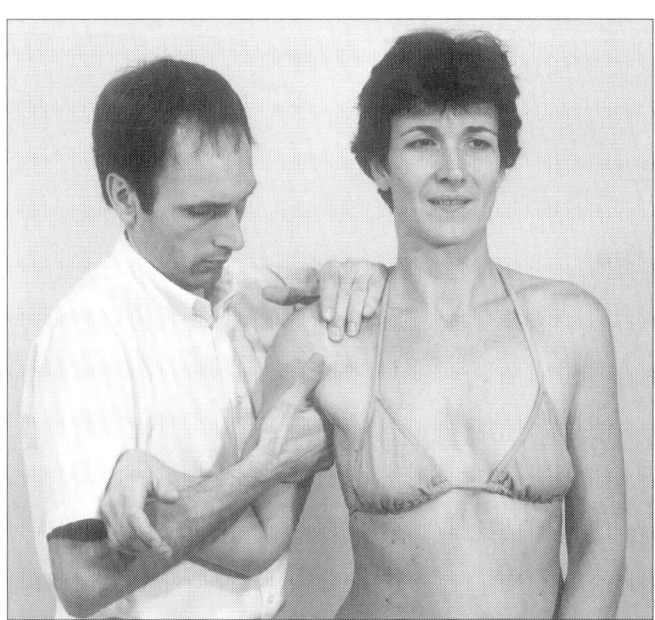

Fig. 19.111

- *Passive mobilization (supine)* (Fig. 19.112)

  1. Have the patient lie supine with her right elbow flexed to 90 degrees. Place a small towel roll beneath her right humeral head.
  2. Stand at her right side, distal to her elbow. Hold her right wrist with your left hand so as to control the amount of external rotation at her shoulder.
  3. With the hypothenar eminence of your right hand, mobilize her right coracoid process posteriorly.
  4. Retest.

- *Home exercise* (Fig. 19.113)—For this exercise, the patient should lie on her right side with her right shoulder protracted and her arm in front of her. Her right elbow should be flexed to 90 degrees, and her forearm should be resting on a pillow.

  Remarks—This exercise could be done while sleeping. The thickness of the pillow determines the amount of external rotation. The weight of the upper body on the mattress provides stabilization to the scapula, while the mattress provides the mobilizing force (the anterior translation). As the shoulder mobility progresses, the pillow can be lowered or the shoulder flexion can be increased (elbow moves away from the body).

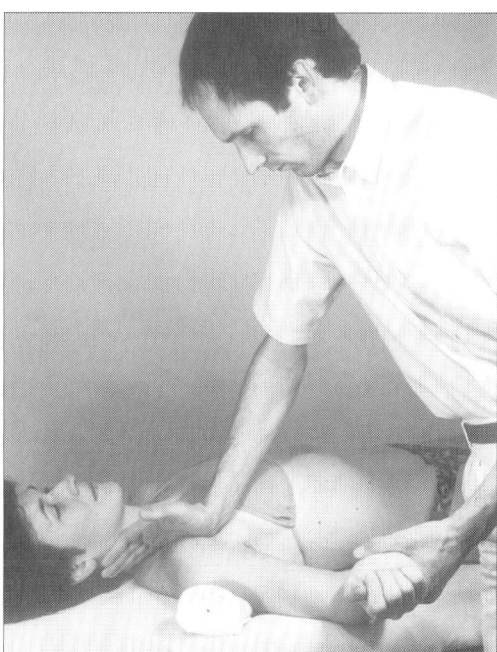

Fig. 19.112

Fig. 19.113

SHOULDER ABDUCTION

During abduction, the humerus moves cranially while the humeral head translates caudally. A restricted glenohumeral abduction will inhibit the abductor muscles when the glenohumeral joint has moved past the inhibiting barrier.

- *Agonists*

    Shoulder abductors:
    supraspinatus (suprascapular nerve; C4,5,6)
    deltoid (axillary nerve; C5,6)

- *Muscle test (abductors)* (Fig 19.114)

    1. With the patient seated, stand behind her right shoulder and place your left hand on her shoulder for stabilization.

    2. Hold her flexed elbow with your right hand and passively abduct her glenohumeral joint in the scapular plane.

    3. When her shoulder is fully abducted, ask her to hold her arm in this position, then reposition your right hand over her lateral epicondyle and gently attempt to adduct her shoulder.

    Remarks—If the test yields weakness, compare the result with a test in which the patient's shoulder is less abducted.

- *Passive mobilization (supine)* (Fig. 19.115)

    1. Have the patient lie supine with her right arm relaxed.

    2. Hold her right wrist and provide traction to her arm in a distal direction.

    3. Oscillate her arm gently in the frontal plane.

    4. Retest.

    Remarks—The advantage of the oscillations is that they make it easy to monitor for relaxation of the shoulder muscles. In addition, they provide the patient's central nervous system with massive amounts of large-diameter nerve fiber input, capable of closing the nociceptive gate.

    This mobilization is most effective when the shoulder has a major restriction and the mobilization needs to be done prior to 45 degrees of abduction.

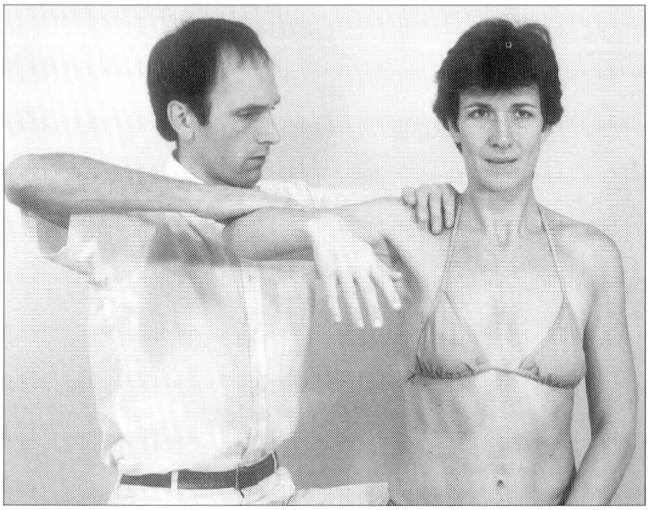

Fig. 19.114

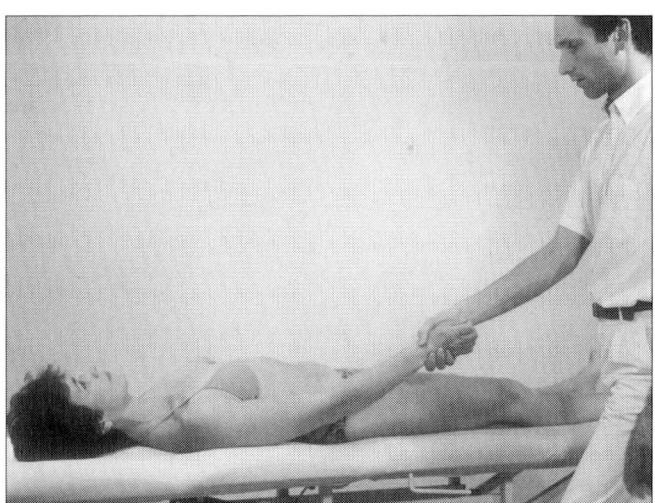

Fig. 19.115

- **Active mobilization (seated)** (Fig. 19.116)

  1. Have the patient be seated with her right shoulder abducted and her elbow flexed.

  2. Have her rest her forearm on your left forearm.

  3. Rest your left wrist in the anterior and lateral part of your right elbow.

  4. Rest your right hand on the patient's deltoid to monitor for deltoid relaxation.

  5. Ask the patient to push her elbow comfortably toward the floor. Then provide unyielding resistance.

  6. Retest.

  Remarks—The patient must have at least 45 degrees of shoulder abduction for this mobilization to be effective. The muscle fibers translating the humeral head caudally are the lateral latissimus dorsi fibers, the lateral pectoralis major fibers, and the teres major.

This exercise can be done as a home exercise. The patient can be seated at a table with her right arm resting on a pillow on top of the table, and should push her elbow into the pillow. As her abduction improves, she can either raise the support beneath her elbow or lower herself in the chair.

When both the abductors and the internal rotators are arthrogenically inhibited by the glenohumeral joint, you can treat both conditions at the same time. Place the patient's arm in abduction and internal rotation and ask her to contract her posterior adductors (latissimus dorsi and teres major) against your unyielding resistance (Fig. 19.117).

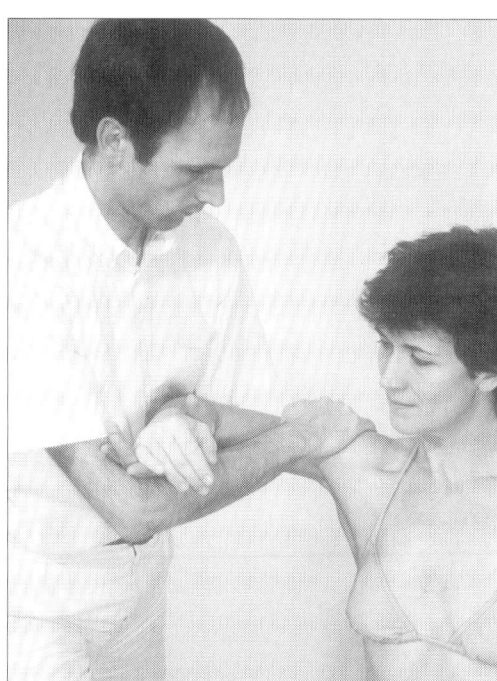

Fig. 19.116

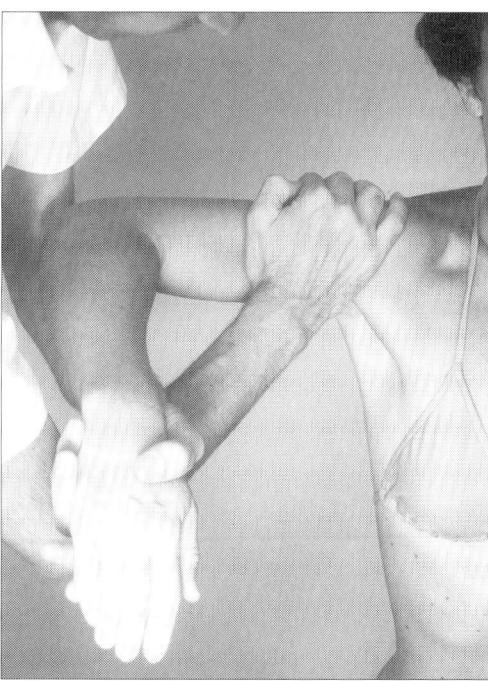

Fig. 19.117

Likewise, when both the abductors and the external rotators are arthrogenically inhibited by the glenohumeral joint, you can again treat both conditions at once. Place the patient's arm in abduction and external rotation and ask her to contract her anterior adductors (pectoralis major) against your unyielding resistance (Fig. 19.118).

- **Active home exercise (supine)** (Fig. 19.119)—Instruct the patient to lie supine with her right shoulder flexed near the limited end range of her glenohumeral joint. She should maintain this position to allow lengthening of the periarticular tissue.

  Remarks—The pectoralis major prevents the arm from moving too far into flexion. Simultaneously, it provides a caudal glide to the humeral head. This home exercise can be used if the patient has more than 90 degrees of flexion available.

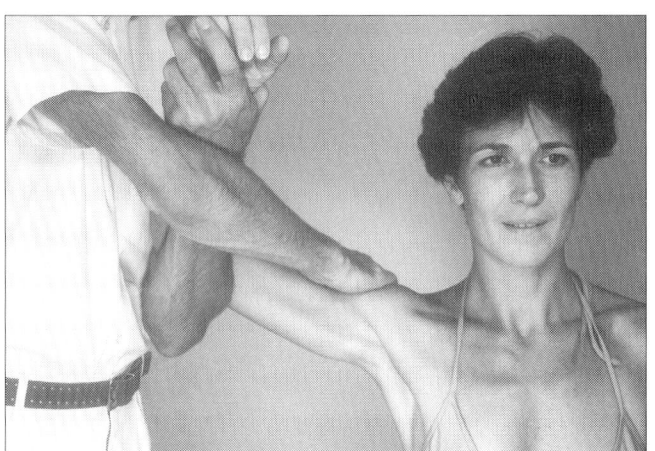

Fig. 19.118

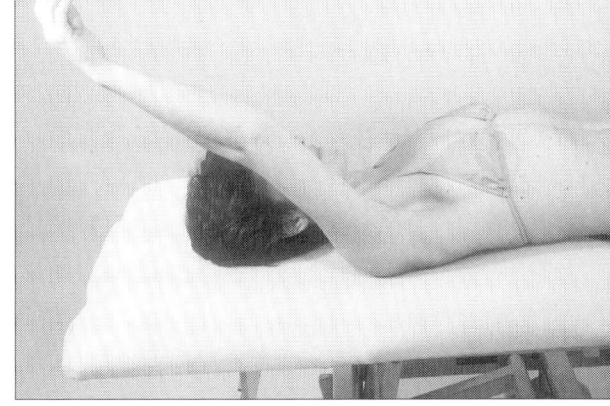

Fig. 19.119

SHOULDER INTERNAL ROTATION

- *Agonists*

    Shoulder internal rotators:
    subscapularis (upper and lower scapular nerves; C5,6,7)
    teres major (lower subscapular nerve; C6,7)
    pectoralis major (lateral and medial pectoral nerves; C5–T1)
    latissimus dorsi (thoracodorsal nerve; C6,7,8)

- *Muscle test* (Fig 19.120)

    1. Have the patient be seated with her right arm slightly abducted and fully internally rotated.
    2. Position yourself in front of her right arm, facing her. Hold her elbow with your left hand.
    3. Hold her wrist with your right hand.
    4. Ask her to hold her arm in place. Then attempt to externally rotate the arm.

    Remarks — Some abduction of the patient's arm is often inevitable when internally rotating the arm.

Therefore, it is good practice to clear abduction first before testing internal rotation.

If this test yields weakness, compare the result with a test in which the patient's arm is less internally rotated.

A limited protraction mobility at the acromioclavicular joint may also inhibit the internal rotators.

- P*assive mobilization (seated)* (Fig. 19.121)

    1. Have the patient be seated with her arm passively internally rotated (her hand resting on the seat behind her back, for instance).
    2. Stand at her right side with your stabilizing left hand on the dorsum of her scapula.
    3. Place your right hand on the anterior aspect of her humeral head and mobilize it posteriorly.
    4. Retest.

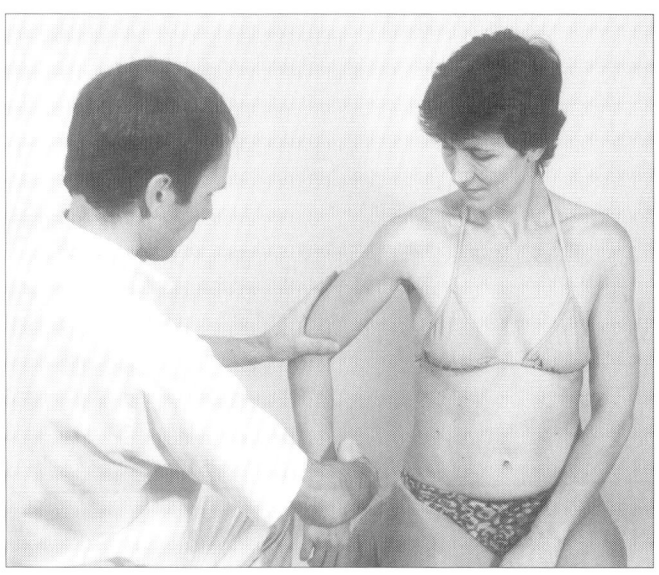

Fig. 19.120

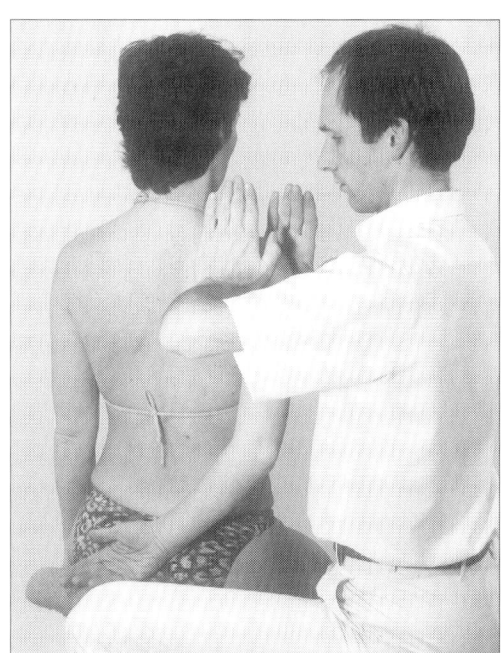

Fig. 19.121

- ***Passive mobilization (supine)*** (Fig. 19.122)

    1. Have the patient lie supine with her right arm slightly abducted.

    2. Stand at her right side, holding her wrist with your right hand.

    3. Place your left hand anteriorly on her humeral head and mobilize it posteriorly.

    4. Maintaining the posterior pressure on the humeral head, internally rotate her arm.

    5. Retest.

    Remarks—The amount of internal rotation depends on the patient's condition. Conditions with high reactivity may be treated just past the dysfunctional barrier. Conditions with low reactivity may be best treated with the glenohumeral joint stretched into the physiological barrier.

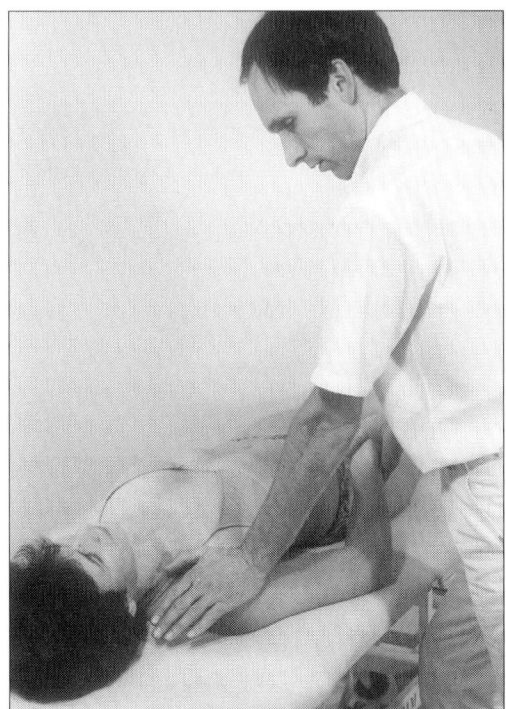

Fig. 19.122

SHOULDER ADDUCTION

- **Agonists**

    Shoulder adductors:
    - pectoralis major (medial and lateral pectoral nerves; C7–T1)
    - latissimus dorsi (thoracodorsal nerve; C6,7,8)
    - teres major (lower subscapular nerve; C6,7)

- **Muscle test** (Fig. 19.123)

    1. Have the patient be seated with her right elbow flexed.
    2. Stand at her right side and place your stabilizing left hand on her iliac crest.
    3. Hold her elbow with your right hand.
    4. Ask her to hold her elbow at her side. Then attempt to abduct the elbow.

    Remarks—Minor adduction dysfunctions might show up only after the patient has elevated her shoulder girdle (Fig. 19.124). During shoulder girdle elevation, the scapula rotates externally. Scapular external rotation creates increased adduction of the humerus at the glenohumeral joint.

Another way to sensitize the adduction test is to push the medial part of the scapula caudally while testing, which will also result in external rotation of the scapula (relative adduction at the glenohumeral joint). The scapula can be externally rotated by placing your stabilizing hand on the patient's right shoulder (Fig. 19.125). Place the base of your hand on the spine of the scapula to push it caudally.

If the adduction test yields weakness, compare the result with a test in which the patient's arm is less adducted.

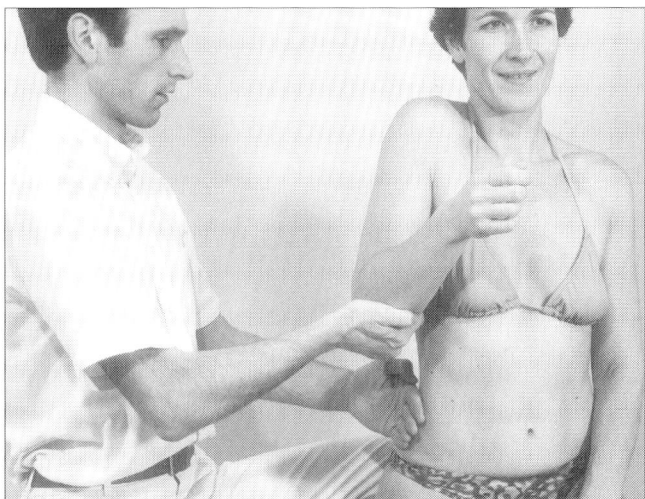

Fig. 19.124

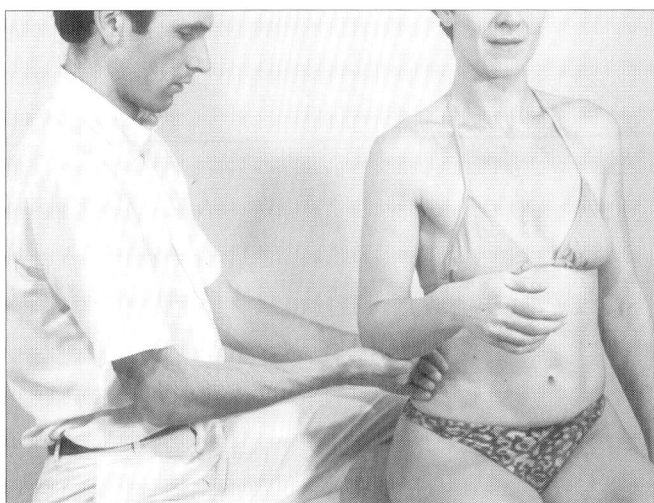

Fig. 19.123

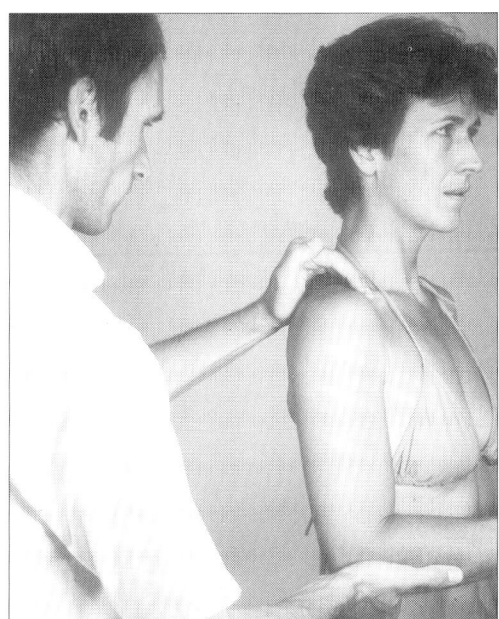

Fig. 19.125

- **Passive mobilization** (Fig. 19.126)

    1. Have the patient be seated with her right elbow flexed and her shoulder adducted.
    2. Sit behind her right shoulder, placing your stabilizing left hand on her shoulder with the base of your hand on the spine of the scapula.
    3. Hold her right elbow with your right hand and mobilize her humerus cranially.
    4. Retest.

- **Active mobilization** (Fig. 19.127)

    1. Have the patient be seated with her right arm adducted and her hand on her thigh.
    2. Ask her to extend her elbow against the resistance of her leg.
    3. Retest.

Remarks—The triceps is the muscle providing the active mobilization. The long head of the triceps originates from the scapula and inserts on the ulna. During a triceps contraction, the scapula and the ulna are pulled closer together, resulting in a cranial force on the humerus at the glenohumeral joint.

This exercise can be given as a home exercise.

The active mobilization can be combined with the passive mobilization. This means that while the patient does the active mobilization, the therapist can reinforce it by performing the passive mobilization.

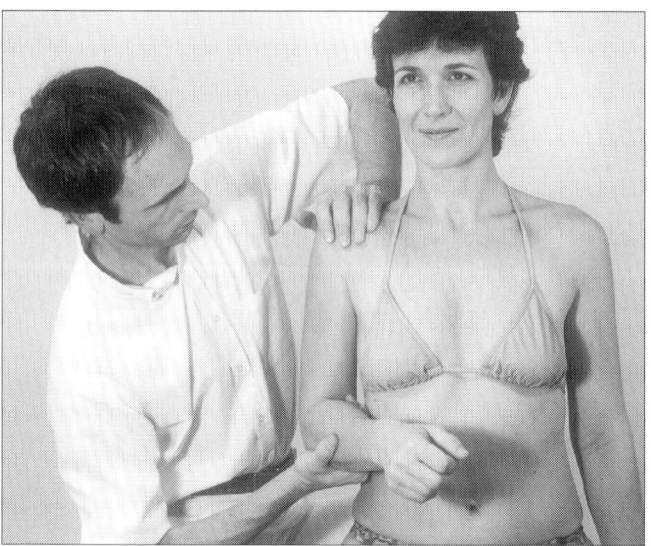

Fig. 19.126

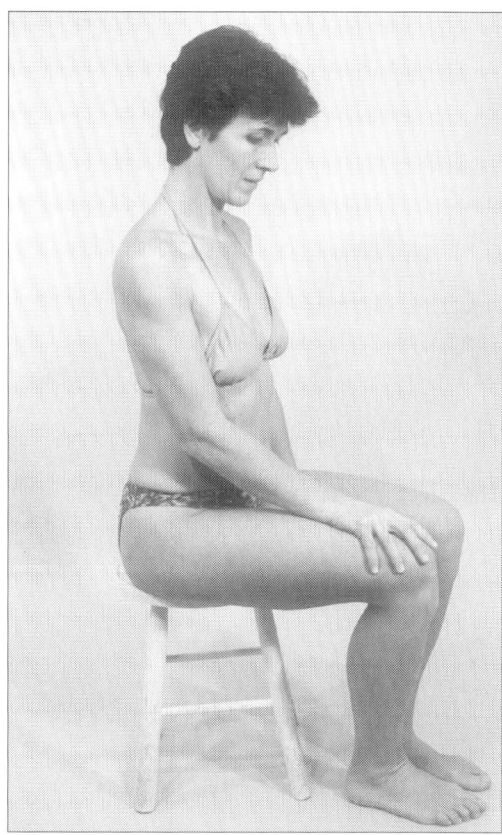

Fig. 19.127

SHOULDER HORIZONTAL ADDUCTION

- *Agonists*

    Shoulder horizontal adductors:
    anterior deltoid (axillary nerve; C5,6)
    pectoris major (medial and lateral pectoral nerves; C5–T1)
    coracobrachialis (musculocutaneous nerve; C5,6,7)

- *Muscle test* (Fig. 19.128)

    1. Have the patient lie supine with her right shoulder flexed to 90 degrees and horizontally adducted (moved to her left). Her humerus should have neutral rotation at the shoulder joint.

    2. Stand at her right side, stabilizing her shoulder joint with your left hand.

    3. Hold her distal forearm with your right hand.

    4. Ask her to keep her arm in place. Then attempt to abduct her arm horizontally (move her arm to her right).

    Remarks—If the test yields weakness, compare the result with one in which the patient's arm is less horizontally adducted.

    The horizontal adductors may also be inhibited by limited protraction mobility at the acromioclavicular joint. In this case, the test will yield more weakness if performed with shoulder girdle protraction and/or elevation (see Shoulder girdle protraction and elevation; Anterior deltoid). If the horizontal adductors are arthrogenically inhibited by the shoulder joint, shoulder girdle protraction increases the strength of these muscles, because protraction reduces horizontal adduction at the glenohumeral joint.

    If the horizontal adductors are inhibited by the upper ribs, a different respiratory phase may alter muscle strength (see Rib cage; Clavicle through rib 6 mobility in a posterior direction; Pectoralis major).

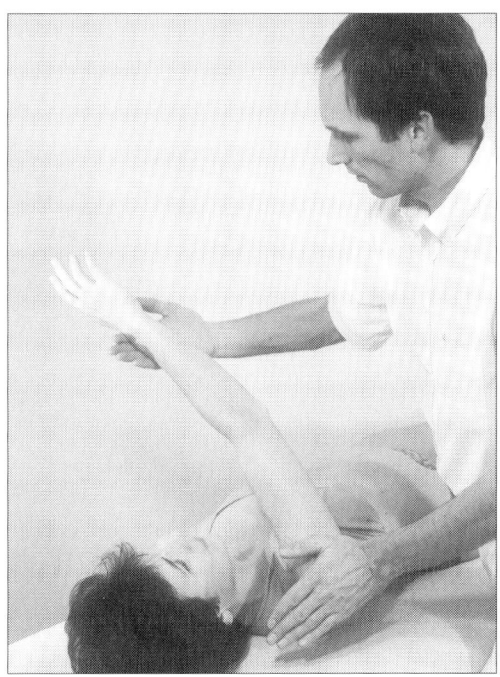

Fig. 19.128

- **Passive mobilization** (Fig. 19.129)
  1. Have the patient lie supine with her right shoulder flexed to 90 degrees and horizontally adducted and her elbow flexed.
  2. Stand at her right side, caudal to her right arm, and place your stabilizing left hand dorsally on her scapula.
  3. Hold her elbow with your right hand and translate her humerus in the direction of your left hand.
  4. Retest.

- **Active mobilization** (Fig. 19.128)—The active mobilization is identical to the evaluation, except the treatment contraction is gentler and sustained. Retest after the mobilization.

  Remarks—The horizontal adductors move the humerus into horizontal adduction and provide posterior translation of the humerus at the shoulder joint.

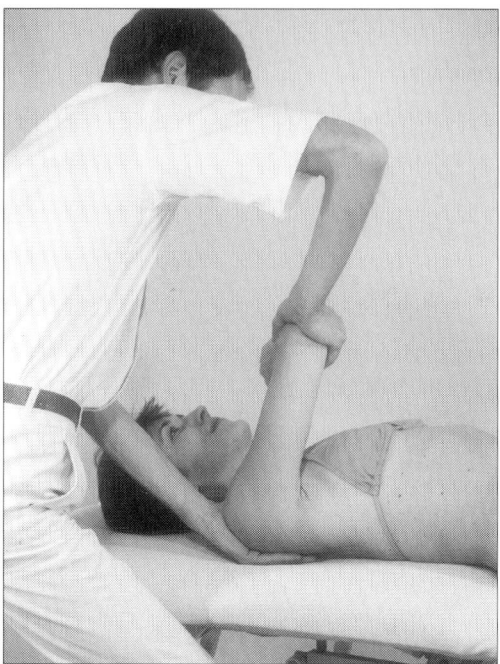

Fig. 19.129

SHOULDER HORIZONTAL ABDUCTION

- *Agonists*

    Shoulder horizontal abductors:
    infraspinatus (suprascapular nerve; C4,5,6)
    posterior deltoid (axillary nerve; C5,6)
    teres minor (axillary nerve; C4,5,6)

- *Muscle test* (Figs. 19.130 and 19.131)

    1. With the patient seated, stand at her right side and place your stabilizing left hand on her shoulder.

    2. Hold her flexed elbow with your right hand.

    3. Abduct her shoulder to 90 degrees, maintaining her shoulder girdle in a neutral position (without elevation) (Fig. 19.130).

    4. Ask her to keep her arm in place, then reposition your left hand to her elbow and your right thenar eminence anteriorly on her coracoid process (Fig. 19.131).

    5. Again ask her to maintain her arm position. Then attempt to adduct her arm horizontally.

Remarks—Since the muscle test involves 90 degrees of glenohumeral abduction, it is good practice to evaluate and treat abduction first.

If the test yields weakness, compare the result with a test in which the patient's glenohumeral joint is less horizontally abducted.

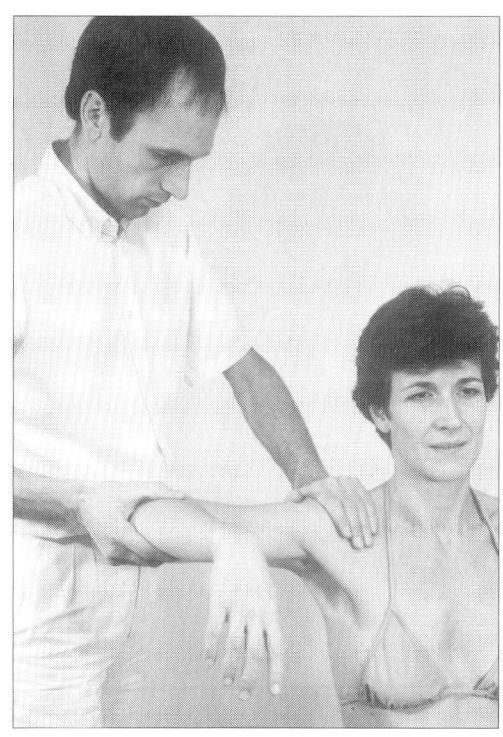

Fig. 19.130

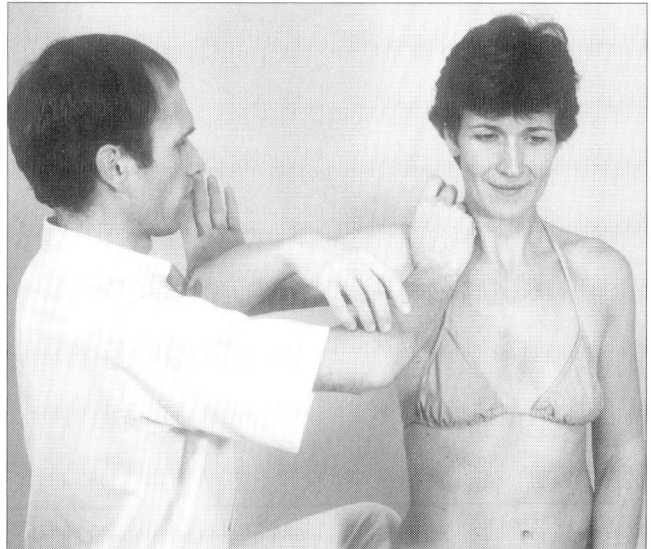

Fig. 19.131

- **Active mobilization** (Fig. 19.132)

  1. Have the patient be seated with her right glenohumeral joint abducted and horizontally abducted.
  2. Stand at her right side. Have her rest her right forearm on your right forearm, and hold her elbow with your right hand.
  3. The web between the thumb and index finger of your left hand should rest on her proximal humerus.
  4. Ask her to gently push her elbow caudally and anteriorly. Then provide unyielding resistance.
  5. Retest.

- **Passive mobilization** (Fig. 19.133)

  1. Have the patient lie supine with her right glenohumeral joint abducted. Place a towel roll beneath her humeral head at a height that moves her right shoulder girdle into protraction.
  2. Stand caudal to her right elbow, stabilizing it with your left hand and maintaining it in a neutral position between internal and external rotation.
  3. Place your right hypothenar eminence on her coracoid process and mobilize it posteriorly.
  4. Retest.

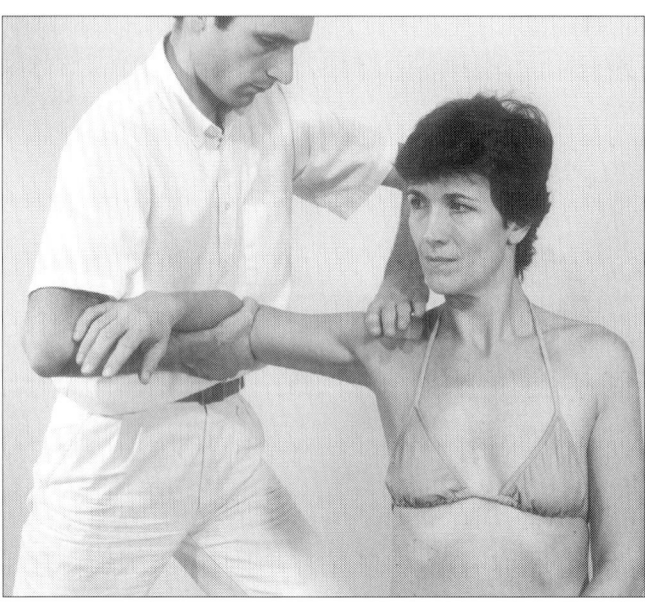

Fig. 19.132

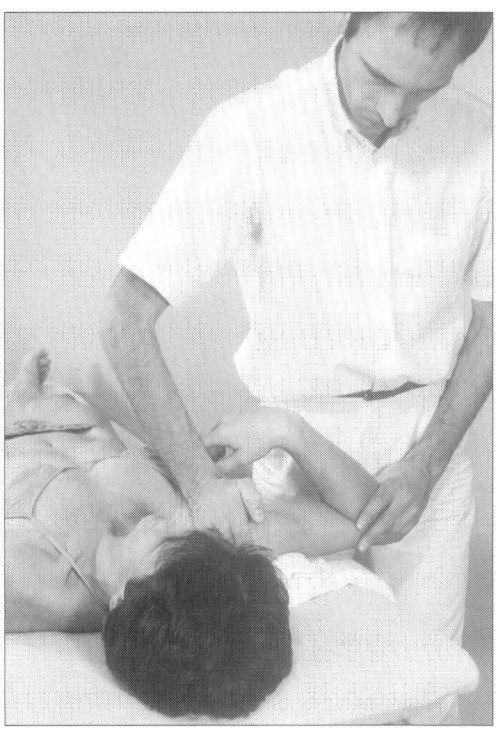

Fig. 19.133

SHOULDER EXTERNAL ROTATION IN 90 DEGREES ABDUCTION

- *Agonists*

    Shoulder external rotators:
    infraspinatus (suprascapular nerve; C4,5,6)
    teres minor (axillary nerve; C4,5,6)

- *Muscle test* (Fig. 19.134)

1. Have the patient be seated with her right arm abducted to 90 degrees and externally rotated.
2. Stand at her right side, fixating her elbow with your right hand.
3. Hold her distal forearm with your left hand.
4. Ask her to maintain this arm position. Then attempt to derotate her arm (internally rotate her arm).

Remarks—Since the muscle test involves 90 degrees of shoulder joint abduction, it is good practice to evaluate and treat abduction first.

If the test yields weakness, compare the result with a test in which the patient's humerus is less externally rotated.

- *Active mobilization* (Fig. 19.135)

1. Have the patient be seated with her right shoulder abducted to 90 degrees and externally rotated, without shoulder girdle elevation.
2. Stand behind her right elbow, supporting her right elbow and lower arm with your right hand.
3. The web between the thumb and index finger of your left hand should rest on her proximal humerus.
4. Ask her to gently pull her arm down and forward (toward her right thigh) while you provide unyielding resistance.
5. Retest.

Remarks—During abduction, the humeral head translates caudally at the glenohumeral joint, and during external rotation, it translates anteriorly. Translation of the humeral head during the combined abduction and external rotation movement is in a caudal and anterior direction. The muscle that translates the humeral head in this direction is the pectoralis major.

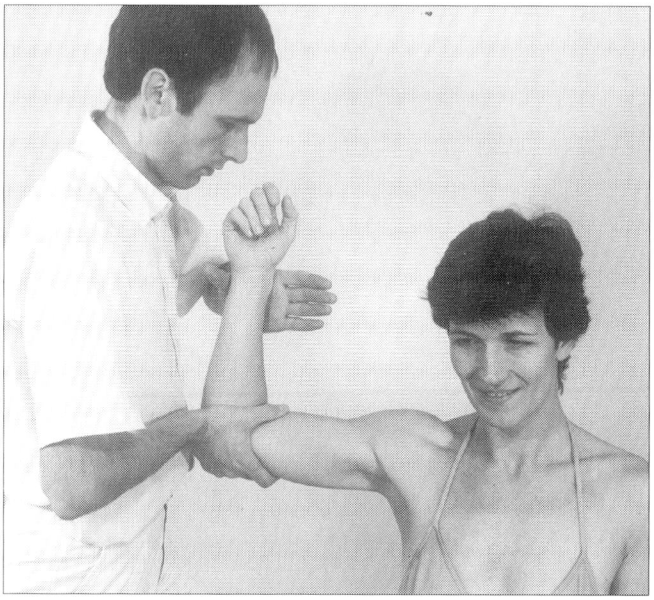

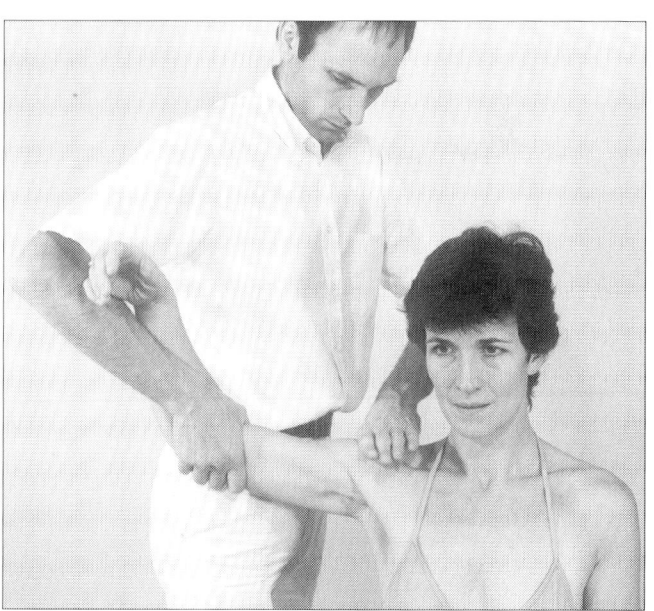

Fig. 19.134

Fig. 19.135

SHOULDER INTERNAL ROTATION IN 90 DEGREES ABDUCTION

- *Agonists*

    Shoulder internal rotators:
    subscapularis (upper and lower subscapular nerves; C5,6,7)
    pectoralis major (medial and lateral pectoral nerves; C5–T1)
    latissimus dorsi (thoracodorsal nerve; C6,7,8)
    teres major (subscapular nerve; C6,7)

- *Muscle test* (Figs. 19.136 and 19.137)

    1. With the patient seated, stand at her right side, stabilizing her shoulder with your left hand.
    2. Hold her elbow with your right hand.
    3. While you maintain shoulder girdle depression, abduct and internally rotate her arm (Fig. 19.136).
    4. Ask her to keep her arm in place, then reposition your left hand to stabilize her elbow (Fig. 19.137).
    5. Hold her distal forearm with your right hand.
    6. Ask her to maintain her arm position. Then attempt to derotate her arm (externally rotate her arm).

Remarks—Since the muscle test involves 90 degrees of glenohumeral abduction, it is good practice to evaluate and treat abduction first.

If the test yields weakness, compare the result with a test in which the patient's humerus is less internally rotated.

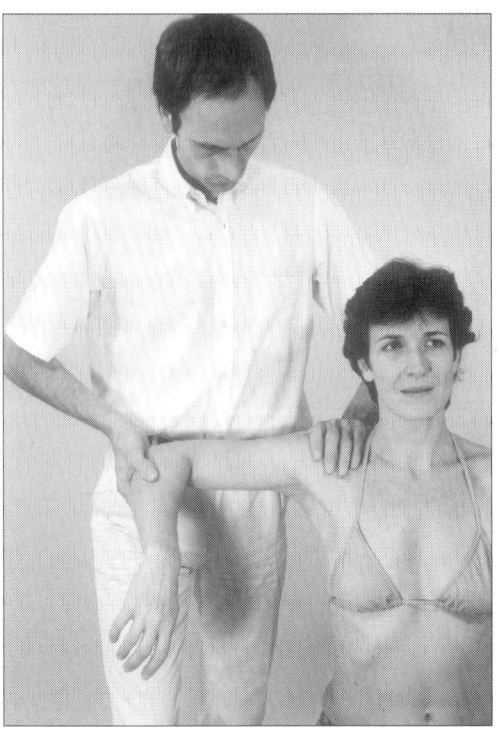

Fig. 19.136

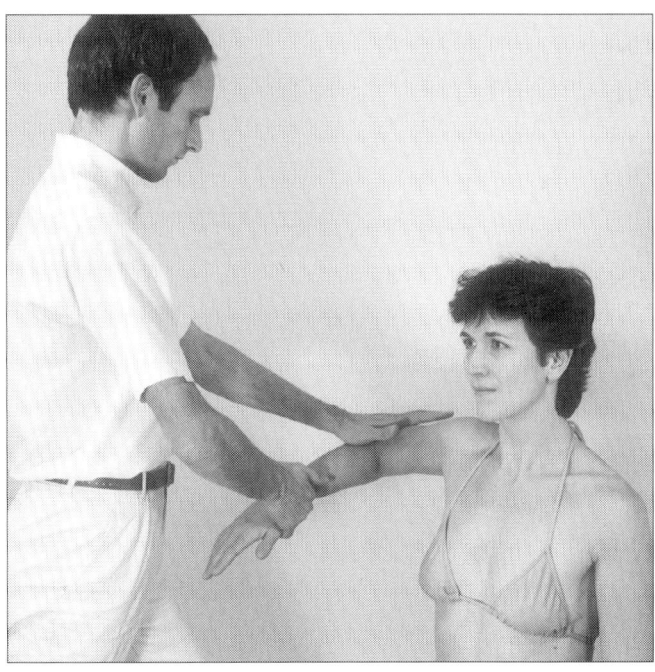

Fig. 19.137

- *Active mobilization* (Fig. 19.138)

    1. Have the patient be seated with her right arm abducted to 90 degrees and internally rotated.

    2. Stand at her right side, supporting her elbow with your left hand.

    3. The web between the thumb and index finger of your right hand should rest on her proximal humerus.

    4. Ask her to gently pull her elbow down and backward (toward her right buttock) while you provide unyielding resistance.

    5. Retest.

    Remarks—During abduction, the humeral head translates caudally at the shoulder joint, and during internal rotation, it translates posteriorly. Translation of the humeral head during the combined abduction–internal rotation movement is in a caudal and posterior direction. The muscles that translate the humeral head in this direction are the latissimus dorsi and teres major.

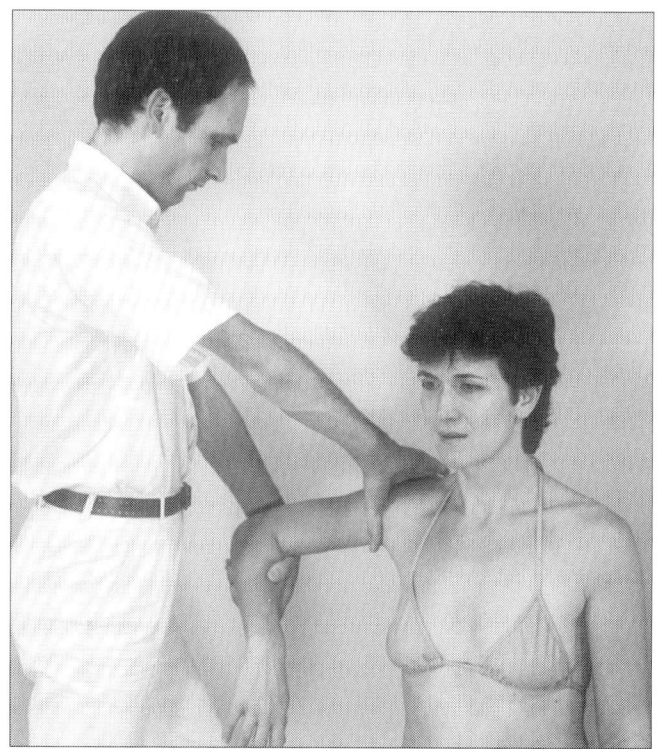

Fig. 19.138

Elbow and Forearm

Flexion and extension of the elbow are movements of the concave ulna and radius on the convex humerus. The contact point of the ulna with the humerus differs from the contact point of the radius with the humerus by about 90 degrees. For instance, with the elbow flexed to 90 degrees, the radius contacts the humerus anteriorly, whereas the ulna contacts the humerus predominately distally.

Both the ulna and the humerus flex and extend at the elbow around a single axis.[10] The axis runs diagonally from medial to lateral and slightly proximal. The diagonal position of the axis for flexion and extension is responsible for the elbow's "carrying angle." The carrying angle is the valgus position of the elbow when extended. Thus, the forearm being laterally in extension is not the result of a changing axis of the elbow joint. To improve elbow extension, it does not help to improve its lateral movement. Nor does stretching the medial collateral ligament improve extension.[16] To improve terminal elbow extension, the radius has to improve its translation dorsally and/or the ulna has to improve its translation proximally on the humerus.

Pronation and supination are movements whereby the radius swings around a stationary ulna. Proximally, the radius is convex on a concave ulna. At the proximal radioulnar joint, supination therefore consists of posterior rolling of the radial head with anterior translation. Distally, the radius is concave on a convex ulna. Distally, supination is a lateral and dorsal movement of the radius on the ulna, with a translation of the radius in the same direction. The reverse occurs during pronation.

- *Capsular pattern*
    - Elbow joint: Flexion is most limited; extension is less limited. In severe cases, pronation and supination are painful.
    - Proximal and distal radioulnar joints: Pain at the end of pronation and supination.

ELBOW FLEXION

- *Agonists*

    Elbow flexors:
    biceps brachii (musculocutaneous nerve; C5,6)
    brachialis (musculocutaneous nerve; C5,6, and the radial nerve; C7)
    brachioradialis (radial nerve; C5,6,7)

- *Muscle test* (Fig. 19.139)

    1. Have the patient be seated with her right elbow fully bent and her forearm in neutral pronation–supination.

    2. Sit in front of her right arm, fixating the anterior aspect of her shoulder with your left hand and holding her distal forearm with your right hand.

    3. Ask her to hold her arm in place. Then attempt to extend her elbow.

    Remarks—This test can be done with the patient's forearm in pronation (emphasis on the brachialis), in supination (emphasis on the biceps), and in a neutral position (emphasis on the brachioradialis). However, if the test is performed in pronation or supination, weakness might also originate from a pronation or a supination dysfunction.

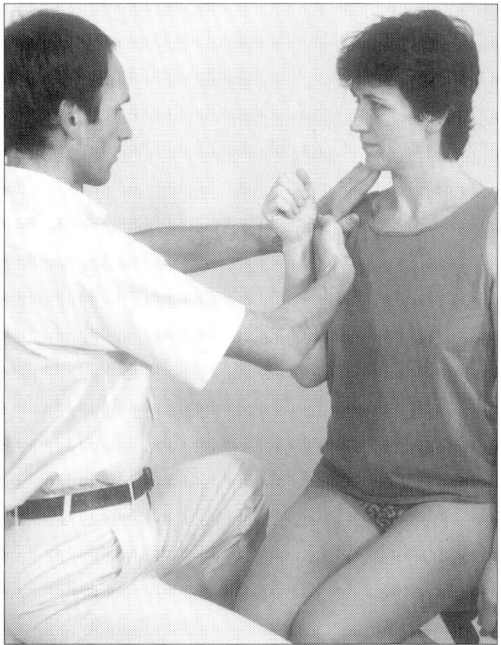

Fig. 19.139

Test pronation and supination to check this. If the test yields weakness, compare the result with a test in which the patient's elbow is less flexed.

The choice of treatment position (pronation, supination, or a neutral position between the two) is based on which position yielded the most weakness. Treatment is performed in the position of maximal weakness.

- **Active mobilization (biceps)** (Fig. 19.139)—The treatment is identical to the testing procedure in supination, except the treatment contraction is gentler and sustained. Retest after the mobilization.

Remarks—At 90 degrees of flexion, the biceps contraction causes a proximal translation of the radial head on the humerus. Near maximal elbow flexion, the biceps contraction causes a distraction of the radiohumeral joint. Both are valuable in mobilizing radiohumeral flexion.

This active mobilization can also be done by the patient as a home exercise. Instruct the patient to hold a weight in her right hand, supinate her forearm, and bend her elbow to the end of her active range (Fig. 19.140), then maintain this position.

- **Active mobilization (brachialis)** (Fig. 19.141)—The treatment is identical to the testing procedure in pronation, except the treatment contraction is gentler and sustained. Retest after the mobilization.

Remarks—The brachialis muscle attaches to the ulna. The contraction of the brachialis is effective when the flexion restriction is located at the ulnohumeral joint and the patient has more then 130 degrees of flexion available.

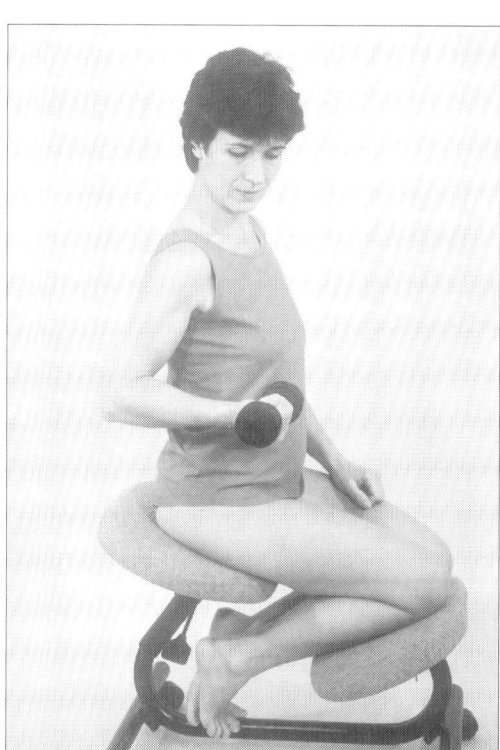

Fig. 19.140

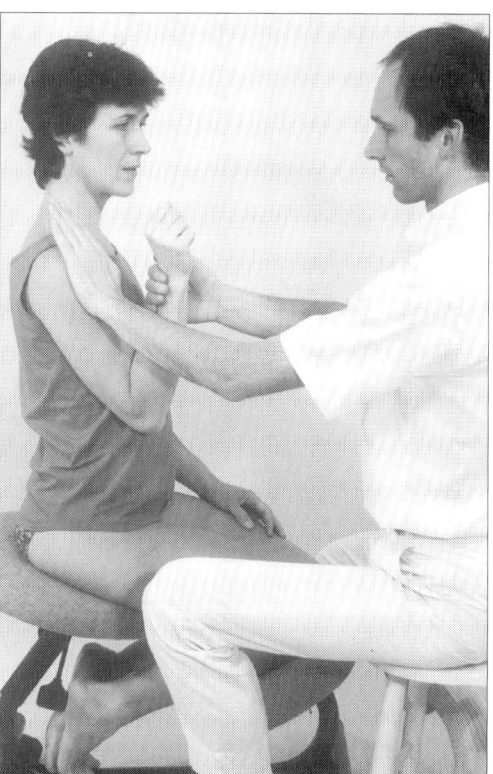

Fig. 19.141

- **Passive mobilization** (Fig. 19.142)

    1. Have the patient be seated.

    2. Sit in front of her right arm, holding her wrist with your right hand, and bend her relaxed right elbow.

    3. Hold her elbow with your left hand, placing your thumb on the dorsal surface of the radial head.

    4. With your left thumb, mobilize her radial head in a dorsal and proximal direction on her humerus.

    5. Retest.

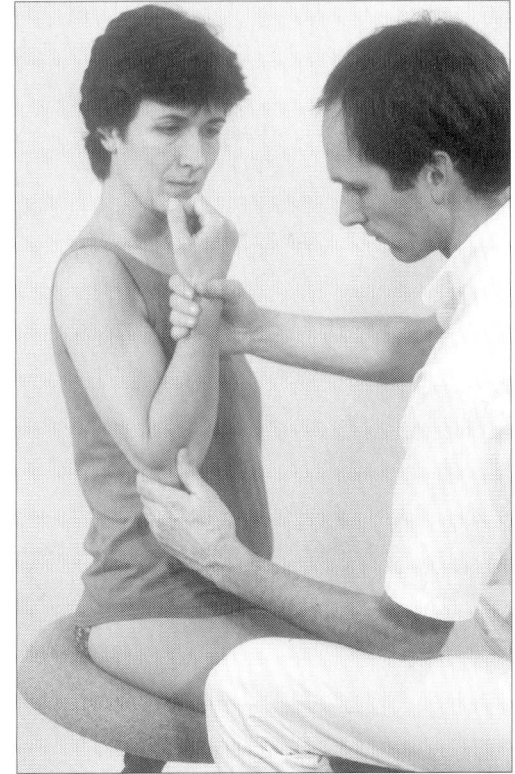

Fig. 19.142

ELBOW EXTENSION

- **Agonist**

  Elbow extensors:
  triceps (radial nerve; C6,7,8)

- **Muscle test** (Fig. 19.143)

  1. Have the patient be seated with her shoulder flexed to 70 degrees, her right elbow fully extended, and her forearm in a neutral pronation–supination position.

  2. Stand in front of her, facing her right arm, and place your stabilizing right hand on the anterior surface of her proximal forearm.

  3. Hold her distal forearm with your left hand.

  4. Ask her to maintain the elbow extension. Then attempt to bend her elbow.

  Remarks—If the test yields weakness, compare the result with a test in which the patient's elbow is less extended.

- **Active mobilization** (Fig. 19.143)—The treatment is identical to the testing procedure, except the treatment contraction is gentler and sustained. The active mobilization is useful for minimal extension dysfunctions. Retest after the mobilization.

- **Passive mobilization** (Fig. 19.144)

  1. Have the patient lie supine with her right shoulder externally rotated and abducted such that her olecranon is just off the table. Her elbow should be extended. A towel roll or sandbag can be placed beneath her distal humerus.

  2. Stand on her right side, medial to her right hand, with your stabilizing left hand holding her wrist, controlling the amount of elbow extension.

  3. Place your right hand on her proximal radius and ulna and mobilize them posteriorly.

  4. Retest.

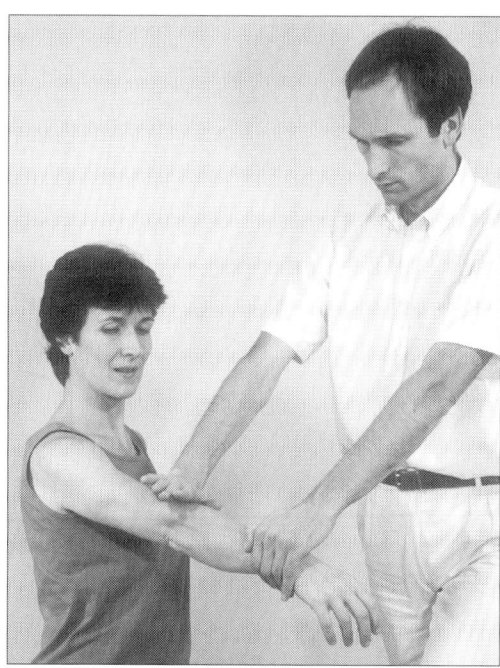

Fig. 19.143

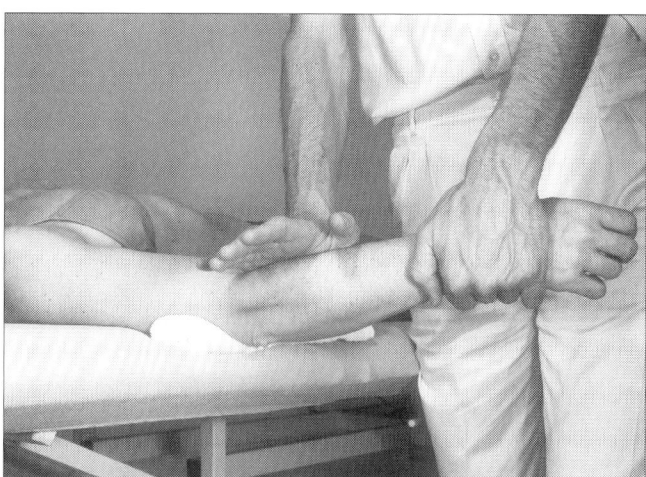

Fig. 19.144

SUPINATION

- **Agonists**

    Supinators:
    supinator (posterior interosseus nerve, deriving from the radial nerve; C5,6)
    biceps (musculocutaneous; C5,6)
    triceps (radial nerve; C6,7,8)

    Remarks—The muscles that provide supination are the supinator and the biceps muscles; however, the biceps also flexes the elbow. The triceps counteracts the elbow flexion of the biceps.

- **Muscle test** (Fig. 19.145)

    1. Have the patient be seated with her right elbow flexed to 90 degrees and her forearm supinated.
    2. Hold her distal forearm such that your hand makes firm contact with the distal radius and the distal ulna.
    3. Cover your fingers with your other hand to reinforce your grip.
    4. Ask her to keep her hand in place. Then attempt to pronate her forearm.

    Remarks—If the test yields weakness, compare the result with a test in which the patient's forearm is less supinated.

    An arthrogenic inhibition of the supinators may result from inhibition from the proximal or distal radioulnar joint. With the forearm supinated, testing the contraction quality of her biceps could emphasize the proximal radioulnar joint (see below). At 90 degrees of elbow flexion, a contraction of the biceps produces an anterior translation of the radial head on the ulna. This is the translation component of supination at this joint. Arthrogenic inhibition of the supinators and the biceps suggests the need for mobilization of the proximal radioulnar joint. Arthrogenic inhibition of the supinators with a strong biceps suggests the need for mobilization of the distal radioulnar joint.

- **Agonist**

    Biceps:
    biceps (musculocutaneous; C5,6)

- **Muscle test** (Fig. 19.146)

    1. Have the patient be seated with her right elbow flexed to 90 degrees and her forearm supinated.
    2. Sit to the right of her arm, stabilizing her elbow with your left hand.
    3. Hold her distal radius and ulna with your right hand.
    4. Ask her to keep her arm in place. Then attempt to extend her elbow.

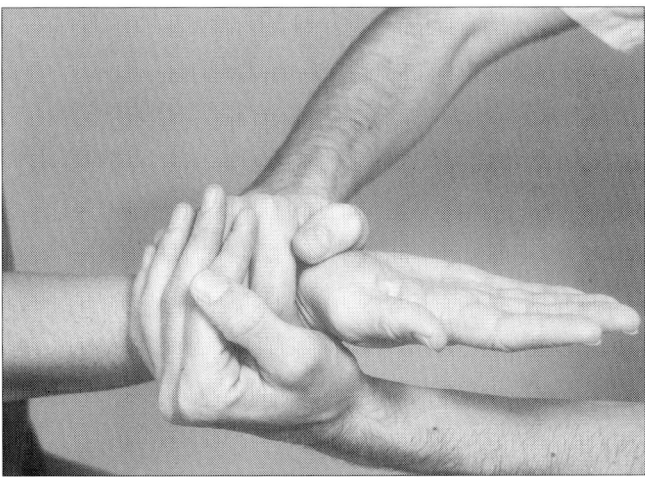

Fig. 19.145

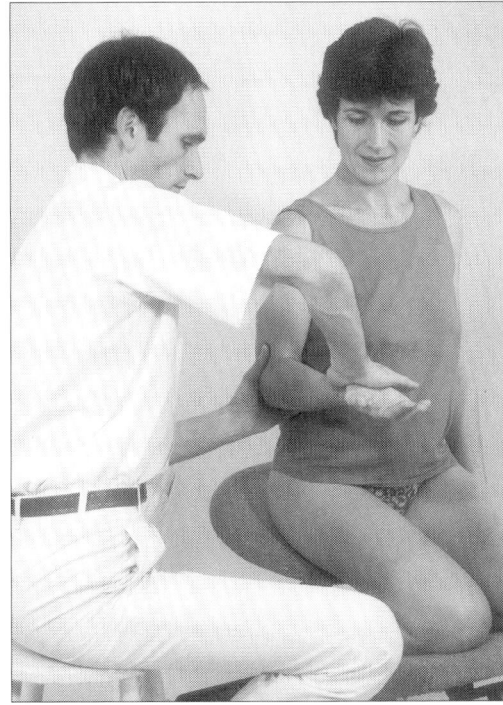

Fig. 19.146

Remarks—If this test yields weakness, compare the result with a test in which the patient's forearm is less supinated.

Arthrogenic inhibition of the biceps when tested in supination suggests the need for a supination mobilization at the proximal radioulnar joint.

- ***Passive mobilization (distal radioulnar joint)*** (Fig. 19.147)

  1. Have the patient be seated with her right elbow flexed to 90 degrees and her forearm supinated.
  2. Sit in front of her with your right hand holding the dorsum of her distal ulna.
  3. Hold her distal radius with your left hand, placing your thumb anteriorly on her wrist and as close to the joint line as possible.
  4. Mobilize her radius dorsally.
  5. Retest.

  Remarks—While mobilizing the distal radioulnar joint, make sure that the mobilization results in a translation (gliding) of the joint only, without increasing the supination movement (angle of the bones). After the mobilization, the supination range and strength can be checked.

- ***Active mobilization (proximal radioulnar joint)*** (Fig. 19.148)

  1. Have the patient be seated with her right elbow flexed to 90 degrees and her forearm supinated.
  2. Hold her distal forearm and ask her to gently attempt to bend her elbow.
  3. Provide unyielding resistance.
  4. Retest.

  Remarks—The active mobilization can also be prescribed as a home exercise. Instruct the patient to hold a weight in her hand, supinate her forearm, and bend her elbow 90 degrees, then maintain this position.

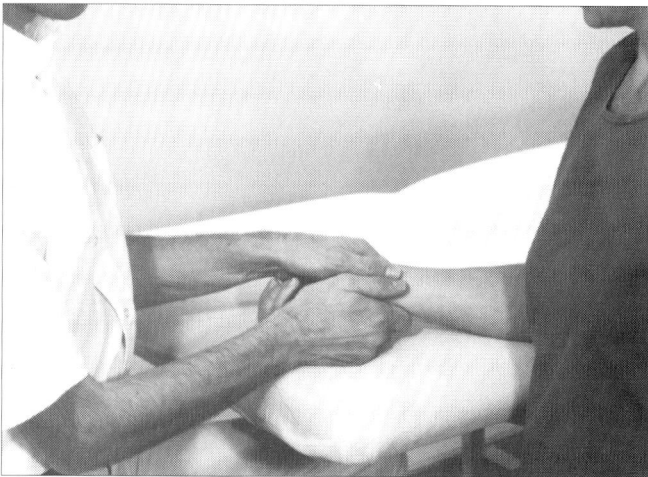

Fig. 19.147

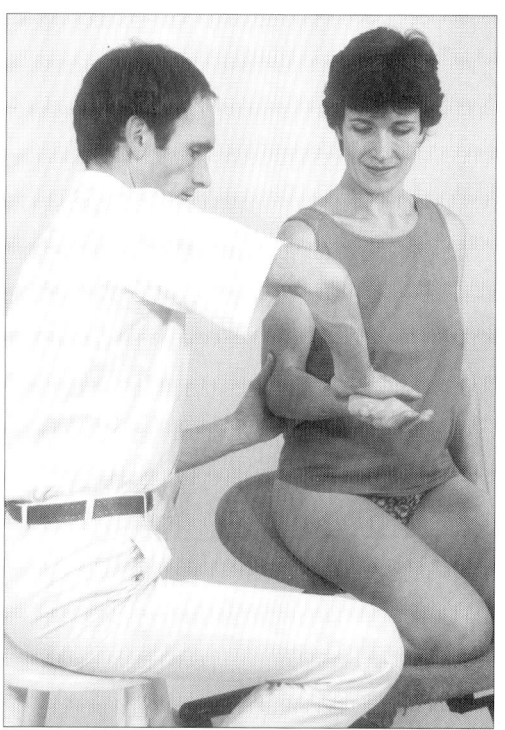

Fig. 19.148

PRONATION

- **Agonists**

    Pronators:
    pronator teres (median nerve; C6,7)
    pronator quadratus (median nerve; C8,T1)

- **Muscle test** (Fig 19.149)

    1. Have the patient be seated with her elbow flexed to 90 degrees and her forearm pronated.
    2. Hold her distal forearm such that your hand is making firm contact with her distal radius and distal ulna.
    3. Cover your fingers with your other hand to reinforce your grip.
    4. Ask her to keep her arm in place. Then attempt to supinate her forearm.

    Remarks—If the test yields weakness, compare the result with a test in which the patient's forearm is less pronated.

- **Passive mobilization (proximal radioulnar joint)** (Fig 19.150)

    1. Have the patient be seated with her elbow and forearm resting on the treatment table. Her forearm should be pronated.
    2. Stand in front of her and, with your left hand, hold her distal forearm in pronation.
    3. Place your right hand anteriorly on her elbow.
    4. Mobilize her proximal radius dorsally on the ulna.
    5. Retest.

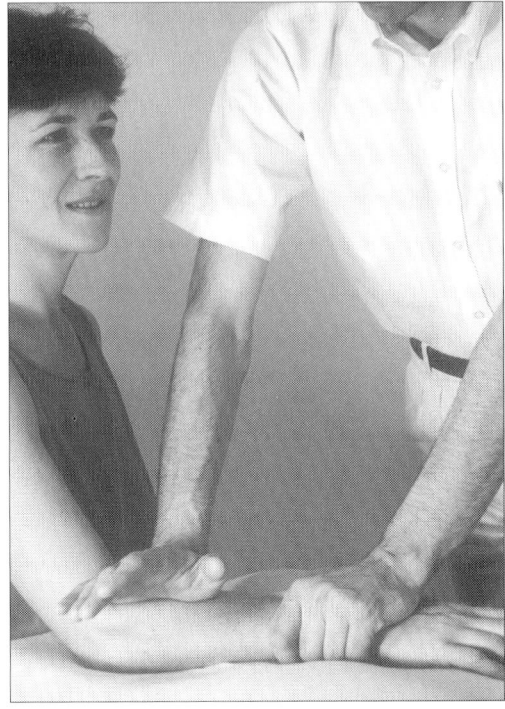

Fig. 19.150

- **Passive mobilization (distal radioulnar joint)** (Fig 19.151)

    1. Have the patient be seated with her elbow flexed and the forearm pronated.
    2. Sit in front of her with your fixating left hand holding her distal ulna anteriorly.
    3. With your mobilizing right hand, hold her distal radius as close to the joint line as possible and mobilize her radius anteriorly.
    4. Retest.

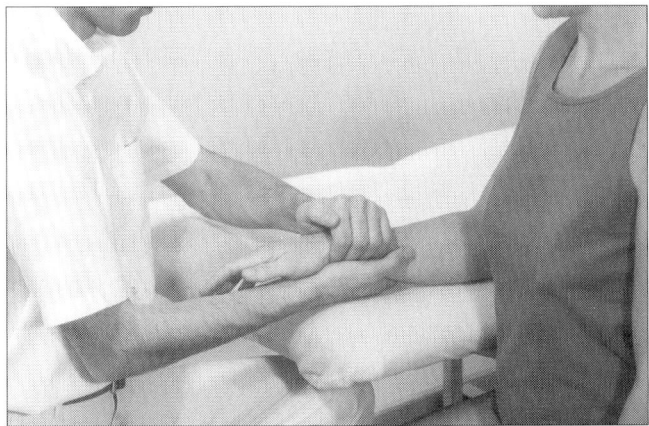

Fig. 19.149

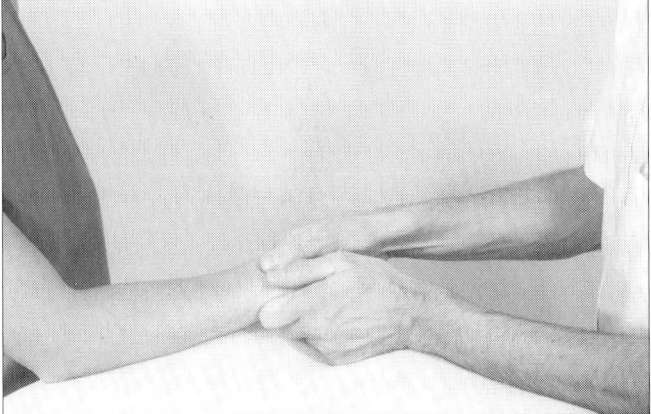

Fig. 19.151

Wrist

Although the mobility of one carpal on another can be restricted and the restriction can cause inhibition of muscles, the individual carpal bones will not be discussed here. Those interested in the evaluation and treatment of the individual carpals are referred to other texts.[25,26,28] As a unit, the carpals function as a convex structure on the distal radius and ulna. Translation of the carpals during wrist movements is always opposite the direction of hand movement. For instance, when the wrist is in palmar flexion, the carpals translate dorsally.

Four movement directions (palmar flexion, dorsiflexion, abduction, and adduction) will be presented; however, if two movement dysfunctions are present, both movements can be combined, provided they are not antagonistic movements. For instance, a combination of limited and weak palmar flexion and ulnar deviation can be evaluated effectively by testing the flexor carpi ulnaris, which flexes and ulnarly deviates the wrist. The dysfunction in both directions can be treated at the same time by a mobilization combining both movements.

- **Capsular pattern**—Palmar flexion and dorsiflexion are equally limited. Ulnar and radial deviation are free.

WRIST DORSIFLEXION

- **Agonists**

    Wrist dorsiflexors:
    extensor carpi radialis longus (radial nerve; C6,7)
    extensor carpi radialis brevis (radial nerve; C7,8)
    extensor carpi ulnaris (posterior interosseous nerve [branch of the radial nerve]; C7,8)

- **Muscle test** (Fig. 19.151)

    1. Have the patient be seated with her elbow flexed and her forearm in a neutral pronation–supination position. Her wrist should be dorsiflexed and her fingers either relaxed or flexed.
    2. Sit in front of her, stabilizing her distal forearm with your right hand.
    3. Place your left hand dorsally on her distal metacarpals.
    4. Ask her to keep her hand in place. Then attempt to palmar flex her wrist.

    Remarks—If this test yields weakness, compare the result with a test in which the patient's wrist is less dorsiflexed.

    To make the muscle test more specific, you can evaluate extension strength over each carpometacarpal joint (CMC), II through V. With your right hand stabilizing the distal forearm, use your left index finger to assess extension strength over the second CMC by pressing dorsally on the distal metacarpal II (Fig. 19.153). Also assess extension strength over the third through fifth metacarpals.

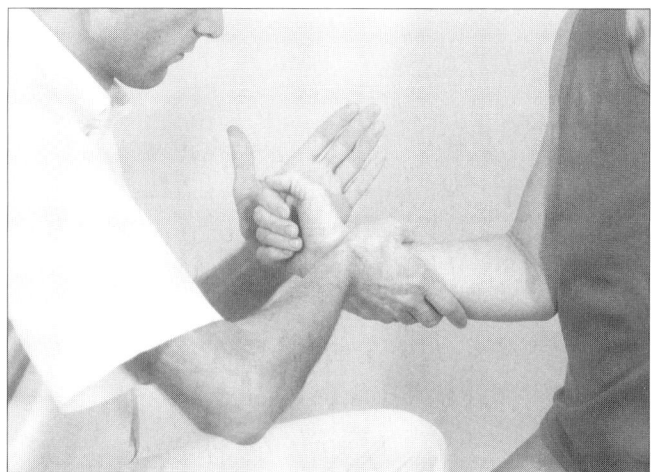

Fig. 19.152

Fig. 19.153

MANUAL THERAPY: IMPROVE MUSCLE AND JOINT FUNCTIONING

- **Passive mobilization** (Fig. 19.154)

  1. Have the patient be seated with her elbow flexed and her forearm in a neutral pronation–supination position. Her wrist should be dorsiflexed with her fingers relaxed.

  2. Facing the patient, stabilize her distal forearm with your right hand.

  3. Place the web between the thumb and index finger of your left hand over the dorsal aspect of her carpals and, with your ring and little fingers, keep her wrist dorsiflexed.

  4. Mobilize her carpals in an anterior direction.

  5. Retest.

- **Compression manipulation** (Fig. 19.155)

  1. With the patient lying supine, stand at her right side, distal to her hand. Hold her wrist between the hypothenar eminences of your hands. Interlace your fingers, with the fingers of your left hand on top.

  2. Ask the patient to relax her arm as you compress her carpals with your left and right hypothenar eminances.

  3. While maintaining the compression, dorsiflex her wrist.

  4. Retest.

- **Active mobilization** (Fig. 19.152)—The active mobilization is identical to the muscle test, except the treatment contraction is gentler and sustained. If extension over a specific CMC tests weak, resistance is given to the related metacarpal. Retest after the mobilization.

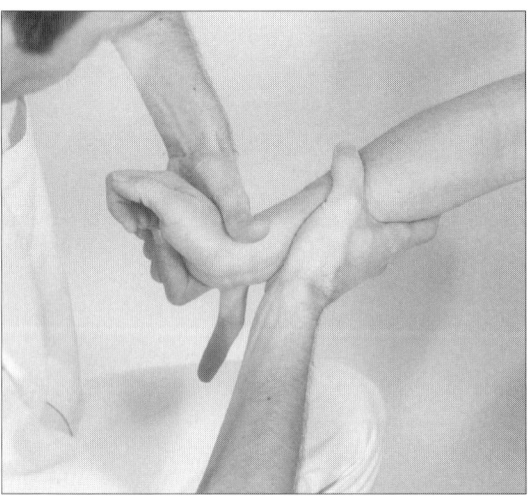

Fig. 19.154

Fig. 19.155

WRIST PALMAR FLEXION

- *Agonists*

    Wrist palmar flexors:
    flexor carpi radialis (median nerve; C6,7)
    palmaris longus (median nerve; C7,8)
    flexor carpi ulnaris (ulnar nerve; C7,8)

- *Muscle test* (Fig. 19.156)

    1. Have the patient be seated with her elbow flexed and her forearm in a neutral pronation–supination position. Her wrist should be palmar flexed, with her fingers relaxed or extended.
    2. Sit in front of her with your left hand stabilizing her distal forearm.
    3. Place your right hand on the palmar side of her distal metacarpals.
    4. Ask her to maintain this hand position. Then attempt to dorsiflex her wrist.

    Remarks—If the test yields weakness, compare the result with a test in which the patient's wrist is less palmar flexed.

    To make the muscle test more specific, you can evaluate flexion strength over each CMC joint (II through V) individually. With your left hand stabilizing the distal forearm, use your right index finger to assess flexion strength over the second CMC by pressing on the palmar aspect of the distal second metacarpal (Fig. 19.157). Also assess flexion strength over the third through fifth metacarpals.

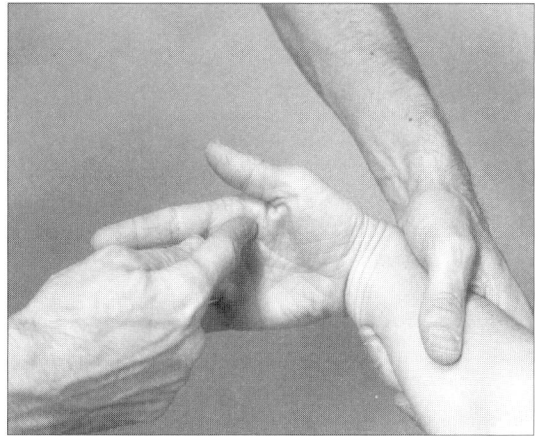

Fig. 19.157

- *Passive mobilization* (Fig. 19.158)

    1. Have the patient be seated with her elbow flexed and her forearm in a neutral pronation–supination position. Her wrist should be palmar flexed with her fingers relaxed.
    2. Facing the patient, stabilize her distal forearm with your left hand.
    3. Place the web between the thumb and index finger of your right hand over the anterior side of her carpals and, with your ring and little fingers, keep her wrist palmar flexed.
    4. Mobilize the carpals in a dorsal direction.
    5. Retest.

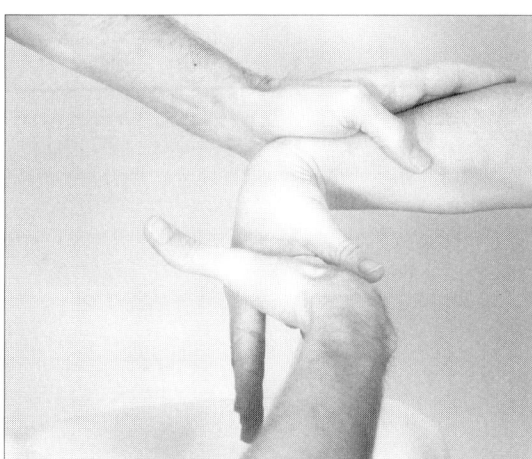

Fig. 19.156

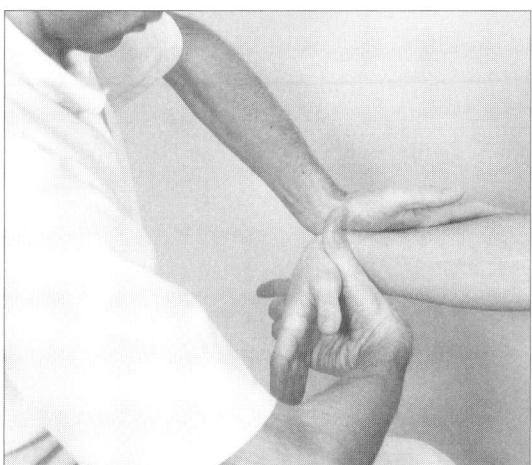

Fig. 19.158

- **_Active mobilization_** (Fig. 19.156)—The treatment is identical to the testing procedure, except the treatment contraction is gentler and sustained. If a specific CMC tested weak, resistance is given to the related metacarpal. Retest after the mobilization.

- **_Compression manipulation_** (Fig. 19.159)

  1. With the patient lying supine, stand at her right side, distal to her hand. Hold her wrist between the hypothenar eminences of your hands. Interlace your fingers, with the fingers of your right hand on top.

  2. Ask the patient to keep her arm relaxed as you compress her carpals with your left and right hypothenar eminances.

  3. While maintaining the compression, palmar flex her wrist.

  4. Retest.

Fig. 19.159

WRIST RADIAL DEVIATION

- **Agonists**

    Wrist radial deviators:
    flexor carpi radialis (median nerve; C6,7)
    extensor carpi radialis longus (radial nerve; C6,7)
    extensor carpi radialis brevis (posterior interosseous nerve [branch of the radial nerve]; C7,8)

- **Muscle test** (Fig. 19.160)

    1. Have the patient be seated with her elbow flexed and her forearm in a neutral pronation–supination position. Her wrist should be radially deviated.
    2. Sit to the right of her hand, stabilizing her distal ulna with your left hand.
    3. With your right hand, hold her hand such that your thumb is on the radial side of her distal second metacarpal and your right index finger is on the proximal fifth metacarpal or the ulnar carpalia.
    4. Ask her to maintain her hand position. Then attempt to ulnarly deviate her wrist.

    Remarks—If the test yields weakness, compare the result with a test in which the patient's wrist is less radially deviated.

- **Passive mobilization** (Fig. 19.161)

    1. Have the patient be seated with her elbow flexed and forearm in a neutral pronation–supination position. Her wrist should be radially deviated.
    2. Sit to the right of her hand, stabilizing her distal ulna with your left hand.
    3. Place the web between the thumb and index finger of your right hand on the radial side of her carpals and, with your ring and little fingers, hold her hand in radial deviation.
    4. With the web between the thumb and index finger of your left hand, mobilize her carpals in an ulnar direction.
    5. Retest.

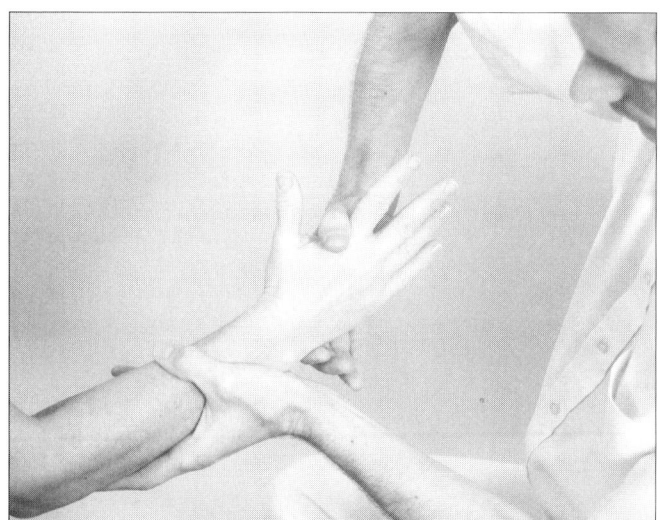

Fig. 19.160

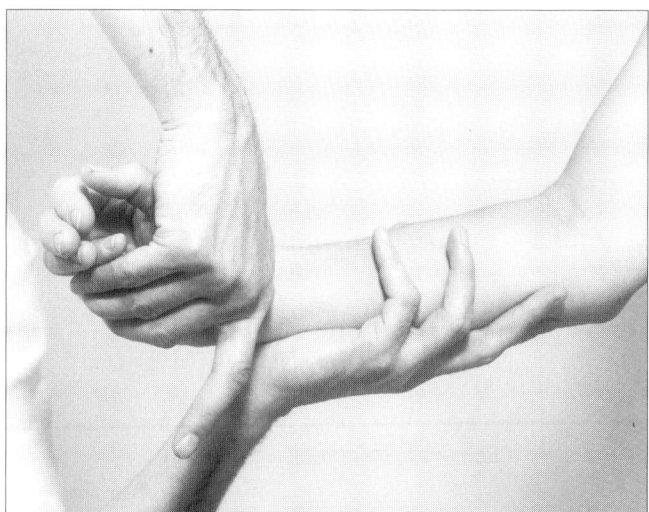

Fig. 19.161

- **Compression manipulation** (Fig. 19.162)

    1. With the patient lying supine, stand at her right side, distal to her hand. Hold her carpals between the hypothenar eminences of your hands and interlace your fingers.

    2. Ask her to keep her arm relaxed as you compress her carpals with your hypothenar eminences.

    3. While maintaining the compression, radially deviate her wrist.

    4. Retest.

Fig. 19.162

WRIST ULNAR DEVIATION

- **Agonists**

  Wrist ulnar deviators:
  flexor carpi ulnaris (ulnar nerve; C7,8)
  extensor carpi ulnaris (posterior interosseus nerve (branch of the radial nerve); C7,8)

- **Muscle test** (Fig. 19.163)

  1. Have the patient be seated with her elbow flexed and her forearm in a neutral pronation–supination position. Her wrist should be in ulnar deviation.
  2. Sit to the right of her hand with your left hand stabilizing her distal radius.
  3. With your right hand, hold her hand such that your middle finger is on the ulnar side of her distal fifth metacarpal and your right thumb is on the proximal part of her second metacarpal.
  4. Ask her to maintain her hand position. Then attempt to radially deviate her wrist.

  Remarks — If the test yields weakness, compare the result with a test in which the patient's wrist is in less ulnar deviation.

- **Passive mobilization** (Fig 19.164)

  1. Have the patient be seated with her elbow flexed and her forearm slightly supinated. Her wrist should be in ulnar deviation.
  2. Sit in front of her with your left hand stabilizing her distal radius.
  3. Place the web between the thumb and index finger of your right hand on the ulnar side of her carpals and, with your ring and little fingers, keep her hand in ulnar deviation.
  4. Mobilize her carpals in a radial direction.
  5. Retest.

- **Compression manipulation** (Fig. 19.165)

  1. With the patient lying supine, stand at her right side, distal to her hand. Hold her wrist between the hypothenar eminences of your hands and interlace your fingers.
  2. Ask her to keep her arm relaxed, then compress her carpals with your hypothenar eminences.
  3. While maintaining the compression, deviate her wrist in an ulnar direction.
  4. Retest.

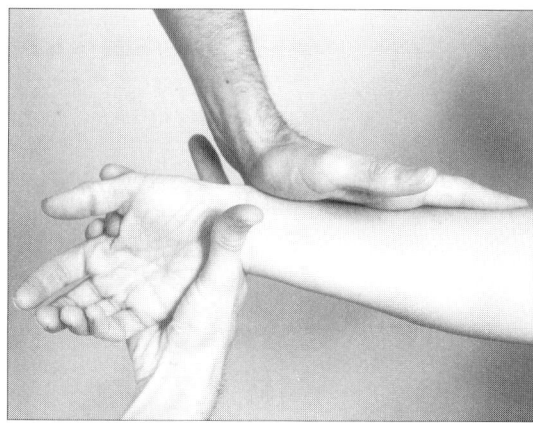

Fig. 19.164

Fig. 19.163

Fig. 19.165

Fingers

Each finger joint consists of a convex proximal bone and a distal concave partner. Translation of the distal bone is in the same direction as the movement.

For evaluation of flexion and extension of the finger joints, the joints are tested in two positions. The first is the "lumbrical" or the "intrinsic–plus" position. This finger position consists of flexion of the MCP joints with extension of the interphalangeal joints. The second is the "claw" position, in which the MCP joints are extended and the IP joints are flexed.

- *Capsular pattern*—Flexion is most restricted. Extension is less restricted.

METACARPOPHALANGEAL (MCP) II–V FLEXION

- *Agonists*

    MCP flexors:
    lumbricals (ulnar two parts of the lumbricals are innervated by the ulnar nerve; C8,T1; medial two parts of the lumbricals are innervated by the median nerve; C8,T1)
    interosseus dorsalis (ulnar nerve; C8,T1)
    interosseus palmaris (ulnar nerve; C8,T1)

- *Muscle test* (Fig 19.166)

    1. Have the patient be seated with her elbow flexed and her forearm in a neutral pronation–supination position. Her wrist should be dorsiflexed, with her fingers in the lumbrical position (MCP flexion, IP extension).

    2. Sit to the right of her hand with your left thumb stabilizing her distal metacarpal II on its dorsal side.

    3. Place your right index finger palmarly on the distal aspect of her proximal phalanx.

    4. Ask her to keep her finger in place. Then attempt to extend her MCP II joint.

    5. Repeat this test with the other MCP joints, comparing the MCP flexor strength of the four fingers.

    Remarks—If any of the four tests yields weakness, compare the result with a test in which the patient's MCP joint is less flexed.

- *Active mobilization* (Fig. 19.166)—The treatment is identical to the testing procedure, except the treatment contraction is gentler and sustained. Retest after the mobilization.

- *Passive mobilization* (Fig. 19.167)

    1. Have the patient be seated with her fingers relaxed.

    2. Sit to the right of her hand. Stabilize her metacarpal with your left hand.

    3. Hold her proximal phalanx with the fingers of your right hand.

    4. Flex her MCP joint and then translate her proximal phalanx in a palmar and proximal direction on her metacarpal.

    5. Retest.

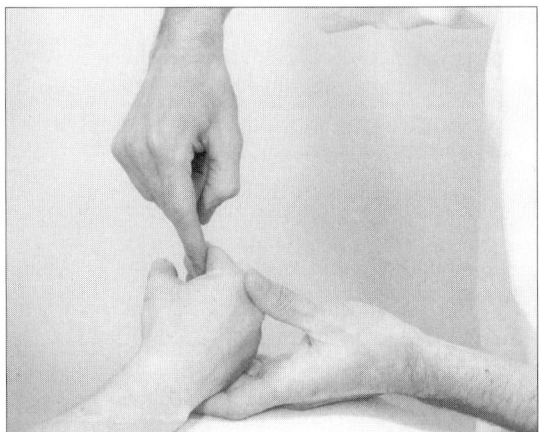

Fig. 19.166

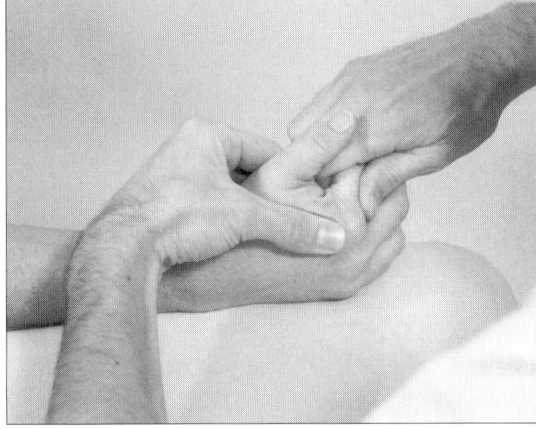

Fig. 19.167

METACARPOPHALANGEAL II–V EXTENSION

- **Agonists**

    MCP extensors:
    extensor digitorum (posterior interosseus nerve [branch of the radial nerve]; C7,8)
    extensor indicis (posterior interosseus nerve [branch of the radial nerve]; C7,8)
    extensor digiti minimi (posterior interosseus nerve [branch of the radial nerve]; C7,8)

- **Muscle test** (Fig. 19.168)

    1. Have the patient be seated with her elbow flexed and her forearm and wrist in a neutral position. Her fingers should be in the claw position (MCP extension, IP flexion).
    2. Sit to the right of her hand, stabilizing the palmar side of her distal metacarpal II with a fingertip of your left hand.
    3. Place your right thumb dorsally on the distal part of her proximal phalanx.
    4. Ask her to keep her finger in place. Then attempt to flex her MCP II joint.
    5. Repeat this test on her other MCP joints, comparing their strength.

    Remarks—If any of the four tests yields weakness, compare the result with a test in which the patient's MCP joint is less extended.

- **Passive mobilization** (Fig 19.169)

    1. Have the patient be seated with her fingers relaxed.
    2. Sit on the palmar side of her hand, stabilizing the distal part of her metacarpal with your right hand.
    3. With the fingers of your left hand, hold her proximal phalanx and extend her MCP joint.
    4. Mobilize her proximal phalanx in a dorsal direction.
    5. Retest.

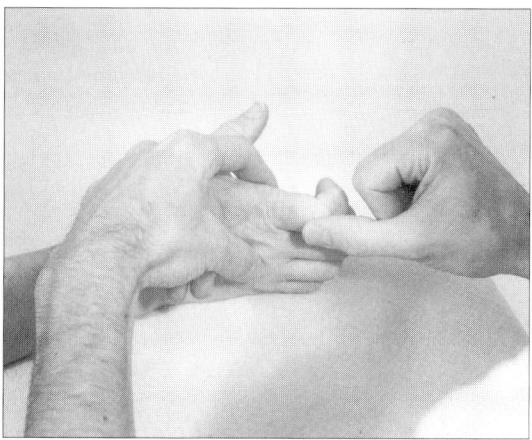

Fig. 19.168

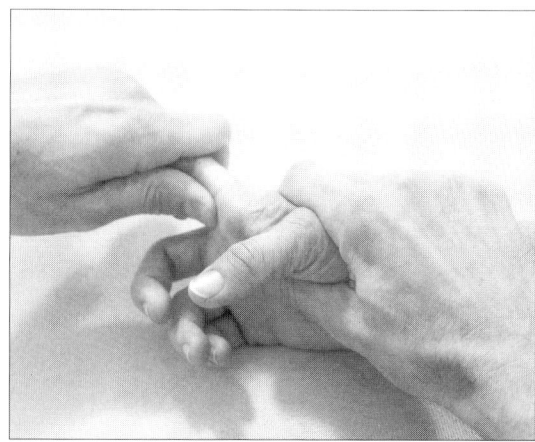

Fig. 19.169

METACARPOPHALANGEAL II–V ABDUCTION

Abduction of the fingers is the movement of the fingers away from an imaginary axis drawn longitudinally through the center of the middle finger. The dorsal interossei abduct by moving the fingers away from this line. Interestingly, both the lateral and medial movements of the middle finger are called abduction.

Similar to the metacarpals being convex for MCP flexion and extension, the metacarpals are convex for MCP abduction and adduction. The proximal phalanx is concave, and translation of the phalanx is in the same direction as the phalanx movement.

- *Agonists*

    MCP abductors:
    dorsal interossei (ulnar nerve, deep branch; C8, T1)
    abductor digiti minimi (ulnar nerve, deep branch; C8, T1)

- *Muscle test* (Fig 19.170)

    1. Have the patient be seated with her elbow flexed and her forearm and wrist in a neutral position. Her MCP joints should be in a loose-packed position for flexion–extension.

    2. With her hand resting in your left hand, stabilize her metacarpal II with your fingertips.

    3. Have her abduct her fingers with her IP joints in a neutral position, then place your right index finger on the radial side of her PIP joint.

    4. Ask her to keep her finger in place. Then attempt to adduct her MCP joint.

    5. Repeat this test on the other fingers. Abduction of the middle finger is tested after the finger has moved in a radial and an ulnar direction.

    Remarks—If any of the tests yields weakness, compare the result with a test in which the patient's MCP joint is less abducted.

- *Passive mobilization* (Fig. 19.171)

    1. Have the patient be seated with her fingers relaxed.

    2. Stabilize her metacarpals with one hand.

    3. With the fingers of your other hand, hold her proximal phalanx and abduct her MCP joint.

    4. Mobilize her proximal phalanx in the direction of the abduction movement.

    5. Retest.

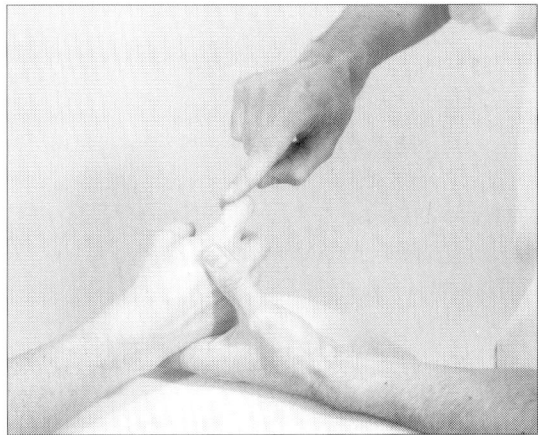

Fig. 19.170

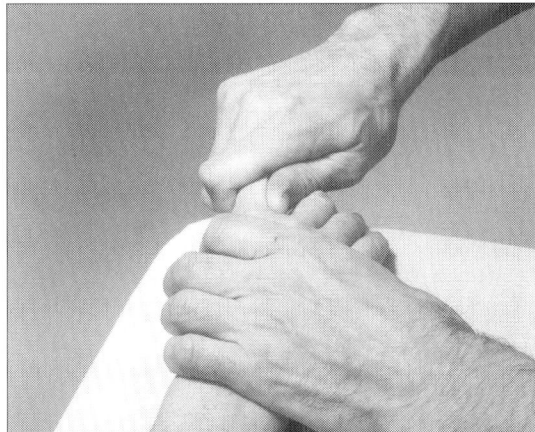

Fig. 19.171

METACARPOPHALANGEAL II–V ADDUCTION

Metacarpal adduction is the movement of the second, fourth, and fifth proximal phalanx toward the middle finger. Translation is in the same direction.

- **Agonists**

    MCP adductors:
      palmar interossei (ulnar nerve, deep branch; C8, T1)

- **Muscle test** (Fig. 19.172)

    1. Have the patient be seated with her elbow flexed and her forearm and wrist in a neutral position. Her MCP joints should be in a loose-packed position for flexion–extension, and her fingers should be in a neutral position.
    2. Stabilize her ulnar three fingers with your left hand.
    3. Hold the DIP of her index finger with your right hand.
    4. Ask her to keep her finger in place. Then attempt to abduct her MCP II joint.
    5. Repeat the test on MCP IV and V.

    Remarks—Adduction of the index finger often is restricted by the presence of the middle finger. The index finger can adduct farther into the range if it adducts in front of or behind the middle finger. Adduction of the ring and little fingers is similarly tested. For maximal adduction, they, too, can move anterior or posterior to their neighboring finger. In addition, adduction of the little finger against the ring finger prevents testing ring finger adduction. To test ring finger adduction, the little finger must be abducted or extended to be out of the way.

    If any of the tests yields weakness, compare the result with a test in which the patient's MCP joint is less adducted. Weakness throughout the range could be the result of radiculopathy, pseudo-radiculopathy, or a PIP joint dysfunction.

- **Passive mobilization** (Fig. 19.173)

    1. Have the patient be seated with her fingers relaxed.
    2. Sit in front of her, stabilizing her metacarpals with one hand.
    3. With the fingers of your other hand, hold her proximal phalanx and adduct her MCP joint in front of or behind its neighboring finger.
    4. Mobilize her proximal phalanx in the direction of the middle finger.
    5. Retest.

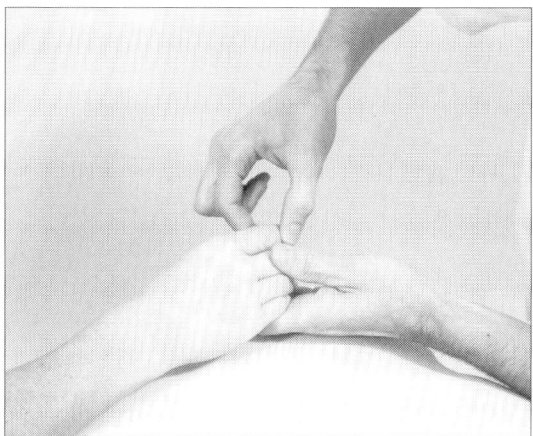

Fig. 19.172

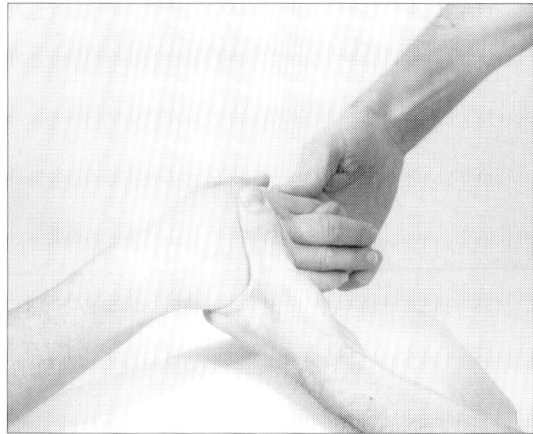

Fig. 19.173

INTERPHALANGEAL (IP) II–V FLEXION

- **Agonists**

    Finger flexors:
    flexor digitorum superficialis (median nerve; C7,8,T1)
    flexor digitorum profundus (median nerve; C8,T1)

- **Muscle test** (Fig. 19.174)

    1. Have the patient be seated with her elbow flexed and her forearm and wrist in a neutral position. She should hold her fingers in the claw position (MCP extension with IP flexion).
    2. Sit to the right of and distal to her hand, stabilizing the distal part of her proximal phalanx with your left thumb.
    3. Place your right index finger on the tip of her finger.
    4. Ask her to keep her finger in place. Then attempt to extend her PIP joint.
    5. Test her other PIP joints in a similar manner.

    Remarks—If any test yields weakness, compare the result with a test in which the patient's PIP joint is less flexed.

    The DIP joints are similarly tested. Stabilize the distal part of the middle phalanx dorsally and, with your testing finger on the tip of the patient's finger, push the DIP joint into extension.

- **Active mobilization** (Fig. 19.174)—The treatment is identical to the testing procedure, except the treatment contraction is gentler and sustained. Retest after the mobilization.

- **Passive mobilization** (Fig. 19.175)

    1. Have the patient be seated with her fingers relaxed.
    2. Stabilize her more proximal phalanx with the fingers of your left hand.
    3. With the fingers of your right hand, hold her more distal phalanx and bring the interphalangeal joint into flexion.
    4. Mobilize the more distal phalanx in a palmar and proximal direction.
    5. Retest.

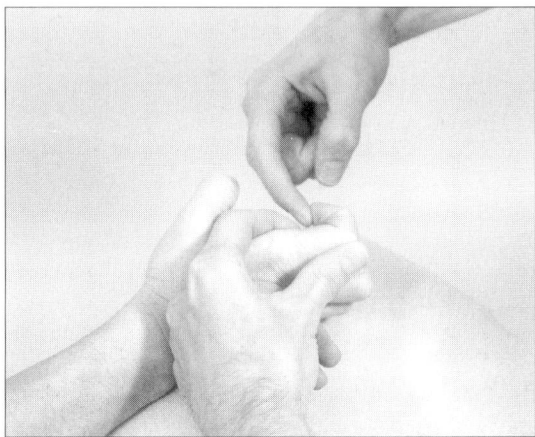

Fig. 19.174

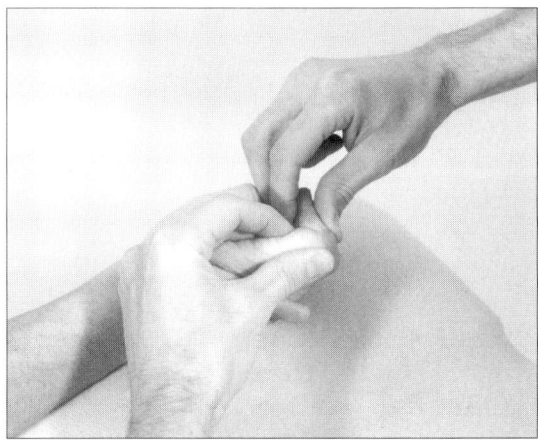

Fig. 19.175

INTERPHALANGEAL II–V EXTENSION

- **Agonists**

    Finger extensors:
    lumbricals (ulnar two lumbricals are
        innervated by the ulnar nerve; C8,T1;
        medial two lumbricals are innervated by
        the median nerve; C8,T1)
    interosseus dorsalis (ulnar nerve; C8,T1)
    interosseus palmaris (ulnar nerve; C8,T1)

- **Muscle test** (Fig. 19.176)

    1. Have the patient be seated with her elbow flexed and her forearm in a neutral position. Her wrist should be slightly dorsiflexed, and her fingers should be in the lumbrical position (MCP flexion, IP extension).

    2. Sit to the right of her hand, stabilizing her more proximal phalanx on its palmar and distal side with one finger of your left hand.

    3. Place your right index finger dorsally and distally on her more distal phalanx.

    4. Ask her to maintain her finger position. Then attempt to flex her IP joint.

    5. Test extension at the other IP joints.

    Remarks—If any of the muscles tests weak, compare the result with a test in which the patient's interphalangeal joint is less extended.

- **Passive mobilization** (Fig. 19.177)

    1. Have the patient be seated with her fingers relaxed.

    2. Sit to the right of her hand, stabilizing her more proximal phalanx with the fingers of your left hand.

    3. Hold her more distal phalanx with the fingers of your right hand.

    4. Passively extend her IP joint.

    5. Mobilize her distal phalanx in a dorsal direction.

    6. Retest.

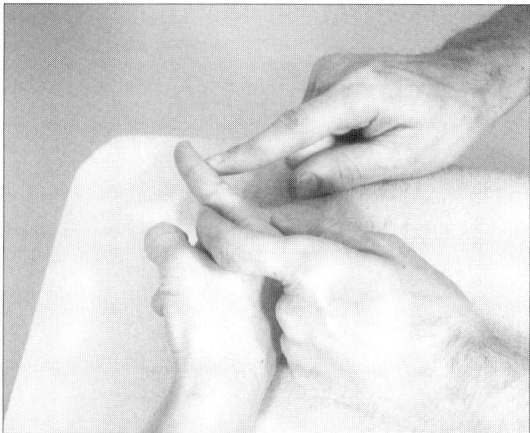

Fig. 19.176

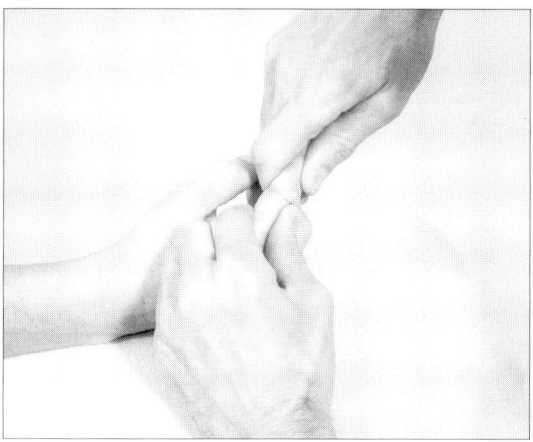

Fig. 19.177

Thumb

Flexion and extension at the carpometacarpal (CMC) I joint are in the same plane as flexion and extension at the other thumb joints. The plane of thumb flexion/extension is almost perpendicular to the plane of finger flexion/extension. For flexion and extension movements of the fingers and the thumb, the distal articular bone is concave. Thus, for flexion and extension movements at the CMC I joint, the metacarpal I is concave as well. Therefore, any mobilization of the distal articular bone in the fingers or thumb to improve flexion or extension is in the same direction as its angular movement.

The CMC I joint is a saddle joint. The proximal part of the first metacarpal is concave for flexion and extension and convex for abduction and adduction. The plane of abduction/adduction at the CMC I joint is perpendicular to its flexion/extension plane. Since the proximal metacarpal I is convex for abduction and adduction movements, translation of the CMC joint is opposite its angular movement. For instance, to mobilize abduction (the thumb moves in a palmar direction), translate the proximal metacarpal in a dorsal direction. To mobilize adduction, translate in a palmar direction.

Opposition of the thumb is a combination of abduction, flexion, and a conjunct rotation of metacarpal I on the carpals.[4,13] During combined abduction and flexion of the thumb, the trapeziometacarpal ligaments tighten and cause the conjunct rotation.[13] Opposition of the thumb is thus not a separate movement that can be tested. Neither is it possible to test the opponens pollicis of the thumb separately. The abductor pollicis brevis and the flexor pollicis brevis are often more active than the opponens pollicis in opposing the thumb.[4,24] Parry demonstrated this by electrical stimulation of the median and ulnar nerves.[21] In two subjects in which (1) the opponens pollicis was innervated by the ulnar nerve and the abductor pollicis and (2) the superficial head of the flexor pollicis were innervated by the median nerve, stimulation of the median nerve produced more opposition of the thumb than stimulation of the ulnar nerve.

- **Capsular pattern**

  - Carpometacarpal I: Extension and abduction equally limited; flexion and adduction free.

  - Other thumb joints: Flexion most restricted; extension less restricted.

CARPOMETACARPAL (CMC) I EXTENSION

- **Agonists**

  Carpometacarpal I extensors:
  extensor pollicis longus (posterior interosseus nerve [branch of the radial nerve]; C7,8)
  extensor pollicis brevis (posterior interosseus nerve [branch of the radial nerve]; C7,8)

- **Muscle test** (Fig. 19.178)

  1. Have the patient be seated with her elbow flexed and her right forearm and wrist in a neutral position. Her CMC I joint should be extended, and her more distal thumb joints should be flexed or in a loose-packed position.

  2. Sit in front of her, stabilizing her metacarpals with your left hand.

  3. Place your right index finger on the distal part of metacarpal I.

  4. Ask her to keep her thumb in place. Then attempt to flex the CMC I joint.

  Remarks—If the test yields weakness, compare the result with a test in which the patient's CMC I joint is less extended.

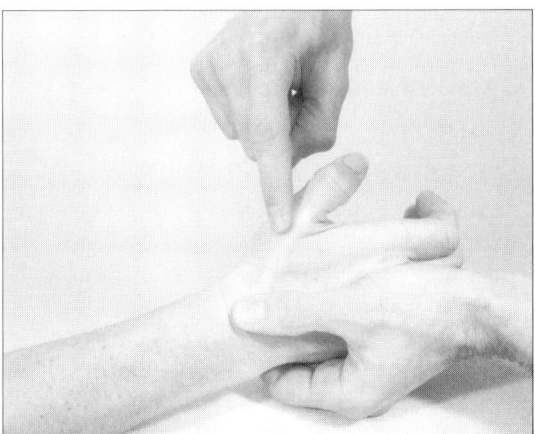

Fig. 19.178

- **Passive mobilization** (Fig. 19.179)

    1. Have the patient be seated with her right forearm supinated and her thumb relaxed.

    2. Sit distal to her hand with your right hand supporting her hand and stabilizing metacarpals II through V.

    3. With your right index finger, stabilize her trapezium on its radial side.

    4. With your left hand, hold metacarpal I and extend it at her CMC joint.

    5. Mobilize her proximal metacarpal with your left thumb in a radial and proximal direction.

    6. Retest.

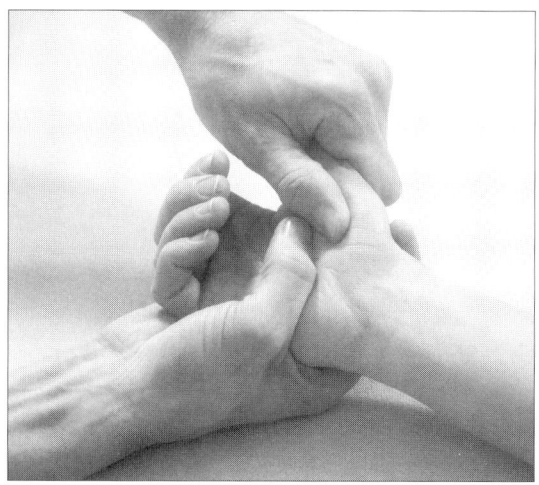

Fig. 19.179

CARPOMETACARPAL I ABDUCTION

- *Agonists*

    Carpometacarpal I abductors:
    abductor pollicis longus (posterior interosseus nerve [branch of the radial nerve]; C7,8)
    abductor pollicis brevis (posterior interosseus nerve [branch of the radial nerve]; C7,8)

- *Muscle test* (Fig. 19.180)

    1. Have the patient be seated with her elbow flexed and her forearm and wrist in a neutral position. Her CMC I joint should be abducted.
    2. Sit in front of her, stabilizing metacarpals II through V with your left hand.
    3. Place your right index finger on the interphalangeal joint of her thumb.
    4. Ask her to keep her thumb in place. Then attempt to adduct it.

    Remarks—If the test yields weakness, compare the result with a test in which the patient's CMC I joint is less abducted. In addition to resulting from a spinal or neurological dysfunction, weakness throughout the range may result from a dysfunction at the MCP I joint. Testing the CMC I abductors with your index finger proximal to the MCP joint may alter strength of the muscles.

- *Passive mobilization* (Fig. 19.181)

    1. Have the patient be seated with her thumb relaxed.
    2. Stabilize metacarpals II through V with your left hand and stabilize the carpals with your left thumb.
    3. Supinate your right arm and hold her first metacarpal with your right hand. Place your right thumb proximal on her metacarpal.
    4. Abduct her thumb and mobilize metacarpal I in a dorsal direction.
    5. Retest.

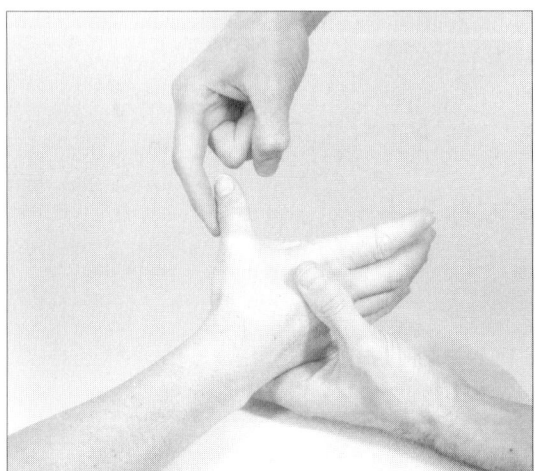

Fig. 19.180

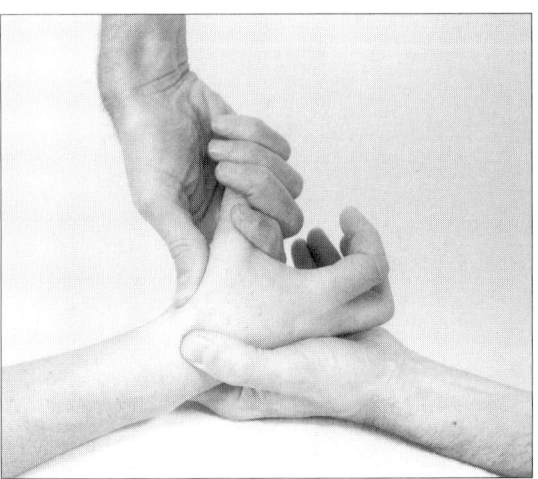

Fig. 19.181

CARPOMETACARPAL I FLEXION

- **Agonists**

    Thumb flexors:
    flexor pollicis longus (median nerve; C8,T1)
    flexor pollicis brevis (superficial head is innervated by the median nerve, C8,T1; the deep head is innervated by the ulnar nerve, C8,T1)

- **Muscle test** (Fig. 19.182)

    1. Have the patient be seated with her elbow flexed and her forearm and wrist in a neutral position. Her CMC I joint should be flexed, and the more distal thumb joints should be extended or in a loose-packed position.
    2. Sit in front of her, stabilizing metacarpals II through V with your left hand.
    3. Place the index finger of your right hand on the palmar and distal part of metacarpal I.
    4. Ask her to keep her thumb in place. Then attempt to extend her CMC I joint.

    Remarks—If the test yields weakness, compare the result with a test in which the patient's CMC I joint is less flexed.

- **Passive mobilization** (Fig. 19.183)

    1. Have the patient be seated with her thumb relaxed.
    2. Stabilize metacarpals II through V with your left hand.
    3. With your right hand, hold metacarpal I and flex her CMC joint.
    4. With your right thumb, mobilize her proximal metacarpal in an ulnar direction.
    5. Retest.

- **Active mobilization** (Fig. 19.182)—The active mobilization is identical to the evaluation, except the treatment contraction is gentler and sustained. Retest after the mobilization.

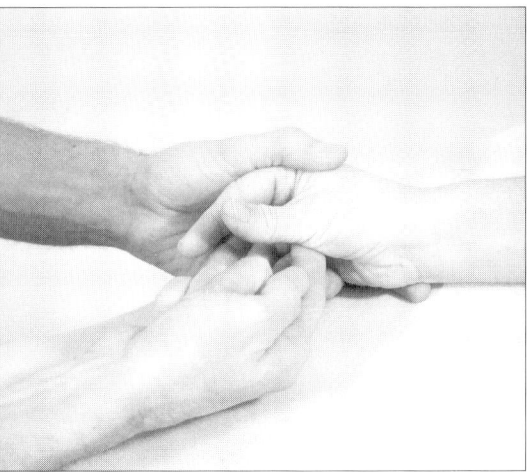

Fig. 19.182

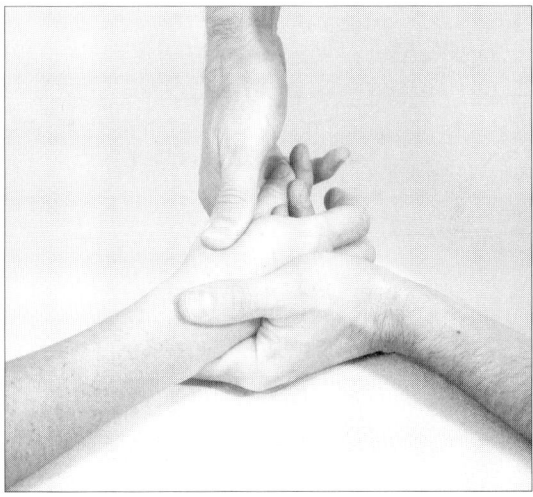

Fig. 19.183

CARPOMETACARPAL I ADDUCTION

- **Agonists**

    Thumb adductors:
    adductor pollicis (ulnar nerve; C8,T1)
    interossei dorsalis I (ulnar nerve; C8,T1)
    extensor pollicis longus (posterior interosseous nerve [a branch of the radial nerve]; C7,8)

- **Muscle test** (Fig. 19.184)

    1. Have the patient be seated with her elbow flexed and her forearm and wrist in a neutral position. The CMC I joint should be slightly extended and fully adducted.
    2. Sit in front of her, stabilizing metacarpal II with your left hand.
    3. Place your right thumb on the interphalangeal joint of her thumb.
    4. Ask her to keep her thumb in place. Then attempt to abduct it.

    Remarks—If the test yields weakness, compare the result with a test in which the patient's CMC I joint is less adducted. Weakness throughout the range might derive from an irritation of the MCP I joint or from a spinal or neurological condition.

- **Passive mobilization** (Fig 19.185)

    1. Have the patient be seated with her thumb relaxed.
    2. Sit to the right of her hand, holding her hand with your left hand and stabilizing the palmar side of her radial carpals with your left index finger.
    3. With your right hand, hold her thumb and adduct it.
    4. With your right thumb, mobilize metacarpal I in a palmar direction.
    5. Retest.

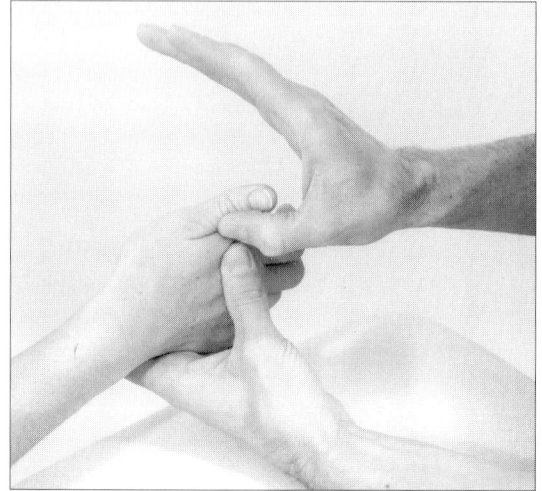

Fig. 19.184

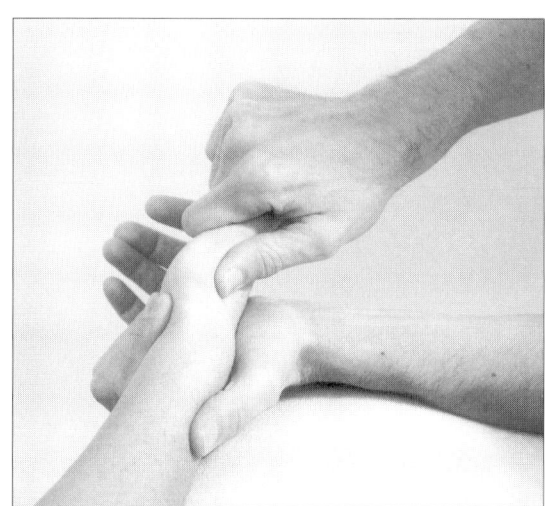

Fig. 19.185

METACARPOPHALANGEAL I AND INTERPHALANGEAL I FLEXION

- *Agonists*

    Thumb flexors:
    flexor pollicis longus (median nerve; C8,T1)
    flexor pollicis brevis (superficial head is innervated by median nerve, C8,T1; deep head is innervated by the ulnar nerve, C8,T1)

- *Muscle test* (Fig. 19.186)

    1. Have the patient be seated with her elbow flexed and her forearm and wrist in a neutral position. Her CMC I joint should be extended, and the more distal thumb joints should be flexed.
    2. Sit to the right of her hand, stabilizing metacarpal I with your left hand.
    3. Place your right index finger on the tip of her thumb.
    4. Ask her to keep her thumb in place. Then attempt to extend her MCP joint by pushing in the direction of her IP joint.

    Remarks—To test the IP joint of her thumb, repeat this test with two modifications. First, stabilize her proximal phalanx with your left hand (Fig. 19.187). Second, place your right index finger on the tip of her thumb and attempt to extend her IP joint.

    If any of the tests yields weakness, compare the result with a test in which the patient's MCP or IP joint are less flexed.

- *Passive mobilization* (Fig. 19.188)

    1. Have the patient be seated with her thumb relaxed.
    2. Sit to the right of her hand, stabilizing her proximal articular bone with the fingers of your left hand.
    3. Hold her distal articular bone with the fingers of your right hand and flex the thumb joint.
    4. With your right hand, translate her distal articular bone in an ulnar (and proximal) direction.
    5. Retest.

- *Active mobilization* (Fig. 19.186)—The treatment is identical to the test procedure, except the treatment contraction is gentler and sustained. Retest after the mobilization.

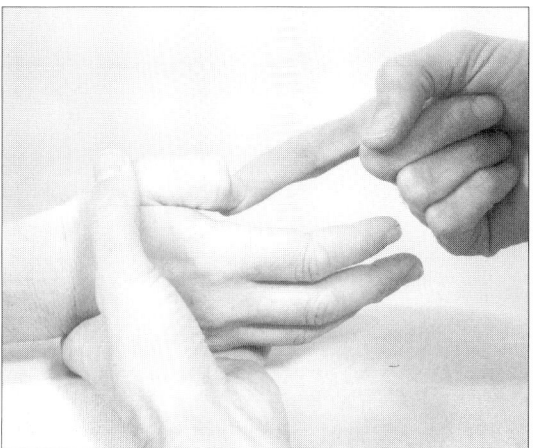

Fig. 19.186

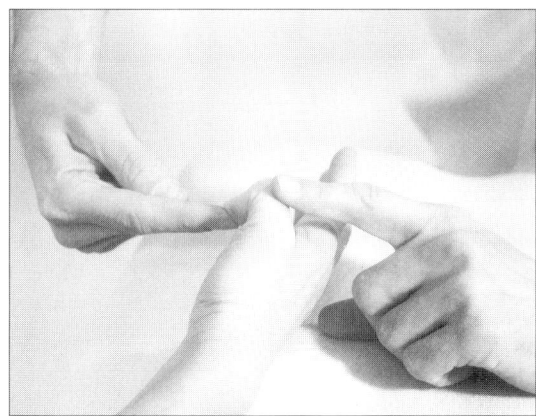

Fig. 19.187

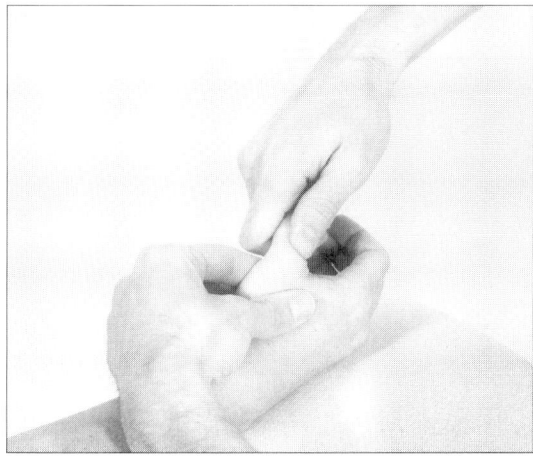

Fig. 19.188

METACARPOPHALANGEAL I AND INTERPHALANGEAL I EXTENSION

- **Agonists**

    Thumb extensors:
    extensor pollicis longus (posterior interosseus nerve [branch of the radial nerve]; C7,8)
    extensor pollicis brevis (posterior interosseus nerve [branch of the radial nerve]; C7,8)

- **Muscle test** (Fig. 19.189)

    1. Have the patient be seated with her elbow flexed and her forearm and wrist in a neutral position. Her CMC I joint should be in a loose-packed position, and the more distal thumb joints should be extended.

    2. Sit in front of her right hand, stabilizing her first metacarpal with the fingers of your right hand.

    3. Place the index finger of your left hand distally on her proximal phalanx.

    4. Ask her to keep her thumb in place. Then attempt to flex her MCP joint.

    Remarks—To test the IP joint of her thumb, repeat this test with two modifications. First, stabilize her proximal phalanx with your right hand (Fig. 19.190). Second, place your left index finger on the tip of her thumb and attempt to flex her IP joint.

    If any of the tests yields weakness, compare the result with a test in which the patient's MCP joint or IP joint is less extended.

- **Passive mobilization** (Fig. 19.191)

    1. Have the patient be seated with her thumb relaxed.

    2. Sit to the right side of her hand, stabilizing her proximal articular bone with the fingers of your left hand.

    3. With the fingers of your right hand, hold her distal articular bone and extend the thumb joint.

    4. Mobilize the distal articular bone in a radial direction.

    5. Retest.

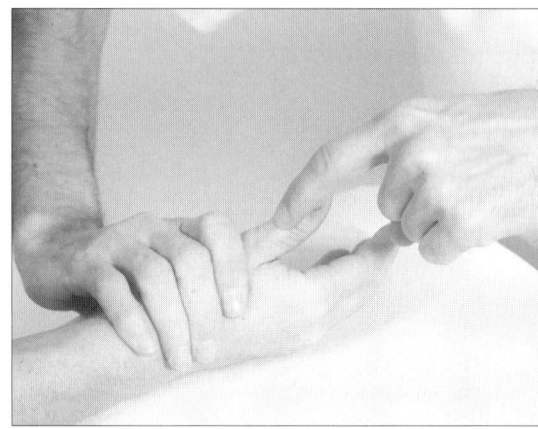
Fig. 19.189

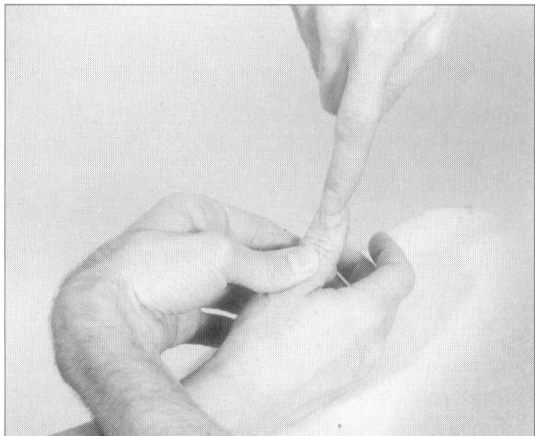
Fig. 19.190

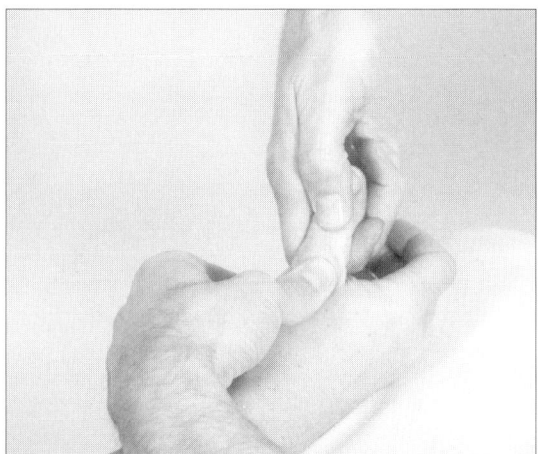
Fig. 19.191

Hip

The hip consists of a convex femur articulating on a concave acetabulum. The translation of the femoral head in the acetabulum is opposite the movement of the femur itself. Since the femur makes an angle at the trochanter, it might be easier to picture trochanter movement in order to visualize the translation in the opposite direction. For instance, during abduction, the trochanter moves cranially while the femoral head translates in a caudal direction. During internal rotation, the trochanter moves anteriorly while the femoral head translates posteriorly.

On testing hip abduction, adduction, internal rotation, and external rotation, hold the patient's ankle when attempting to move the leg. This provides forces over the knee. Knee disorders such as medial or lateral collateral ligament sprains can cause weakness in these tests. In such cases, the weakness is throughout the range of the hip joint. However, altering the knee angle may improve muscle strength, suggesting that the knee was the cause of the inhibition.

- ***Capsular pattern***—Internal rotation is most limited. Flexion, abduction, and extension are less limited. External rotation and adduction are least restricted.

Hip Internal Rotation

- ***Agonists***

    Internal rotators:
    gluteus medius, anterior part (superior gluteal nerve; L5,S1)
    gluteus minimus, anterior part (superior gluteal nerve; L5,S1)
    tensor fascia latae (superior gluteal nerve; L4,5)

- ***Muscle test*** (Fig. 19.192)

    1. Have the patient lie prone with her right hip in a neutral position between abduction and adduction. Her right knee should be flexed to 90 degrees, and her hip should be internally rotated.

    2. Stand distal to her right leg with your left hand stabilizing her medial knee.

    3. Hold her ankle with your right hand.

    4. Ask her to keep her leg in place. Then attempt to externally rotate her leg (push her ankle medially).

Remarks—If the test yields weakness, compare the result with a test in which the patient's hip is less internally rotated. If the internal rotators are arthrogenically inhibited by the hip joint, a clear inhibiting barrier can be found. Thus, slightly beyond the barrier the muscle will test weak, whereas slightly before the barrier the muscle will test fully strong.

Internal rotator weakness might also result from an innominate with limited external rotation at the sacroiliac joint ("inflare"). In this case, internal rotation is dysfunctional (weak), but no inhibiting barrier can be found. As the internal rotators are retested in gradually less internally rotated positions, the muscles gradually test stronger. More about the sacroiliac joint inhibiting the internal rotators can be found under the headings "Innominate external rotation on S1" and "Innominate external rotation on S3."

A pseudoradiculopathy or radiculopathy can cause weakness throughout the full hip range. This weakness is dependent on the position of the spine.

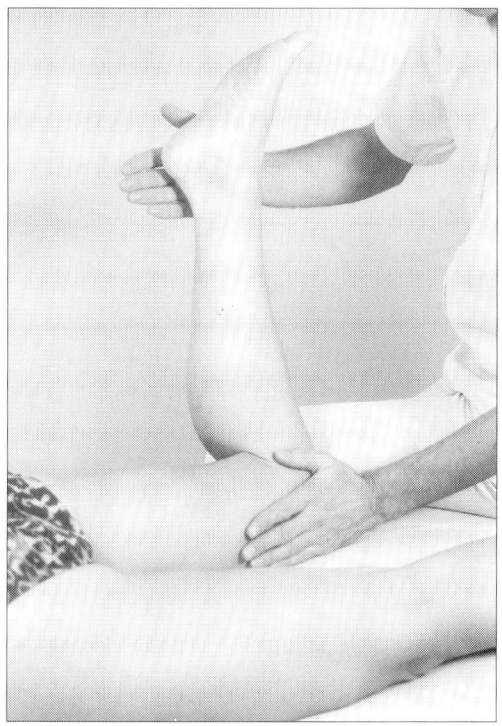

Fig. 19.192

Combinations of the above dysfunctions are also possible. You should treat the most obvious dysfunction first.

- **Passive mobilization** (Fig. 19.193)

    1. Have the patient lie prone with her right knee flexed to 90 degrees and her right hip internally rotated. Place a towel roll between her right trochanter and the table to stabilize her femur.

    2. Stand at her right side with your left hand holding her ankle, maintaining the internal rotation of her leg.

    3. Place your right hand on her right innominate (acetabulum) and mobilize it anteriorly to stretch her posterior hip capsule.

    4. Retest.

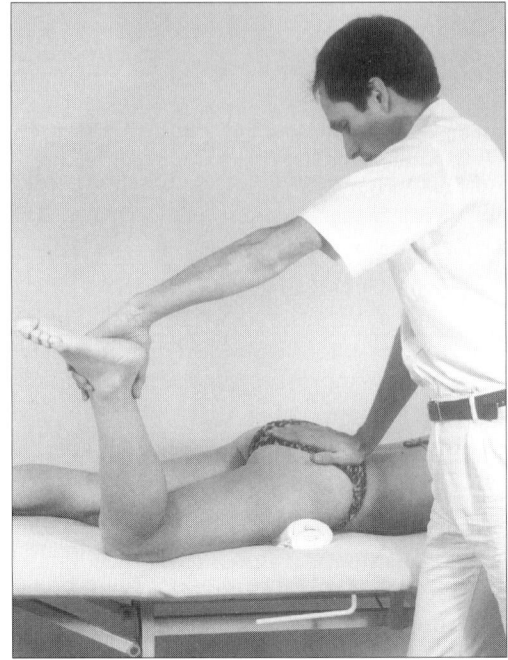

Fig. 19.193

Hip Flexion

- *Agonists*

    Hip flexors:
      psoas major (lumbar plexus; L1–3)
      iliacus (femoral nerve; L2,3)

- *Muscle test* (Fig. 19.194)

    1. Have the patient lie supine with her right hip fully flexed.

    2. Stand at her right side, stabilizing her left ASIS with one hand, placing your other hand on her knee.

    3. Ask her to keep her leg in this position. Then attempt to extend her hip (push her knee caudally).

    Remarks—If the test yields weakness, compare the result with a test in which the patient's hip is less flexed. Dysfunctions of the lumbar spine and the sacroiliac joints may also cause inhibition of the psoas major and iliacus. However, if the hip is responsible for hip flexor inhibition, a clear inhibiting barrier can be found in the flexion range of the hip. In addition, testing the iliacus and the psoas separately will yield an identical pattern of weakness of both muscles (same inhibiting barrier).

- *Passive mobilization* (Fig. 19.195)

    1. Have the patient lie supine with her right hip fully flexed.

    2. Sit on the treatment table distal to her right leg and brace her right foot against your chest or right shoulder.

    3. Interlace the fingers of both your hands and place them anterior and proximal on her femur.

    4. Mobilize her proximal femur in a caudal direction.

    5. Retest.

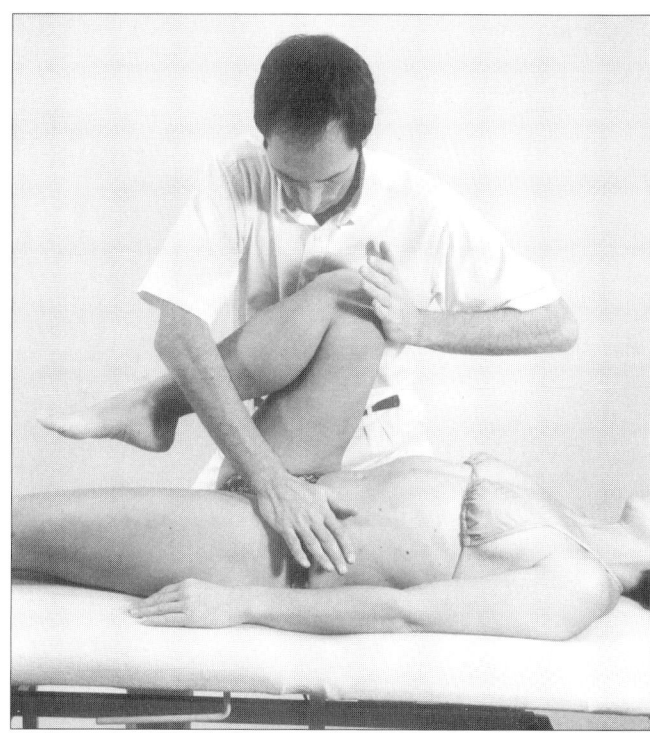

Fig. 19.194

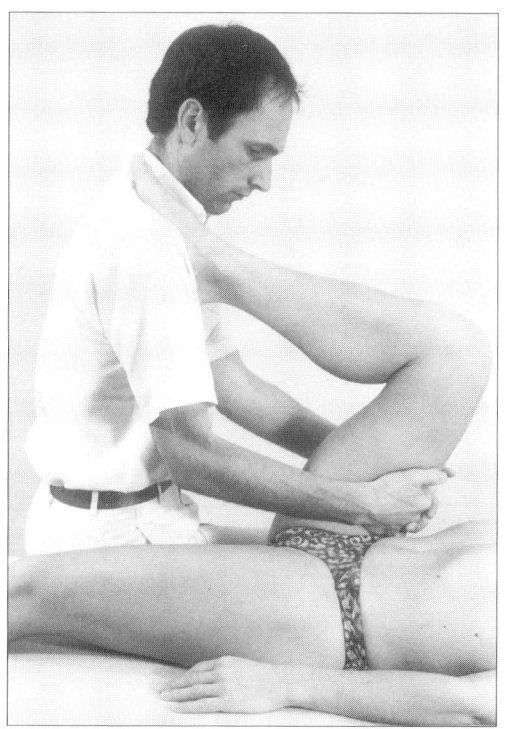

Fig. 19.195

Hip Abduction

- **Agonists**

    Hip abductors:
    gluteus medius (superior gluteal nerve; L5,S1)
    gluteus minimus (superior gluteal nerve; L5,S1)
    gluteus maximus, cranial fibers (inferior gluteal nerve; L5–S2)
    tensor fascia latae (superior gluteal nerve; L4,5)

- **Muscle test** (Fig. 19.196)

    1. Have the patient lie supine with her legs abducted.
    2. Stand at her feet holding one of her ankles in each hand.
    3. Her left leg should rest on the table; her right leg should be raised off the table and supported by your hand.
    4. Ask her to maintain her right leg position. Then attempt to adduct her leg.

    Remarks—If the test yields weakness, compare the result with a test in which the patient's hip is less abducted.

- **Passive mobilization** (Fig. 19.197)

    1. Have the patient lie supine with both legs abducted.
    2. Hold her right ankle with both hands.
    3. Mobilize her relaxed right leg in a distal direction to achieve an inferior glide (stretch the medial inferior hip capsule).
    4. Retest.

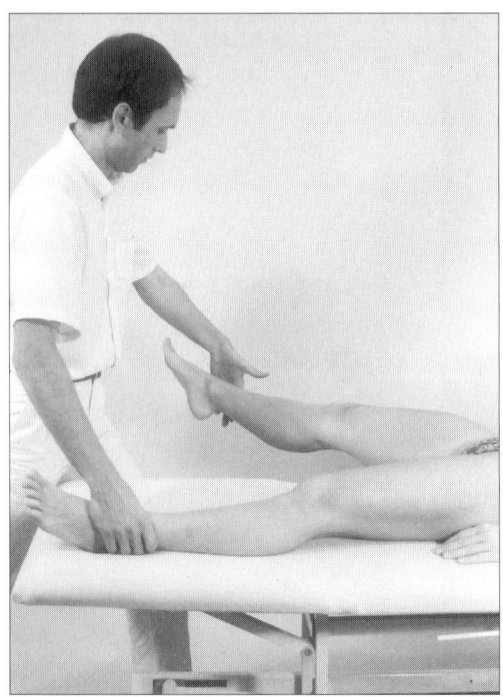

Fig. 19.196

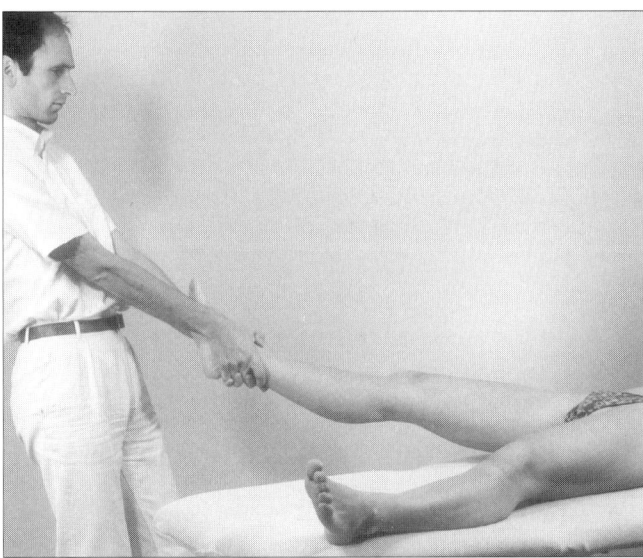

Fig. 19.197

Hip Extension

- **Agonists**

    Hip extensors:
    gluteus maximus (inferior gluteal nerve; L5–S2)
    gluteus medius, posterior fibers (superior gluteal nerve; L5,S1)

- **Muscle test (prone)** (Figs. 19.198 and 19.199)

    1. Have the patient lie prone with her right knee flexed to 80 degrees.
    2. Stand at her right side with your right hand stabilizing her left iliac crest.
    3. Hold her right ankle with your left hand and assist her in lifting her right knee off the table (Fig. 19.198).
    4. Ask her to keep her right hip extended. Then reposition your left hand to hold her right calcaneus (Fig. 19.199).
    5. Ask her to keep her knee lifted off the table. Then attempt to push the knee closer to the table (flex her hip). The pressure on her calcaneus should be directed to her right knee.

Remarks—Instead of pushing on the right calcaneus, your left hand can also push on the distal thigh. However, the disadvantage of pushing on the distal thigh is that you are pressing on soft tissues (hamstrings) and not on bone. A fluctuating tone in the hamstrings, for example, could be misinterpreted as leg movement.

If the test yields weakness, compare the result with a test in which the patient's hip is less extended.

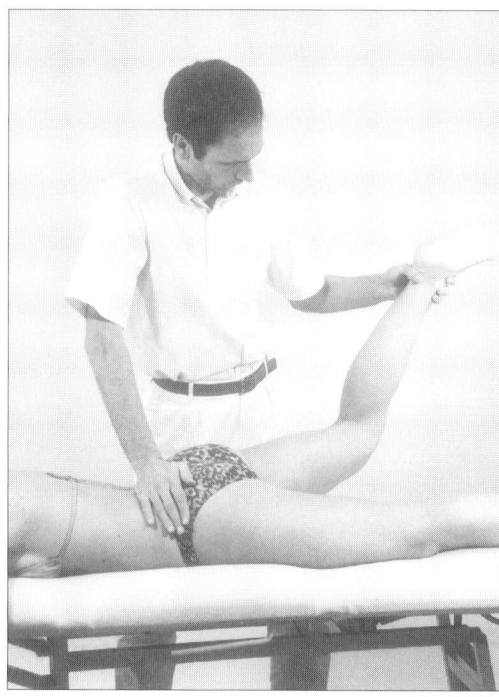

Fig. 19.198

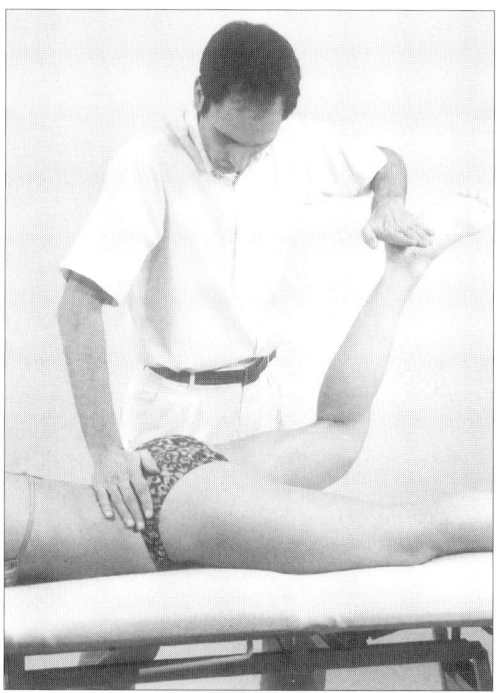

Fig. 19.199

- **Muscle test (supine)** (Fig. 19.200)

  1. Have the patient lie supine, holding her flexed left knee close to her chest with both hands.

  2. Position yourself at her right side, stabilizing her right ASIS with your left hand.

  3. Hold the posterior part of her right femoral condyles with your right hand.

  4. Ask her to keep her leg in this position. Then attempt to flex her right hip (lift her right knee).

  Remarks—If this test yields weakness, compare the result with a test in which the patient's hip is less extended.

- **Passive mobilization** (Fig. 19.201)

  1. Have the patient lie prone and place a towel roll under her right ASIS. Her right leg should be extended and supported by pillows or towels under her knee, so that the knee is comfortably bent.

  2. Position yourself at her right side, with your left hand stabilizing her right ankle to maintain neutral hip rotation.

  3. Place your right hand on her femur, just distal to her buttock.

  4. Mobilize her femur anteriorly.

  5. Retest.

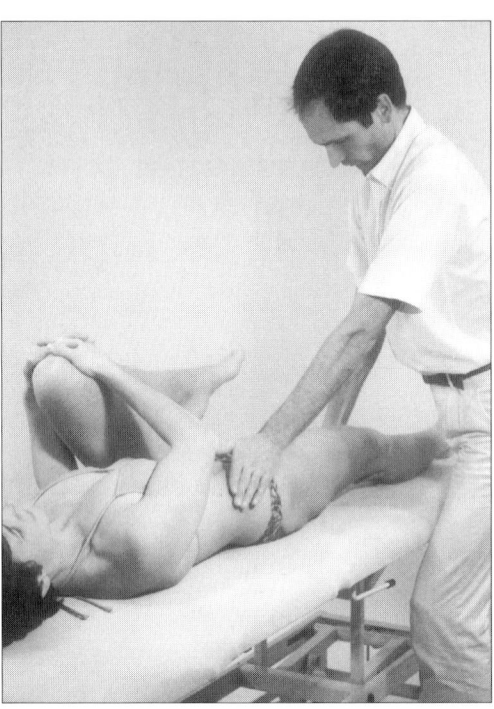

Fig. 19.200

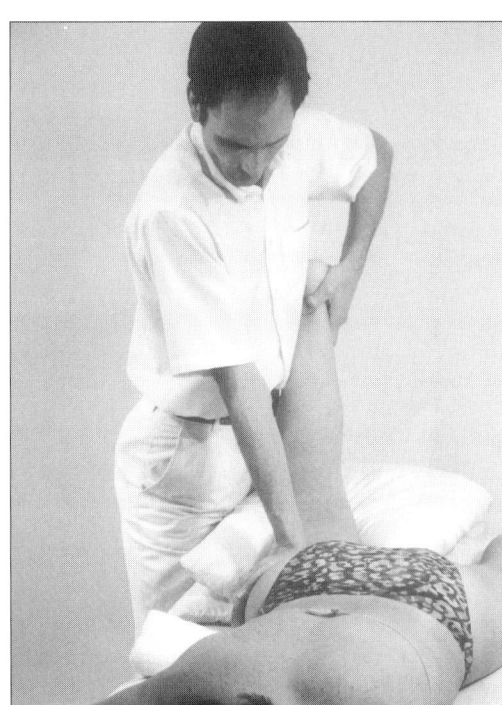

Fig. 19.201

Hip External Rotation

- **Agonists**

    Hip external rotators:
    - gluteus maximus (inferior gluteal nerve; L5–S2)
    - gluteus medius, posterior part (superior gluteal nerve; L5,S1)
    - gluteus minimus, posterior part (superior gluteal nerve; L5,S1)
    - piriformis (L5–S2)
    - obturator internus (nerve to the obturator internus; L5,S1)
    - obturator externus (obturator nerve; L3,4)
    - gemelli superior and inferior (nerve to the quadratus femoris; L5,S1)
    - quadratus femoris (nerve to the quadratus femoris; L5,S1)

- **Muscle test** (Fig. 19.202)

    1. Have the patient lie prone with her right leg in a neutral position between abduction and adduction. Her knee should be flexed to 90 degrees and her hip externally rotated.

    2. Position yourself distal to her right knee, stabilizing her lateral knee with your right hand.

    3. Hold her ankle with your left hand.

    4. Ask her to keep her leg in this position. Then attempt to internally rotate her leg (push her ankle laterally).

    Remarks—If the test yields weakness, compare the result with a test in which the patient's leg is less externally rotated. If the external rotators are arthrogenically inhibited by the hip joint, a clear inhibiting barrier can be found. Slightly beyond the barrier, the external rotators will test weak, whereas slightly before the barrier, the muscles will test fully strong.

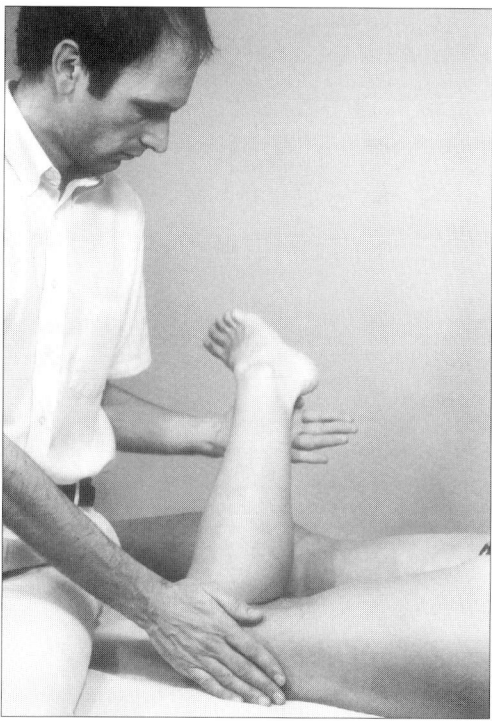

Fig. 19.202

External rotator weakness may also result from an innominate with limited internal rotation at the sacroiliac joint (an "outflare"). In this case, external rotation is dysfunctional (weak), but no inhibiting barrier can be found. As the external rotators are tested in gradually less externally rotated positions, the muscles gradually test stronger. More about the sacroiliac joint inhibiting the external rotators can be found under the headings "Innominate internal rotation on S1" and "Innominate internal rotation on S3."

A pseudoradiculopathy or radiculopathy can also cause weakness throughout the full hip range. In these cases, weakness is dependent on the position of the spine.

- **Passive mobilization** (Fig. 19.203)

    1. Have the patient lie prone with her right knee flexed to 90 degrees and her right hip externally rotated.

    2. Stand on her right side with your left hand holding her right ankle, maintaining the externally rotated position of her leg.

    3. Place your right hand on her greater trochanter and push it anteriorly.

    4. Retest.

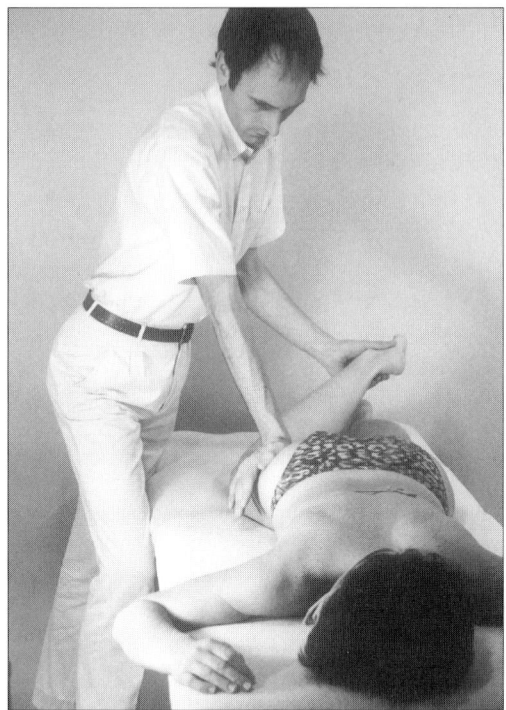

Fig. 19.203

Hip Adduction

- *Agonists*

    Hip adductors:
    pectineus (femoral nerve; L2,3, and accessory obturator; L3)
    adductor longus (obturator nerve; L2,3,4)
    gracilis (obturator; L2,3)
    adductor brevis (obturator; L2,3,4)
    adductor magnus (obturator nerve; L2,3,4, and sciatic nerve; L2,3,4)

- *Muscle test* (Fig. 19.204)

    1. With the patient supine, stand at her feet, facing her. Fixate her left ankle on the table with your right hand while holding her right ankle with your left hand.

    2. Lift her right leg and adduct it in front of her left leg.

    3. Ask her to keep her right leg in this position. Then attempt to abduct her leg.

    Remarks—If the test yields weakness, compare the result with a test in which the patient's hip is less adducted. A limited mobility of the right pubis in a caudal direction may also inhibit the right adductors. This condition causes weakness throughout much of the hip range without an inhibiting barrier.

- *Passive mobilization* (Fig. 19.205)

    1. Have the patient lie supine with her right leg adducted across her left leg.

    2. Stand at her feet with your hands holding her right ankle and foot.

    3. Provide an axial compression at her heel (translate her femoral head cranially).

    4. Retest.

- *Active mobilization* (Fig. 19.204)—The treatment is identical to the testing procedure, except the treatment contraction is gentler and sustained. Retest after the mobilization.

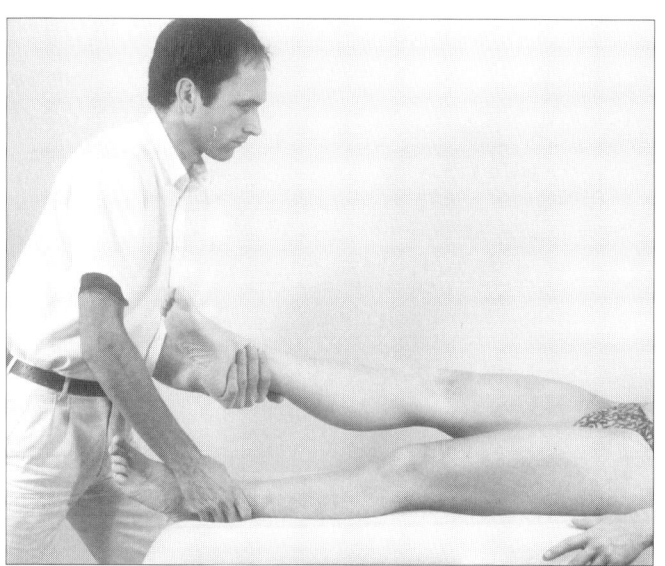

Fig. 19.204

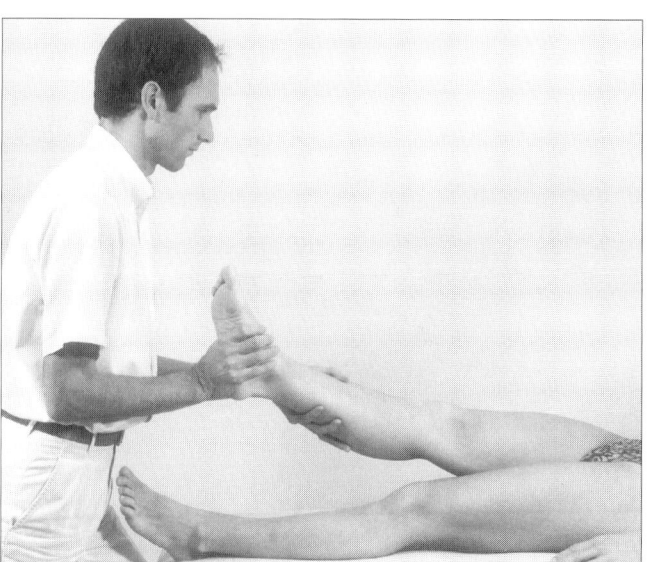

Fig. 19.205

Hip Horizontal Adduction

Horizontal adduction of the hip is the adduction movement of the hip after it has flexed to 90 degrees.

- **Agonists**

    Hip adductors:
    pectineus (femoral nerve; L2,3, and accessory obturator; L3)
    adductor longus (obturator nerve; L2,3,4)
    gracilis (obturator; L2,3)
    adductor brevis (obturator; L2,3,4)
    adductor magnus (obturator nerve; L2,3,4, and sciatic nerve; L2,3,4)

- **Muscle test** (Fig. 19.206)

    1. Have the patient lie supine with her right hip flexed to 90 degrees and horizontally adducted. Her right knee should be passively flexed.
    2. Stand at her right side with your left hand stabilizing her left ASIS.
    3. Hold the medial aspect of her right knee with your right hand.
    4. Ask her to keep her knee in this position. Then attempt to push the knee laterally.

    Remarks—If this test yields weakness, compare the result with a test in which the patient's right hip is less horizontally adducted.

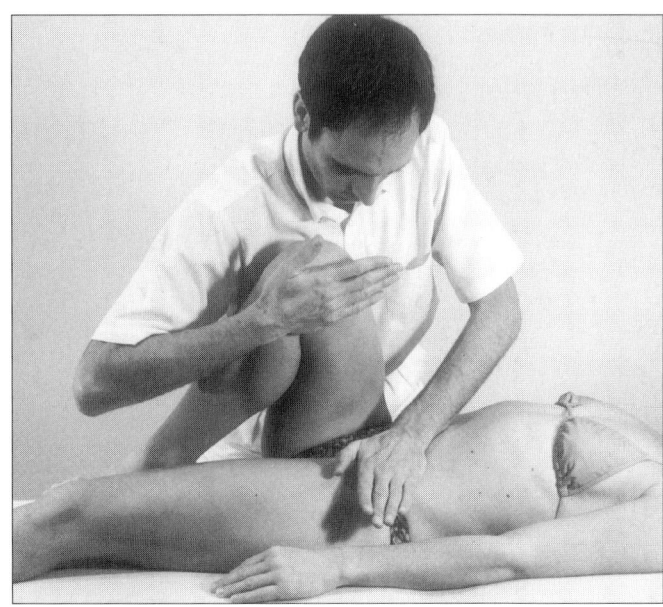

Fig. 19.206

Limited mobility of the right pubis in an anterior direction and/or limited external rotation mobility of the innominate over the sacrum may also inhibit the adductors. The active mobilization described hereunder can correct all three conditions.

- **Active mobilization** (Fig. 19.206)—The treatment is identical to the muscle test, except the treatment contraction is gentler and sustained. Retest after the mobilization.

Knee

The tibia is concave under the femoral condyles. During flexion and extension, the proximal tibia translates in the same direction as the osteokinematic movement of the tibia. Internal rotation of the tibia is a posterior translation of the medial tibial plateau with an anterior translation of the lateral tibial plateau. The posterior translation could be called a flexion movement of the medial tibia. The anterior translation of the lateral tibial plateau could be described as a translation of the lateral tibial plateau into extension.

- *Capsular pattern*

    - Knee joint: Flexion is grossly limited. Extension is much less limited. Rotations are affected only at the end stage of arthritis.

    - Tibiofibular joints: Pain may be present on stretching the joint.

Knee Flexion

The knee is actively flexed by the hamstrings, which attach, among other places, to the proximal tibia and fibula. Strong uninhibited hamstrings are especially important when the restraint to anterior tibial translation by the anterior cruciate ligament is lost.[27] The hamstrings can substitute for lost function of the anterior cruciate ligament by providing the posterior translation of the tibia.

- *Agonists*

    Knee flexors:
    hamstrings (sciatic nerve; L5–S2)

- *Muscle test* (Fig. 19.207)

    1. Have the patient lie supine with her right knee fully flexed.

    2. Stand distal to her right leg, stabilizing her knee with your left hand.

    3. Hold her right ankle with your right hand.

    4. Ask her to keep her knee in this position. Then attempt to extend it.

    Remarks—During flexion of the knee, the tibia translates posteriorly (and/or proximally) on the femur and the patella translates distally (and/or posteriorly) on the femoral condyles. Limited mobility of either structure could inhibit the hamstrings. The distal translation mobility of the patella can be tested by testing the hamstrings while the patella is manually mobilized in a distal direction (Fig. 19.208). Compare the result of this test with one in which you do not translate the patella distally.

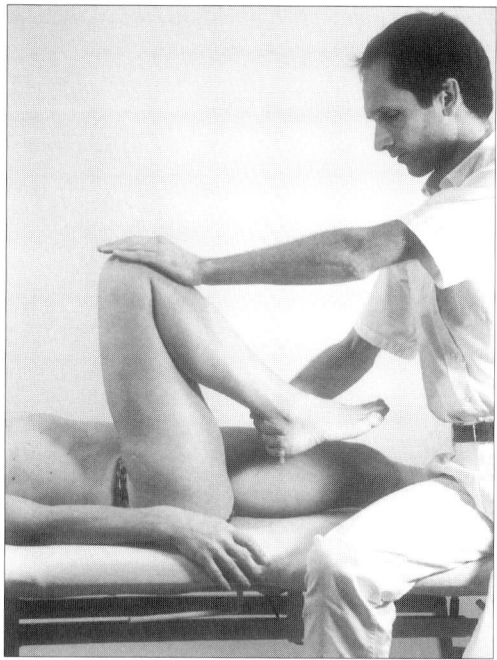

Fig. 19.207

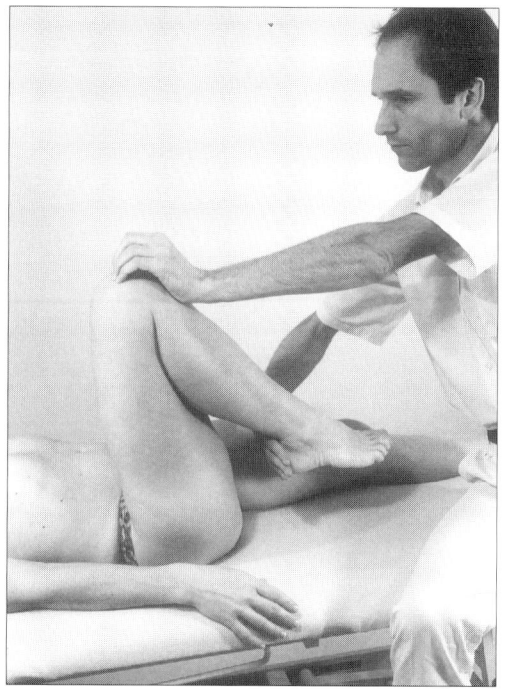

Fig. 19.208

While in flexion, you can add internal or external rotation of the lower leg. During internal rotation of the tibia, the medial tibial plateau translates farther into flexion, whereas the lateral tibial plateau loses some of its flexion translation. Asking the patient to maintain knee flexion after the knee has internally rotated specifically tests flexion of the medial knee compartment (Fig. 19.209). The semitendinosus and semimembranosus muscles are primarily responsible for maintaining knee flexion while the knee is internally rotated. External rotation of the tibia while the knee is flexed puts emphasis on flexion of the lateral compartment of the knee and tests predominantly the biceps femoris (Fig. 19.210). If any of the above three knee flexion tests (neutral rotation, internal rotation, external rotation) yields weakness, compare the result with a test in which the patient's knee is less flexed.

During knee flexion, the anterior structures of the knee are stretched. Internal rotation puts emphasis on the anteromedial knee capsule. External rotation emphasizes the anterolateral knee capsule. In addition, the different parts of the medial and lateral patellar retinaculum are emphasized. Parts of the patellar retinacula can be further emphasized by applying manual translation of the patella.

- **Active mobilization** (Fig. 19.207)—The treatment is identical to the testing procedure, except the treatment contraction is gentler and sustained. Place her knee in the position where it tested maximally weak. Retest after the mobilization.

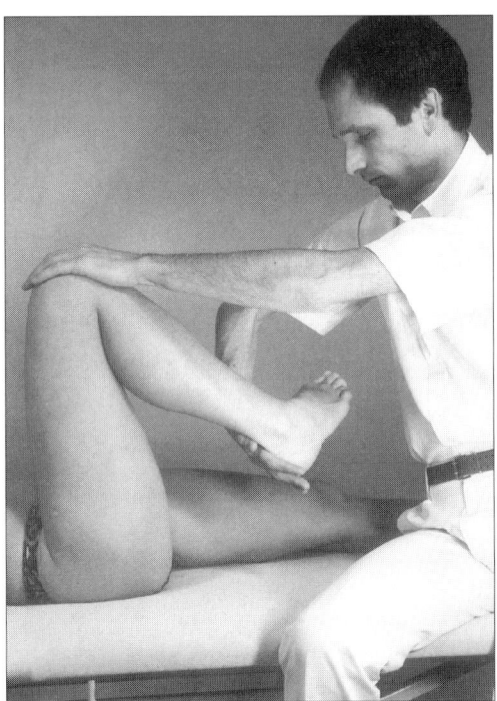

Fig. 19.209

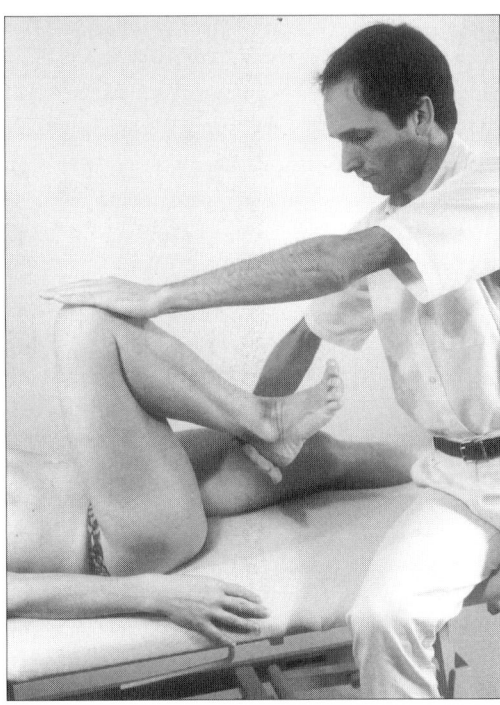

Fig. 19.210

- **Passive mobilization** (Fig. 19.211)

  1. Have the patient lie supine with her right knee fully flexed and rotated into the position of maximal weakness (neutral rotation, internal rotation, or external rotation).
  2. Stand her foot on the table and stabilize it with your right leg.
  3. Stabilize her distal femur with your left hand.
  4. With your right hand, mobilize her tibial plateau posteriorly.
  5. Retest.

  Remarks — If the medial knee has insufficient flexion, you can mobilize the medial tibial plateau posteriorly with your mobilizing hand. If the lateral part of the knee has insufficient flexion, you can mobilize the lateral tibial plateau posteriorly.

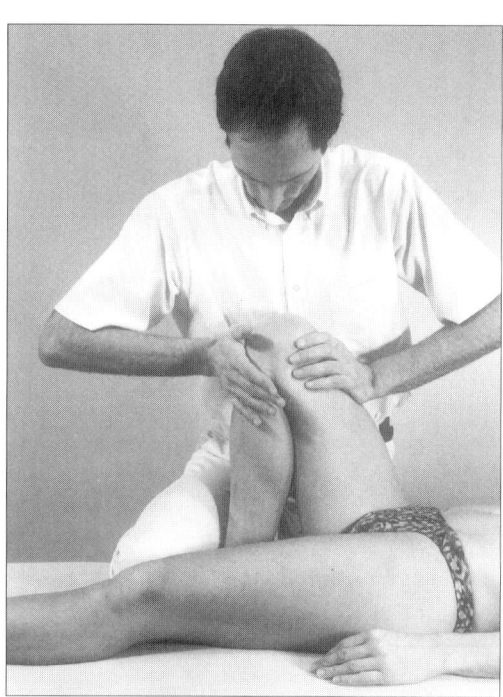

Fig. 19.211

- **Home exercise** (Fig. 19.212) — If the patient tolerates kneeling on the floor and has more than 90 degrees of flexion available in her hips and knees, the following exercise can improve flexion mobility of her knee.

  1. Instruct the patient to kneel on the floor and bend forward onto her forearms, with her elbows close to her knees and her feet plantar flexed.
  2. She should slowly and comfortably lower her buttocks toward her heels.

  Remarks — The mobilizing effect is caused by the floor pushing the tibial tubercle posteriorly while part of the body weight pushes the femur in the opposite direction. The effect of this exercise can be enhanced by having the patient tense her right hamstrings by making her foot rest more lightly on the floor. A towel roll can be placed under her tibial tubercles. Internal or external rotation of the foot emphasizes flexion of the medial or lateral part of the knee.

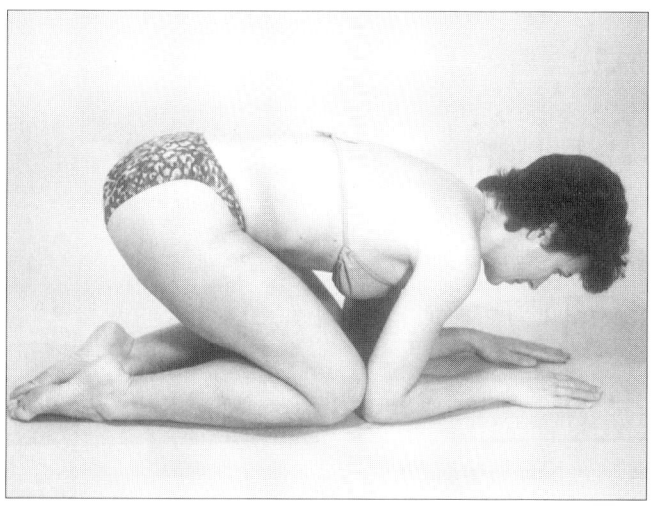

Fig. 19.212

Knee Extension

Three muscle tests are available to evaluate knee extension: (1) a quadriceps test, (2) an adductor test, and (3) an abductor test.

- *Agonists*

    Knee extensor:
    quadriceps femoris (femoral nerve; L2,3,4)

- *Muscle test (quadriceps)* (Fig. 19.213)

    1. With the patient supine, stand at her right side.
    2. Place the ulnar side of your left hand on her proximal tibia to stabilize her knee posteriorly.
    3. With your right hand, hold her ankle (at the malleoli).
    4. Ask her to fully straighten her knee and to maintain the knee extension.
    5. With your arms straight (the movement coming from your shoulders and your trunk), gradually and gently attempt to flex her knee.

    Remarks—If the test yields weakness, compare the result with a test in which the patient's knee is less extended. If the quadriceps tested weak in extension and tests strong in less extension, the cause could be the posterior knee capsule (limiting tibiofemoral extension) or the patellar retinacula (limiting patellofemoral movement). The patellofemoral joint can be evaluated using the hamstrings (see "Patellar mobility"). To further evaluate for the possibility of the posterior capsule inhibiting the quadriceps, the abductors and adductors could be tested.

- *Agonists*

    Hip adductors:
    pectineus (femoral nerve; L2,3, and accessory
        obturator; L3)
    adductor longus (obturator nerve; L2,3,4)
    gracilis (obturator; L2,3)
    adductor brevis (obturator; L2,3,4)
    adductor magnus (obturator nerve; L2,3,4,
        and sciatic nerve; L2,3,4)

- *Muscle test (adductors)* (Fig. 19.214)

    1. Have the patient lie supine with her legs in a neutral position between abduction and adduction.
    2. Stand at the patient's feet, facing her. Fixate her left ankle on the table with your right hand.
    3. Support her right ankle, which is lifted off the table, with your left hand.
    4. Ask her to straighten her right knee completely and to maintain the position of her leg. Then attempt to abduct her right leg.

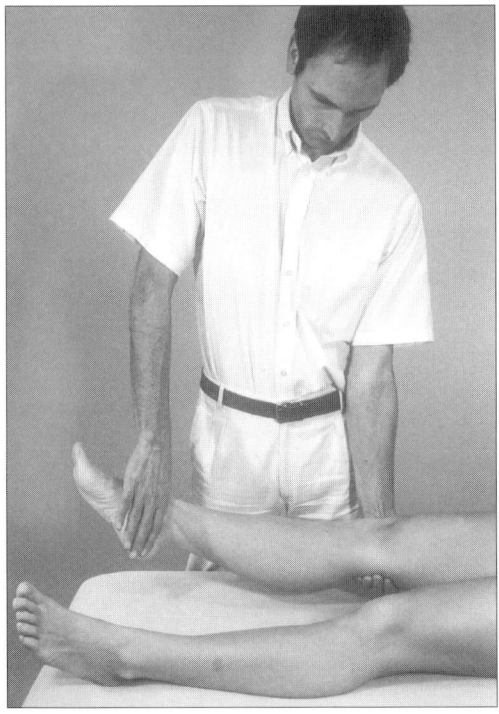

Fig. 19.213

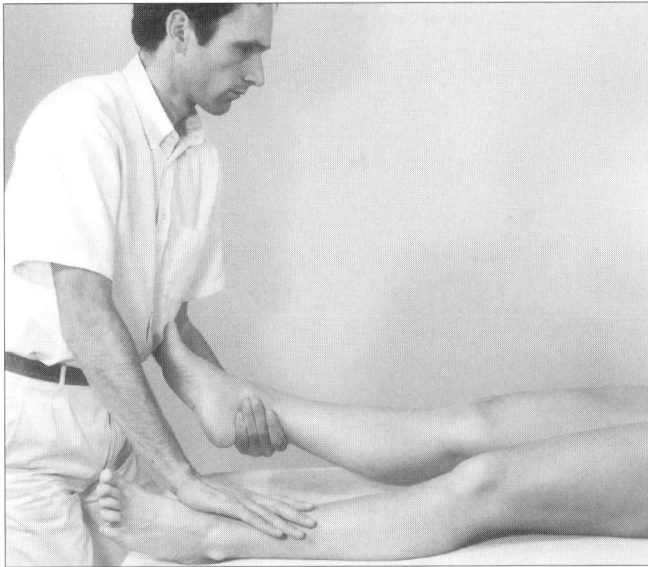

Fig. 19.214

Remarks—If the test yields weakness, compare the result with a test in which the patient's knee is less extended. Often, a mere 5 degrees of knee flexion will remove the inhibition of the adductors. This test emphasizes the medioposterior capsule of the knee joint. The medioposterior knee joint capsule is stretched with extension and with knee valgus in extension. Resisted adduction while the knee is extended stretches the medioposterior part of the knee capsule. Thus, with the knee extended, limited extensibility of the medioposterior capsule inhibits both the quadriceps and the adductors.

If the quadriceps and the adductors are weak in extension but become strong with a slight amount of knee flexion, an extension mobilization of the medial knee compartment could correct the dysfunction.

- *Agonists*

    Hip abductors:
    gluteus medius (superior gluteal nerve; L5,S1)
    gluteus minimus (superior gluteal nerve; L5,S1)
    gluteus maximus, cranial fibers (inferior gluteal nerve; L5–S2)

- *Muscle test (abductors)* (Fig. 19.215)

    1. Have the patient lie supine with her legs in a neutral position between abduction and adduction.
    2. Stand at the patient's feet, facing her. Fixate her left ankle on the table with your right hand.
    3. Support her right ankle, which is lifted off the table, with your left hand.
    4. Ask her to fully straighten her knee, maintaining the position of her leg. Then attempt to adduct her right leg.

Remarks—If the test yields weakness, compare the result with a test in which the patient's knee is less extended. Often, a mere 5 degrees of knee flexion will remove the abductor inhibition. This test emphasizes the lateroposterior capsule of the knee joint. The lateroposterior knee joint capsule is stretched with extension and with knee varus in extension. Resisted abduction while the knee is extended stretches the lateroposterior part of the knee capsule. Thus, with the knee extended, limited extensibility of the lateroposterior capsule inhibits both the quadriceps and the abductors.

If the quadriceps and the abductors are weak in extension but become strong with a slight amount of knee flexion, an extension mobilization of the lateral knee compartment could correct the dysfunction.

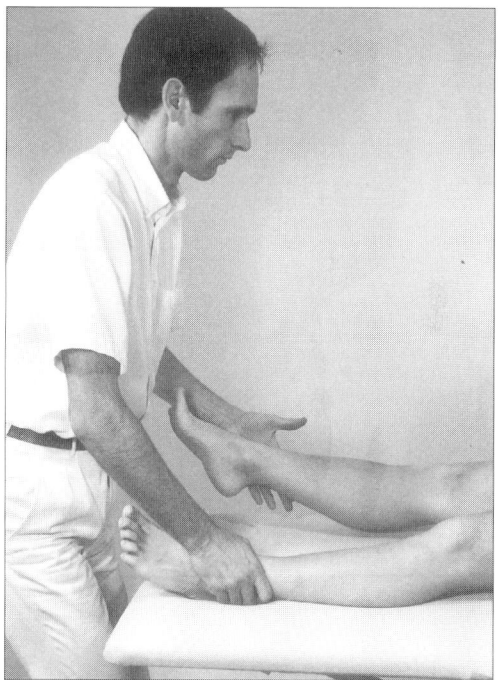

Fig. 19.215

- **Combined active and passive mobilization** (Fig. 19.216)

  1. Have the patient lie prone with her feet off the edge of the table. Position a small towel roll just proximal to her right patella.
  2. Stand at her right side with your right hand on her proximal tibia.
  3. Hold her ankle with your left hand.
  4. Ask her to extend her knee comfortably (bring her foot closer to the floor). Then apply a light counterpressure at her ankle to resist knee extension.
  5. Continue to apply pressure to the back of her proximal tibia with your right hand for stabilization and for passive mobilization.
  6. Retest.

Remarks—If her abductors tested weak, mobilize her lateral tibial plateau anteriorly with your right hand. If her adductors tested weak, mobilize her medial tibial plateau anteriorly. If both muscle groups test weak, both the medial and the lateral tibial plateau are mobilized. To prevent pressure on her popliteal artery, it is best to cup your right hand, which puts pressure on the lateral and medial side of her tibia.

- **Passive mobilization (medial compartment)** (Fig. 19.217)

  1. Have the patient lie supine with her right ankle resting on a towel roll.
  2. Stand at her right side with your stabilizing left hand anteriorly on her distal femur (proximal to her patella).
  3. Place the fingers of your right hand posteriorly and medially on her proximal tibia.
  4. Position her knee in comfortable extension.
  5. While stabilizing her femur with your left hand, pull her medial tibia anteriorly with your right hand.
  6. Retest.

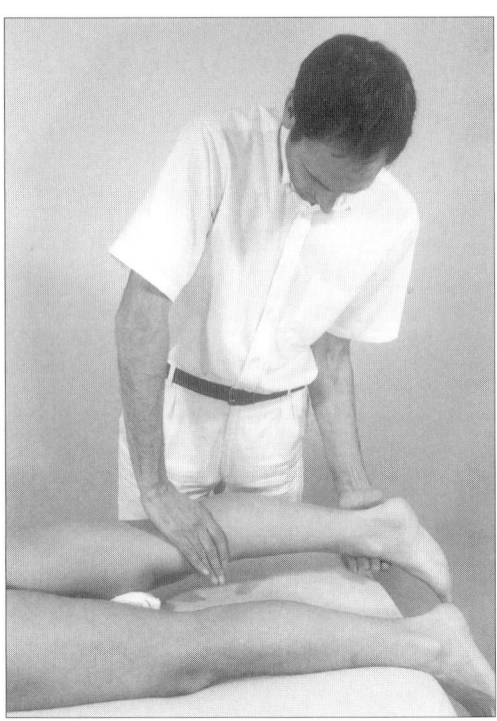

Fig. 19.216

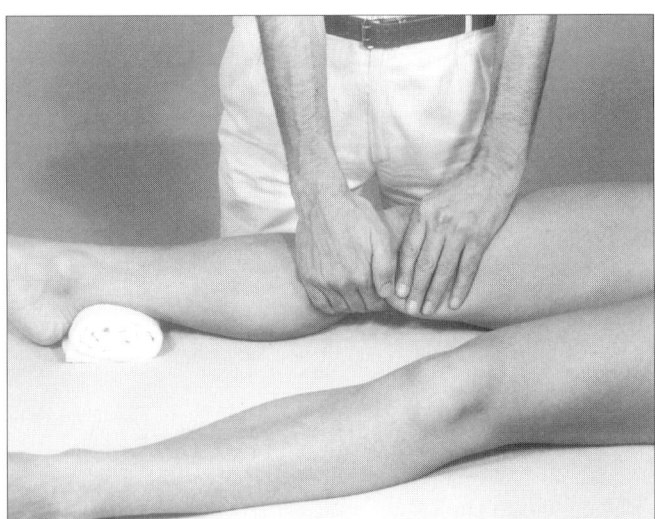

Fig. 19.217

- *Passive mobilization (lateral compartment)* (Fig. 19.218)

  1. Have the patient lie supine with her right ankle resting on a towel roll.
  2. Stand at her left side with your stabilizing right hand anteriorly on her distal femur (proximal to her patella).
  3. Place the fingers of your left hand posteriorly and laterally on her proximal tibia.
  4. Position her knee comfortably in extension.
  5. While stabilizing her femur with your right hand, pull her lateral tibia anteriorly with your left hand.
  6. Retest.

  Remarks—If both the medial and lateral tibial plateau need an extension mobilization, your mobilizing hand could be substituted by a firm towel roll.

- *Active home exercise* (Fig. 19.219)

  1. Instruct the patient to lie supine with a towel roll under her right proximal tibia.
  2. The patient should do "quad sets" by lifting her heel off the floor and pushing her tibia into the towel roll. A weight can be wrapped around her distal tibia (at the ankle).

  Remarks—Doing quad sets with a towel roll under the distal femur to strengthen the quadriceps has been a popular exercise among rehabilitation professionals. I doubt it has much value, though, because a weak quadriceps should be strengthened by addressing the cause of the inhibition. If the weakness is believed to be due to lack of use, many other functional exercises, such as walking, running, or bicycling, would be much better. In addition, if the weakness is due to inhibition from the knee joint, the mobilizing force from the inhibited quadriceps working against gravity is questionable. However, if this exercise must be used, its effectiveness could be improved by placing the towel roll under the proximal tibia instead of the femur. This creates an anterior translation of the tibia. The quad set exercise becomes easier with the towel roll under the tibia because the quadriceps has less weight to lift. This decreased workout by the quadriceps can be countered by increasing the weight around the ankle.

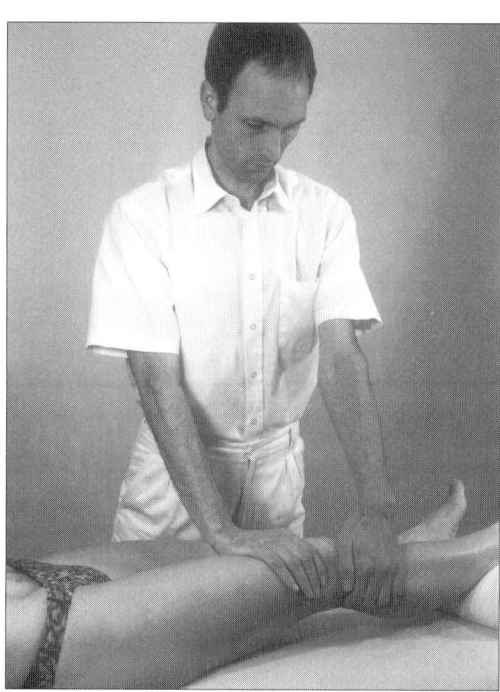

Fig. 19.218

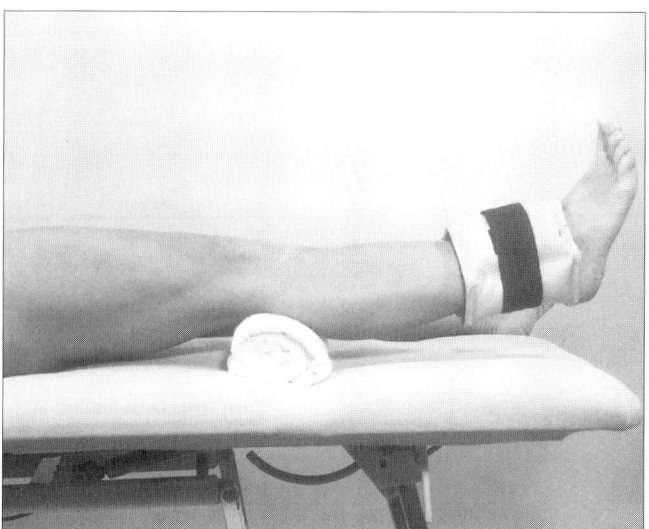

Fig. 19.219

Patellar Mobility

The medial and lateral patellar retinaculum aid in passively restraining patellar mobility. These ligaments attach to the patella, fanning out from there like rays of the sun. Limited mobility of (part of) a retinaculum can cause patellar tracking problems. In addition, limited mobility of a retinaculum can inhibit parts of the quadriceps. For instance, limited mobility of the distal fibers (if the right patella were the center of a clock, the seven and eight o'clock fibers) of the lateral retinaculum might inhibit the vastus medialis.

- *Agonists*

    Knee flexors:
        hamstrings (sciatic nerve; L5–S2)

- *Muscle test (medial patellar retinaculum, proximal part)* (Fig. 19.220)

    1. Have the patient lie supine with her right knee fully flexed and her lower leg externally rotated.
    2. Stabilize her knee with your left hand, translating her patella laterally and distally with your index finger.
    3. Hold her ankle with your right hand.
    4. Ask her to keep her leg in place. Then attempt to extend her knee.

- *Muscle test (medial patellar retinaculum, middle part)* (Fig. 19.221)

    1. Have the patient lie supine with her right knee fully flexed and her lower leg externally rotated.
    2. Stabilize her knee with your right hand, pushing the patella laterally with the base of your hand.
    3. Hold her ankle with your left hand.
    4. Ask her to keep her leg in place. Then attempt to extend her knee.

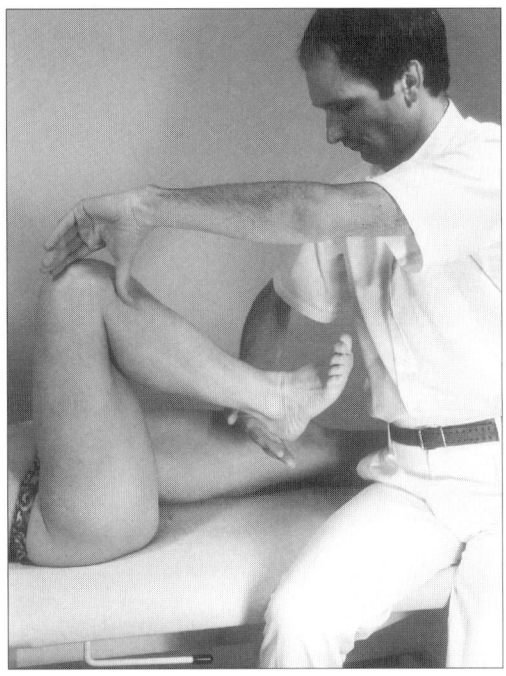

Fig. 19.220

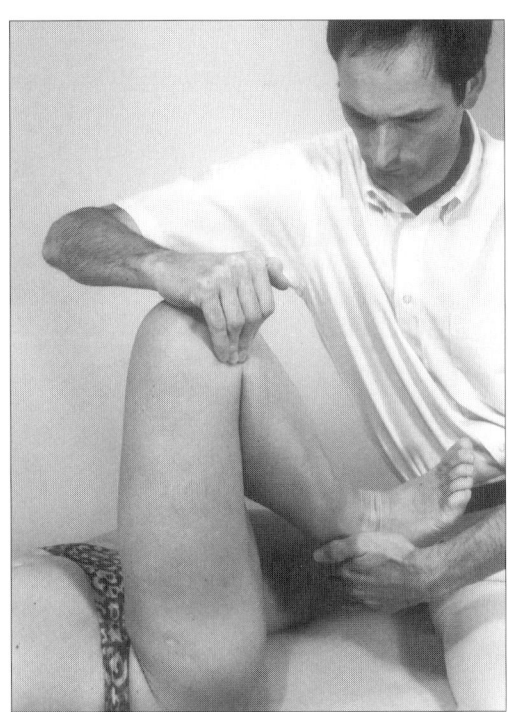

Fig. 19.221

TECHNIQUE PROCEDURES

- *Muscle test (medial patellar retinaculum, distal part)* (Fig. 19.222)

  1. Have the patient lie supine with her right knee fully flexed and her lower leg internally rotated.

  2. Stabilize her knee with your right hand, translating the patella laterally and proximally with the base of your hand.

  3. Hold her ankle with your left hand.

  4. Ask her to keep her leg in place. Then attempt to extend her knee.

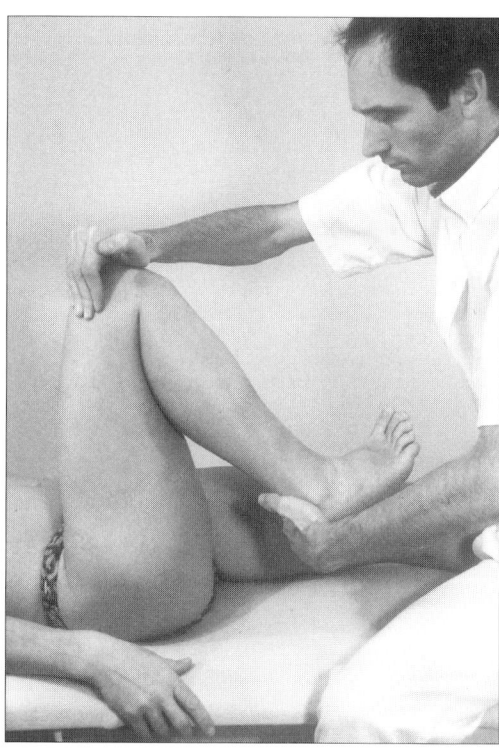

Fig. 19.222

- *Muscle test (lateral patellar retinaculum, proximal part)* (Fig. 19.223)

  1. Have the patient lie supine with her right knee fully flexed and her lower leg internally rotated.

  2. Stabilize her knee with your right hand, with your index finger translating her patella medially and distally.

  3. Hold her ankle with your left hand.

  4. Ask her to keep her leg in place. Then attempt to extend her knee.

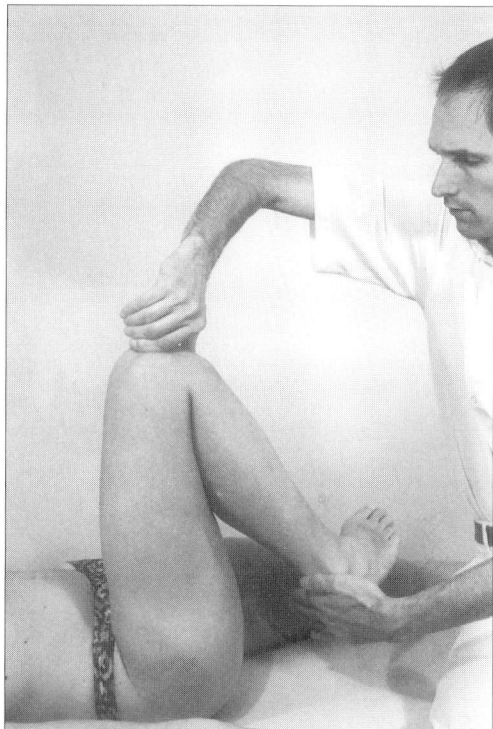

Fig. 19.223

- **_Muscle test (lateral patellar retinaculum, middle part)_** (Fig. 19.224)

  1. Have the patient lie supine with her right knee fully flexed and her lower leg internally rotated.
  2. Stabilize her knee with your left hand, translating the patella medially with the base of your hand.
  3. Hold her ankle with your right hand.
  4. Ask her to keep her leg in place. Then attempt to extend her knee.

- **_Muscle test (lateral patellar retinaculum, distal part)_** (Fig. 19.225)

  1. Have the patient lie supine with her knee fully flexed and her lower leg externally rotated.
  2. Stabilize her knee with your left hand, translating the patella medially and proximally with the base of your hand.
  3. Hold her ankle with your right hand.
  4. Ask her to keep her leg in place. Then attempt to extend her knee.

Remarks—If any of the above tests yields weakness, compare the result with a test in which you do not push against the patella or you push the patella in the opposite direction.

Weakness of the quadriceps is commonly caused by limited mobility of the posterior knee capsule or the patellar retinaculum, preventing sufficient cranial movement of the patella. A quadriceps testing weak in extension with strong abductors and adductors suggests limited patellar mobility. Limited mobility of the patella in a proximolateral direction may inhibit the vastus lateralis, whereas limited mobility in a proximomedial direction may inhibit the vastus medialis. Limited extensibility of any part of the patellar retinaculum can cause patellar tracking dysfunctions.

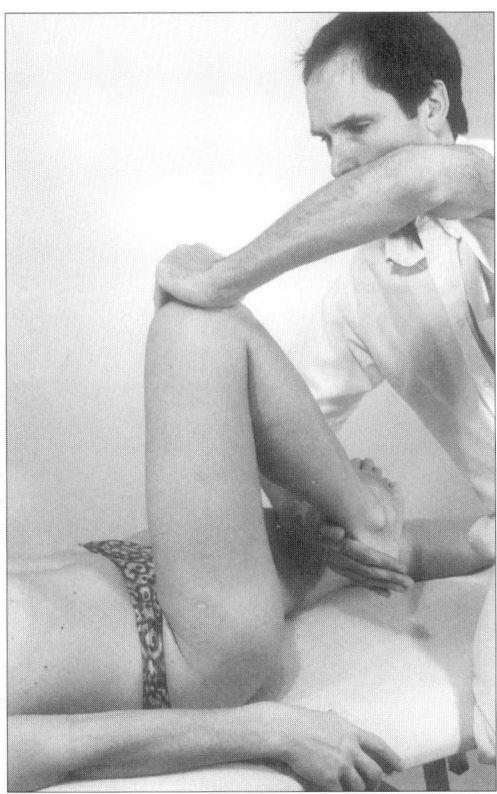

Fig. 19.224

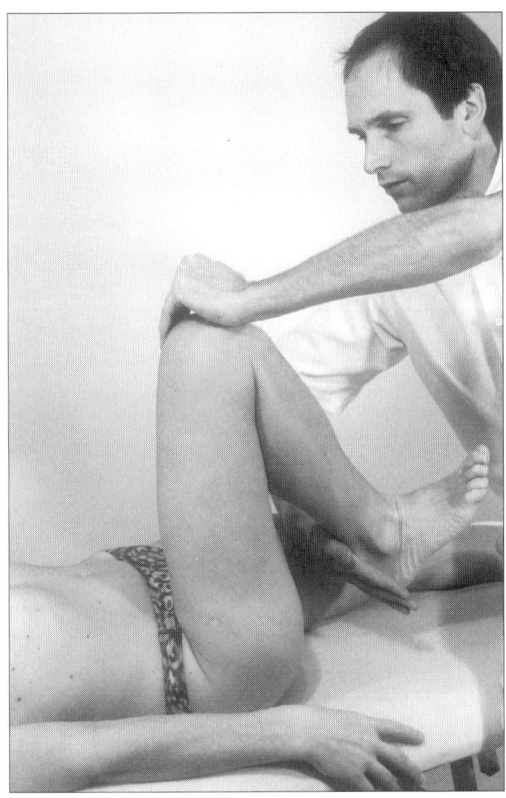

Fig. 19.225

- *Passive mobilization (patellar retinaculum)*
  (Fig. 19.226)

  1. Have the patient lie supine with her right knee bent and her foot in standing position on the treatment table. Her lower leg should be internally or externally rotated, depending on the position in which weakness was found.

  2. Place your right knee on the treatment table to stabilize her right foot.

  3. Mobilize her patella in the direction that previously caused inhibition of the hamstrings. Stabilize her knee with your other hand by pushing the femoral condyles in the opposite direction of the mobilizing hand.

  4. Retest.

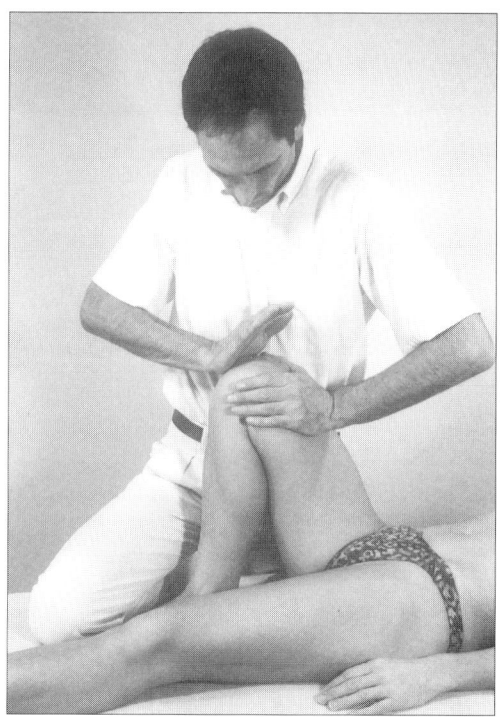

Fig. 19.226

Proximal Tibiofibular Joint

The proximal and distal tibiofibular joints are joints without an active range. However, muscles that primarily function over neighboring joints do influence the tibiofibular joints. For instance, depending on the amount of knee flexion, the biceps femoris, attaching to the fibular head, may pull the fibular head proximally, posteriorly, or distally.

- **Agonists**

    Biceps:
        biceps femoris (sciatic nerve; L5–S2)

- **Muscle test (proximal translation of fibular head)** (Fig. 19.227)

    1. With the patient supine, stand at her feet, holding one of her ankles in each hand.
    2. Stabilize her left leg on the table.
    3. Lift her right leg off the table and externally rotate the leg.
    4. Ask her to dorsiflex and evert her foot.
    5. Ask her to maintain this leg position. Then attempt to move the leg anteriorly and medially.

Remarks—If the test yields weakness, compare the result with a test in which the patient's foot is fully plantar flexed and inverted. This will bring her fibula in a distal direction.

If limited proximal translation of the fibula is responsible for the inhibition, a slight amount of knee flexion or hip adduction should cause no difference in the test result. However, if these positions do alter the outcome of the test, the inhibition is deriving from the hip or knee.

- **Muscle test (dorsal translation of fibular head)** (Fig. 19.228)

    1. Have the patient lie supine with her right knee flexed to 90 degrees and her foot lifted off the table. Her lower leg should be externally rotated.
    2. Position yourself distal to her right leg, supporting her right knee with your left hand.
    3. Hold her ankle with your right hand.
    4. Ask her to maintain this leg position. Then attempt to extend her knee.

Remarks—For this test, the position of the foot is not important. If the test yields weakness, compare the result with a test in which the patient's knee is fully flexed and externally rotated or to one in which her knee is extended.

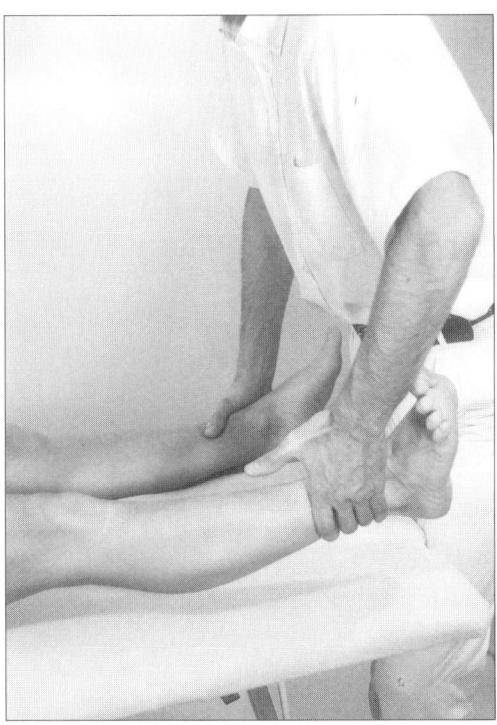

Fig. 19.227

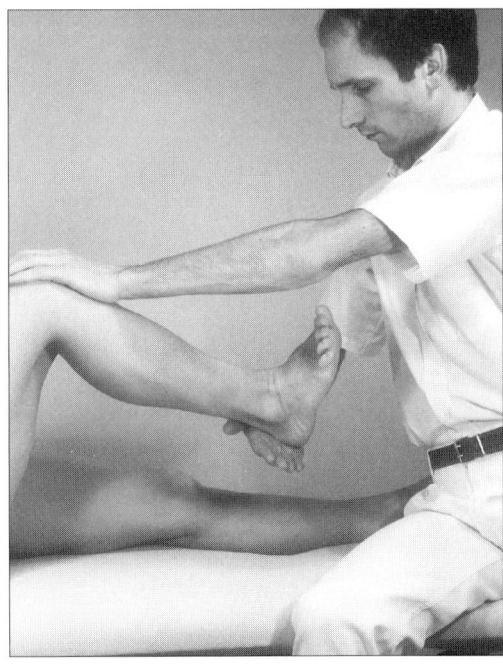

Fig. 19.228

- ***Muscle test (distal translation of fibular head)*** (Fig. 19.229)

    1. Have the patient lie supine with her right knee fully bent and her foot lifted off the table.

    2. Position yourself distal to her right leg, supporting her right knee with your left hand.

    3. Ask her to externally rotate her right lower leg and to plantar flex her foot.

    4. Hold her ankle with your right hand.

    5. Ask her to maintain this leg position. Then attempt to extend her right knee.

    Remarks—If the test yields weakness, compare the result with a test in which the patient's foot is dorsiflexed and everted. Dorsiflexion and eversion reduce the distal translation mobility of the fibula.

- ***Active mobilization of the fibular head***—The treatment procedure for each test is identical to the test, except the treatment contraction is gentler and sustained. Retest after the mobilization.

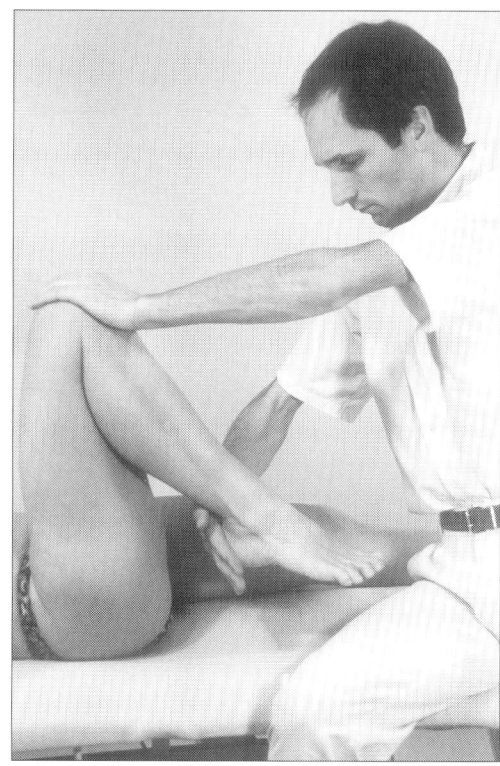

Fig. 19.229

Ankle and Tarsal Joints

Plantar and dorsiflexion occur between the convex talus and the concave distal tibia and fibula. During plantar flexion, the convex talus translates anteriorly. During dorsiflexion, the talus translates posteriorly between the malleoli.

The subtalar joint allows inversion and eversion. The subtalar joint consists of two parts: anterior and posterior. The posterior part of the joint consists of a convex calcaneus on a concave talus. The anterior part of the joint has the reverse shape; the calcaneus is concave on the talus. During inversion, the anterior part of the calcaneus, along with the forefoot, swings medially. The direction of anterior calcaneus translation is medial as well. The posterior part of the calcaneus translates laterally during inversion. The reverse occurs during eversion; the anterior calcaneus translates laterally while the posterior calcaneus translates medially.

Function of the midtarsal joints can be assessed by testing plantar flexion, dorsiflexion, inversion, and eversion. In addition, the midtarsal bones can be evaluated and treated with passive angular movements. For instance, each of the five tarsal rows can be passively moved. For evaluation and treatment, it is not necessary to know which tarsal bone is convex and which is concave.

- ***Capsular pattern***

    – Ankle joint: Plantar flexion is most limited; dorsiflexion is less limited.

    – Subtalar joint: Inversion is grossly limited; eversion is less limited.

    – Midtarsal joints: Equal limitation of supination, plantar flexion, and adduction. Extension is less limited. Abduction and pronation are free.

Ankle Plantar Flexion

- ***Agonists***

    Ankle plantar flexors:
    gastrocnemius (tibial nerve; S1,2)
    soleus (tibial nerve; S1,2)
    plantaris (tibial nerve; S1,2)
    tibialis posterior (tibial nerve; L4,5)
    flexor hallucis longus (tibial nerve; S2,3)
    flexor digitorum longus (tibial nerve; S2,3)
    peroneus longus (superficial peroneal nerve; L5–S2)
    peroneus brevis (superficial peroneal nerve; L5–S2)

- ***Muscle test*** (Fig. 19.230)

    1. Have the patient lie supine with her heel off the edge of the table and her right ankle plantar flexed.

    2. Stand to the right of her ankle, fixating her distal tibia with your left hand.

    3. Place the fingers of your right hand on the plantar surface of her distal metatarsals and your right thumb on the dorsal side of her tarsals.

    4. Ask her to maintain the plantar flexion. Then attempt to dorsiflex her ankle.

    Remarks—If the test yields weakness, compare the result with a test in which the patient's foot is less plantar flexed.

    In addition to a spinal or nerve tissue disorder, weakness throughout the range could be caused by limited flexion mobility of the midtarsals or toes.

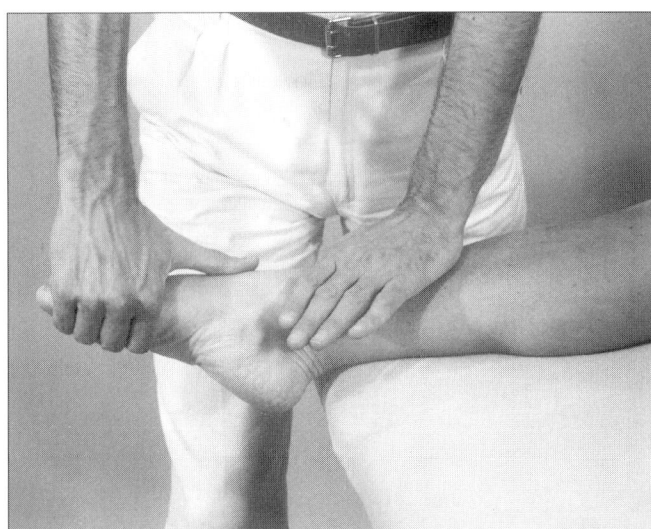

Fig. 19.230

- *Passive mobilization* (Fig. 19.231)

1. Have the patient lie supine with her right knee bent and her foot in standing position on the table. The amount of knee flexion should be such that her ankle is maximally plantar flexed.
2. Stand distal to her foot, stabilizing it with your right hand.
3. With your left hand, mobilize her distal tibia and fibula posteriorly on her talus.
4. Retest.

Remarks—If the plantar flexors and invertors tested weak, simply mobilizing the fibula posteriorly could simultaneously improve the strength of both muscle groups. If the plantar flexors and evertors tested weak, mobilizing the tibia posteriorly could improve strength in both muscle groups.

Limited flexion mobility at the midtarsal joints may also inhibit the plantar flexors. The evaluation for a tarsometatarsal restriction is to passively flex the individual rays, comparing the end-feel of one ray to another. Actively testing flexion strength with your index finger on the plantar and distal part on each metatarsal is another way to evaluate the individual rows (Fig. 19.232). Treatment is a passive mobilization of the restriction.

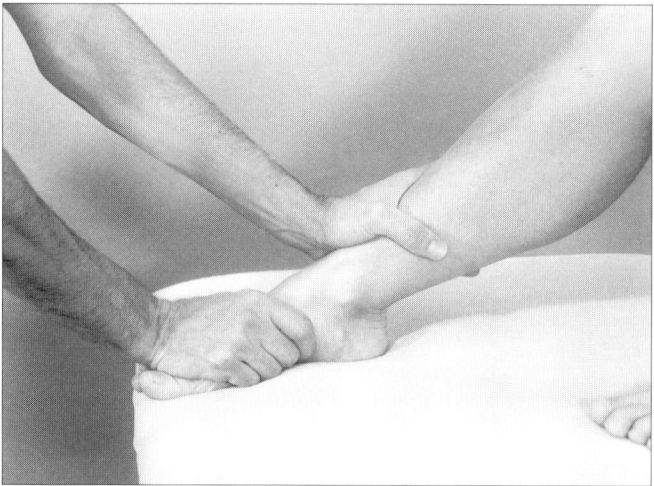

Fig. 19.231

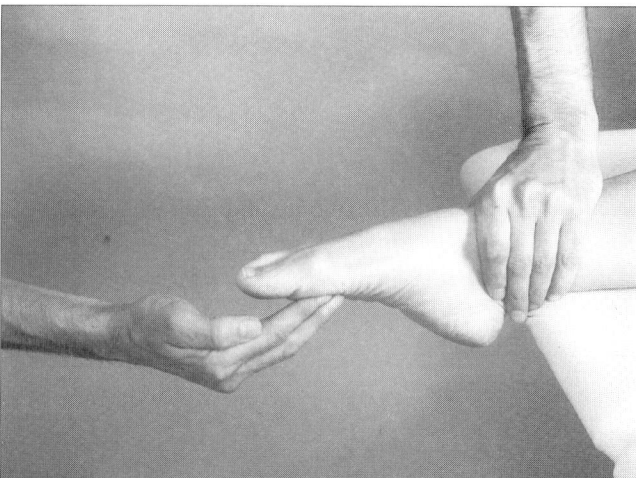

Fig. 19.232

- **Passive home exercise** (Fig. 19.233)—The following exercise will not be possible for all patients, but if it can be done, it can significantly reduce treatment time.

  1. Instruct the patient to sit on her heels with the dorsum of her feet on the floor and a towel roll under her distal tibiae. The height of the towel roll should be determined by the amount of plantar flexion available at her ankles.

  2. The patient should bear the weight of her trunk on her calcanei. If limited knee flexion prevents her from sitting on her calcanei, she can place a pile of folded towels between her buttocks and her heels.

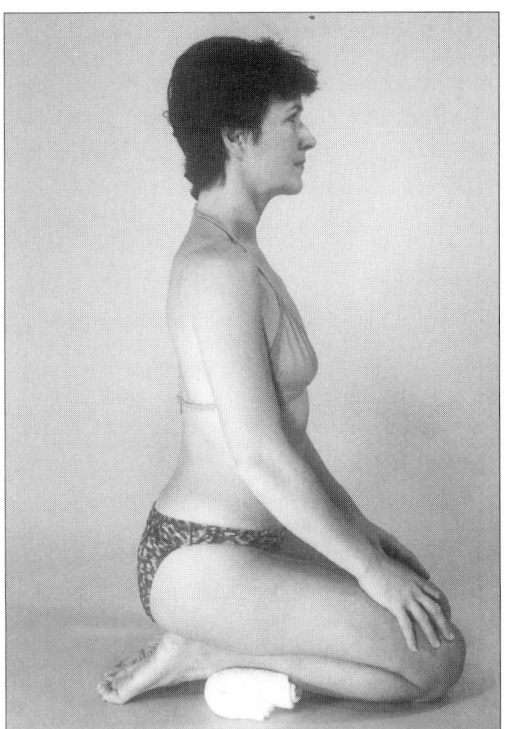

Fig. 19.233

Ankle Dorsiflexion

- **Agonists**

    Ankle dorsiflexors:
    anterior tibialis (deep peroneal nerve; L4,5)
    extensor hallucis longus (deep peroneal nerve; L5,S1)
    extensor digitorum longus (deep peroneal nerve; L5,S1)
    peroneus tertius (deep peroneal nerve; L5,S1)

- **Muscle test** (Fig. 19.234)

    1. Have the patient lie supine with her right knee flexed and her right ankle fully dorsiflexed.

    2. Stand at her right side, cupping her heel with your right hand.

    3. Place your left hand on the dorsum of her distal metatarsals.

    4. Ask her to keep her foot dorsiflexed. Then attempt to plantar flex her ankle.

    Remarks—If the test yields weakness, compare the result with a test in which the foot is less dorsiflexed.

    In addition to a spinal or a nerve tissue lesion, weakness throughout the range could be caused by a limited extension mobility at the midtarsal or toe joints. The evaluation for midtarsal restrictions is to passively extend the individual rays, comparing the end-feel of one ray to another. Actively testing extension strength with your fingers on the dorsal and distal part of each metatarsal is another way to evaluate the individual rows (Fig. 19.235). The treatment is a passive mobilization of the restriction.

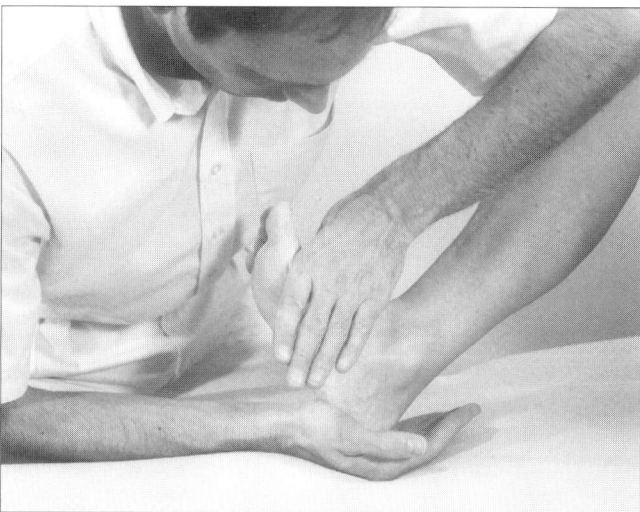

Fig. 19.234

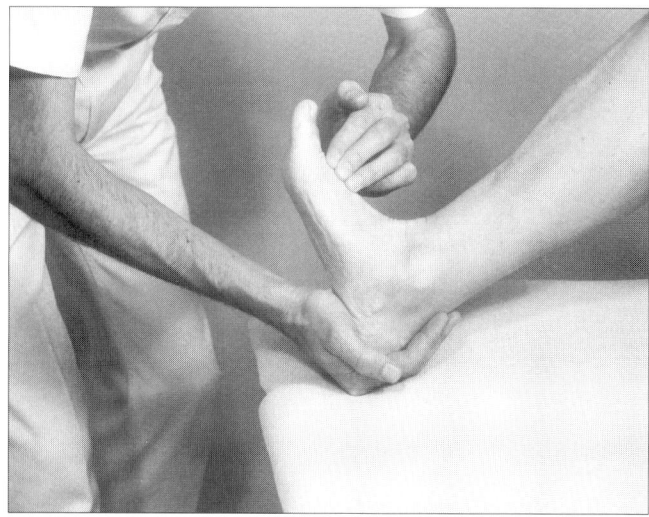

Fig. 19.235

- *Passive mobilization* (Fig. 19.236)

  1. Have the patient stand with her left foot in front of her right foot, holding onto the treatment table in front of her. Her right knee should be flexed and her ankle fully dorsiflexed while her heel remains on the floor.

  2. Kneel behind her right foot and place your thumbs, metacarpals, or the base of your hands on the posterior aspects of her medial and lateral malleoli.

  3. Mobilize the malleoli anteriorly.

  4. Retest.

  Remarks—If the dorsiflexors and invertors tested weak, simply mobilizing the tibia anteriorly could simultaneously strengthen both muscle groups. If the dorsiflexors and evertors tested weak, mobilizing the fibula anteriorly could simultaneously strengthen both muscle groups.

  To enhance the effect of the mobilization, the patient can attempt to lift her toes. This will add an active mobilization.

- *Home exercise* (Fig. 19.237)

  1. Instruct the patient to squat (sit on her heels with her ankles dorsiflexed), maintaining her balance by holding onto a table or chair.

  2. She then actively dorsiflexes her right foot.

  Remarks—A comfortable knee flexion mobility is a prerequisite to performing this exercise.

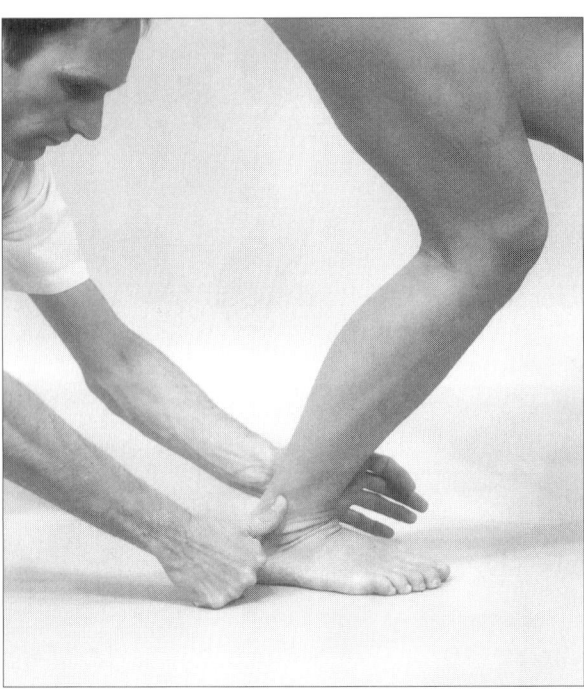

Fig. 19.236

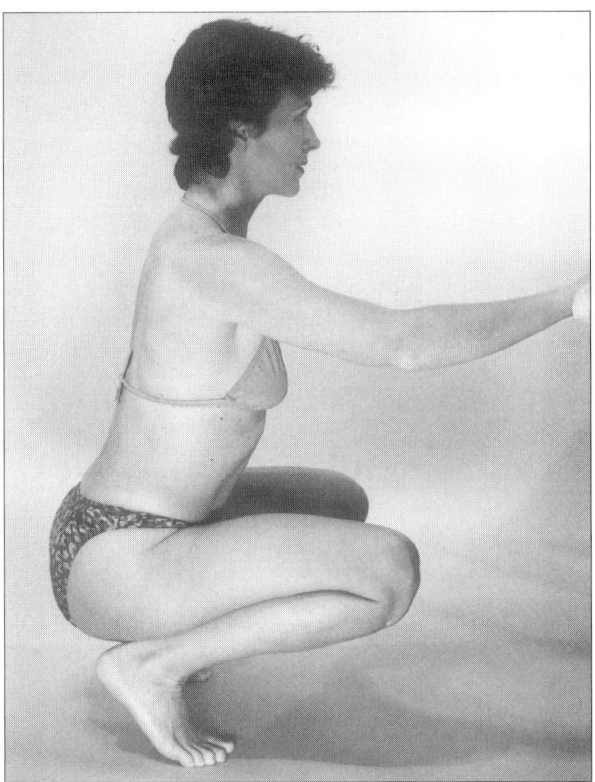

Fig. 19.237

Ankle Inversion

- **Agonists**

    Ankle invertors:
    anterior tibialis (deep peroneal nerve; L4,5)
    posterior tibialis (tibial nerve; L4,5)
    flexor hallucis longus (tibial nerve; S2,3)
    flexor digitorum longus (tibial nerve; S2,3)

- **Muscle test** (Fig. 19.238)

    1. Have the patient lie supine with her right heel off the edge of the table. Her foot should be plantar flexed and inverted.

    2. Position yourself distal to her foot, fixating her lateral malleolus with your left hand.

    3. Place the web between the thumb and index finger of your right hand distally on her metatarsal I.

    4. Ask her to maintain this foot position. Then attempt to evert her foot.

Remarks—If the test yields weakness, compare the result with a test in which the foot is less inverted. If the invertors test strong in less inversion, inhibition from the subtalar joint is likely. Such patients often have a limited passive inversion excursion as well. To differentiate between inhibition from the anterior and posterior parts of the subtalar joint, these parts can be stressed separately during the muscle test.

To stress the anterior part of the subtalar joint while testing the invertors, do the following. While stabilizing the lateral malleolus with your left hand, press with your left thumb on the anterior part of the calcaneus in a medial direction (Fig. 19.239). This places emphasis on the anterior part of the subtalar joint. Then test the strength of the invertors.

Similarly, while stabilizing the lateral malleolus with the base of your left hand, use your left fingers to pull the posterior part of the calcaneus in a lateral direction (Fig. 19.240). This places emphasis on the posterior part of the subtalar joint. Then test the strength of the invertors. In addition to using muscle testing to determine if the posterior and/or anterior parts of the subtalar joint have a limited inversion excursion, you can also passively assess the end-feel of the two joints.

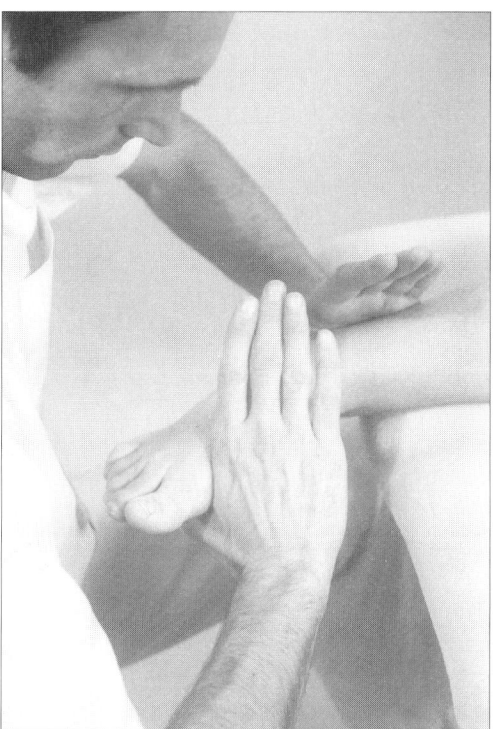

Fig. 19.238

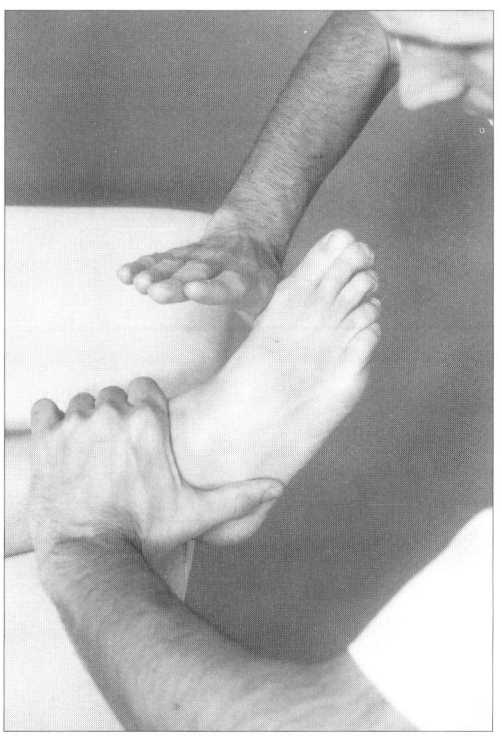

Fig. 19.239

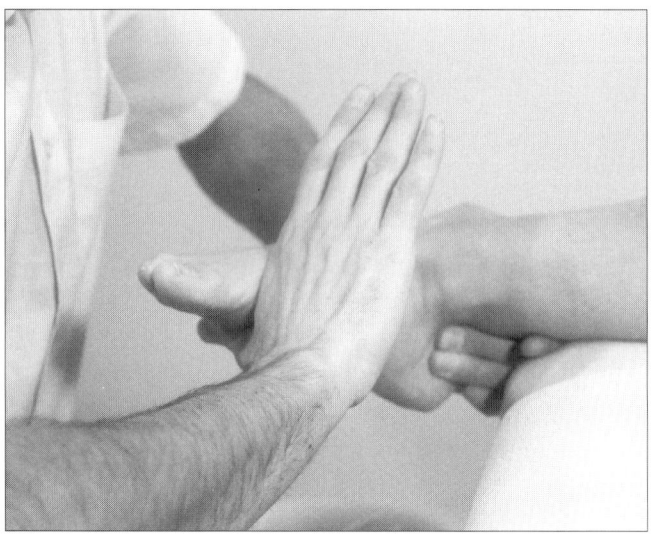

Fig. 19.240

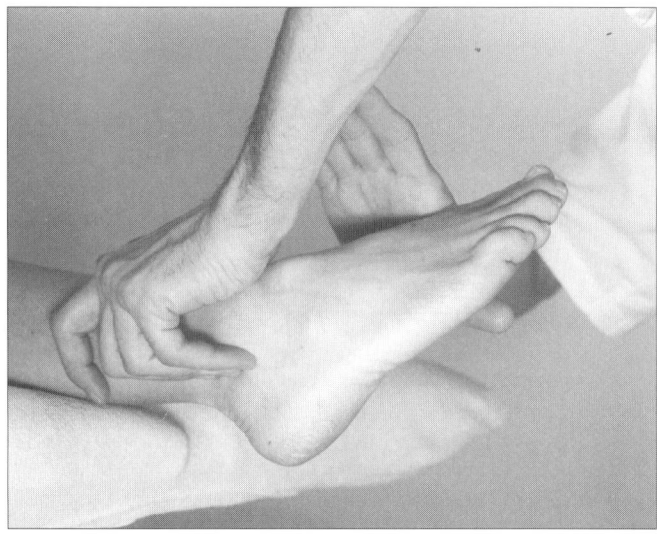

Fig. 19.241

If the invertors tested weak throughout their range and the weakness is not due to a spinal or nerve tissue disorder, the cause could be located in the distal tibiofibular joint. During the muscle test, either in maximal inversion or in less inversion, the invertor tendons push the distal tibia anteriorly. Thus, a limited anterior translation of the distal tibia on the fibula can inhibit the invertors. Such patients might have a normal passive inversion excursion.

To differentiate between the distal tibiofibular joint and other causes of invertor inhibition throughout the range, you could eliminate any possible inhibition from the distal tibiofibular joint. This is done in the following manner. While stabilizing the lateral malleolus with your left hand, hook your fingertips onto the lateral malleolus to pull it anteriorly (Fig. 19.241). With your thumb or the base of your hand, push the distal tibia posteriorly. If this maneuver removes inhibition from the invertors, a limited anterior mobility of the distal tibia on the fibula is likely.

Invertor weakness throughout the range may also result from abduction stiffness of the metatarsophalangeal (MTP) I joint. To check for inhibition from the MTP I joint, repeat the invertor test, keeping the MTP I joint passively abducted with your thumb (Fig. 19.242). If limited hallux abduction is responsible for invertor weakness, the invertors will test weaker when the toe is abducted and stronger when the toe is adducted.

Limited inversion mobility at the midtarsal joints can also inhibit the ankle invertors. Weakness will be present in a large part of the range. The evaluation for a midtarsal restriction is done by passively moving the forefoot into inversion, comparing the end-feel of one ray to another. The treatment is a passive mobilization of the restriction.

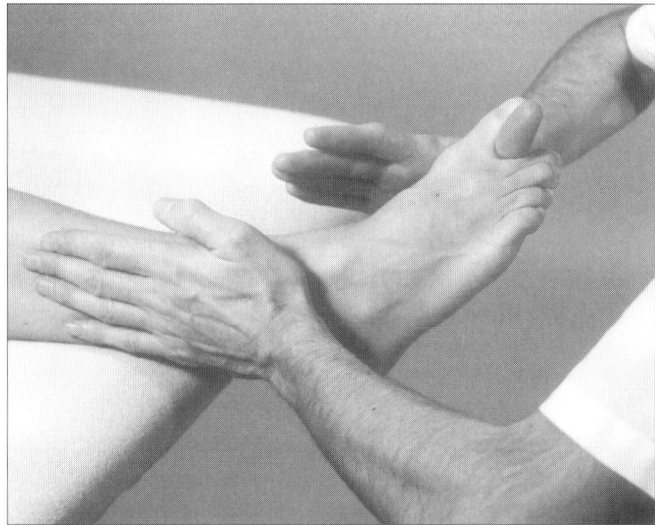

Fig. 19.242

- *Combined active and passive mobilization (distal tibiofibular joint)* (Fig. 19.243)—The treatment is similar to the testing procedure, except the treatment contraction is gentler and sustained. With the hand (your left) fixating the ankle, assist the mobilization by passively pulling the distal tibia anteriorly with your fingertips while pushing the distal fibula posteriorly with your thumb. Retest after the mobilization.

  Remarks—If the invertors are arthrogenically inhibited by the distal tibiofibular joint (weakness throughout the range), and the plantar flexors are arthrogenically inhibited as well (weakness in plantar flexion only), simply mobilizing the fibula posteriorly on a fixed, plantar flexed foot may simultaneously improve the strength of both muscle groups.

  If the invertors are arthrogenically inhibited by the distal tibiofibular joint (weakness throughout the range), and the dorsiflexors are arthrogenically inhibited as well (weakness in dorsiflexion only), mobilizing the tibia anteriorly on a fixed, dorsiflexed foot may improve the strength of both muscle groups simultaneously.

- *High-velocity thrust maniuplation (distal tibiofibular joint)* (Fig. 19.244)

  1. Have the patient lie supine with her right heel just off the edge of table.
  2. Stand distal to her right foot, supporting her ankle with your right hand, with your fingertips on the outside of her lateral malleolus. Place your right thenar eminence posterior to her medial malleolus (thumb pointing proximally).
  3. Place the base of your left hand anteriorly on her lateral malleolus.
  4. Lift her leg, making sure it is relaxed
  5. While maintaining your hand position, gently drop her leg to the table.
  6. Retest.

  Remarks—As the leg drops, the impact of your right hand with the table quickly thrusts the medial malleolus anteriorly. The inertia of your left arm and the fibula will quickly thrust the lateral malleolus posteriorly.

  Good knee extension is a prerequisite for this technique.

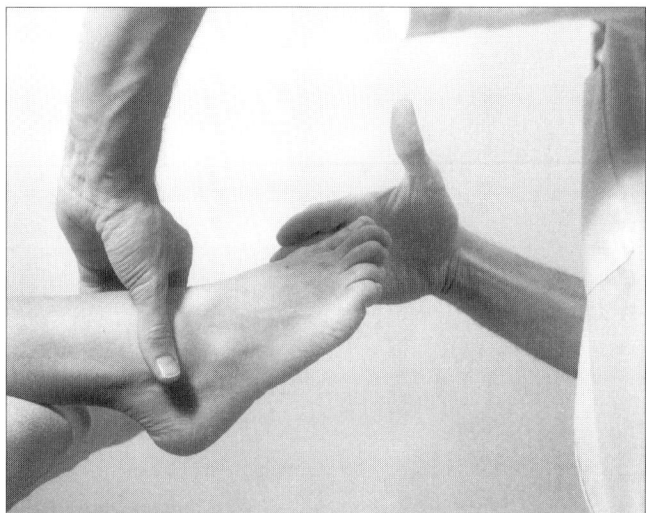

Fig. 19.243

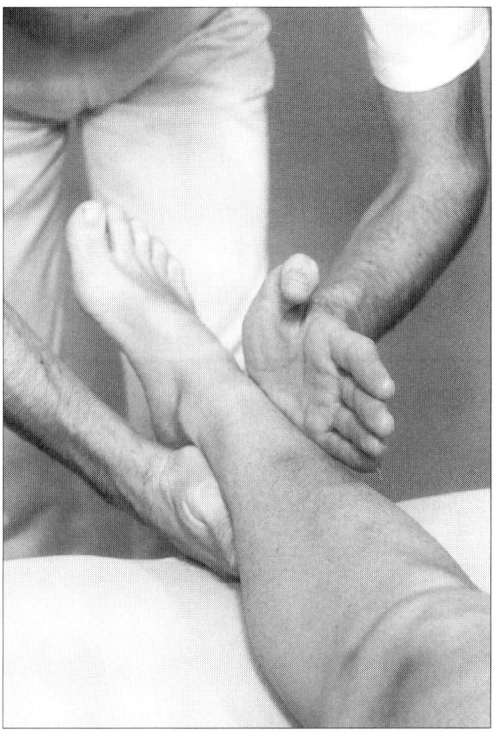

Fig. 19.244

- *Passive mobilization (anterior part of subtalar joint)* (Fig. 19.245)

    1. Have the patient lie supine with her right heel off the edge of the table.
    2. Sit distal to her right foot, fixating her medial malleolus with your right hand.
    3. Hold her midtarsals with your left hand, placing your pisiform on the distal part of her calcaneus.
    4. Invert her foot with your left hand.
    5. With the pisiform of your left hand, mobilize the anterior part of her calcaneus medially.
    6. Retest.

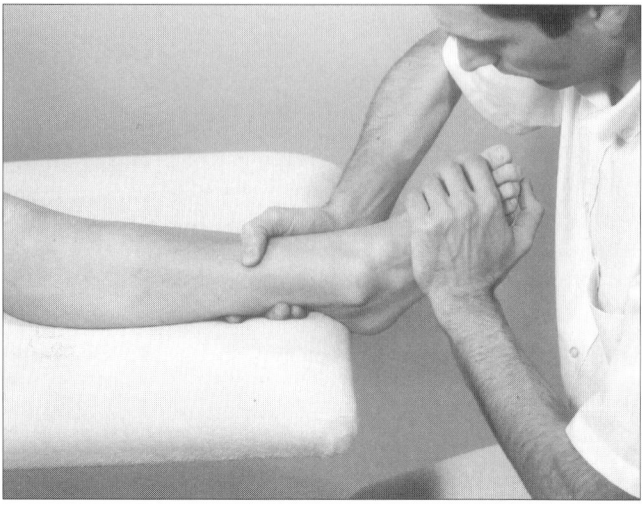

Fig. 19.245

- *Passive mobilization (posterior part of subtalar joint)* (Fig. 19.246)

    1. Have the patient lie supine with her right heel off the edge of the table.
    2. Sit distal to her right foot, fixating her lateral malleolus with your left hand.
    3. Hold her calcaneus with your right hand, placing your thenar eminence on the posterior and medial part of her calcaneus.
    4. Invert her foot with your right hand.
    5. With the thenar eminence of your right hand, mobilize the dorsal part of her calcaneus laterally.
    6. Retest.

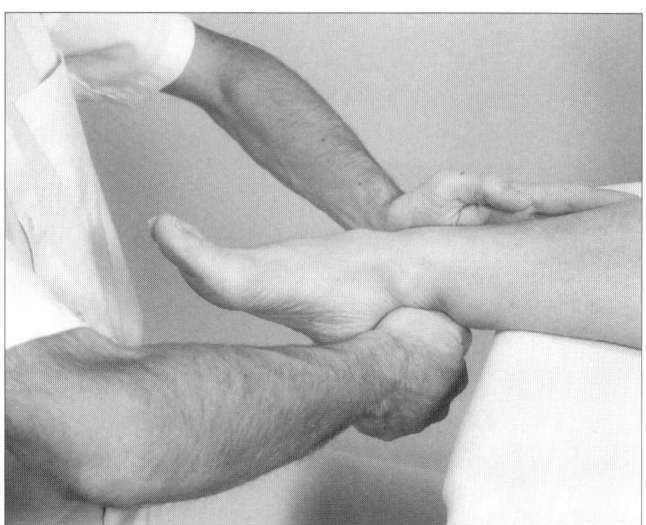

Fig. 19.246

- *Hallux (MTP I) abduction* (Fig. 19.247)

    1. With the patient supine, sit at her right side, stabilizing her metatarsals with your left hand.
    2. With the fingers of your right hand, hold the proximal phalanx of her hallux and abduct the hallux.
    3. Translate the hallux in a medial direction.
    4. Retest.

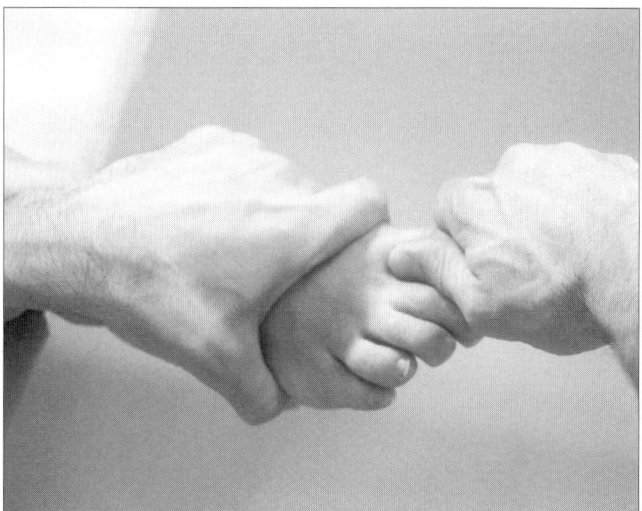

Fig. 19.247

Ankle Eversion

- **Agonists**

    Ankle evertors:
    peroneus longus (superficial peroneal nerve; L5–S2)
    peroneus brevis (superficial peroneal nerve; L5–S2)

- **Muscle test** (Fig. 19.248)

    1. Have the patient lie supine with her right heel off the edge of the table. Her right foot should be plantar flexed and everted.

    2. Fixate her medial malleolus with your right hand.

    3. Place the web between the thumb and index finger of your left hand distally on her metatarsal V.

    4. Ask her to keep her foot everted. Then attempt to invert her foot.

Remarks—If the test yields weakness, compare the result with a test in which the patient's foot is less everted. If the evertors test strong in less eversion, an inhibition from the subtalar joint is likely. Patients with arthrogenic inhibition of the evertors deriving from the subtalar joint often have a limited passive eversion excursion as well. To differentiate between inhibition from the anterior and the posterior parts of the subtalar joint, stress each part separately during the muscle test.

To stress eversion at the anterior part of the subtalar joint, do the following. While stabilizing the medial malleolus with your right hand, press the anterior part of the calcaneus in a lateral direction with your left thumb (Fig. 19.249). Then test the strength of the evertors.

To stress eversion at the posterior part of the subtalar joint, do the following. While stabilizing the medial malleolus with your right hand, pull the posterior and proximal part of the calcaneus in a lateral direction with your right fingertips (Fig. 19.250). Then test the strength of the evertors. In addition to using muscle testing to determine if the posterior and/or anterior parts of the subtalar joint have a limited eversion excursion, you can also passively assess the end-feel of the two joints.

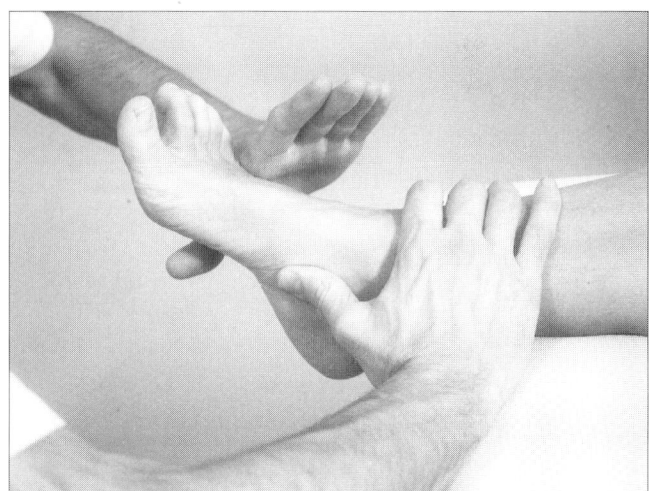

Fig. 19.249

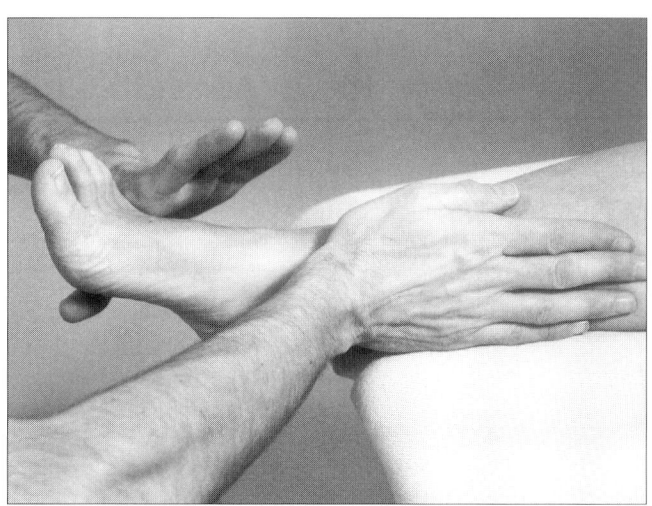

Fig. 19.248

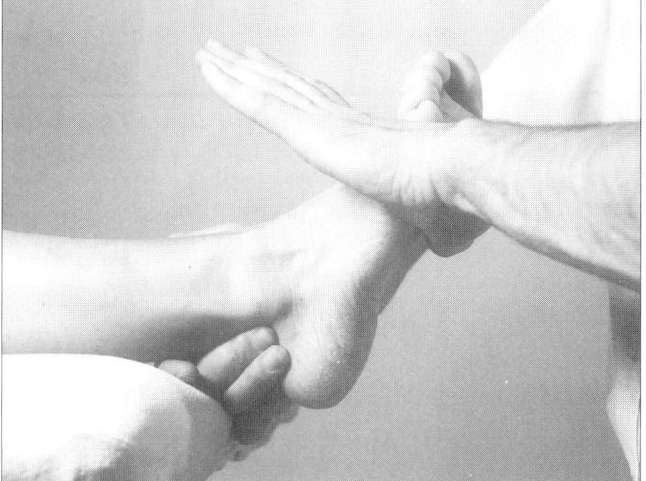

Fig. 19.250

If the eversion test yields weakness and testing in less eversion does not improve this condition, the cause may be located in the distal tibiofibular joint. During the muscle test, whether in full eversion or in less eversion, the peroneal tendons push the lateral malleolus anteriorly. Limited anterior mobility of the distal fibula on the tibia can inhibit the evertors. Patients with a stiff distal tibiofibular joint may have a normal passive eversion excursion.

To differentiate between the distal tibiofibular joint and other causes of evertor inhibition throughout the range, the therapist can eliminate inhibition from the distal tibiofibular joint. While stabilizing the medial malleolus with your right hand, push the lateral malleolus posteriorly with your fingertips (Fig. 19.251) and push the tibia anteriorly with your thumb (Fig. 19.252). In case of inhibition from the distal tibiofibular joint, this procedure eliminates inhibition and restores evertor strength.

- **Combined active and passive mobilization (distal tibiofibular joint)** (Fig. 19.253)—The treatment procedure is similar to the testing procedure, except the treatment contraction is gentler and sustained. With the hand (your right) fixating the ankle, assist the mobilization by passively mobilizing the distal fibula anteriorly with your fingertips while pushing the tibia posteriorly with the base of your hand. Retest after the mobilization.

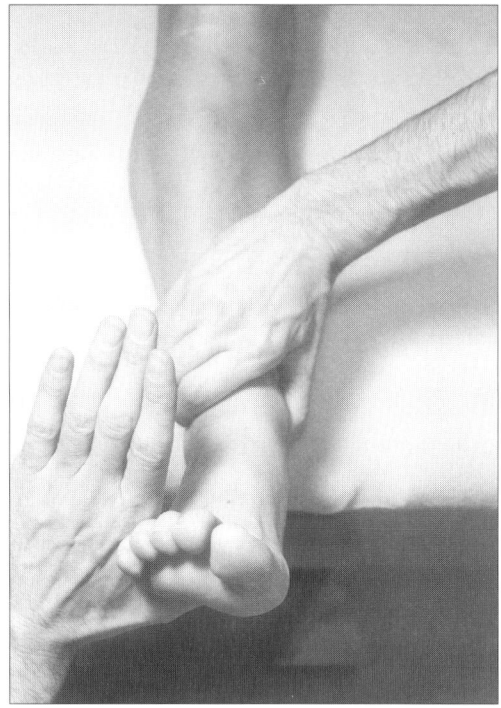

Fig. 19.252

Remarks—If the evertors are arthrogenically inhibited by the distal tibiofibular joint (weakness throughout the range) and the plantar flexors are arthrogenically inhibited by the ankle joint (weakness in plantar flexion only), simply mobilizing the tibia posteriorly while the foot is plantar flexed might simultaneously improve the strength of both muscle groups.

Fig. 19.251

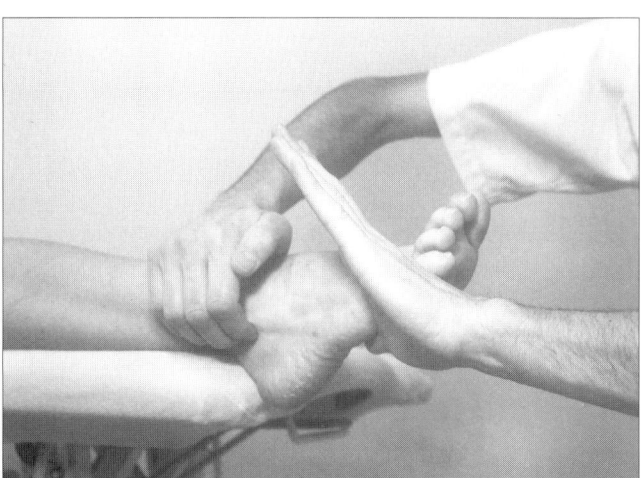

Fig. 19.253

If the evertors are arthrogenically inhibited by the distal tibiofibular joint (weakness throughout the range) and the dorsiflexors are arthrogenically inhibited by the ankle joint (weakness in dorsiflexion only), mobilizing the fibula anteriorly while the foot is stabilized in dorsiflexion might simultaneously improve the strength of both muscle groups.

- *High-velocity thrust manipulation (distal tibiofibular joint)* (Fig. 19.254)

  1. Have the patient lie supine with her right heel slightly off the table.
  2. Place your left hand posterior on her ankle with your fingers on the medial side. Place your thenar eminence behind her lateral malleolus, with your left thumb pointing cranially.
  3. Rest your right hand anteriorly on her distal tibia.
  4. Lift her right leg and ensure that the leg is relaxed.
  5. Maintaining your hand position, drop the leg to the table.
  6. Retest.

  Remarks—The patient needs to have good extension function of her knee.

The manipulating force is produced by your left hand landing on the table, thrusting the fibula anteriorly. The counterforce is produced by the inertia of your right hand and the tibia, translating the tibia posteriorly.

Limited eversion mobility at the midtarsal joints can also inhibit the ankle evertors. The evaluation for a midtarsal restriction is to evert the midfoot passively and evaluate mobility. The treatment is a passive mobilization of the restriction.

- *Passive mobilization (anterior part of subtalar joint)* (Fig. 19.255)

  1. Have the patient lie supine with her right heel off the edge of the table and her right foot relaxed.
  2. Position yourself distal to her right foot, stabilizing her lateral malleolus with your left hand.
  3. Hold her foot with your right hand, placing your right pisiform medially on the anterior part of her calcaneus.
  4. Evert her foot with your right hand and, with your right pisiform, mobilize her calcaneus laterally.
  5. Retest.

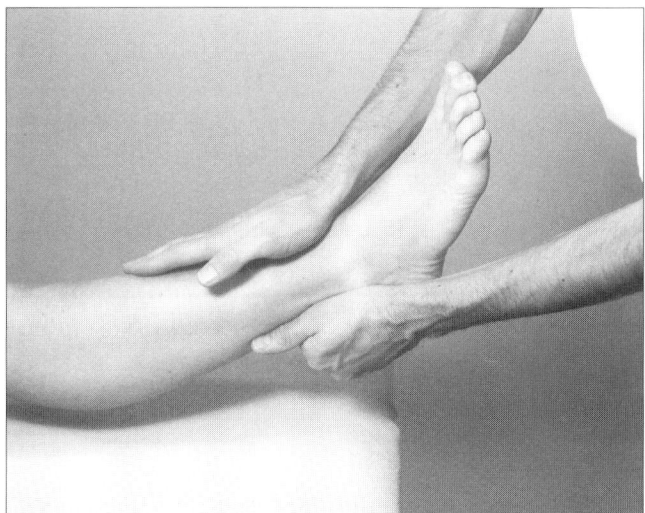

Fig. 19.254

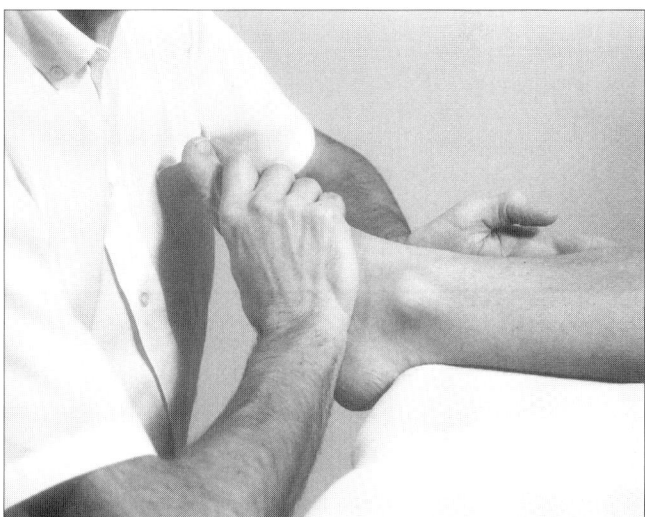

Fig. 19.255

- **_Passive mobilization (posterior part of subtalar joint)_** (Fig. 19.256)

  1. Have the patient lie supine with her right heel off the edge of the table and her right foot relaxed.

  2. Position yourself distal to her right foot, stabilizing her medial malleolus with your right hand.

  3. Hold her posterior calcaneus with your left hand, placing the thenar eminence on the posterior and proximal part of her calcaneus.

  4. Evert her calcaneus with your left hand and, with your thenar eminence, translate her posterior calcaneus in a medial direction.

  5. Retest.

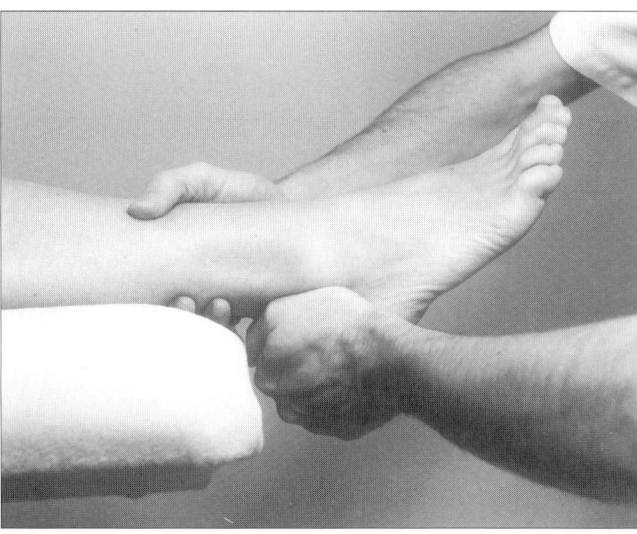

Fig. 19.256

Toes

The distal articular bones in the toes are concave on their more proximal articular partners. Therefore, the direction of translation is always in the direction of movement of the distal bone. The MTP I joint follows these same rules for abduction and adduction movements as well. During hallux abduction, translation of the proximal phalanx is medial, and during adduction, the translation is lateral.

- **Capsular pattern**

  - Metatarsophalangeal I joint: Extension is markedly limited; flexion is slightly limited

  - Metatarsophalangeal II–V joints: Flexion and extension are variably limited, but in the end stage, flexion is more limited.

  - Interphalangeal joints: Extension is often more limited than flexion.

Toe Extension

- **Agonists**

  Toe extensors:
  extensor hallucis longus (deep peroneal nerve; L5,S1)
  extensor hallucis brevis (deep peroneal nerve; L5,S1)
  extensor digitorum longus (deep peroneal nerve; L5,S1)
  extensor digitorum brevis (deep peroneal nerve; S1,S2)

- **Muscle test** (Fig. 19.257)

  1. Have the patient lie supine with her right foot moderately plantar flexed. Her toes should be extended.
  2. Position yourself distal to her foot.
  3. Fixate the distal and plantar part of her metatarsal I with your left thumb.
  4. Place your right index finger on the dorsal and distal aspect of her proximal phalanx I.
  5. Ask her to keep her toes extended. Then attempt to flex her MTP I joint.
  6. Test MTP extension of the other four rows in a similar manner, comparing strength.

Remarks—If any of the tests yields weakness, compare the result with a test in which the patient's toes are less extended.

Extensor strength over the interphalangeal joints is tested similarly. Stabilize the proximal phalanx and test the more distal phalanx (Fig. 19.258).

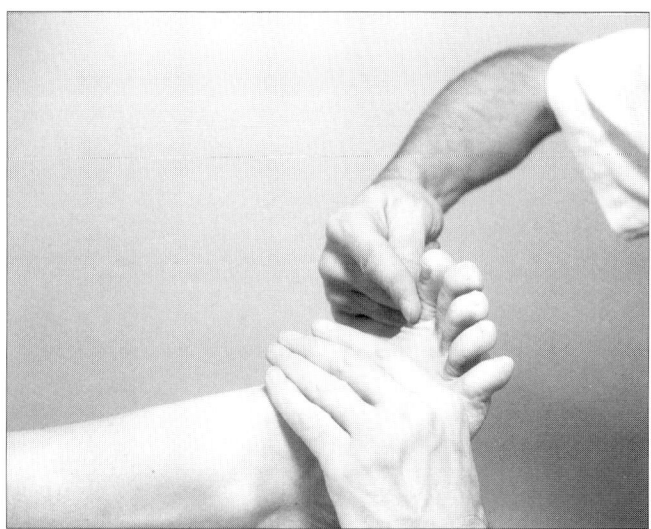

Fig. 19.257

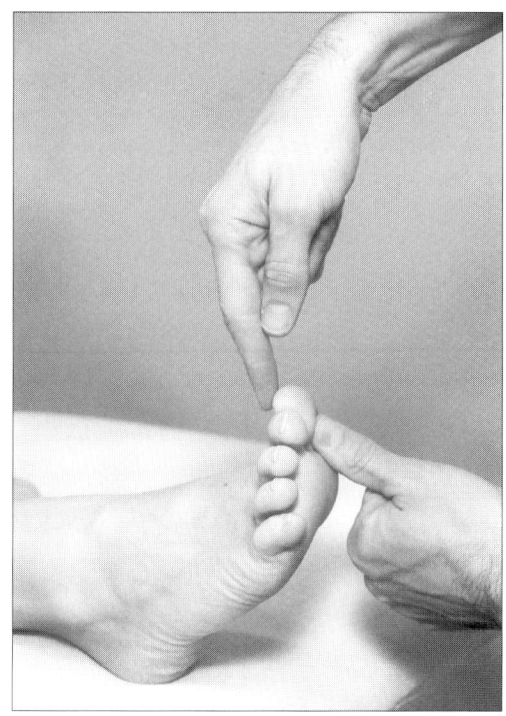

Fig. 19.258

- ***Passive mobilization*** (Fig. 19.259)

    1. Have the patient lie supine with her right foot relaxed.

    2. Stand to the right of her foot.

    3. Fixate the proximal articular bone (metacarpal or phalanx) with the fingers of your left hand. Your left thumb should be on the dorsal side of the foot, close to the joint.

    4. With the fingers of your right hand, hold the more distal phalanx and passively extend it.

    5. Mobilize the distal phalanx dorsally.

    6. Retest.

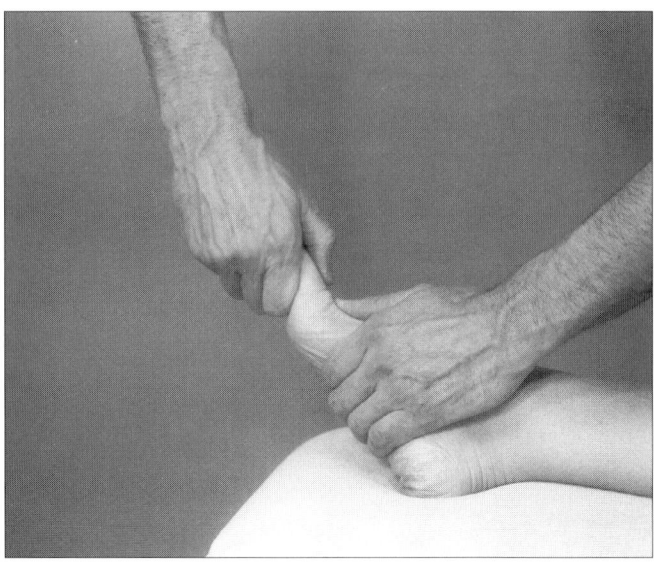

Fig. 19.259

Toe Flexion

- **Agonists**

    Toe flexors:
    flexor hallucis longus (tibial nerve; S2,3)
    flexor digitorum longus (tibial nerve; S2,3)
    flexor hallucis brevis (medial plantar nerve, a branch from the tibial nerve; S2,3)
    flexor digitorum brevis (medial plantar nerve, a branch from the tibial nerve; S2,3)
    flexor digitorum accessorius (lateral plantar nerve, a branch from the tibial nerve; S2,3)
    flexor digiti minimi brevis (lateral plantar nerve, a branch from the tibial nerve; S2,3)
    dorsal and plantar interossei (lateral plantar nerve, a branch from the tibial nerve; S2,3)

- **Muscle test** (Fig. 19.260)

    1. Have the patient lie supine with her right foot dorsiflexed and her toes flexed.
    2. Sit at her right side.
    3. Fixate the distal and dorsal aspect of her metatarsal I with your left thumb.
    4. Place your right index finger on the plantar and distal aspect of her toe (hallux).
    5. Ask her to keep her toes flexed. Then attempt to extend her MTP I joint.
    6. Test MTP flexor strength over the other four rows in a similar manner, comparing strength.

    Remarks—If any of the tests yields weakness, compare the result with a test in which the patient's toes are less flexed.

    Flexor strength over the interphalangeal joints is tested in a similar manner. Stabilize the phalanx directly proximal to the joint tested with your left index finger. Place your right index finger on the plantar tip of her toe. Direct the pressure of your right index finger toward extending the joint being tested.

- **Passive mobilization** (Fig. 19.261)

    1. Have the patient lie supine with her right foot relaxed.
    2. Fixate the proximal articular bone (metatarsal or phalanx) with your left hand.
    3. With your right hand, hold the phalanx distal to the joint that was tested and passively flex the joint.
    4. Mobilize the distal articular bone in a plantar direction.
    5. Retest.

- **Active mobilization** (Fig. 19.260)—The active treatment is identical to the muscle test, except the treatment contraction is gentler and sustained. Retest after the mobilization.

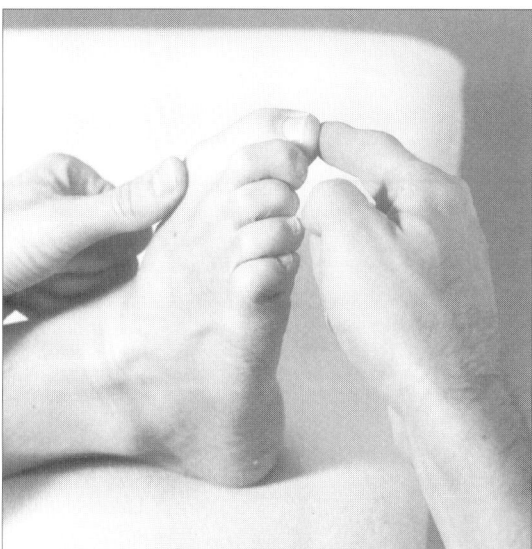

Fig. 19.260

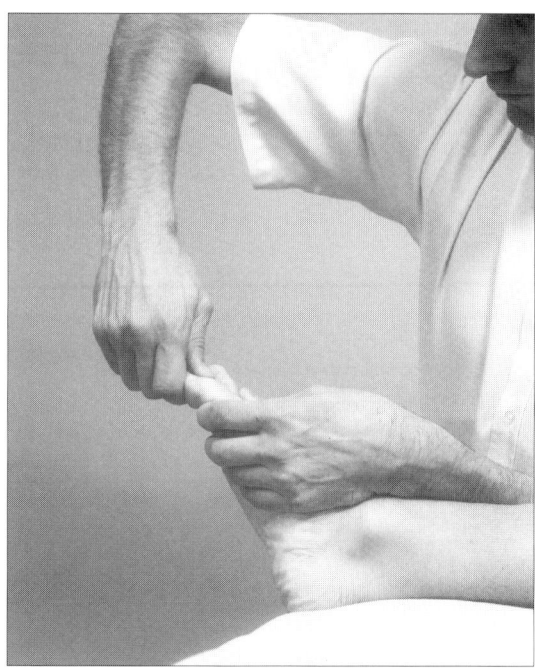

Fig. 19.261

Radiculopathy and Pseudoradiculopathy

Testing the extremity muscles can provide information regarding the function of the spine and the level of any possible dysfunction. In addition to evaluating for spinal dysfunctions, peripheral muscle tests can be used to evaluate for a spinal component to peripheral dysfunctions.

The extremity muscles are tested in ipsilateral facet flexion and extension. The listing of muscle groups serves as an example only. The list coincides with Peck's algorithm for the innervation of peripheral muscles.[22] Any extremity muscle or muscle group can be tested for the influence of spinal position on the muscle or muscles.

The position of the patient's extremity joint is not critical, as a spinal cause of extremity muscle weakness causes weakness throughout the muscle's entire range. However, a midrange position of the muscle may be preferable to any shortened position to eliminate picking up arthrogenic inhibition from the extremity joint the muscle crosses.

Treatment of upper or lower extremity muscle weakness caused by radiculopathy or pseudoradiculopathy involves treating the cervical or lumbar spine or their respective nerve roots. For treatment of radiculopathy or pseudoradiculopathy, refer to chapter 17 of this text.

Cervical Facet Extension (foraminal closing)

The performance of unilateral cervical extension, whether as a muscle test or as a treatment, consists of cervical movements that are similar to the vertebral artery test. Be aware of the possibility of vertebral artery symptoms.

- **Agonists**

    shoulder girdle:
    – elevators; C3,C4
    – depressors; C7,C8,T1

    shoulder:
    – abductors; C5,C6
    – adductors; C7,C8,T1

    elbow:
    – flexors; C5,C6
    – extensors; C6,C7,C8

    wrist:
    – flexors; C7,C8,T1
    – extensors; C6,C7

    thumb:
    – flexors; C8,T1
    – extensors; C7,C8

    fingers:
    – flexors; C8,T1
    – extensors; C8,T1

- **Muscle tests** (Fig. 19.262)

    1. Have the patient be seated with her cervical spine in right facet extension (ask her to bring her right ear toward her right shoulder blade).
    2. Stand next to her right arm.
    3. Test the above muscle groups on her right side.

    Remarks—If any test yields weakness, compare the result with a test in which the patient's right cervical spine is in a neutral position or flexed.

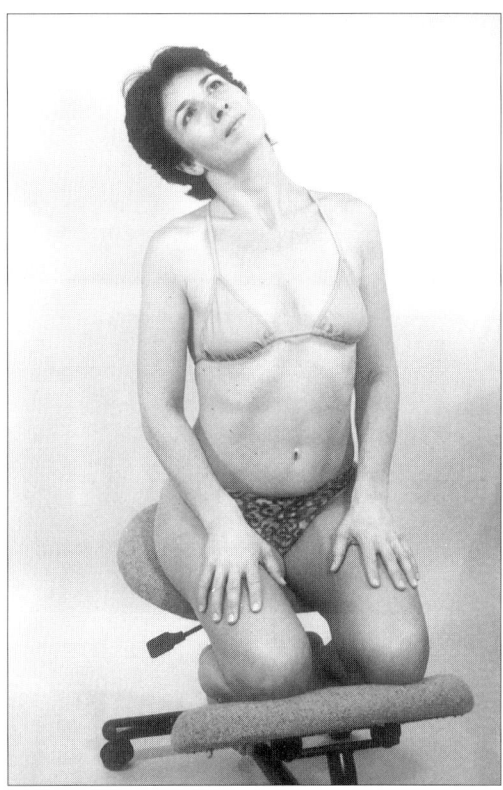

Fig. 19.262

Cervical Facet Flexion (foraminal opening)

- **Agonists**

    shoulder girdle:
    – elevators; C3,C4
    – depressors; C7,C8,T1

    shoulder:
    – abductors; C5,C6
    – adductors; C7,C8,T1

    elbow:
    – flexors; C5,C6
    – extensors; C6,C7,C8

    wrist:
    – flexors; C7,C8,T1
    – extensors; C6,C7

    thumb:
    – flexors; C8,T1
    – extensors; C7,C8

    fingers:
    – flexors; C8,T1
    – extensors; C8,T1

- **Muscle tests** (Fig. 19.263)

    1. Have the patient be seated with her cervical spine in right facet flexion (ask her to look at her left hip).

    2. Stand next to her right arm.

    3. Test the above right muscle groups.

    Remarks—If any test yields weakness, compare the result with a test in which the patient's right cervical spine is in a neutral position or extended.

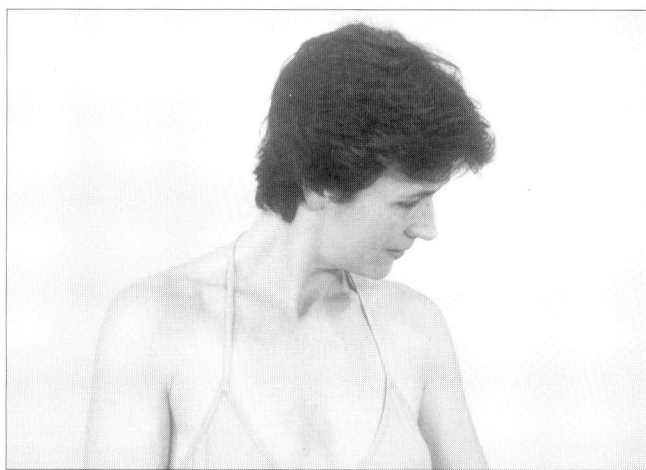

Fig. 19.263

Lumbar Facet Extension (foraminal closing)

- *Agonists*

    hip:
    - flexors; L2,L3
    - extensors; L5,S1,S2

    knee:
    - extensors; L3,L4
    - flexors; L5,S1,S2

    ankle:
    - dorsiflexors; L4,L5
    - plantar flexors; S1,S2

    toes:
    - extensors; L5,S1
    - flexors; S1,S2

- *Muscle test* (Fig. 19.264)

    1. Have the patient lie supine with both legs to her right (right lumbar facet extension).
    2. Stand next to her right leg.
    3. Test the above right muscle groups.

    Remarks—If any test yields weakness, compare the result with a test in which the patient's right lumbar spine is in a neutral or flexed position.

    Be aware that testing the hip flexors in right hip flexion may reduce the patient's lordosis. This can be countered by adducting the left leg more after the right hip is flexed.

Lumbar Facet Flexion (foraminal opening)

- *Agonists*

    hip:
    - flexors; L2,L3
    - extensors; L5,S1,S2

    knee:
    - extensors; L3,L4
    - flexors; L5,S1,S2

    ankle:
    - dorsiflexors; L4,L5
    - plantar flexors; S1,S2

    toes:
    - extensors; L5,S1
    - flexors; S1,S2

- *Muscle test* (Fig. 19.265)

    1. Have the patient lie supine with her left hip and knee bent and holding her left knee with both hands.
    2. Stand at her right and side bend her lumbar spine to the left (bring her pelvis to her left).
    3. Test the above right muscle groups.

    Remarks—If any test yields weakness, compare the result with a test in which the patient's right lumbar spine is in a neutral or extended position.

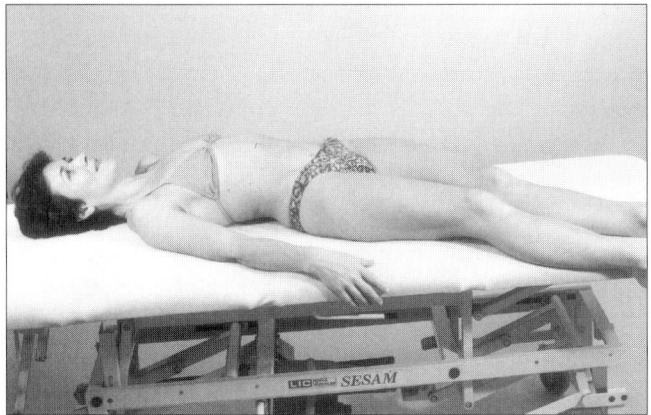

Fig. 19.264

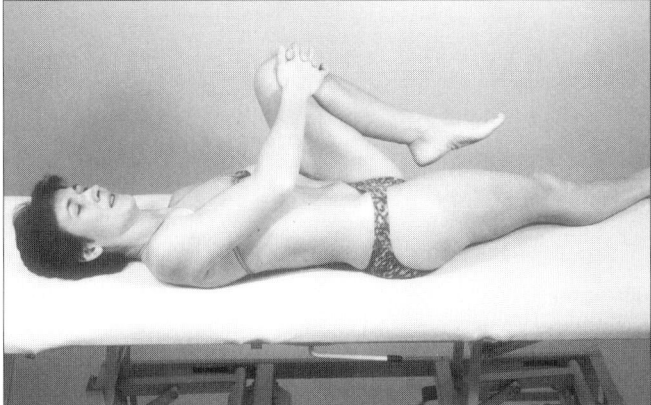

Fig. 19.265

References

1. Bakker A, Enschede, Netherlands, Praktijk voor Manuele Geneeskunde, Nov. 18, 1994, in a personal communication.
2. Biederman H: Kinematic imbalances due to suboccipital strain in newborns. J Manual Medicine 6: 151-156, 1992
3. Bijl Gvd, Graaf CG de, Ridder PA de: Actief en Passief Bewegen in de Gewrichten der Extremiteiten. Lochem, The Netherlands, De Tijdstroom, 1975
4. Brandsma JW: Intrinsieke parese-paralyse van de hand, in het bijzonder met betrekking tot de leprapatiënt. Ned Tijdschr Fysiotherapie 7/8: 187-193, 1976
5. Cibulka M, Koldehoff R: Evaluating chronic sacroiliac joint dysfunction. Clinical Management 6: 12-15, 1986
6. Cibulka MT, Rose SJ, Delitto A, et al: Hamstring muscle strain treated by mobilizing the sacroiliac joint. Physical Therapy 66: 1220-1223, 1986
7. Dorman TA, Brierly S, Fray J, et al: Muscles & pelvic clutch: Hip abductor inhibition in anterior rotation of the ilium. Journal Manual & Manipulative Therapy 3: 85-90, 1995
8. Grant R: Vertebral artery insufficiency: A protocol for pre-manipulative testing of the cervical spine. In Boyling JD, Palastanga N: Grieve's Modern Manual Therapy: The Vertebral Column, ed 2. New York, Churchill Livingstone, 1994, pp 371-380
9. Greenman PE: Principles of Manual Medicine, ed 2. Baltimore, Williams & Wilkins, 1996
10. London JT: Kinematics of the elbow. J Bone Joint Surg 63-A: 529-535, 1981
11. McKenzie RA: The Lumbar Spine: Mechanical Diagnosis and Therapy. Lower Hutt, New Zealand, Spinal Publications Ltd., 1986
12. McNeill T, Warwick D, Andersson G, et al: Trunk strength in attempted flexion, extension, and lateral bending in healthy subjects and patients with low-back disorders. Spine 5: 529-538, 1980
13. Mink AJF, Veer HJ ter, Vorselaars JACTh: Extremiteiten: Functie-Onderzoek en Manuele Therapie. Houten, The Netherlands, Bohn Stafleu Van Loghum, 1993
14. Mitchell FL: Elements of Muscle Energy technique. In Basmajian JV, Nyberg R (eds): Rational manual therapies. London, Williams & Wilkins, 1993
15. Mitchell FLjr, Mitchell PKG: The Muscle Energy manual: Concepts & mechanisms, the musculoskeletal system, cervical region evaluation and teatment. East Lansing, Michigan, MET Press, 1995, vol 1
16. Morrey BF, Kai-nan A: Articular and ligamentous contributions to the stability of the elbow joint. Am J Sports Med 11: 315-319, 1983
17. Oostendorp RAB, Bernards ATM, Meldrum HA, et al: Neurologie en manuele therapie: De vertebrobasilaire insufficiëntie (deel III). Nederlands Tijdschrift voor Manuele Therapie 4: 18-39, 1985
18. Oostendorp RAB, Bernards ATM, Querido C, et al: Neurologie en manuele therapie: De vertebrobasilaire insufficiëntie (deel I, II). Nederlands Tijdschrift voor Manuele Therapie 3: 48-73, 1984
19. Oostendorp RAB, Hagenaars LHA, Meldrum HA, et al: Neurologie en manuele therapie: De vertebrobasilaire insufficiëntie (deel IV). Nederlands Tijdschrift voor Manuele Therapie 5: 16-45, 1986
20. Paris SV: S3 Course Notes. St. Augustine, Florida, Institute Press, 1988, p 61
21. Parry CB: Rehabilitation of the Hand, ed 3. Boston, Butterworths, 1977, p 9
22. Peck D, Brower T: Algorithms for the segmental motor innervation of the extremities. The American Surgeon 53: 270-273, 1987
23. Pettman E: Stress tests of the craniovertebral joints. In Boyling JD, Palastanga N (eds): Grieve's Modern Manual Therapy: The Vertebral Column. New York, Churchill Livingstone, 1994, pp 529-537
24. Schreuders TAR, Brandsma JW: Het onderzoek van de spierkracht van de hand. FysioPraxis 5(8): 14-19, 1996
25. Van Bachum A, Elkhuizen J, Tilstra S: Epicondyalgie lateralis. Nederlands Tijdschrift Manuele Therapie 2: 2-20, 1983
26. Warmerdam A: Arthrokinetische Therapie: Spierfunctie verbeteren door gewrichtsmobilisaties. Ned T Fysiotherapie 101: 222-228, 1991
27. Warmerdam A: Arthrokinetic Therapy[SM]: Improving muscle performance through joint manipulation. Proceedings of the 1992 Conference of the International Federation of Orthopaedic Manipulative Therapists, Vail, Colorado, pp 204-207
28. Warmerdam A: Arthrokinetische Therapie: Naschrift. Ned T Fysiotherapie 102: 58, 1992
29. Williams PL, Warwick R (eds): Gray's Anatomy, ed 36. New York, Churchill Livingstone, 1986
30. Wyke B: Articular neurology and manipulative therapy. In Glasgow E, Twomey L (eds): Aspects of Manipulative Therapy. Edinburgh, Scotland, Churchill Livingstone, 1985, pp 72-77

MANUAL THERAPY: IMPROVE MUSCLE AND JOINT FUNCTIONING

20. Case Studies

Five case studies are presented in this chapter. The last three cases include examples of notation. The capital L, R, or B preceding the name of a muscle or muscle group denotes, respectively, left, right, or bilateral. Immediately following the name is a description of the test position, abbreviated as s, n, and l. These letters designate whether the muscle or muscle group was tested in a shortened position, in a neutral or loose-packed position of its joint, or in a lengthened position. More information on notation can be found in chapter 18.

Case 1, Median Nerve Paresis

A 30-year-old Dutchman presented with a median nerve paresis. He reported that he had gone skiing in Switzerland six months earlier and had dislocated his right shoulder during a fall. The orthopedist at the ski resort repositioned his shoulder, but it immediately dislocated again. During this office visit, the shoulder dislocated two more times before it remained in place. The orthopedist put his arm in a sling and told him to see an orthopedist when he returned to his home country after vacation. A week later, the patient returned to the Netherlands, where he visited an orthopedist close to his home town. This orthopedist put his arm and shoulder girdle in a cast, due to the multiple dislocations he had suffered. A day later, on Saturday, the patient noticed tingling in his hand, but it did not bother him. On Sunday, he noticed he could not move his fingers anymore. The patient waited until a regular business day (Monday) to call the orthopedist's office but was informed that the orthopedist's schedule was booked for that day and could not see him until Tuesday. On Tuesday, the cast was removed, leaving the patient with a median nerve paresis. The orthopedist referred him to a physical therapist.

The patient reported that his shoulder was very stiff and that he could not turn his arm out (external rotation) or lift his arm. He had constant tingling in his hand, almost to the point of numbness. He barely had enough strength to hold a sheet of paper between his thumb and index finger. In testing his grip strength on a bathroom scale, he found he could squeeze only 1.5 kilograms with his right hand, versus 30 kilograms with his other hand.

A week after the visit to the orthopedist, the patient began therapy with a local physical therapist. The patient reported that the physical therapist mobilized his shoulder and performed electrical stimulation to improve the muscle strength in his hand. Two months later, the patient was discharged with good shoulder range, but he still could not use his hand properly. I saw the patient six months after his discharge from physical therapy.

FIRST VISIT

The patient's main complaint was a weak "key grip." Although the tingling in his hand had disappeared a few months earlier, my evaluation revealed weakness of the muscles innervated by the median nerve: pronators, palmar flexors, thumb abductors, and flexors. His other arm muscles tested strong. His right forearm flexors and right thenar eminence showed diminished muscle bulk. Thus far, my findings confirmed the diagnosis of a median nerve lesion.

I decided to examine his pronators further. If the median nerve was regenerating, his pronator teres, being the most proximal muscle innervated by the median nerve, would be the first to receive restored innervation. Surprisingly, his pronators were not weak throughout the full pronation–supination range, but only in the last 30 degrees of pronation. This suggested that the pronators were arthrogenically inhibited by the distal and/or proximal radioulnar joint. Examining the end-feel of the radioulnar joints, I found that the distal radioulnar joint was stiff for pronation. A pronation mobilization restored pronator strength throughout the full pronation–supination range. I was pleased with this progress.

Next I examined his palmar flexors, and they too turned out to be arthrogenically inhibited. The flexors tested weak beyond 15 degrees of wrist flexion and tested strong in a loose-packed position of the wrist. A palmar flexion mobilization of the radial carpals moved the inhibitive barrier from 15 to 30 degrees of palmar flexion.

The patient's chances of recovery increased greatly when I determined that his thumb flexors were also arthrogenically inhibited by limited MCP I flexion. A passive flexion mobilization improved thumb flexor strength.

SECOND VISIT

Three days later, I saw the patient a second time. The day after his first visit, he had felt muscle soreness in his thenar eminence on flexing his thumb. I suspected that the removal of inhibition had made his thumb flexors more active in ADL activities and had created muscle soreness.

In assessing the effect of the previous treatment, I found that pronation and wrist flexion were weak in a shortened position. The pronators were weak in the last 5 to 10 degrees of pronation, and the wrist flexors were weak in the last 10 degrees of wrist flexion. I suspected that the patient needed only a little more mobilization to restore full pronation and wrist flexion function. I repeated the pronation mobilization of his distal radioulnar joint, and I mobilized his ulnar carpals to correct a flexion restriction. After the mobilization, the pronators and wrist flexors tested strong in a maximally shortened position.

When I checked the influence of the patient's cervical spine on the pronators, I was confused at first. The pronators tested weak in right cervical facet flexion and extension. I doubted my previous finding that the pronators had been fully strengthened. On retesting my findings, however, I found the pronators strong in a neutral cervical position but weak in right cervical facet flexion and extension. Extension caused more weakness of the pronators than flexion, so I evaluated right cervical facet extension first. In right cervical facet extension, the right cervical facet extensors were weak, as was the left lesser rhomboid. These muscles tested strong in a cervical neutral position. Right cervical facet extension was pain free, and the patient did not exhibit any signs of a radiculopathy. Thus, he probably had a right C7 pseudoradiculopathy. The active mobilization using the left rhomboid restored right cervical facet extensor and left rhomboid strength.

Next, I evaluated right cervical facet flexion. The right cervical facet flexors were weak in a shortened position, and the right C5 and C7 facet joints seemed responsible. A passive mobilization of these two segments restored strength of the right cervical facet flexors, and the pronators now tested strong in right cervical facet flexion and extension.

It is possible that during the first visit, I had completely restored pronation, but that pronation weakness and stiffness had partly returned because of the inhibiting influence of the cervical spine.

THIRD VISIT

Four days later, the patient returned for his third visit. On examination, I found that his pronators and wrist flexors had retained their strength in a maximally shortened position and in right cervical facet flexion and extension.

Next, I examined his thumb abductors and flexors, which tested weak in a shortened position but strong in a loose-packed position of the thumb joints. More specific muscle testing revealed inhibition from the CMC I joint affecting the thumb abductors and flexors. The MCP I joint was also still inhibiting the thumb flexors. A passive abduction and flexion mobilization of the CMC I joint and a passive flexion mobilization of the MCP I joint restored strength of the thumb abductors and flexors.

FOURTH VISIT

The patient was seen once more a week later. All muscles still tested strong. The visible atrophy of the right forearm and thenar eminence muscles was still present; however, since these muscles now had good contraction quality, I expected that the forearm muscles would automatically increase in circumference with daily use and time.

I called the patient a year later while writing this case presentation, and he told me that his hand had been functioning well. He said that the muscle bulk in his forearm was identical to the other side but that his thenar eminence might still be somewhat diminished in bulk.

CONCLUSION

This case is interesting, though not unique, in that symptoms of a seemingly paretic nerve were resolved with joint mobilization. The patient probably had suffered a true median nerve entrapment, causing weakness of the muscles in the median nerve's innervation territory. During the paretic period, the weak muscles atrophied, causing reduced forearm circumference and thenar eminence atrophy. The weak muscles failed to move their constituent joints through adequate range, resulting in stiffness of the distal radioulnar joint for pronation, the wrist for palmar flexion, and the thumb joints for flexion and abduction. After the median nerve regenerated, the affected muscles found themselves inhibited by the stiff joints they crossed. Thus, whereas during the paretic period the muscles had been weak throughout their range, they were now weak only in their shortened range. A mobilization of the joints restored normal joint mechanics and muscle strength.

The dysfunctions in the patient's cervical spine may have been coincidentally present but may also have been developed in response to excessive bombardment of the C6 and/or C8 cord segments by nociceptive afference from the right arm. The nociceptive bombardment of the central nervous system may have caused local changes in the C5 and C7 spinal segments. The cervical dysfunctions prevented recovery of the lower arm, because the cervical dysfunctions inhibited the already inhibited median nerve muscles.

Case 2, Median Nerve Laceration

FIRST VISIT

This patient, a healthy 36-year-old female, who had sustained a laceration injury to her right forearm two years earlier, was referred to me for a right median nerve injury. Her main complaint was numbness and weakness of her hand, which was evidenced by her suddenly dropping objects she was holding. An EMG test of her median nerve, however, had been negative, and she reported no neck pain.

Inspection revealed a scar on the flexor side of her forearm, about a third of the way down its length. The grip strength of her right hand was 7 kilograms versus 28 for her left hand. Surprisingly, her palmar flexors were weak in a shortened position only. The palmar flexors tested strong in a loose-packed position of her wrist, which meant that the palmar flexors were probably arthrogenically inhibited and the weakness was not neurogenic (median nerve). I passively mobilized palmar flexion, convinced of the simplicity of the case. After the mobilization, the flexors had good contraction quality in a shortened position, but grip strength had only improved to 11 kilograms.

On evaluation of the influence from the cervical spine, I found that flexion of the patient's right cervical facets inhibited her right palmar flexors. Right cervical facet extension restored palmar flexor strength. The patient had a cervical pseudoradiculopathy, which explained why her grip strength had only partially improved.

Further evaluation of the patient's cervical spine revealed that her right cervical facet flexors were strong in right facet extension (lengthened position) but became weak before her neck achieved a neutral position. The segmental evaluation revealed a major flexion restriction (more than 50 percent of the range) of C6 on the right. I passively mobilized the right C6 facet joint into flexion, which improved her grip strength from 11 to 21 kilograms.

SECOND VISIT

Seven days later, when the patient returned for her second visit, the grip strength of her right hand was 19 kilograms. I found a right C7 facet flexion stiffness, and mobilizing it improved grip strength to 22 kilograms.

THIRD VISIT

At her third visit a week later, the patient had no complaints, however, evaluation revealed a minimal flexion restriction for C6 on the right. The right cervical facet flexors tested weak in a shortened position with pressure on the lamina of C6 on the right but tested strong with pressure on the lamina of C7 on the right.

The patient also had an inhibition of her finger flexor due to a stiff MCP II joint. After mobilization of the two dysfunctions, grip strength on the right side improved to 27 kilograms versus 28 for the left. Since the patient again said she had not experienced problems during the past week and felt she had accomplished her goal, she was discharged.

CONCLUSION

The sudden weakness the patient had experienced, causing her to drop objects from her hand, may have been directly related to the position of her cervical spine. Her wrist flexors may have been strong if she picked something up in cervical extension, but when she flexed her right C6 and C7 facet joints (e.g., rotated her head to the left), her wrist flexors would instantly be inhibited, possibly to the point of dropping objects from her hand. It is also interesting that, despite the gross cervical dysfunction, the patient had not been aware that anything was wrong with her neck.

A median nerve lesion may have been present following the forearm laceration. The median nerve seemed to function well at the time of discharge and was probably fine at the initial visit as well. The numbness the patient complained of at her initial visit may simply have been a memory. She did not mention it again during the time that I worked with her. The cervical C7 and C8 pseudoradiculopathy (from the C6 and C7 spinal segments) mimicked a median nerve dysfunction. The cervical changes may have developed in response to chronic nociception from her arm (due to the laceration).

Case 3, Lyme Disease

FIRST VISIT

The patient was a 25-year-old female with Lyme disease, a disease transmitted by the deer tick, which is thought to carry the bacterium. She had contracted the disease some 13 years earlier but had gone undiagnosed for 10 years until testing positive on a blood test for Lyme disease.

Lyme disease can cause generalized arthritis with neurological and cardiac complications. The patient reported that her heart was affected, since she could not tolerate any exertion. She moved very slowly and deliberately, saying "I have no stamina." On the initial evaluation form, she indicated pain in her neck, both upper arms, her lower back, both upper legs, and both knees (Fig. 20.1). When I asked her if she currently felt pain in these areas, she reported that she felt pain in her left upper arm, her lower back, and her right thigh and knee. She requested that I first work on her lower back, since this caused her the most pain.

The patient's lower back pain was reproduced with forward bending and squatting. In evaluating lumbar flexion, I found the following:

Weak	Strong
	L lumbar flexors s
R lumbar flexors s	n
in R lumbar fl: R hip flexors n	in lumbar extension
R hamstrings n	in lumbar extension
R hip ext n	in lumbar extension
	in R lumbar fl: R quadriceps n
	R plantar flexors n
	R dorsiflexors n

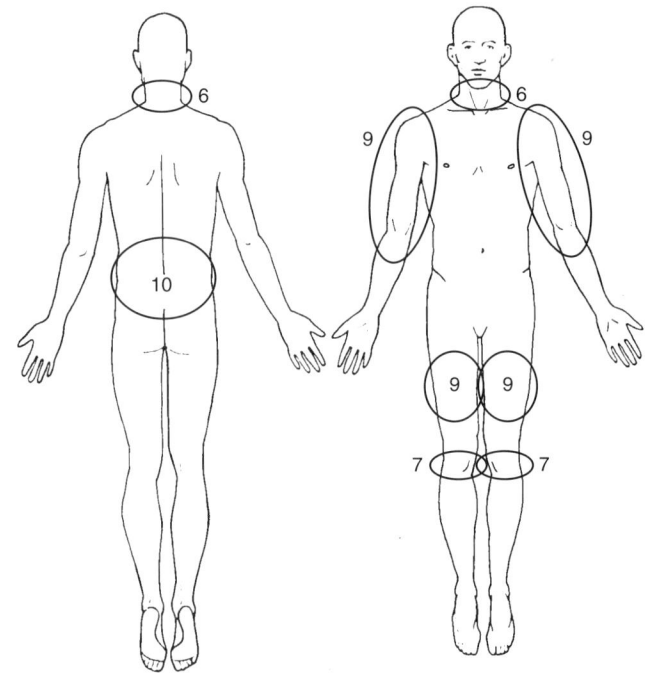

Fig. 20.1 Case 3 pain diagram.

Strength tests revealed that the patient's left lumbar flexors were strong. Her right lumbar flexors tested strong in a neutral position of the lumbar spine but became weak in a shortened position (in right lumbar facet flexion). This suggested a minor flexion dysfunction of one or more right lumbar facet joints.

When I tested her right leg muscles in right lumbar facet flexion, the weakness of her hip flexors, hamstrings, and hip extensors becoming strong in a lumbar neutral position suggested the segmental levels of her lumbar dysfunctions. The hip extensors and hamstrings are innervated by L5 through S2. Since her plantar flexors (S1,2) tested strong, the L5 segment seemed the cause of inhibition. The second dysfunctional level was evidenced by her having weak hip flexors (L2,3) with strong quadriceps (L3,4). This suggested the L2 level as the dysfunctional segment. Motion palpation of her lumbar spine confirmed limited right facet flexion of L2 and L5.

I passively mobilized L2 and L5 in a side-lying position. After the treatment, the lumbar right facet flexors tested strong in a shortened position. Forward bending was pain free. The patient still felt pain with squatting, but only in her right knee. She said that if she stayed in this position, her lower back would get stiff.

Since bilaterally the patient's lumbar spine had good flexion (strong flexors), it was possible that the sacroiliac joint produced the feeling of stiffness. I evaluated the sacroiliac joints and found the following:

Weak	Strong
L iliacus s, n	R iliacus s
	B psoas major s
	B adductors s
L quadratus lumborum s, n, l	R quadratus lumborum s

The weakness of the left iliacus throughout most of its range in conjunction with a strong left psoas major made me suspect a left posterior innominate. The iliacus weakness was not likely spinal in origin, since other muscles that shared the same segmental innervation (psoas major and adductors) were strong. The patient's hip flexion did not inhibit her iliacus either, since her psoas major, also a hip flexor, was strong. This left a posteriorly rotated innominate as the cause. The weakness of the left quadratus lumborum throughout its lumbar range confirmed a left posteriorly rotated innominate.

I mobilized the left innominate anteriorly. Following the mobilization, the iliacus and quadratus lumborum tested strong throughout their complete range. Squatting and forward bending produced no pain in the patient's lower back, but she still felt right knee pain on squatting.

I evaluated her right knee flexion and found:

Weak	Strong
R hamstrings s	n

Her right hamstrings tested weak in full knee flexion but were strong when tested in less flexion. The treatment was an active flexion mobilization. Approximately 3 minutes of active flexion mobilization restored the hamstrings' strength. The patient reported that squatting felt fine.

When I asked her about her left arm, she reported that internal rotation and elevation reproduced her pain and demonstrated this. The muscle tests revealed the following:

Weak	Strong
L external rotators s	n
L abductors s	n
L internal rotators s, n	l
L adductors s	n

Initially, I tested the external rotators, abductors, internal rotators, and adductors in a shortened position. All four muscle groups tested weak. This provided no information on where to start treatment. When I retested her shoulder in a loose-packed position, only the internal rotators tested weak, and they became strong in a lengthened position. I initially performed the passive internal rotation mobilization in the maximally loose-packed position. As the patient's internal rotation improved, I performed the internal rotation mobilization with her shoulder in more internally rotated positions. I continued the treatment until her internal rotators tested strong in a maximally shortened position. After the treatment, I asked her to internally rotate and elevate her arm, and she reported that these movements were pain free.

SECOND VISIT

Seven days later, the patient returned for her second visit, reporting that her shoulder was less painful. Shoulder evaluation revealed that protraction of her left shoulder girdle reproduced the pain. The muscle tests revealed:

Weak	Strong
L protractors s	n
L anterior deltoid protraction	

Her protractors were weak in a shortened position and were strong in less protraction. This suggested arthrogenic inhibition from her acromioclavicular or sternoclavicular joint. Weakness of the left anterior deltoid suggested that her acromioclavicular joint was responsible for the inhibition. I passively mobilized her lateral clavicle anteriorly, and after the mobilization, her protractors and anterior deltoid tested strong. Protraction was pain free.

The patient reported that her right knee had bothered her while walking. Evaluation of her right knee revealed:

Weak	Strong
	R hamstrings s
R quadriceps s	10° knee flexion
R abductors knee extension	10° knee flexion
	R adductors knee extension

The patient's hamstrings were still strong from the previous week's treatment, but her quadriceps and abductors were weak in knee extension. These muscles tested strong in 10 degrees of knee flexion. I mobilized her lateral knee compartment into extension until the muscles tested strong.

The patient also reported that her lower back still bothered her, although less. Evaluation of the lower back revealed:

Weak	Strong
	B quadratus lumborum s
	L iliacus s
	R lumbar flexors s
L Lumbar flexors s	n
in L lumbar flex: L hamstrings	lumbar n
L hip extensors	lumbar n
	in L lumbar fl: L hip flexors
	L quadriceps
	L dorsiflexors
	L plantar flexors
B internal rot. in abduction s, n	l

First, I checked the results of the previous week's treatment. Both quadratus lumborum muscles were still strong, as were the iliacus and right lumbar flexors. However, the left lumbar flexors tested weak in a shortened position, suggesting a left facet flexion dysfunction. In left facet flexion, the hamstrings and hip extensors became weak. Thus, the patient needed more flexion mobility of her left L5 facet joint. I passively mobilized the L5 segment.

Next, both internal rotators tested weak in hip abduction throughout most of the range. This suggested a bilaterally limited S1 nutation, so I passively improved bilateral S1 nutation. After the mobilization, the internal rotators tested strong. Additionally, she reported that her back pain was gone.

THIRD VISIT

On the patient's third visit, again a week later, she reported that she was less tired and that all pain was gone except in her right lower back. Her right lower back pain was reproduced with lumbar extension.

Evaluation of her lower back revealed the following:

Weak	Strong
R quadratus lumborum s, n, l	
R iliacus s, n	R psoas s
R latissimus dorsi 90° abduc.	adduction

I suspected a right lumbar facet extension dysfunction, so I first tested the right quadratus lumborum. However, this muscle was not only weak in its shortened position but throughout its range. I then assumed a downslip or a posteriorly rotated innominate was inhibiting the quadratus lumborum. I needed to correct this before the quadratus lumborum could reveal any possible lumbar extension dysfunction.

The weakness of the right iliacus throughout most of the hip range in conjunction with a strong psoas suggested the patient needed an anterior rotation mobilization of her innominate. This was confirmed by latissimus dorsi weakness. I mobilized the innominate anteriorly. After the treatment, the muscles tested as follows:

Weak	Strong
	R iliacus
	R latissimus dorsi 90° abduc.
R quadratus lumborum s	n
in R lumbar ext: R quadriceps n	in lumbar n
	in R lumbar ext: R hip flexors
	R dorsiflexors
	R plantar fl
	R hamstrings
	R hip ext

Retesting the latissimus dorsi, iliacus, and quadratus lumborum showed that the posterior innominate had been corrected. Inhibition of the right quadratus lumborum suggested a minor right lumbar facet extension dysfunction.

The patient's quadriceps became weak when her right lumbar facet joints were extended, suggesting levels L3 and L4 were involved. With motion palpation, I found limited extension mobility at the level of L3. A passive L3 extension mobilization removed the inhibition from the quadriceps and quadratus lumborum. After retesting all lumbar and pelvic muscles, the only weakness I found was in the left hamstrings during full hip flexion. This hamstrings weakness in full hip flexion suggested that the left innominate needed a posterior rotation mobilization. The mobilization restored hamstrings strength.

The patient left pain free, indicating she would call if she needed further treatment.

FOURTH VISIT

Five weeks later, I saw the patient again. She reported that she had been doing well, but that two days earlier her back and right leg had "gone out." She complained of pain in front of her right thigh and continuing down to her right patella. Examination revealed the following:

Weak	Strong
	R hamstrings s
R hamstrings full hip flexion	less hip flexion
R hip extensors s, n	
R quadriceps s, n, l	

The right hamstrings weakness in full hip flexion and the hip extensor weakness in a large part of the range suggested that the patient's innominate needed a posterior mobilization. I actively mobilized her right innominate posteriorly, which restored the strength of her hamstrings and hip extensors. Finding that her quadriceps was still weak, I evaluated the quadriceps and found the following:

Weak	Strong
R quadriceps s, n, l	in R lumbar ext: R quadriceps s

The quality of the quadriceps contraction was influenced by the position of the lumbar spine. The quadriceps became weaker in flexion and got stronger in extension. L4 felt stiff for right facet flexion. After the mobilization, the quadriceps still tested weak, but less so. I repeated the mobilization, and this made the quadriceps fully strong. The patient had no more complaints and again indicated that she would call if she needed further treatment.

CONCLUSION

Patients with rheumatoid Lyme disease or arthritis have sensitized joints. A nonrheumatoid person could have minor joint dysfunctions without being aware of them. The sensitization of joints in people with rheumatoid arthritis could cause a similar minor joint dysfunction to be a debilitating condition with pain, as well as strong neurological reflexes and muscle tone changes.

This patient responded well to therapy, which is not unusual for cases of rheumatoid arthritis without joint destruction. Often only minor mobilizations are needed to improve joint function. I believe the patient's joint dysfunctions (not her rheumatoid arthritis) were minor, but the expressions of those joint dysfunctions in her activity level were major.

Case 4, Drop Foot

The patient was a 60-year-old male with a long history of low back pain with sciatica. He had recently tripped a few times and injured himself as a result. He had seen his neurologist and mentioned his "balance" problem. The neurologist referred him for physical therapy with the diagnosis "lumbar disc disease with left foot drop."

FIRST VISIT

The evaluation revealed that the patient could stand for 5 seconds on his right leg but could not stand for even 1 second on his left. He was unable to walk on his left heel. A straight leg raise with dorsiflexion of that leg was limited to 30 degrees and reproduced his sciatic pain. Prone knee bending (femoral nerve) was slightly limited and did not produce any pain. Muscle tests presented a confusing picture. The examination revealed the following:

Weak	Strong
	L hip flexors s
L quadriceps s	
L dorsiflexors s	
L plantar flexors s	L evertors s
	L invertors s
	L hamstrings s
L hip extensors s	

Muscles with innervation from multiple segmental levels were weak, and there seemed to be no mechanical or neurological relationship between the weak muscles. I thought the patient would likely have multiple dysfunctions and decided to evaluate his lower lumbar spine in more detail. I first looked for a flexion dysfunction in the left lumbar spine and discovered the following:

Weak	Strong
L lumbar flexors s	n
with pressure on L L5	with pressure on L S1

His left lumbar flexors were strong in a neutral position of the lumbar spine but became weak in left facet flexion. The L5 left facet felt stiff for flexion. The left lumbar facet flexors tested weak with posterior–anterior pressure on the left transverse process of L5 (pushing L5 farther into left facet flexion) and tested strong with the same pressure on the left of S1. Again, this suggested a limited flexion mobility of the left L5 facet joint. A postisometric stretch technique to improve flexion of the left L5 segment restored strength of the left lumbar flexors in a shortened position.

I tested the left hip extensors (innervated by L5) to see if the L5 mobilization had improved their strength. They were still weak over a large part of their range, raising the question of whether the patient had a limited posterior rotation of the innominate. When I tested his strong hamstrings in maximal hip flexion, they became weak, confirming limited posterior rotation of the left innominate. The ensuing posterior rotation manipulation restored strength in both muscle groups.

The left quadriceps still tested weak but tested stronger in left lumbar extension, suggesting that the L3 or L4 left facet joints had a flexion dysfunction. Passive motion palpation confirmed a left L4 facet flexion restriction. After I mobilized the L4 left facet with a postisometric stretch technique, I felt more strength on the manual muscle test of the quadriceps; however, the quadriceps still tested weak.

Next, I worked on the plantar and dorsiflexors, which were both weak throughout their working range. Lumbar flexion and extension did not influence their strength, so I mobilized the sciatic nerve with straight leg raises. A subsequent assessment revealed that the straight leg raise had improved by 10 degrees and the plantar and dorsiflexors had increased strength.

As a result of this first treatment, the patient reported that his back felt a little better and that it was easier to put on his socks.

SECOND VISIT

When the patient returned for his second visit the following day, he reported that everything felt better. A muscle strength evaluation with his lumbar spine in a neutral position revealed:

Weak	Strong
	L hip flexors s
L quadriceps s	
L dorsiflexors s	L plantar flexors s
	L invertors s
	L evertors s
	L hamstrings s
	L hip extensors s
in L lumbar fl: L quadriceps s	in L lumbar ext: little stronger

The plantar flexors and hip extensors had remained strong after the previous day's treatment, but the quadriceps and dorsiflexors still needed work. The quadriceps became slightly stronger in left lumbar extension, suggesting that one of the left lumbar facets needed more flexion mobility. I mobilized the L4 left facet into flexion but failed to notice any increase in quadriceps strength afterward.

The patient's straight leg raise was 40 degrees. Mobilizing his sciatic nerve for 5 minutes resulted in increased quadriceps strength when tested manually, so I prescribed a home exercise to mobilize the sciatic nerve.

THIRD VISIT

During his third visit seven days later, the patient reported that he was doing well. He noted that his foot came up more while walking. I repeated the L5 flexion mobilization and the sciatic nerve mobilization.

FOURTH VISIT

A week later, during his fourth visit, the patient again reported that his leg was doing well; forward bending was easier. His quadriceps and dorsiflexors were the only muscles in his leg that were still weak. A closer evaluation of his quadriceps revealed that it was weak in knee extension only. The treatment consisted of an extension mobilization of the lateral part of the knee, after which the quadriceps tested strong in knee extension.

I also repeated the left tibial nerve stretch. Following the treatment, the dorsiflexors tested stronger but still weak throughout the full ankle range. His dorsiflexion range was limited with the knee straight and bent, so I prescribed a dorsiflexion mobilization as a home exercise, since I expected an arthrogenic inhibition to be present as well. The neurogenic inhibition of his dorsiflexors had undoubtedly created arthrogenic inhibition from the ankle (stiff dorsiflexion).

FIFTH VISIT

During his fifth visit a week later, the patient reported that his balance was getting better. He could balance on his left leg and could pull his pants on while standing, something he had been unable to do before treatment. His quadriceps still tested strong in knee extension. An evaluation of the sciatic nerve revealed a limited extensibility of the peroneal nerve with less involvement of the other parts of the sciatic nerve. I mobilized the peroneal nerve, and for the first time, his dorsiflexors tested strong in a lengthened position; however, they were still weak in a shortened position (dorsiflexion). During the remainder of the treatment, dorsiflexion range was mobilized.

SIXTH VISIT

During the patient's sixth visit two weeks later, I repeated the peroneal nerve stretch and the dorsiflexion mobilization.

SEVENTH VISIT

During his final visit a week later, the patient reported that his balance was normal. He demonstrated this by standing on his left leg for 6 seconds while showing off by kicking his other leg up and down. His dorsiflexors tested strong. I repeated the peroneal nerve stretch and the dorsiflexion mobilization, and then discharged the patient.

CONCLUSION

The patient's problem could be summarized as a dysfunction of the sciatic nerve that weakened his plantar and dorsiflexors. His quadriceps became stronger in response to, among other things, mobilization of the sciatic nerve. The improved extensibility of the posterior leg's nerve tissue (sciatic nerve) may have reduced the inhibition it created on the muscle that stretched the nerve tissue.

The weakness of the other muscles in the patient's left leg may originally have been due to lack of proper use of the extremity and spine. At the initial evaluation, the weakness of multiple muscles in the leg gave me a baseline of his condition but failed to indicate where to begin treatment. In this jungle of signs, I took the first path that offered a possible way out. I began by examining lumbar left facet flexion mobility, but I could have started with any of the dysfunctions in the left lower back or leg. My treatments were guided by testing the muscles and finding weakness or strength through application or removal of tension from antagonistic or spinal structures.

Case 5, Low Back Pain

FIRST VISIT

The patient was a healthy 42-year-old woman who was referred for physical therapy with a diagnosis of myositis–radiculitis. She complained of low back pain radiating to both hips and her right groin (Fig. 20.2). The pain had started six months earlier for no apparent reason. Until that time, she had worked out two hours a day in a gym. Sitting and standing were painful for her, and walking sometimes made it worse. Lying down relieved the pain. She used pain medication daily and had stopped her workouts about five months earlier secondary to the pain. On a scale of 0 to 10 (10 being the worst possible pain), the patient rated her pain as 5. She reported that X-rays had shown arthritis of her spine, and MRI had shown a swollen disc. Her goal for therapy was to work out again for two hours a day.

The evaluation showed that both forward and backward bending were painful, while standing eased the

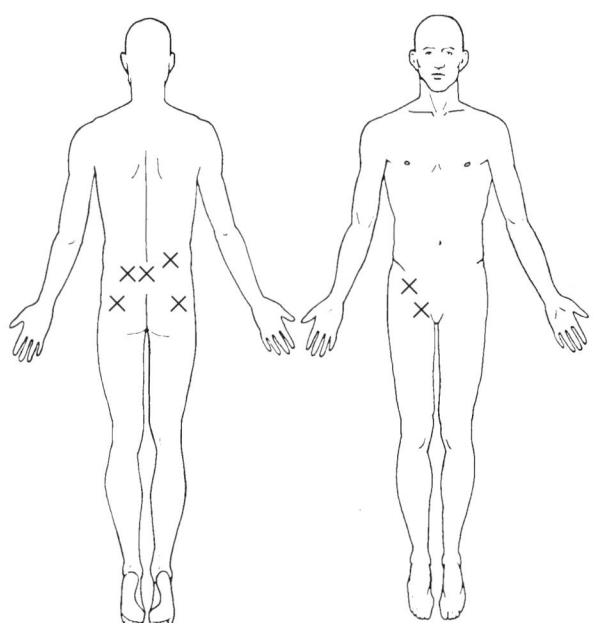

Fig. 20.2 Case 5 pain diagram.

pain. On forward bending, the patient could easily put her hands on the floor. The left Kemp test (left lumbar facet extension) reproduced the pain she felt on extension. The right Kemp test was negative. The muscle tests revealed:

Weak	Strong
R lumbar flexors s	n
	L lumbar flexors s
B quadratus lumborum s, n	l
L hamstrings L lumbar extension	L lumbar flexion
B adductors at 40° flexion s, n	s
(reproduced pain)	
B internal rotation in abduction n, l	

The patient's greatest lumbar dysfunction was extension (bilateral quadratus lumborum weakness in a loose-packed position of her spine). Her right lumbar flexors were only weak in a shortened position. On the left side, the extension dysfunction was probably located at L5 (hamstrings). She likely also had bilateral stiffness of S1 for nutation (adductors and internal rotators in abduction).

Since the patient did not experience any L5 radicular symptoms (e.g., foot pain, numbness, or limited straight leg raise), the hamstring weakness was probably due to a pseudoradiculopathy. I decided to mobilize left L5 extension, which would later assist in correcting the limited S1 nutation. I used a postisometric stretch technique to improve L5 left facet extension, after which the left hamstrings and left quadratus lumborum tested strong in lumbar left side bending (left facet extension).

Next, the patient actively mobilized bilateral nutation of S1 using a contraction of her adductor muscles in 40 degrees of hip flexion. After the active mobilization, the adductors tested somewhat stronger. She reported that backward bending and the left Kemp test felt fine.

I prescribed the active S1 nutation mobilization as a home exercise and asked the patient if she would be comfortable starting a daily walking program. My rationale for the walking program was that it would assist in maintaining the increased extension mobility of L5, mobilize sacral nutation, and start her doing daily exercises. She told me she felt she could start walking.

SECOND VISIT

The patient returned four days later. She had walked twice for a two-hour period and reported feeling much better. The evaluation revealed:

Weak	Strong
	B quadratus lumborum s
R lumbar flexors s	n
	in R lumbar fl: hip flexors
	knee extensors
	dorsiflexors
	plantar flexors
	hamstrings
	hip extensors

The patient felt no pain when bending backward from a standing position, but flexion produced pain on the right side. Lumbar extension remained pain free (bilateral quadratus lumborum), but the right lumbar flexors still needed more strength and tested the same as on the previous visit. When I tested the right leg muscles in lumbar right facet flexion, all muscles tested strong. This meant that the cause of the lumbar flexion dysfunction was located outside the L1–L5 region. On passive motion palpation, I found T12 to be stiff for right facet flexion. A passive flexion mobilization restored the strength of the right lumbar flexors. Forward bending from a standing position was pain free after the treatment.

The adductors still tested weak throughout most of their range and when tested at 40 degrees of hip flexion (S1 nutation); however, the contraction did not reproduce pain, as it had on the previous visit. The adductors also tested weak in 90 degrees of hip flexion (posterior symphysis pubis). The treatment consisted of an active mobilization in 40 and 90 degrees of hip flexion. I asked the patient to continue her home walking exercises and the adductor contraction exercise.

THIRD VISIT

The patient returned three days later for her third visit. She reported that on the day after the treatment her pain had increased, but on the day after that and the day of the visit, it had improved again. She only felt pain in her left lower back. Evaluation revealed that forward and backward bending were pain free, but I found a minor extension dysfunction in her left lumbar spine. The muscle tests revealed:

Weak	Strong
L quadratus lumborum s	n
L hip flexors in L lumbar extension	lumbar flexion

The quadratus lumborum tested weak in left facet extension but strong in a lumbar neutral position. The level of the dysfunction was probably L2 or L3, since hip flexor strength was reduced in lumbar extension. Intervertebral motion palpation revealed an extension stiffness at L2. The treatment consisted of a postisometric stretch technique to improve L2 left facet extension. After the treatment, the left quadratus lumborum and left hip flexors were strong in left lumbar extension.

FOURTH VISIT

The patient again returned three days later for her fourth visit. She had resumed her gym workout and had felt only minor discomfort in her lower back. Backward bending caused pain across her buttocks. I found weakness of the muscles associated with limited S1 nutation and mobilized both sacroiliac joints, after which backward bending became pain free.

FIFTH VISIT

Returning for her fifth visit three days later, the patient reported that her daily two-hour walk had been comfortable. On the previous day, she had decided to start running, but this caused her pain to return. Backward bending reproduced her pain, whereas forward bending was pain free. The muscle tests showed:

Weak	Strong
	R quadratus lumborum s
L quadratus lumborum s	n
L hip flexors L lumbar ext.	L lumbar flexion

She had a minor lumbar left facet extension dysfunction at the level of L2 or L3. I passively mobilized extension at the level of L2, which removed the inhibition from the two muscles. Backward bending was again pain free.

SIXTH VISIT

A week later, the patient had increased her daily workout to 15 minutes of running plus walking or her gym routine. On the previous day, she had felt some pain in her right buttock. The day of the visit was her best day so far.

Backward bending reproduced the pain in her left buttock. I found the same pattern of weakness as on the previous visit, although weakness was only present in maximal left facet extension. I repeated the previous week's treatment, and the patient reported that backward bending was again pain free. The muscles tested strong.

SEVENTH VISIT

On her seventh and final visit a week later, the patient reported that she had had a good week. She had worked out in the gym for two hours a day, or had walked or run three miles, and had felt no pain. Examination of her lumbar muscles revealed:

Weak	Strong
	B quadratus lumborum s
	B lumbar flexors s
	B internal rotation in abduction

The previous dysfunctions were corrected, and I found no additional lumbar or sacral dysfunction. In evaluating her lower thoracic spine, I found some dysfunctions and corrected them.

CONCLUSION

This patient was extremely motivated to return to her daily workouts. Whereas most patients need coaching in their exercise program, she coached herself and reported the increases in her workout schedule to me on each visit. Surprisingly, despite the rapid increases in her workout, she never complained of muscle soreness. Her increasing workouts caused dysfunctions to surface that would have been undetectable in someone with a sedentary lifestyle.

The patient's most prominent dysfunction was limited bilateral sacral nutation. During her first visit, applying a force to improve nutation mobility (the adductor contraction) reproduced her pain. Normalizing her S1 nutation mobility required an unusual amount of time.

Courses

Courses that teach the concepts presented in this book are conducted regularly in both Europe and the United States. Those interested in attending or sponsoring a course can contact the author at:

Bellmore-Wantagh Physical Therapy
2947 Jerusalem Avenue
Wantagh, NY 11793-2020
USA

or

Almere Motion
Kweekgrasstraat 60
1313 BT Almere
The Netherlands

Information about courses and this book can be obtained at http://www.almotion.com.

Index

Abdominal reflex, 14
Acetabulum, 229–30, 293–94
Acetylcholine, 12
Achilles tendon, 169
Achilles tendon reflex, 73
ACL, 10, 16, 76–77, 79–80, 90
Acromegaly, 8
Actin filament, 9
Action potential, 9, 12, 32, 38, 49, 67, 94, 98
Active positioning, 49, 95
Adams, 11, 19, 21, 47, 50
Adenosine triphosphate, 136
Adipose tissue, 31, 33, 35
Aging, 3, 7, 18, 21, 26, 89–90, 132
Agonists, 14–15, 36, 73, 76, 102–4, 125, 177–78, 182–84, 186, 189, 191, 193–194, 197, 199–200, 202–4, 206, 208, 210, 213–16, 220, 222–23, 225–29, 231–32, 234–38, 243, 246, 248, 250, 252, 255, 257, 259, 261, 263–64, 266, 270, 272–73, 275, 277, 279–86, 288–93, 295–97, 299, 301–3, 306–7, 310, 314, 316, 319, 321, 325, 329, 331–34
Alcohol, 9
Alpha motoneuron, 10, 14, 16–17, 48–49, 57–63, 65, 71, 76–77, 82, 84–85, 94–95, 100, 139, 141
Alpha-1 motoneuron, 62–66
Alpha-2 motoneuron, 62–66
Altered movement patterns, 7
Anatomical barrier, 135
Anesthesia, 50, 71–72, 77, 101, 103, 124
Anesthetic, 10, 47–49, 72, 83, 94–96, 118, 124, 166
Anesthetic injection, 47–49, 72, 83, 94–96, 118, 124, 166
Ankle sprain, 77–78, 80, 91
Antagonists, 10, 14–15, 61, 73, 78, 84, 101–4, 125, 162, 164, 175–76, 178
Antalgic gait, 10, 58
Anterior cruciate ligament, 10, 23, 25–26, 30–32, 35, 40–43, 46, 48, 50–51, 59, 67–69, 74, 76–77, 79–80, 90–91, 109–10, 126, 303
Anterior horn, 11–12, 31, 38, 57, 94, 116, 118
Antigravity muscles, 11
Appendicitis, 116
Arterial pulses, 6
Arthritic atrophy, 123, 125

Arthritis, 8–9, 23, 40, 68, 89, 93, 96–104, 106–7, 109, 117, 137, 170–72, 177, 303, 340, 343, 345
Arthrogenic atrophy, 123
Arthrogenic inhibition, 111, 123, 128, 131–32, 134–39, 142–43, 146–48, 151–52, 154–55, 157, 163, 175, 178, 220, 270–71, 325, 332, 338–39, 341, 345
Arthrogenic muscle dysfunction, 89, 123, 125, 169
Arthrogenic muscle weakness, 136, 139, 151, 175
Arthrogenous muscle wasting, 65, 123
Arthrogenous muscle weakness, 68, 107, 111, 123, 126, 170–71, 174
Arthrosis, 89, 93, 99, 101–2, 104, 177
Articular receptor, 17, 27, 29, 31–34, 36–39, 42–43, 45, 47–50, 52–54, 57–59, 64–66, 68, 71–73, 75–77, 79, 81–84, 90, 94–95, 98–101, 103–4, 109, 119, 123, 125, 151, 158–60, 165–67, 171
Articular reflexes, 15–16, 21, 51, 57–58, 64, 67, 72–73, 75, 78–80, 82, 85, 89–90, 95, 103–4, 110, 125, 164–67, 172
Articulo-muscular protective reflex, 1, 71, 74–78, 81–82, 89, 90, 101, 104, 109, 123, 125
Arvidsson, 10, 19
Assessment of muscle strength, 131–132, 143, 175, 344
Atherosclerosis, 6
Atrophic articular paralysis, 123
Atrophy, 3–9, 12, 18–21, 26, 29, 65, 96–98, 100, 104–5, 109, 123, 125, 137, 157, 338–39
Autoimmune disease, 12
Axon reflex, 37, 98
Axoplasmic flow, 98, 118

Babinski's sign, 11, 14
Bacterial, 9, 177
Bakker, 233, 335
Baxendale, 19, 21, 26, 50, 66, 68, 78, 83–85, 87, 95, 104–5, 107–8, 174
Bed rest, 4–5, 19, 24
Berg, 4–5, 19
Blood flow, 7, 12–13, 30, 112
Brachial plexus tension test, 15, 19
Bracing, 5
Brain stem, 11, 59, 181
Buerger's disease, 6
Butler, 20, 25, 113, 156, 172, 174

Calcium, 9
Candida tropicalis, 9
Capillaries, 12, 31, 111–12
Capsular pattern, 100–104, 123, 127, 137, 181, 191, 198, 210, 219, 237, 239, 250, 266, 273, 280, 286, 293, 303, 316, 329
Cardiac failure, 8
Carpal tunnel syndrome, 9
Cartilage, 30–31, 37, 48, 109, 209
Casting, 5, 13
Cauda equina compression, 119, 121, 173
Central lumbar stenosis, 147, 150, 154
Central nervous system (CNS), 6, 25, 31, 45, 53, 57, 61, 63, 74, 84, 99, 117, 124, 155, 157, 160, 165, 172, 174, 252, 339
Centralization of pain, 115
Cessation of smoking, 7, 18
Charcot, 96–97, 100, 105, 109, 125
Charcot's joint, 109
Chloroform, 101, 103–104
Chondromalacia patella, 117
Chronic anterior cruciate ligament, 10
Chronic obstructive pulmonary disease, 9
Cibulka, 166, 172, 335
Cimetidine, 9
Circulation, 6, 13, 18, 99, 112–113, 168, 181
Circulatory disorders, 3, 18
Clofibrate, 9
Close-packed position, 45
CNS (central nervous system), 6, 11–12
Codman's exercise, 54
Colchine, 9
Colebatch, 11, 20
Collagen fiber, 32–33, 35–36, 58, 77, 89, 104, 158
Compression, 12–13, 18, 30, 72, 95, 111–13, 118–19, 121, 124, 139, 150, 152, 161, 168, 274, 276, 278–79, 301
Compression manipulation, 274, 276, 278–79
Connective tissue, 8–9, 14–15, 30–37, 40, 58, 61, 74, 111, 157
Connective tissue massage, 14
Contractures, 8, 74, 100, 170
Contraindications, 136, 167–68, 170–71
Cortex, 6, 9, 11, 38, 47–49, 53–54, 57, 89, 96, 155
Corticosteroids, 9, 12, 20
Costotransverse joint, 30
Costovertebral joint, 30, 32, 36, 42, 57, 68, 116, 198
Counterforce, 131–32, 163, 327
Creatine phosphate, 136
Cremaster reflex, 14
Cross bridges, 5
Cross-sectional area (CSA), 4, 157
Crossover point, 134, 137
Cryostat sections, 5
CSA (cross-sectional area), 4–5
Cushing's syndrome, 8
Cutaneous afferent, 19, 48, 51–52, 66, 82, 86

Cutaneous receptor, 13, 37, 47–48, 50–51, 54, 59, 64, 81–82
Cutaneous reflex, 20, 24, 81
Cyriax, 2, 93, 103, 105, 118, 133–34, 143

D-penicillamine, 9
Davies, 6, 20, 24
De Quervain's syndrome, 145
DeAndrade, 20, 85, 95, 105, 110, 125
Decreased skin temperature, 6
Deep tendon reflex, 61, 63, 72–74, 82, 94, 118
Deep tissue, 84
Deep tissues, 18
Degree of freedom, 43–45
Demyelinative disease, 12
Dermatome, 115–16, 118, 120, 146, 152
Dermatomyositis, 8
Diabetes mellitus, 9, 109
Diarrhea, 8–9, 20
Diplopia, 12
Discharge rate, 7, 32–34, 44–45, 49, 84, 95, 159–60
Disease, 6–9, 12, 20, 22, 25, 48, 50, 54, 65, 74, 90, 93, 95, 100, 109, 117–18, 120–21, 125, 154–55, 181, 340, 343
Disorders of nerve conduction, 11, 18, 132
Distal muscles, 7–8, 11–12, 153
Distending fluid, 94–95
Dixon, 20, 23, 85–86, 94–95, 105–6
Dorsal horn, 38, 85, 98, 116
Double vision, 12
Downslip, 218–220
Drop foot, 343
Drug-induced myopathy, 7, 9
Duchateau, 6, 20–21
Duchenne's muscular dystrophy, 7–8
Dynamic gamma motoneuron, 62–64, 66
Dynamometer, 4–5, 14
Dysfunctional barrier, 130, 134–38, 158–59, 162, 175
Dysphagia, 8

Edema, 8, 99, 111
Effusion, 16, 38, 40, 45, 48, 51, 85, 93–97, 99, 103–107, 111, 126, 177
Elbow, 5–6, 11, 19–20, 31, 41, 73, 77–78, 80–81, 83–85, 102, 123–25, 127, 130, 161–62, 164, 173, 266–72, 335
Elderly, 7, 21, 27, 89
Electrolyte imbalance, 9
Electromyography (EMG), 22, 26, 39, 51, 67, 71, 77–80, 83, 85–86, 94, 96, 98, 107, 110, 117, 124–125, 142, 157, 166, 174, 339
EMG (electromyography), 4–6, 10, 15–16, 71, 77–78, 80, 94, 96, 157, 166, 339
Endocrine myopathy, 7–8
Endoneural blood flow, 13
Energy storage, 6, 157
Enzyme activity, 4
Enzymes, 8
Epidural injection, 10
Epineurium, 111

Eripheral nerve, 156
Evaluation, 2, 18–19, 23, 25, 46, 52, 68, 119–20, 128, 132–33, 139–40, 142–43, 145–47, 150–51, 154–56, 159, 167–71, 173, 175–78, 184, 225, 228–29, 260, 273, 280, 289, 316–17, 319, 322, 327, 337, 339–46
Excessive joint movement, 15, 83
Exercise, 1, 3–7, 14–15, 18, 19, 21–26, 54, 96, 132, 139–41, 157, 169–72, 305, 309
   active, 139, 141, 169, 254, 309
   passive, 169, 318
   home, 169, 172, 229, 251, 253–54, 258, 267, 271, 305, 309, 318, 320, 345–46
   strengthening, 1, 3–4, 6–7, 14–15, 18, 77, 132, 157, 170–71
Eyelids, 8

Facet extension, 102, 145–49, 151–52, 154–55, 167, 180–82, 186–88, 191, 193–96, 210, 212–15, 220, 228, 239, 242, 332, 334, 338–39, 342–43, 346–47
Facet flexion, 118, 145–49, 151, 154–55, 167–68, 180, 183, 188–91, 196, 199, 210, 216, 221, 231, 332–34, 338–46
Facet joint, 30, 33, 35, 37–42, 44, 46, 68, 91, 97, 106, 110–11, 116–19, 126, 128–29, 139, 145–52, 154–55, 167–68, 171, 180, 184–85, 187–88, 191, 194, 196, 199, 210, 214–15, 220–21, 338–40, 342–47
Facilitated segment, 1, 116–18, 120, 150
Fascioscapulohumeral muscular dystrophy, 8
Fast-twitch fiber, 5
Fatigue, 9, 64
Ferrell, 19, 21, 26–27, 46, 54, 66, 68, 76, 78, 83–85, 87, 90–91, 95–96, 100, 104–8, 125, 172–74
Fever, 9
Fiatarone, 7, 21
Fibrous capsule, 30–33, 35, 37, 95
Flexion receptor, 32–33, 37, 65
Flexion reflex, 58, 68, 78, 81, 85–87, 105
Foot drop, 13, 343
Foramen, 111–13, 119, 147, 150, 152–53, 167–68, 180
Foraminal closing, 113, 147–50, 332, 334
Foraminal opening, 112–13, 147–50, 167–68, 333–34
Foraminal stenosis, 147, 150
Force generation, 4–5
Force-time diagram, 131–132
Fortin, 119–20, 156
Fractures, 9, 145, 170–71, 177
Freeman, 16, 21–22, 32, 35, 39, 54, 58–59, 67, 71–73, 77–79, 85, 90, 110, 120, 125
Functional skills, 8
Functional Technique, 17, 170, 173
Fungal infections, 9
Furcal nerve, 152, 156

Gait, 7, 10, 18, 42, 52, 58, 68, 91
Gamma bias, 16, 61–62, 64–65, 137, 159, 179
Gamma loop system, 48–50, 101
Gamma motoneuron, 13–14, 16–17, 34, 36, 39, 48–49, 58–62, 64–65, 72–73, 75–77, 100, 159
Gamma-spindle loop, 78

Gamma-spindle system, 60, 62
Ganglion, 53, 98, 111
Gate-control theory, 10, 24, 26, 38, 53–55, 58, 160, 172
Gating of nociception, 10–11, 18, 38, 53–54, 89, 160, 167, 172, 252
Generalized muscle weakness, 139, 151
Generator potential, 31, 35
Germain, 5, 22
Giving way, 77, 95, 109
Gold therapy, 48, 99
Golgi tendon organ, 32–33, 36–37, 41, 49, 82, 139
Grading muscle strength, 178
   qualitative, 130, 140
   quantitative, 129, 140
Grant, 20, 39, 85, 105, 110, 125, 181, 335
Greenman, 22, 46, 172, 181, 335
Grieve, 20–21, 23, 25, 39, 86, 119–21, 125, 156, 172, 174, 335
Guarding, 10, 29, 53, 103–4, 130, 163, 165, 167
Guillain-Barré, 12

Hamstrings, 15–16, 29, 48–49, 62, 74, 76–80, 85, 103, 117, 128–29, 158, 160–61, 163, 166, 175–77, 218–19, 225–27, 231, 297, 303, 305–6, 310, 313, 340–44, 346
Heart attack, 115
Hemarthrosis, 38
Hemiparesis, 11–12, 20
Hemiplegia, 11–13, 19, 98–99, 105, 107
Hemochromatosis, 9
Herniated nucleus pulposus, 150, 154, 168
Herniation, 112, 118–19, 150, 156, 167–68, 173
Herpes zoster, 12
High-velocity thrust manipulation, 165–67, 171, 221–23, 225–26, 323, 327
Hilton, 30, 40, 74, 79, 100–101, 103–4, 106
Hilton's law, 30, 74, 101
HIV, 9–10
Hoffmann reflex, 94, 106
Home exercise, 169, 172, 229, 251, 253–54, 258, 267, 271, 305, 309, 318, 320
Hormone-replacement therapy, 7, 25
Hormones, 9
Human immunodeficiency virus, 9
Hurley, 3, 22
Hyperactive tendon reflexes, 11
Hyperparathyroidism, 8
Hyperthyroidism, 8
Hypertonic muscle, 133, 162, 164, 168–169
Hypertrophy, 5, 24
Hypocalcemia, 9
Hypokalemia, 9, 20
Hypothyroidism, 8
Hysteresis, 65–66, 69, 158–60

Immobilization, 5–6, 18, 20, 22, 24, 36, 103–105, 132, 139, 157, 173
Immunization, 12

Inactivity, 4–5, 7, 18, 132
Increased muscle tone, 11, 14, 29, 76, 103, 167
Indications, 76, 170–172
Infectious myopathy, 7, 9
Inflammation, 8–9, 22, 83, 85, 93, 97–101, 103–6, 109, 113, 150, 167–68, 170, 173
Inflammatory disorders, 7–8
Inflare, 218, 293
Influenza, 9
Inhibition, 1, 3–5, 10, 14–21, 23–24, 26, 29, 36–37, 39, 53, 61–62, 65, 67, 72–73, 75–76, 85–87, 90, 94–98, 101, 103–4, 106–8, 110–11, 117–18, 120, 123–25, 139, 146–47, 151, 154–55, 157, 165–66, 168–69, 175, 178, 188, 191–92, 195–96, 205, 209, 212, 216, 219–21, 223, 228, 232, 237, 242–43, 246, 270–71, 273, 293, 295, 307, 309, 313–14, 321–22, 325–26, 332, 335, 338–43, 345, 347
Inhibitory interneurons, 10
Innervation territory, 117, 339
Inspiration receptor, 32
Instability, 54, 77–78, 89–90, 109–10, 173, 181
Insulin, 9
Intermittent claudication, 6, 21
Internal rotation receptor, 32
Interneuron, 17, 34, 36–38, 57, 64, 67, 81–82, 84–85, 87, 116
Intertester reliability, 140–141
Intima, 31, 33, 37
Intraarticular fluid, 94, 159
Intraarticular injection, 118
Intraarticular pressure, 93–97, 103–4, 125, 159, 173
Intratester reliability, 140, 141
Ischemia, 12, 112, 150
Ischemic ulcers, 7
Isometric strength testing, 96
Isometric strengthening, 129, 132, 141

Janda, 3, 10, 14, 23, 86, 124–26, 143
Jayson, 23, 42, 55, 86, 91, 94–95, 106
Johansson, 19, 23, 25–26, 40–41, 43, 46, 66–68, 74–80, 90–91, 126, 159, 173
Joint damage, 15, 22, 43, 73–74, 98–99, 101, 109, 135, 171
Joint effusion, 16, 26, 93–96, 103–4, 107
Joint nutrition, 109, 167
Joint play, 124
Joint position, 1, 29, 33–34, 36–37, 43–45, 47–49, 54, 57, 65–66, 83, 100, 105, 124, 142, 160
Joint protective reflex, 1, 71, 74–78, 81–82, 89–90, 101, 103–4, 109, 123
Joint stability, 26, 52, 68, 71, 73–75, 77–78, 80, 86, 89, 93, 101, 335
Joints without active range, 104, 132, 137–39, 143, 154–55
Jones, 23, 25, 96, 106, 121, 156, 173

Kemp test, 346
Kendall, 13, 23, 156
Kennedy, 23, 40, 75, 79, 90, 106, 110
Kidney disorders, 9

Kinesthesia, 41, 46–49, 51–52, 65
Kinesthetic sensation, 47–48
Krauspe, 23, 79, 90–91

Lateral epicondylitis, 123
Lateral foraminal stenosis, 150, 154, 167–68
Lateral horn, 116
Lateral spinal stenosis, 147
LeBlanc, 4, 24
Length-tension curve, 166
Leprosy, 109
Lever arm, 78, 163
Lidocaine, 75
Ligamento-muscular protective reflex, 19, 25, 75, 78, 80, 103
Ligamentum flavum, 111
Lignocaine, 83, 95–96
Local reflexes, 14
Loose-packed position, 16, 33–34, 37, 43, 49, 78, 95, 97, 101, 103–4, 128, 131–33, 138, 146, 151, 159, 169, 175–76, 178, 282–283, 286, 289, 292, 337–39, 341, 346
Loss of developmental milestones, 7
Loubert, 13, 24, 55, 166, 173
Low back pain, 15, 23, 42, 52, 55, 69, 91, 126, 134, 139, 151, 170–73, 176, 210, 340, 342–43, 345
Lower motoneuron lesions, 11
Lyme disease, 340, 343

MacDougall, 5, 24
Magnetic resonance imaging, 4, 23
Maitland, 156
Malaise, 12
Malnutrition, 7, 9
Manipulation, 34, 41, 45, 59, 68, 86, 117–18, 124, 142–143, 165–67, 170–74, 221–23, 225–26, 274, 276, 278–79, 335
Manipulative technique, 17
Massage, 11, 14, 54
Maximal-intensity exercises, 5
Maximally close-packed position, 44–45
Maximally loose-packed position, 43–45, 134–36, 169, 341
McKenzie, 115, 121, 168, 173, 335
Mechanoreceptor, 10–11, 15–16, 21, 24–25, 27, 32–34, 36–43, 46, 48–51, 53–54, 57–59, 64–69, 71–73, 75–83, 85, 89–91, 95–96, 98, 100, 105–106, 108–10, 123, 125–26, 159–60, 172, 174
Median nerve lesion, 337, 340
Median nerve paresis, 337
Medical workup, 7, 18
Membrane potential, 9
Meningitis, 116
Mennell, 124, 126
Menopause, 7
Metabolic activity, 7
Metabolic alkalosis, 9
Metabolic myopathy, 7, 9
Midline neutral point, 134–36
Milner-Brown, 6, 24

Mitchell, 24, 46, 68, 173, 335
Mobilization, 1, 17–18, 29, 34, 36–37, 54, 59, 68, 123–24, 137, 142, 145, 152, 155, 158–72, 175–76, 178, 338–39, 341–46
   active, 158–63, 169, 171, 178, 182–83, 187–88, 191–94, 199, 213, 225–26, 228–29, 237–38, 241, 244–45, 247, 253, 258, 260, 262–63, 265, 267, 269, 271, 274, 276, 280, 284, 289, 291, 301–2, 304, 315, 320, 331, 338, 341, 343, 346
   oscillatory, 160, 171
   passive, 137, 159–62, 169, 178, 185, 197, 201–2, 204–5, 207, 209, 211–13, 217, 227, 232–36, 247, 250–52, 255–56, 258, 260, 262, 268–69, 271–72, 274–75, 277, 279–85, 287–92, 294–96, 298, 300–301, 305, 308–9, 313, 317, 319–20, 322–24, 326–28, 330–31, 338–39, 341–43, 346–47
   positonal, 158
   sustained, 160
Morphine, 99
Motor cortex, 6, 11, 82
Motor unit, 3, 6–7, 20–22, 24–26, 34, 36, 59, 66, 71–73, 76, 78, 85–86, 106, 157, 172–73
MRI, 4–5, 345
Multijoint muscle, 128–29, 132, 137
Multinutrient supplementation, 7
Multiple sclerosis, 9, 139
Muscle Energy, 18, 24, 46, 68, 159, 173, 178, 335
Muscle fiber size, 3, 6
Muscle fibers
   extrafusal, 3, 48–49, 60–61, 63
   intrafusal, 48–49, 60–61
Muscle receptors, 33, 36–37, 47, 49–51, 59, 66–67, 76, 78, 82
Muscle spasm, 103, 116, 121, 125
Muscle spindle, 14, 23, 26, 48–51, 60, 62, 67–68
Muscle testing, 12
   concentric, 129
   isometric, 129, 141
Muscle wasting, 26, 65, 123
Muscle weakness, 1, 3–13, 15, 17–20, 23, 25, 95–96, 132, 139, 157–59, 164–65, 167–72, 174, 337, 345
Muscle-specific factors, 4–5, 18
Muscular dystrophy, 7–8, 26
Myasthenia gravis, 12, 25
Myofascial release, 14
Myometer, 11
Myopathy, 3, 7–9, 18, 132, 139
Myosin filament, 9
Myosin head, 9
Myotatic reflex, 61, 63, 72–74, 82, 94, 118
Myotome, 12, 115–16, 118, 120, 151, 156
Myotonia, 8
Myotonic muscular dystrophy, 7–8

Nausea, 12
Needle biopsies, 5
Neostigmine, 12

Nerve compression, 111–13, 118, 124, 139, 150, 152, 156, 168
Nerve conduction, 3, 11, 18, 30, 132
Nerve deformation, 35, 113
Nerve hyperactivity, 118
Nerve root, 1, 97, 109, 111–13, 118–19, 142, 146, 150, 152–53, 156, 172, 332
Neural tissue, 13, 15, 89
Neurogenic arthropathy, 109
Neurogenic inflammation, 98–100, 106
Neurological disease, 109
Neurological disorders, 13, 109
Neuromuscular disorders, 9, 21
Neuromuscular junction, 11–12
Neuropathic arthritis, 109
Neuropathic arthropathy, 109
Neuropathy, 48, 95, 99
Neuropeptide, 40, 98, 106, 112, 118, 173
Neuropraxia, 112
Neurotransmitter, 98–100, 104, 109, 112, 118, 167
Nicholas, 14, 24, 86
Nociception, 3, 10–11, 18, 38, 53–54, 58, 65, 89, 98–100, 104, 109, 115, 118, 120, 124, 132–33, 139, 150, 155, 160, 166–67, 172, 340
Nociceptive afference, 53, 58, 103, 339
Nociceptive gate, 54, 89, 252
Nociceptive memory, 133
Nociceptive reflex, 10, 15, 57–58, 68, 75, 81, 87
Non-weight-bearing activities, 4
Notation, 175–76
Novocain, 101
Noxious stimulation, 14, 22
Numbness, 6, 150, 337, 339–40, 346
Nursing home residents, 7
Nutritional supplements, 7

Occlusion, 6, 90, 102
Oostendorp, 24, 41, 68, 79, 91, 181, 335
Oscillations, 36, 54, 159–60, 171, 211, 233, 252
Oscillatory joint movements, 11
Osteoarthritis, 39–40, 48, 50, 66, 68, 78, 90, 93, 106–8, 173–74
Osteoarthrosis, 177
Osteomalacia, 9, 25
Osteophytes, 111, 168
Osteoporosis, 96, 171
Outflare, 218, 299

Pain, 6–11, 15, 18, 21–24, 26, 29, 31, 38, 40–42, 52–55, 57–59, 65, 68–69, 79, 81, 85–87, 89–91, 93, 95, 97–108, 112–13, 115–20, 123–27, 130–31, 133–35, 137, 139, 145–46, 150–51, 155–56, 166–67, 169–78, 181, 210, 219, 237, 239, 266, 303, 340–43, 345–47
Paralysis, 9, 11, 20, 27, 96–97, 99, 123, 128
Parasitic, 9
Paresis, 11, 24, 124, 337
Paresthesia, 6, 12, 119–20, 146, 181

Paris, 21, 25, 41, 55, 112, 116, 121, 156, 166, 173, 182, 335
Partridge, 25, 74, 80, 107
Passive movements, 11–12, 25, 34–35, 37, 46, 51, 54, 59, 105, 126, 172
Passive positioning, 48–49, 95, 186
Pathological barrier, 135
Payr, 74, 80, 101, 107
Peak tension, 13
Peck, 46, 153, 156, 332, 335
Periarthritis, 9
Periarticular tissue, 5, 30–32, 59, 109, 165–166, 254
Peripheral compression neuropathies, 13
Peripheral nerve, 11–13, 18, 26, 53, 111–13, 118, 132, 154–55
Peripheralization of pain, 115
Persistent exertion, 6
Pettman, 181
Phasic alpha motoneuron, 62–66
Physiological barrier, 135, 137, 178, 225, 231, 256
Pins and needles, 118
Plasma flow, 112
Poliomyelitis, 12–13, 99, 105
Polymyositis, 8
Position sense, 45, 47–48, 50–52, 54, 75, 90, 105
Positional distraction, 13, 168
Positioning
    active, 49, 95
    passive, 48–49, 95, 186
Posterior horn, 31, 57, 81, 133
Postmobilization weakness, 142, 164–65, 172
Postural sensation, 48–49, 58
Potassium, 9, 172
Preganglionic neurons, 116
Progressive arterial vascular diseases, 6
Proprioception, 10, 22–23, 40, 46, 48–51, 67, 89–91, 104, 124, 173
Protective reflex, 1, 15, 71, 74–78, 80–82, 89–90, 101, 103–4, 109, 123, 125
Proximal muscles, 7–8, 11–12, 337
Pseudohypertrophy, 8
Pseudoparesis, 14, 23, 86, 124–125
Pseudoradicular syndrome, 118, 120
Pseudoradiculopathy, 84, 104, 115, 117–20, 131–32, 139–40, 147–48, 150–52, 154–55, 166–69, 171, 175, 221, 225, 231–32, 283, 293, 299, 332, 338–40, 346
Ptosis, 12
Pyridostigmine, 12

Quadriceps lag, 16, 94

Rabies, 9
Radial nerve lesion, 13
Radiculopathy, 12, 15, 30, 111–12, 115, 118–20, 131–32, 134, 139, 141–42, 146–48, 150–52, 154–55, 167–69, 171–72, 175, 177, 232, 283, 293, 299, 332, 338
Radiological density, 4
Range of motion, 4, 7–8, 36, 62, 97, 109, 129–130, 135, 165

Raymond, 97, 107
Raynaud's disease, 6
Raynaud's phenomenon, 8
Receptive field, 31
Receptor potential, 31, 38
Reciprocal inhibition, 61, 73, 124, 162, 169
Reciprocal innervation, 61–62
Reduced extensibility, 3, 7, 13, 17–18, 93, 104, 151
Reduction of compression, 6, 13, 18
Referred pain, 115–16, 118–20, 151, 156
Reflex atrophy, 97, 123
Reflex inhibition, 36, 68, 85, 94, 96, 105–7, 123, 125–26, 166, 174
Reflex neurogenic immobilization, 103
Reflex reversal, 83
Relative importance of afferent input, 18
Reliability, 140–41
Renal failure, 9
Renal tubular acidosis, 9, 27
Repeated contractions, 6
Resistive exercises, 3
Respiratory failure, 8
Respiratory muscle weakness, 8
Respiratory muscles, 12, 202, 204
Response frequency, 32
Restrictive barrier, 135–36, 148, 151, 163, 178–79, 187, 190, 211, 216–17
Rheumatoid arthritis, 9, 23, 40, 48, 51, 86, 98–100, 104–7, 171, 177, 343
Rhythmic voluntary contractions, 11
Rickets, 9
Rubella, 9

Sacroiliac joint, 1, 85, 101, 115, 117, 119, 124, 128, 137–38, 143, 155–56, 166, 172–74, 177, 205, 218–20, 227–30, 237, 293, 295, 299, 335, 341, 347
Sarcopenia, 7, 19, 21
Scleroderma, 8, 20, 23
Sclerotome, 115–16, 118, 120
Segmental innervation, 115–16, 146, 152–53, 341
Sensation, 7, 47–51, 58, 86, 89, 95, 109, 120, 164–65, 181
Sensation, loss of, 109
Sensitization, 68, 107, 207, 343
Sensory disturbances, 12, 115
Sensory gating, 10–11
Sequencing treatment, 169, 171
Sherrington's law of reciprocal innervation, 14
Shingles, 12
Skeletal deformities, 9
Skin, 7–8, 13–14, 17–18, 29–30, 40, 71, 81–86, 100–101, 115, 117
Skin receptors, 14, 37, 47–48, 50–51, 54, 59–60, 64, 81–82
Skin tightness, 8, 14
Slow-twitch fiber, 5
Slump test, 15, 21
Smoking, 7, 18

Space flight, 4
Spasm, 10, 101, 103, 116, 121, 125
Spastic muscles, 12
Spasticity, 11–12, 24
Spencer, 26, 96, 107, 126
Spin movement, 44–45
Spinal cord, 10–11, 17, 38, 53–54, 57, 66, 82–83, 97–99, 101, 116–18, 120, 133, 150–53, 157, 165, 167, 181
Spinal motoneurons, 6, 59
Spindle cell, 60–66, 73, 82, 84, 94, 137, 139, 157, 159
Spinothalamic tract, 116
Splinting, 75, 103
Spondyloarthropathy, 9
Stability of a joint, 16, 71, 73–75, 77–78, 80, 86, 89, 93, 101, 335
Stair-climbing power, 7
Static alpha motoneuron, 62
Static gamma motoneuron, 62, 64, 66
Straight leg raise, 15, 19, 22, 118, 343–344, 346
Strain-Counterstrain, 17, 23, 156, 170
Stratford, 16, 26, 107
Strengthening exercises, 1, 3–4, 6–7, 14–15, 18, 77, 132, 157, 170–71
Stretch weakness, 13
Stretching, 7, 13–15, 17, 19, 23, 31, 74–78, 80, 102–3, 123, 125, 158, 160, 162, 166, 170–71, 266, 303
Strong muscle, 177, 223; defined, 130
Subintima, 31, 35
Submaximal loads, 4–5
Substance P, 37, 39, 98–100, 104–5, 107, 112
Superficial tissues, 18, 115
Supraspinal, 6, 17, 24, 59, 64, 82, 85
Supraspinal nuclei, 59, 64, 82, 85
Surgery, 10, 12, 89–90
Sympathetic nervous system, 116, 118, 120, 171
Symphysis pubis, 101–2, 137, 178, 227, 237, 346
Synchronization ratio, 6
Synergistic muscles, 17, 61, 64, 82, 127
Synergists, 14, 61, 76, 124
Synovial membrane, 30–33, 35, 41, 93, 95, 102–3
Syphilis, 109
Systemic sclerosis, 8, 20

t cells, 10
Tabes dorsalis, 109
Tactile stimulation of the skin, 14, 24, 81, 86
Target organ, 112, 152
Technique procedures, 177–334
    Ankle and tarsal joints, 316–28
        Ankle dorsiflexion, 319
        Ankle eversion, 325
        Ankle inversion, 321
        Ankle plantar flexion, 316
    Cervical spine, 180–90
        C1 right rotation, 184
        C2 through T4 right facet extension, 186
        C2 through T4 right facet flexion, 189
        C7 and T1 right facet extension, 188
        Occiput right facet extension, 182
        Occiput right facet flexion, 183
        Subcranial region, 160
    Elbow and forearm, 266–72
        Elbow extension, 269
        Elbow flexion, 266
        Pronation, 272
        Supination, 270
    Fingers, 280–85
        Interphalangeal II–V extension, 284
        Interphalangeal flexion, 258
        Metacarpophalangeal II–V abduction, 282
        Metacarpophalangeal II–V adduction, 283
        Metacarpophalangeal II–V extension, 281
        Metacarpophalangeal II–V flexion, 285
    Glenohumeral joint, 250–65
        Shoulder abduction, 252
        Shoulder adduction, 257
        Shoulder external rotation, 250
        Shoulder external rotation in 90 degrees abduction, 263
        Shoulder horizontal abduction, 261
        Shoulder horizontal adduction, 259
        Shoulder internal rotation, 255
        Shoulder internal rotation in 90 degrees abduction, 264
    Hip, 293–302
        Hip abduction, 296
        Hip adduction, 301
        Hip extension, 297
        Hip external rotation, 299
        Hip flexion, 295
        Hip horizontal adduction, 302
        Hip internal rotation, 293
    Knee, 303–15
        Knee extension, 306
        Knee flexion, 303
        Patellar mobility, 310
        Proximal tibiofibular joint, 314
    Lumbar spine, 210–17
        Flexion T6 through L5, 215
        Flexion T10 through L5, 216
        L1 through L5 facet extension, 213–14
        L1 through L5 right facet extension, 210
        T6 through L5 right facet extension, 210
    Radiculopathy and pseudoradiculopathy, 332–34
        Cervical facet extension (foraminal closing), 332
        Cervical facet flexion (foraminal opening), 333
        Lumbar facet extension (foraminal closing), 334
        Lumbar facet flexion (foraminal opening), 334
    Rib cage, 198–209
        Clavicle through rib 6 mobility in a posterior direction, 208
        Ribs 1 and 2 inhalation, 199

Ribs 1 through 5 mobility in a posterior direction, 206
Ribs 4 through 12 inhalation, 202
Ribs 6 through 12 inhalation, 203
Ribs 7 through 9 inhalation, 200
Ribs 7 through 12 exhalation, 203
Ribs 9 through 12 exhalation, 204
Twelfth rib inhalation, 205
Sacroiliac joints, 218–36
Innominate anterior rotation, 223
Innominate caudal translation, 227
Innominate cranial translation on the sacrum, 220–22
Innominate external rotation on S1 (S1 nutation), 229–33
Innominate external rotation on S3 (S3 counternutation), 234
Innominate internal rotation on S1 (S1 counternutation), 235
Innominate internal rotation on S3 (S3 nutation), 236
Innominate posterior rotation, 225–26
S1 nutation, 228
Shoulder girdle, 239–49
Shoulder girdle depression, 248
Shoulder girdle elevation, 246
Shoulder girdle protraction, 243
Shoulder girdle protraction and elevation, 245
Shoulder girdle retraction, 239–42
Symphysis pubis, 237–38
Anterior translation of the right pubis/posterior translation of the left pubis, 238
Caudal translation of the right pubis/cranial translation of the left pubis, 237
Thoracic spine, 191–97
C7 through T7 facet extension, 193
C7 through T12 right facet extension, 191
T6 through L5 flexion, 197
T7 through L5 right facet extension, 194–95
T7 through L5 right facet flexion, 196
Thumb, 286–92
Carpometacarpal I abduction, 288
Carpometacarpal I adduction, 290
Carpometacarpal I extension, 286
Carpometacarpal I flexion, 289
Metacarpophalangeal I and interphalangeal I extension, 292
Metacarpophalangeal I and interphalangeal I flexion, 291
Toes, 329–31
Toe extension, 329
Toe flexion, 331
Wrist, 273–79
Wrist dorsiflexion, 273
Wrist palmar flexion, 275
Wrist radial deviation, 277
Wrist ulnar deviation, 279

Temporomandibular joint, 30, 36, 39–40, 46, 67, 90–91
Tenderness, 8–9, 18, 117
Tendon organ, 33, 41, 82, 139
Tendon reflexes, 11–12, 61, 72–74, 118
Tesch, 4, 19
Thalamus, 59, 116
Thoracic outlet compression syndrome, 6, 13
Thromboangiitis obliterans, 6
Tightness, 3, 8, 14, 18, 123–24, 139, 169
Tinel's sign, 12
Tonic alpha motoneuron, 63–66
Torque, 4–5, 11, 21, 78, 166
Toxin-induced myopathy, 7, 9
Toxoplasmosis, 9
Training, 3–7, 14, 21–26, 157
Trapezius, 14–15, 19, 128, 138, 145, 163, 169, 186, 191–92, 194, 239, 242, 246
Troponin, 9

Uncinate process, 111
Unloading, 4–5, 19
Upper motoneuron lesions, 11–12
Upslip, 218–19, 225, 227

Validity, 140–42
Van Meeteren, 13, 26
Vasoconstrictive effect, 7
Vasotome, 115
Venules, 12, 112
Vertebral artery insufficiency syndrome, 13, 181, 335
Viral, 9, 12, 177
Viscerotome, 115
Vitamin D deficiency, 9
Voluntary contraction, 4, 8, 16, 25, 60, 78, 84, 87, 94, 164, 174
Vomiting, 9
Von Uexküll's law, 83–84

Wall, 24, 26–27, 55, 84, 86–87, 115, 117, 121–22
Weak muscle, 10, 13–14, 29, 130, 132, 134, 138–39, 142, 151, 157, 171, 175, 339, 344; defined, 130
Weight bearing, 4–5, 59
Weight lifting, 6
Weight loss, 8
Weightlessness, 4
Wheelchair, 8
Withdrawal reflex, 1, 15, 58, 68, 78, 81–87, 89–90, 100, 105, 123
Wood, 21, 26–27, 66, 85, 87, 94–95, 105, 108, 174
Woolf, 26–27, 84, 86–87, 115, 117, 122
Wrist drop, 13
Wyke, 16, 21–22, 27, 32–33, 35–42, 55, 58–59, 64, 67–69, 71–73, 78–80, 85, 90–91, 108, 110, 120–21, 125, 174, 335

Xylocaine, 118

NOTES